Lecture Notes in Comput

5

Commenced Publication in 1973
Founding and Former Series Editors:
Gerhard Goos, Juris Hartmanis, and Jan van

Editorial Board

David Hutchison
  Lancaster University, UK
Takeo Kanade
  Carnegie Mellon University, Pittsburgh, PA, USA
Josef Kittler
  University of Surrey, Guildford, UK
Jon M. Kleinberg
  Cornell University, Ithaca, NY, USA
Friedemann Mattern
  ETH Zurich, Switzerland
John C. Mitchell
  Stanford University, CA, USA
Moni Naor
  Weizmann Institute of Science, Rehovot, Israel
Oscar Nierstrasz
  University of Bern, Switzerland
C. Pandu Rangan
  Indian Institute of Technology, Madras, India
Bernhard Steffen
  University of Dortmund, Germany
Madhu Sudan
  Massachusetts Institute of Technology, MA, USA
Demetri Terzopoulos
  University of California, Los Angeles, CA, USA
Doug Tygar
  University of California, Berkeley, CA, USA
Moshe Y. Vardi
  Rice University, Houston, TX, USA
Gerhard Weikum
  Max-Planck Institute of Computer Science, Saarbruecken, Germany

Phillip B. Gibbons   Tarek Abdelzaher
James Aspnes   Ramesh Rao (Eds.)

Distributed Computing in Sensor Systems

Second IEEE International Conference, DCOSS 2006
San Francisco, CA, USA, June 18-20, 2006
Proceedings

 Springer

Volume Editors

Phillip B. Gibbons
Intel Research
4720 Forbes Avenue, Suite 410, Pittsburgh, PA 15213, USA
E-mail: phillip.b.gibbons@intel.com

Tarek Abdelzaher
University of Illinois at Urbana-Champaign, Department of Computer Science
Urbana, IL 61801, USA
E-mail: zaher@cs.uiuc.edu

James Aspnes
Yale University, Department of Computer Science
51 Prospect Street, New Haven, CT 06520-8285, USA
E-mail: aspnes@cs.yale.edu

Ramesh Rao
University of California at San Diego
9500 Gilman Drive, La Jolla, CA 92093-0436, USA
E-mail: rrao@ucsd.edu

Library of Congress Control Number: 2006927240

CR Subject Classification (1998): C.2.4, C.2, D.4.4, E.1, F.2.2, G.2.2, H.4

LNCS Sublibrary: SL 5 – Computer Communication Networks and
Telecommunications

ISSN   0302-9743
ISBN-10   3-540-35227-9 Springer Berlin Heidelberg New York
ISBN-13   978-3-540-35227-3 Springer Berlin Heidelberg New York

This work is subject to copyright. All rights are reserved, whether the whole or part of the material is concerned, specifically the rights of translation, reprinting, re-use of illustrations, recitation, broadcasting, reproduction on microfilms or in any other way, and storage in data banks. Duplication of this publication or parts thereof is permitted only under the provisions of the German Copyright Law of September 9, 1965, in its current version, and permission for use must always be obtained from Springer. Violations are liable to prosecution under the German Copyright Law.

Springer is a part of Springer Science+Business Media

springer.com

© Springer-Verlag Berlin Heidelberg 2006
Printed in Germany

Typesetting: Camera-ready by author, data conversion by Scientific Publishing Services, Chennai, India
Printed on acid-free paper   SPIN: 11776178   06/3142   5 4 3 2 1 0

Message from the General Chair

Welcome to DCOSS 2006 – the second version of the meeting series. DCOSS focuses on distributed computing issues in large-scale networked sensor systems, including systematic design techniques and tools, algorithms, and applications.

I am indebted to the Program Chair, Phil Gibbons, for his efforts in handling the review process and composing the technical program. I appreciate his leadership in putting together a strong and diverse Technical Committee to address various aspects of this interdisciplinary area. I would also like to thank him for his input in resolving a number of meeting-related issues.

I would like to thank all of the authors who submitted papers, our invited speakers, the external referees we consulted, the Vice Chairs and the members of the Program Committee.

I would like to thank Sotiris Nikoletseas for his efforts as the Workshop Chair for DCOSS 2006.

Several volunteers assisted me in putting together the meeting. I would like to thank Jim Reich for handling the poster session, Wendi Heinzelman for publicizing the event, Amol Bakshi for handling Web-based publicity, Loren Schwiebert for handling the student scholarships, Jie Wu for interfacing with IEEE TCDP for student scholarships and Yang Yu for his assistance in putting together these proceedings. Special thanks go to Amol Bakshi for his invaluable input in deciding the meeting focus, format and local arrangements.

I would like to thank Jose Rolim, DCOSS Steering Chair for inviting me to be the General Chair. Indeed, it was a pleasure working with him and with Jie Wu, Vice General Chair. Their invaluable input in putting together the meeting program and in shaping the meeting series is gratefully acknowledged.

I would like to acknowledge support from the IEEE Technical Committee on Distributed Processing and from the Centre Universitaire d'Informatique of the University of Geneva.

Rosine Sarafian, our administrative coordinator, deserves special thanks for her assistance with local arrangements.

The field of networked sensor systems is rapidly evolving. It is my continued hope that this meeting series serve as a forum for researchers from various aspects of this interdisciplinary field to interact and in particular to offer opportunities for those working in algorithmic, theoretical and high-level aspects to interact with those addressing challenging issues in complementary areas such as wireless networks, communications and systems composed of these underlying technologies.

I hope you enjoy the technical sessions as well as San Fransisco.

June 2006   Viktor K. Prasanna

Message from the Program Chair

This volume contains the 33 full papers presented at the Second IEEE International Conference on Distributed Computing in Sensor Systems (DCOSS 2006), which took place in San Francisco, California, during June 18–20, 2006. These papers were selected by the Program Committee from 87 submissions received in response to the call for papers. Submissions were received from 18 countries across 5 continents, and directed to one of three tracks: algorithms, applications, or systems. Each track had its own Program Committee that reviewed the papers and recommended either "accept", "reject", or "accept if room". In a joint meeting between the Vice Chairs and myself we reviewed and discussed this latter category of papers to arrive at the final program.

DCOSS 2006 presentations were arranged into seven sessions, ranging from Data Aggregation and Dissemination to Programming Support and Middleware to Lifetime Maximization. Papers from the three tracks were intermixed within the sessions. Other highlights of the conference included keynote talks by Leo Guibas and Bill Kaiser, two workshops and a poster session.

I would like to add my thanks to Viktor's to all the DCOSS organizers, the authors, the external reviewers, and the Program Committee members. I am especially indebted to the Program Vice Chairs Tarek Abdelzaher, James Aspnes, and Ramesh Rao for their efforts in forming and running the three track Program Committees. The 44 Program Committee members are at universities and research labs from 12 different countries, further evidence that DCOSS is truly an international conference. The quality of the program reflects positively on the expertise and dedication of the Vice Chairs and Program Committee members.

Finally, it was a pleasure working with Viktor Prasanna, General Chair, and José Rolim, Steering Committee Chair, who both worked tirelessly to ensure the success of DCOSS 2006.

June 2006   Phillip B. Gibbons

Organization

General Chair

Viktor K. Prasanna — University of Southern California, USA

Vice General Chair

Jie Wu — Florida Atlantic University, USA

Program Chair

Phillip B. Gibbons — Intel Research, Pittsburgh, USA

Program Vice Chairs

Algorithms
James Aspnes — Yale University, USA

Applications
Ramesh Rao — University of California at San Diego and Calit2, USA

Systems
Tarek Abdelzaher — University of Illinois, Urbana Champaign, USA

Steering Committee Chair

Jose Rolim — University of Geneva, Switzerland

Steering Committee

Sajal Das — University of Texas at Arlington, USA
Josep Diaz — UPC Barcelona, Spain
Deborah Estrin — University of California, Los Angeles, USA
Phillip B. Gibbons — Intel Research, Pittsburgh, USA
Sotiris Nikoletseas — University of Patras and CTI, Greece
Christos Papadimitriou — University of California, Berkeley, USA
Kris Pister — University of California, Berkeley, and Dust, Inc., USA
Viktor Prasanna — University of Southern California, Los Angeles, USA

Poster Chair

Jim Reich   Palo Alto Research Center, USA

Workshops Chair

Sotiris Nikoletseas   University of Patras and CTI, Greece

Proceedings Chair

Yang Yu   Motorola Labs, USA

Publicity Co-chairs

Wendi Heinzelman   University of Rochester, USA
Amol Bakshi   University of Southern California, USA

Finance Chair

Germaine Gusthiot   University of Geneva, Switzerland

Student Scholarships Chair

Loren Schwiebert   Wayne State University, USA

Sponsoring Organizations

IEEE Computer Society Technical Committee on Parallel Processing
  (TCPP)
IEEE Computer Society Technical Committee on Distributed Processing
  (TCDP)

Held in Cooperation with

ACM Special Interest Group on Computer Architecture (SIGARCH)
ACM Special Interest Group on Embedded Systems (SIGBED)
European Association for Theoretical Computer Science (EATCS)
IFIP WG 10.3

Program Committee

Costas Busch	Rensselaer Polytechnic Institute, USA
Edgar Chavez	University of Michoacana, Mexico
Bogdan Chlebus	University of Colorado at Denver, USA
Shlomi Dolev	Ben-Gurion University of the Negev, Israel
Alfredo Ferro	University of Catania, Italy
Stefan Fischer	University of Luebeck, Germany
Mohamed Gouda	University of Texas at Austin, USA
Tian He	University of Minnesota, USA
Wendi Heinzelman	University of Rochester, USA
Jennifer Hou	University of Illinois, Urbana Champaign, USA
Anura Jayasumana	Colorado State University, USA
Dariusz Kowalski	University of Liverpool, UK
Bhaskar Krishnamachari	University of Southern California, USA
Phil Levis	Stanford University, USA
Jie Liu	Microsoft Research, USA
Julia Liu	Palo Alto Research Center, USA
Chenyang Lu	Washington University in St. Louis, USA
Haiyun Luo	University of Illinois, Urbana Champaign, USA
Rajeev Motwani	Stanford University, USA
C. Siva Ram Murthy	IIT Madras, India
Radhika Nagpal	Harvard University, USA
Suman Nath	Microsoft Research, USA
Sotiris Nikoletseas	University of Patras and CTI, Greece
Boaz Patt-Shamir	Tel-Aviv University, Israel
Pino Persiano	University of Salerno, Italy
John Regehr	University of Utah, USA
Andrea Richa	Arizona State University, USA
Kurt Rothermel	University of Stuttgart, Germany
Andreas Savvides	Yale University, USA
Christian Scheideler	Technical University of Munich, Germany
Maria Jose Serna	UPC Barcelona, Spain
Devavrat Shah	Massachusetts Institute of Technology, USA
Vikram Srinivasan	National University of Singapore, Singapore
Mani Srivastava	University of California, Los Angeles, USA
Jack Stankovic	University of Virginia, USA
Ivan Stojmenovic	University of Ottawa, Canda
Gaurav Sukhatme	University of Southern California, USA
Violet R. Syrotiuk	Arizona State University, USA
Nalini Venkatasubramanian	University of California, Irvine
Chieh-Yih Wan	Intel Research, USA
Stephen Wicker	Cornell University, USA
Peter Widmayer	ETH Zurich, Switzerland
Yinyu Ye	Stanford University, USA
Ying Zhang	Palo Alto Research Center, USA

Referees

Rida Bazzi
Karthik Dantu
Hen Fitoussi
Yinnon Haviv
Ronen Kat
Philip Kuryloski
Michael Margaliot
Pedro Marron
Darryl Morrel
Melih Onus
Sameer Pai

Rami Puzis
Hui Qu
Marina Sadetsky
Elad Schiller
Allon Shafrir
Christina Tavoularis
Hector Tejeda
Nir Tzachar
Donglin Xia
Reuven Yagel
Xin Zhang

Limor Lahiani
Olga Brukman
Bodhi Priyantha
Ioannis Chatzigiannakis
Tassos Dimitriou
Athanassios Kinalis
Dennis Pfisterer
Young-ri Choi
Maria Blesa

Table of Contents

Evaluating Local Contributions to Global Performance in Wireless Sensor and Actuator Networks

Christopher J. Rozell and Don H. Johnson*

Department of Electrical and Computer Engineering
Rice University, Houston, TX 77025-1892
{crozell, dhj}@rice.edu

Abstract. Wireless sensor networks are often studied with the goal of removing information from the network as efficiently as possible. However, when the application also includes an actuator network, it is advantageous to determine actions in-network. In such settings, optimizing the sensor node behavior with respect to sensor information fidelity does not necessarily translate into optimum behavior in terms of action fidelity. Inspired by neural systems, we present a model of a sensor and actuator network based on the vector space tools of *frame theory* that applies to applications analogous to reflex behaviors in biological systems. Our analysis yields bounds on both absolute and average actuation error that point directly to strategies for limiting sensor communication based not only on local measurements but also on a measure of how important each sensor-actuator link is to the fidelity of the total actuation output.

1 Introduction

Recent interest in wireless sensor networks has fueled a tremendous increase in the study of signal and information processing in distributed settings. Energy conservation is very important for most interesting applications, which generally translates into minimizing the communication among sensors to preserve both individual node power and total network throughput. Consequently, recent sensor network research has primarily focused on adapting well-known signal processing algorithms to distributed settings where individual nodes perform local computations to minimize the information passed to distant nodes (e.g., [1,2,3]).

The goal of many proposed sensor network algorithms has been to get the information *out* of the network (via a special node connected directly to a more traditional data network) with a good trade-off between fidelity and energy expended. However, in many applications the implicit assumption is that the information coming out of the network will be used to monitor the environment and take action when necessary. A significant and natural extension to the sensor network paradigm is a wireless sensor and actuator network (WSAN). A WSAN consists of a network of sensor nodes that can measure stimuli in the environment

---

* This work was supported by the Texas Instruments Leadership University Program.

P. Gibbons et al. (Eds.): DCOSS 2006, LNCS 4026, pp. 1–16, 2006.
© Springer-Verlag Berlin Heidelberg 2006

and a network of actuator nodes capable of affecting the environment. While one possible strategy summarizes information for a system outside the network to determine actuator behaviors, greater efficiency should be achieved by determining actions through in-network processing. A more subtle issue is that processing and communication strategies optimizing sensor data fidelity may not yield the best results when actuation performance fidelity is the desired metric.

While WSANs are often discussed, quantitative analysis of their performance has not received much attention. Existing work can be found in areas such as software development models for WSANs [4] and heuristic algorithms for resource competition based on market models [5]. Other recent work [6] has used techniques from causal inference to evaluate specific actuation strategies. Most relevant is the recent work of Lemmon et al. [7] analyzing distributed control systems while considering the underlying communication network. A control system approach is certainly appropriate for some WSAN application models, but may use more communication resources (especially from actuators to sensors) and may require the sensors and actuators to operate in the same signal space.

Merging sensed information directly into actions without centralizing the information and decision making has rarely been considered in man-made systems. Fortunately, we have examples from biology that demonstrate the effectiveness of this strategy. Neural systems perform a chain of tasks very similar to the needs of WSANs: sensing, analysis, and response. Furthermore, evidence indicates that neural systems represent and process information in a distributed way (using groups of neurons) rather than centralizing the information and decision making in one single location. This shrewd strategy avoids creating a single point of vulnerability, so the system can function in the presence of isolated failures.

In neural systems, two types of behaviors exist, depending on whether there is "thinking" involved, which we call *conscious* and *reflex* behaviors. In conscious behavior, biological systems gather sensory information, make inferences from that information about the structure of their environment, and generate actions based on that inferred structure. In reflex behavior, a sensed stimulus directly generates an involuntary and stereotyped action in the peripheral nervous system before the brain is even aware of the stimulus [8]. An obvious example of a reflex behavior is the knee-jerk reaction achieved by a doctor's well-placed tap below the kneecap. A more subtle example is the eye position correction that allows our vision to stay focused on an object even when our head is moving.

WSAN applications have an analogous division, which we call *object-based* and *measurement-based* network tasks. For example, the canonical target tracking scenario is an object-based task because it involves using sensory measurements to infer information about objects in the environment. On the other hand, an application such as agricultural irrigation is a measurement-based task because sensor measurements directly contain all the necessary information — there is no underlying environmental object to try and infer. In this work we consider models of measurement-based WSAN applications. While measurement-based systems are simpler and possibly more limited than object-based systems, they provide an entry point for analyzing and designing WSAN algorithms.

WSANs are complex systems with many interacting layers of operation. There are significant communication and networking challenges in these systems that are the focus of current research efforts. While the biological reflex systems described earlier do not appear to adaptively change their communication strategy on short time scales, the nature of wireless networking may necessitate dynamic decisions to employ different communication strategies based on current network conditions. Networking strategies to limit communication in the system must weigh the cost of executing individual communication links against the detrimental effect of performing suboptimal information processing. The role of our present research is to analyze a distributed WSAN model for a broad class of applications. We want to determine the optimal information processing strategy and to quantify the effects of suboptimal strategies resulting from eliminating communication links. As a simple starting place for our analysis, we will use vector space methods to model sensors and actuators, leveraging the notion of *frame theory* to analyze systems of nodes with overlapping influence.

2   Sensors and Actuators

As an example reflex behavior that will shape our thinking about WSANs, we consider the crayfish visual system. The crayfish has a dorsal light reflex [9] where light movement in the visual field elicits predictable reflex movement in the eyestalk that attempts to keep a constant orientation of the visual field. The main visual representation (in neurons called "sustaining fibers") is comprised of sensory elements that sum light activity in overlapping spatial regions. All of the information available to the creature about the light stimulus is contained in this collection of sustaining fiber responses.

The crayfish eyestalk movement is controlled by a set of motorneurons, which send signals to several small muscles. Each muscle generates movement in one specific direction. As with the sensory units, the muscle movement directions also overlap in the movement space (i.e., muscle movements are not "orthogonal"). Most importantly, the activity in each motorneuron is determined directly from a processed combination of some sustaining fiber inputs. Though all of the motorneurons have to be coordinated to produce the desired total action, their distributed individual responses are generated directly from the distributed sustaining fiber representation and without a centralized decision-making structure. Previous research has shown that even in this critical behavior, the contributions of each sensory unit to the total action are simple and essentially linear [10].

Our WSAN model will follow the principles seen in this example from the crayfish. Though the constraints facing biological systems are different from the constraints imposed by wireless networking, neural systems must also be very resource efficient and try to minimize communication (each neural signal generated means expending more metabolic energy). Biological systems must have solutions that do a good job (some would even argue optimal) at trading-off performance and efficiency, and we use them as a rough guide.

In our model, a collection of sensors measuring overlapping spatial regions gather information about a stimulus field. A collection of actuators have individual environmental effects that overlap and must be coordinated. Each actuator determines its individual contribution to a behavioral goal through a combination of the sensor measurements. We start with the simplest scenario where only this direct sensor-to-actuator communication is allowed. By eliminating inter-sensor and inter-actuator communication, we also eliminate the communication overhead necessary for such a scenario. It may be possible to improve system performance by allowing additional communication and cooperation, depending on the specific networking model and communication costs involved.

A major goal in any information processing strategy for WSANs is retaining good performance in the total actuation while reducing the communication burden from the sensors to the actuators. To analyze the performance of a WSAN under different design decisions, we use mathematical models based in the familiar tools and terminology of vector spaces.

2.1   Vector Space Models of Sensors and Actuators

Sensor network models often begin with a collection of sensors distributed over a 2-D spatial field limited to the spatial domain W (e.g., $W = [0, 1]^2$). Sensors are indexed by $k \in K$, and are located either irregularly or on a regular grid. The spatial region being sensed contains a stimulus field, denoted by $x(w)$, where $w \in W$ is a vector indicating location in the field.

Sensor measurement models often consist of averaging the stimulus field over non-overlapping spatial regions surrounding each sensor [11]. We generalize that notion by representing each sensor by a receptive field $s_k(w)$ over W that performs a weighted average over a spatial region. The sensor receptive fields are defined by the physics of the devices and could indicate sensors that are directional or have varying sensitivity over a region. Sensor measurements of the field are therefore given by

$$m_k = \int_W x(w)s_k(w)dw. \tag{1}$$

We will not assume any particular arrangement or shape of the sensor fields; in general we expect sensors to be irregularly spaced and have highly overlapping receptive fields. The measurement form given in equation (1) includes the special case of sensors averaging the field over disjoint local regions.

Recasting equation (1), the sensor measurements can be written as an inner product over the field W, $m_k = \langle x, s_k \rangle$. This vector space view of the sensor measurements indicates that with no further processing the measurements can represent any stimulus signal in the space $\mathcal{H}_x = \text{span}(\{s_k\})$. The space $\mathcal{H}_x$ represents a restricted class of fields that is consistent with the resolution of the sensors. For example, $\mathcal{H}_x$ may be a space of spatially bandlimited functions over W. The actual stimulus field in the environment may not be in $\mathcal{H}_x$, but the sensors have a limited resolution (depending on design and placement of the sensors) that precludes them from sensing an unrestricted class of signals.

Therefore, we assume that $x \in \mathcal{H}_x$, though in reality x only represents the component of the true environmental field within the sensing resolution of the network.

Just as individual sensors have local but overlapping regions of sensitivity, actuator networks are composed of individual actuators that each affect the environment through (possibly overlapping) local regions of influence. Actuators are indexed by $l \in L$, and again are located either irregularly or on a regular grid. Whereas each sensor is represented by a receptive field, each actuator is represented by a influence field over W, denoted by a function $a_l(w)$. An actuator's influence field depends on the physics of the specific problem, and again may indicate actuators that are directional or have varying influence over a region.

Each actuator responds with an intensity that indicates how strongly it acts on the environment. We will model an actuator's intensity d_l as weighting its influence function. The resulting total actuation field over W is $y = \sum_{l \in L} d_l a_l$, where, for simplicity (and to emphasize the vector space view), we drop the explicit notation of spatial location $w \in W$ from the actuator influence function $a_l(w)$ and the total actuation field $y(w)$. The collection of actuators can therefore cause any actuation field y in the space $\mathcal{H}_y = \text{span}(\{a_l\})$. The space $\mathcal{H}_y$ represents a restricted class of fields that is consistent with the resolution and placement of the actuators (e.g., a class of spatially bandlimited signals, etc.).

It is critical to note here that the collection of sensors $\{s_k\}$ and actuators $\{a_l\}$ do not share many characteristics; they can have different numbers of elements at different locations over W. Most importantly, individual sensor and actuator functions can have different shapes and even involve different modalities (e.g., temperature sensors and water delivery actuators). Consequently, $\mathcal{H}_x$ and $\mathcal{H}_y$ can be *very* different functions spaces, and using general vector space definitions allows us to connect sensed inputs to actuation outputs.

In order to design effective communication strategies between sensors and actuators, we need methods to analyze the relationship between individual node activity (m_k and d_l) and the resulting impact on signals in $\mathcal{H}_x$ and $\mathcal{H}_y$. The analysis is complicated because of the overlap between both individual sensor receptive fields and actuator influence fields; in short, the representational elements are not orthogonal. We appeal to the tools of *frame theory* to analyze systems of linearly dependent sensor and actuator functions.

2.2 Frame Theory

In section 2.1 we described the sensor measurement process as a projection of a stimulus field onto a collection of sensor representation functions. Similarly, we described actuators generating an effect as a weighted sum of individual actuator representation functions. In both the collections of sensors and actuators, the basic functions form a representation for a signal space ($\mathcal{H}_x$ and $\mathcal{H}_y$, respectively). The notion of representing a signal in terms of a collection of orthonormal basis (ONB) vectors is one of the most fundamental ideas in signal processing. Though the situation here is more complicated than an ONB, the collections of sensors and actuators are vectors that form a similar representation for their associated signal spaces. In this section, we will consider a general collection of vectors $\{\phi_j\}$ indexed over J. Fundamental results about this generic collection of vectors will be applied to the sensor and actuator representations in section 3.

An orthonormal basis has the property that any energy represented by the projection onto one vector will not be present in the projections onto any other vectors. As a consequence, reconstructing the signal from the projections is trivial; the projection coefficients simply weight the same vectors in the reconstruction. However, in general, collections of sensor receptive fields and actuator influence fields will not be orthogonal. In fact, in the most general case, these collections of functions may be linearly dependent and no longer form a basis.

A collection of M vectors $\{\phi_j\}$ forms a *frame* [12] for $\mathcal{H}$ if there exist constants $0 < A \leq B < \infty$ so that Parseval's relation is bounded for any $x \in \mathcal{H}$,

$$A \left\|x\right\|^2 \leq \sum_{j \in J} |\langle \phi_j, x \rangle|^2 \leq B \left\|x\right\|^2.$$

In general, there will be more vectors than are necessary to represent $\mathcal{H}$ ($M > N$, where $N = \dim(\mathcal{H})$), meaning that the frame is redundant. When the frame vectors are normalized $\left\|\phi_j\right\|^2 = 1$ (which we assume here), the frame bounds measure the minimum and maximum redundancy of the system and satisfy $A \leq \frac{M}{N} \leq B$. Frames were originally introduced in 1952 in the context of nonharmonic Fourier series [13] and later played a key role in wavelet theory [14]. They have recently been used in many other areas, including filterbanks [15], image processing [16], communications [17], coding [18] and machine learning [19].

The frame condition given above guarantees that the analysis coefficients obtained from projecting a signal onto the frame vectors contains all of the information necessary to synthesize (or reconstruct) the signal. Mathematically, the analysis coefficients are generated through the frame analysis operator $\Phi : \mathcal{H} \to l^2$, which is given by $(\Phi x)_j = c_j = \langle \phi_j, x \rangle$. In vector notation, the collection of all analysis coefficients is given by $c = \Phi x$. For finite dimensional frames (as in practical systems), the operator Φ is a matrix multiplication.

The adjoint of the frame analysis operator is the frame synthesis operator, $\Phi' : l^2 \to \mathcal{H}$, given by $\Phi' c = \sum_{j \in J} c_j \phi_j$. Because of the dependency present between frame vectors, the same set of vectors cannot generally be used for both analysis and synthesis. Even though Φ' and Φ are inverse operations in an ONB, in general Φ will not have a unique inverse. Therefore, the usual reconstruction will not work, $x \neq \Phi' \Phi x = \sum_{j \in J} \langle x, \phi_j \rangle \phi_j$. Instead, the pseudoinverse operator $\Phi^* = (\Phi' \Phi)^{-1} \Phi'$ is used for reconstruction, $x = \Phi^* \Phi x = (\Phi' \Phi)^{-1} \sum_{j \in J} \langle x, \phi_j \rangle \phi_j$. Equivalently, we can view the reconstruction as using a different set of vectors $\{\tilde{\phi}_j\}$ called the dual set, $x = \sum_{j \in J} \langle x, \phi_j \rangle \tilde{\phi}_j$. While there are an infinite number of sets of dual vectors that will work, the canonical dual set is given by $\tilde{\phi}_j = (\Phi' \Phi)^{-1} \phi_j$. These dual vectors are also a frame for $\mathcal{H}$, with lower and upper frame bounds $\left(\frac{1}{B}, \frac{1}{A}\right)$, respectively. Importantly, the frame and dual set are interchangeable in the reconstruction equation,

$$x = \sum_{j \in J} \langle \phi_j, x \rangle \tilde{\phi}_j = \sum_{j \in J} \langle \tilde{\phi}_j, x \rangle \phi_j.$$

The frame bounds are related directly to the eigenstructure induced by the frame vectors: $A = \lambda_{\min}$ and $B = ||\Phi'\Phi|| = \lambda_{\max}$, where $\{\lambda_i\}$ are the eigenvalues of $(\Phi'\Phi)$. When a collection of vectors has frame bounds that are equal, $A = B = \frac{M}{N}$, it is called a *tight frame*. When a frame is tight, the dual vectors are simply rescaled versions of the frame vectors, $\widetilde{\phi}_j = \frac{1}{A}\phi_j$. A collection of vectors is an orthonormal basis if and only if it is a tight frame with $A = B = 1$.

In an ONB, perturbing a measurement coefficient (including removing it entirely) has a proportional impact on the reconstruction — the energy in the reconstruction error is the same as the energy in the perturbation. The redundancy present in a frame can provide a measure of robustness to perturbations that is not present in orthonormal systems, but it also makes the effect of such perturbations harder to analyze. When we apply frame theoretic models to the analysis of sensor and actuator networks, we want to know the impact of reducing communication costs by using approximate coefficients in the reconstruction.

Stated generally, we need to calculate a bound on the maximum error when a perturbation p_j is added to each frame coefficient c_j in the reconstruction, $\hat{x} = \sum_{j \in J} (c_j + p_j)\widetilde{\phi}_j$. Perturbations may include removing the coefficient from the reconstruction, $p_j = -(c_j)$. The error resulting from these perturbations is

$$||x - \hat{x}||^2 = \left|\left|\sum_{j \in J} p_j \widetilde{\phi}_j\right|\right|^2. \tag{2}$$

We recall that the dual set $\{\widetilde{\phi}_j\}$ is also a frame for $\mathcal{H}_x$, and we denote the analysis operator for the dual frame to be $\widetilde{\Phi}$. Note that the error signal recast in matrix notation is $(x - \hat{x}) = \widetilde{\Phi}'p$, where p is the perturbation vector $p = [p_1 p_2 \ldots p_{|J|}]'$. Linear algebra can yield a bound on the error,

$$\left|\left|\widetilde{\Phi}'p\right|\right|^2 = \langle p, \widetilde{\Phi}\widetilde{\Phi}'p\rangle \leq \left|\left|\widetilde{\Phi}\widetilde{\Phi}'\right|\right| \cdot ||p||^2.$$

Note that because the singular values of $\widetilde{\Phi}$ are the square roots of the eigenvalues of both $\left(\widetilde{\Phi}\widetilde{\Phi}'\right)$ and $\left(\widetilde{\Phi}'\widetilde{\Phi}\right)$, it follows that $\left|\left|\widetilde{\Phi}\widetilde{\Phi}'\right|\right| = \left|\left|\widetilde{\Phi}'\widetilde{\Phi}\right|\right|$. Because the dual set is a frame for $\mathcal{H}$ with upper frame bound $\left(\frac{1}{A}\right)$ and because of the relationship between the eigenvalues of $\left(\widetilde{\Phi}'\widetilde{\Phi}\right)$ and the frame bounds, we can finally write a useful bound (alluded to in [20]) on the reconstruction error

$$||x - \hat{x}||^2 \leq \frac{||p||^2}{A}. \tag{3}$$

In words, the perturbation energy is reduced in the reconstruction by at least the minimum redundancy in the set of frame analysis vectors $\{\phi_j\}$. The upper bound in equation (3) is consistent with probabilistic robustness results when stochastic noise is added to frame coefficients [18].

3   Connecting Sensors to Actuators

Following our example of reflex behavior, actuators must generate activity using received sensors measurements without communicating with other actuators. The overlapping actuator influence fields prevent a purely greedy approach where each actuator generates the locally optimal activity. Nearby actuators could be nearly identical and wildly overcompensate their actions in a greedy approach. Sensors must coordinate behavior (without communication) to account for the the action field components covered by the other sensors.

3.1   Generating Optimal Actuation

To formalize this notion of coordination, we draw on our discussion of frame theoretic models for sensors and actuators in section 2.2. We assume that the collection of sensors represented by $\{s_k\}$ form a frame for $\mathcal{H}_x$ with frame bounds (A_s, B_s) and with dual functions given by $\{\tilde{s}_k\}$. Similarly, we assume that the collection of actuators represented by $\{a_l\}$ form a frame for $\mathcal{H}_y$ with frame bounds (A_a, B_a) and with dual functions given by $\{\tilde{a}_l\}$. Note that the dual sets $\{\tilde{s}_k\}$ and $\{\tilde{a}_l\}$ aren't realized directly in physical systems. For example, the sensor receptive field dual functions $\{\tilde{s}_k\}$ may have spatial characteristics that would be impossible to build into any type of real-world sensor.

To generate coordinated behavior in the actuator network, we must necessarily start with the ideal solution for generating actions. Each WSAN has an application specific goal that defines its existence. For example, a system might use sensed rainfall to order the diversion of floodwater or the delivery of irrigation to meet specified conditions. Though the actions necessary to achieve the goal depend on the specific observed stimulus, the goal itself is stimulus independent. To quantify this application goal, we assume that for any measured stimulus field x there is a mapping $T : \mathcal{H}_x \to \mathcal{H}_y$ that defines the ideal action field response, $y = Tx$. The mapping T would be determined as a design specification for the WSAN in advance. While it may be possible to reconfigure a WSAN to perform a different application (with a different goal) on long time scales, we assume that the goal (as quantified by T) stays fixed.

An ideal actuator network would have each node determine action coefficients $\{d_l\}$ to generate the optimal response $Tx = \sum_{l \in L} d_l a_l$. Drawing on the frame theory results from section 2.2, the coefficients weighting the action influence field vectors are given by the inner products between the action *dual* vectors and the action signal that we are trying to generate,

$$d_l = \langle \tilde{a}_l, Tx \rangle. \tag{4}$$

To determine the optimal action coefficients, consider first the reconstruction equation for the stimulus field based on the sensor measurements,

$$x = \sum_{k \in K} m_k \tilde{s}_k. \tag{5}$$

Substituting equation (5) into equation (4), the optimal action coefficients are

$$d_l = \langle \tilde{a}_l, T \sum_{k \in K} m_k \tilde{s}_k \rangle = \sum_{k \in K} m_k \langle \tilde{a}_l, T \tilde{s}_k \rangle. \tag{6}$$

The conversion from sensor measurements $m = [m_1, m_2, \ldots, m_{|K|}]'$ to actuator intensity coefficients $d = [d_1, d_2, \ldots, d_{|L|}]'$ in matrix form is $d = Vm$, where

$$V = \begin{bmatrix} \tilde{a}'_1 T \tilde{s}_1 & \tilde{a}'_1 T \tilde{s}_2 & \cdots & \tilde{a}'_1 T \tilde{s}_{|K|} \\ \tilde{a}'_2 T \tilde{s}_1 & \ddots & & \vdots \\ \vdots & & & \\ \tilde{a}'_{|L|} T \tilde{s}_1 & \cdots & & \tilde{a}'_{|L|} T \tilde{s}_{|K|} \end{bmatrix}.$$

The expression in equation (6) (or equivalently the entries of V) illuminate the form of the actuator intensity coefficients necessary to generate the optimal total action Tx. Unfortunately, each coefficient d_l is a sum including sensor measurements s_k over all $k \in K$; each individual actuator would require knowledge of *every* sensor measurement in order to generate an optimal actuation intensity.

A scenario where every sensor in the network communicates its measurement to every actuator would present an unreasonable communication burden on the network — approximately $|K| \cdot |L|$ communication links would be necessary. While a portion of this burden could be reduced through broadcast communication, some sensor-to-actuator links may involve several communications in a multi-hop routing scheme. Any realistic networking scheme will have to eliminate some of these communication links based on their communication cost and their contribution to the total actuation performance. Intuitively, some sensor measurements will be more important than others in determining an actuators behavior. For example, a moisture sensor spatially located a long distance away from the influence field of a specific irrigation actuator will likely have very little relevance on that actuator's optimal behavior coefficient. Using the frame theory results presented in section 2.2 along with the vector space model of sensor and actuator networks, we have tools for analyzing the effects of eliminating communication links on the total actuation performance.

3.2 Limiting Communication Costs

Each entry of the matrix V indicates a communication link from a sensor to an actuator. Before blindly reducing communications, a networking scheme must know the importance of each possible communication. In a sensor network, performance is often judged by assessing the fidelity of the information removed from the network at representing the original sensor measurements (or the underlying stimulus field). However, the only performance metric of any consequence in a WSAN is the fidelity of the resulting total action.

To quantify the importance of individual communications, we must determine how the total actuation performance is affected when a communication is not executed. We quantify this notion of importance through the results described in equation (3). Consider the case where for actuator l, a subset of sensor nodes $E_l \subset K$ do not transmit their measurement coefficient to this actuator. Instead of optimal actuator intensity coefficients (see equation (6)), actuators form approximate intensity coefficients using the received sensor measurements

$$\hat{d}_l = \sum_{k \in (K \setminus E_l)} m_k \langle \tilde{a}_l, T\tilde{s}_k \rangle. \tag{7}$$

The approximate actuator intensities generate a total action field approximating the desired optimal action Tx,

$$\hat{y} = \sum_{l \in L} \hat{d}_l a_l.$$

Generating a total action field with the approximate coefficients $\{\hat{d}_l\}$ is equivalent to performing a frame reconstruction with perturbed coefficients, as described in section 2.2. Subtly, the actuator frame vectors are performing synthesis, meaning that dual vectors (with lower frame bound $\frac{1}{B_a}$) are now the analysis set. Therefore, equation (3) relates the fidelity of the approximate actuator intensity coefficients to the fidelity of the resulting total action field,

$$\|Tx - \hat{y}\|^2 \leq B_a \sum_{l \in L} |d_l - \hat{d}_l|^2.$$

Using equations (7) and (6), we can write the total action field error in terms of individual sensor coefficients *not* communicated to actuator nodes

$$\|Tx - \hat{y}\|^2 \leq B_a \sum_{l \in L} \left| \sum_{k \in E_l} m_k \langle \tilde{a}_l, T\tilde{s}_k \rangle \right|^2 \tag{8}$$

$$\leq B_a \sum_{l \in L} \sum_{k \in E_l} |m_k \langle \tilde{a}_l, T\tilde{s}_k \rangle|^2. \tag{9}$$

As we see in equation (9), the networking strategy for sensor node k can use the value of $|m_k \langle \tilde{a}_l, T\tilde{s}_k \rangle|^2$ to quantify the maximum contribution it would make to the total action error by *not* communicating its measurement to actuator l. The bound in equation (9) can be used to set a threshold γ guaranteeing an absolute upper limit on the actuation error.

Importantly, the form of the error bound in equation (9) isolates each communication link as an independent term so that no communication overhead is required to determine the absolute worst actuation error that can be incurred by eliminating a communication link[1]. In applications where a WSAN must respond quickly to critical but rare events (e.g., a fire suppression system), an

---

[1] We are assuming that the setup phase of the WSAN has given nodes information about the relative locations of their neighboring nodes that can be used to calculate the necessary inner product.

absolute bound on the actuation error computed locally is probably appropriate. To ensure that the actuation error is within an absolute tolerance, the active communication links between sensors and actuators will necessarily change depending on the input signal. While this dynamic decision making doesn't impose a large computational burden on the sensor nodes, the underlying communications network must be able to handle large fluctuations in demand for resources.

Because the sensor and actuator fields overlap and form a frame (instead of an orthonormal basis), the contributions from two different sensor measurements to an actuator coefficient could, in effect, "cancel" each other. Because the error bound provided in equation (9) is expressly written in terms of local sensor node measurements, this bound favors a conservative interpretation rather than accounting for these interactions. Given a specific communication and networking scenario, it may or may not be advantageous to allow sensors to explicitly communicate to calculate a tighter error estimate (based on the original error expression in equation 2) and coordinate their communication accordingly. While the frame theoretic analysis paradigm introduced here would allow such an analysis, it would necessarily be specific to the application details (particularly the communication and networking scenario).

In many settings, designing around an absolute error constraint results in a system that is too conservative in its average behavior. To analyze the average actuation error one must assume a stochastic model for the measurements, such as assuming that the sensor measurements have zero mean ($\mathcal{E}[m] = 0$) and covariance matrix Γ_m. The covariance matrix Γ_m will be determined by a combination of the the sensor receptive field properties and the distribution assumed on x within the signal space $\mathcal{H}_x$. Only the first two moments of the distribution on m are relevant, so we need not assume Gaussian distributions.

Average WSAN performance is much easier to calculate if we recast equation (8) using matrix notation. We first need to write approximate actuator coefficients in equation (7) in terms of a perturbation of V, which captures the ideal transformation from sensor measurements to actuator coefficients. Let the approximate actuator coefficient be given by $\hat{d} = \left(V + \widetilde{V}\right) m$, where the matrix $\widetilde{V}$ is defined to remove inactive communication links:

$$\left(\widetilde{V}\right)_{k,l} = \begin{cases} -\left(\tilde{a}'_1 T \tilde{s}_1\right) & \text{if } k \in E_l \\ 0 & \text{if } k \in (K \setminus E_l). \end{cases}$$

Incorporating this definition into equation (8) and taking the expectation of both sides lets us bound the average error

$$\mathcal{E}\left[||Tx - \hat{y}||^2\right] \leq B_a \text{Tr}\left[\widetilde{V}\Gamma_m\widetilde{V}'\right], \tag{10}$$

where $\text{Tr}[\cdot]$ is the trace operator.

A system designer could use equation (10) to characterize (on average) how important a communication link between a specific sensor and actuator pair is to generating the total actuation field. Using this information, a WSAN design could choose *a priori* which communication links between sensors and actuators will be active in the network. Such a scheme has the disadvantage that it may not react well to events that are large deviations from the usual behavior. The advantages to this type of non-adaptive communication scheme in a WSAN are that the communication resources are used more efficiently most of the time, the network can count on a limited communication burden for any stimulus field, and the real cost of executing individual communication links (through a possibly multi-hop network) can be easily integrated into generating an optimal strategy. Also, it is worth noting that the bound in equation (10) is tighter than the bound in equation (9) (because it is based directly on equation (8)), reflecting the fact that all of the communication links can be considered jointly when designing the system for average error performance.

4   An Example WSAN System

As an illustrative example, consider a WSAN operating a fire suppression system in an office building with four research labs. Each lab contains expensive equipment, so there is a strong desire to localize the fire suppression to minimize water damage to adjacent labs. The building space is covered with a network of 21 temperature sensors (modeled with radially symmetric, exponentially-decaying receptive fields) and 13 actuators (modeled with an oriented and exponentially decaying influence field), all illustrated in Fig. 1. This WSAN has 273 possible communication links from the sensor nodes to actuator nodes. In this example we assume an equal communication cost for each link (i.e., we would like to use as few links as possible regardless of which links are in use).

We specified a function T mapping the temperature inputs to an imaginary desired fire suppression output. To illustrate that this mapping may be spatially varying, we note that fire activity in all labs will induce fire suppression activity along a path to the main exit. We used two sample temperature fields indicating a fire in different labs areas (shown in Fig. 2, along with optimal responses). As discussed in section 3, the quantity $|m_k\langle \tilde{a}_l, T\tilde{s}_k\rangle|$ determines the importance of each communication link (sorted and plotted in Fig. 3 for these test signals). In these signals, a threshold of $\gamma = .2$ allows approximately 15 of the 273 possible

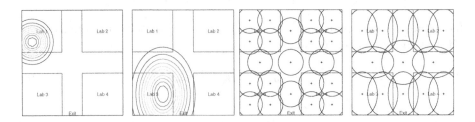

Fig. 1. Contour plot of example sensor *(Far left)* and actuator *(Middle left)* nodes. Layout and shape of the sensor *(Middle right)* and actuator *(Far right)* nodes.

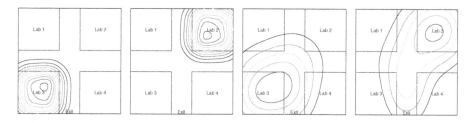

Fig. 2. Contour plots of sample temperature fields for test signal 1 indicating a fire in lab 3 *(Far left)* and test signal 2 indicating a fire in lab 2 *(Middle left)*. Contour plots of optimal actuation responses to the two test scenarios *(Middle right and far right, respectively)*. Different spatial response characteristics keep the main exits clear.

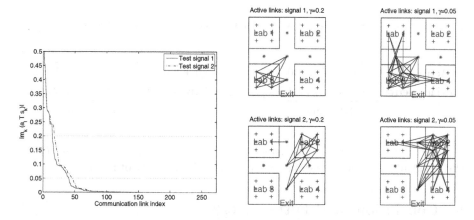

Fig. 3. *Left:* The importance measurements of each communication link ($|m_k \langle \tilde{a}_l, T\tilde{s}_k \rangle|$) are sorted and plotted for the two test signals. *Right:* Connection diagrams for the two test signals under the two thresholds in the example system. Sensor nodes are marked with a blue (+) and actuator nodes are marked with a red (*). Active connections from a sensor to an actuator are denoted by a blue line.

communication links to be active, and $\gamma = .05$ allows approximately 40 active communication links. The resulting active communication links are shown in Fig. 3. Close examination of the connection diagrams shows that some communication choices are non-obvious; the most important sensor to a particular actuator is not always the one with heavily overlapping influence functions.

The actuation response is generated for both test signals using threshold values of $\gamma = .2$ and $\gamma = .05$, and the resulting total actuation fields are plotted in Fig. 4. The reduced communication scheme based on the thresholds resulted in the number of active communication channels and associated percentage errors given in Table 1. The principles discussed in section 3 allow the WSAN to generate excellent approximations to the optimal actuation field by using local rules to activate only a fraction of the communication links. Interestingly, if we acti-

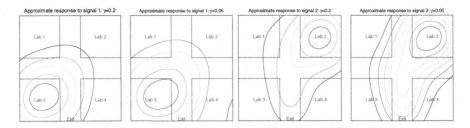

Fig. 4. Contour plots of actuation responses when using only a subset of possible communication links (determined by thresholding each link's importance to the total actuation). Approximate responses to test signal 1 are shown when using 14 and 40 communication links (*Far left and middle left*). Approximate responses to test signal 2 are shown when using 17 and 45 communication links (*Middle right and far right*).

Table 1. Results from the example WSAN fire suppression system

	$\gamma = .2$	$\gamma = .05$		$\gamma = .2$	$\gamma = .05$
Active links	14	40	Active links	17	45
Relative error	2.22%	0.04%	Relative error	2.46%	0.15%
	Test signal 1			Test signal 2	

vate the same number of links using the more intuitive measure $|m_k \langle a_l, T s_k \rangle|$, the resulting actuation error increases by roughly an order of magnitude.

5   Conclusions and Future Work

WSANs are often discussed as a logical extension to sensor networks, but there is little research investigating sensor and actuator systems working in concert together. While algorithms that reduce communications and ensure data fidelity for sensor measurements are important for many applications, they are not the ultimate arbiter for obtaining good actuation performance. The total system must be designed and managed with the final actuation goal in mind. Our frame-theoretic WSAN model illustrates one strategy for taking such a holistic information management view with actuation fidelity as the relevant metric.

The analytic tools we present characterize the effect of eliminating an individual communication link between a sensor and an actuator, both in terms of absolute (for specific sensor measurements) and average actuation error. Choosing a networking strategy for eliminating communication links is both difficult and non-intuitive. While intuition would indicate that the relationship between the activation fields of a sensor and an actuator are the relevant quantity characterizing the importance of the communication between those two nodes, our work shows that it is the relationship between the mathematical *duals* of the activation fields that captures this inherent importance. It is through these dual functions that the relationship of the whole sensor network to the whole actuator network can be accounted for in local communications between pairs of nodes. Characterizing the importance of individual communication links to the overall goal points directly to how a networking strategy could weigh the costs and benefits of each communication link to achieve the desired balance between performance and energy efficiency. The value of our analysis is highlighted in an example WSAN system where link activations based on the sensor and actuator duals performed an order of magnitude better than activations based on the simple overlap of the sensor and actuator receptive field functions.

Today we are only seeing the beginning of work in information management in WSANs. In this work, we have given explicit upper bounds on actuation error that can be determined locally with no cooperation between the sensors. We have also indicated how this analysis framework could be used in a specific application and networking scenario to investigate the benefits of allowing local sensor coordinate their communications to an actuator. Finally, we have also derived analogous average error bounds that could be used to design static networking strategies for applications where that approach is more appropriate.

We are currently working on many extensions to this work. We have considered the case where perfect (analog) coefficients are sent on active communication links. While real systems would have to use quantized coefficients, we believe that typical quantization schemes would have only a second order effect relative to other actions taken to limit communication (such as eliminating communication links). However, it is more interesting to consider a variable rate communication scheme where some links could send coefficients with variable fidelity. Such variable rate schemes could be particularly interesting as we consider incorporating information about the variable networking costs of different communication links. We are working to more tightly integrate the costs and benefits of individual communication links to find optimal strategies for determining which links to activate dynamically and with minimal overhead.

Finally, our system model considers a single actuation response to a set of sensor measurements. This is something of an open-loop system because the sensors don't necessarily receive any direct feedback from the actuators. This generality is appealing in many senses; our model allows sensors and actuators to live in separate signal spaces and it may be possible that actuation is not directly observable by the sensors. However, in many practical applications, future sensor measurements will be affected by actuator behavior even when they operate in different signal spaces (e.g., fire suppression actions will reduce the temperature measured by sensors). We are working on methods for extending this work to consider the dynamic properties of such an implicit feedback system.

References

1. Nowak, R.: Distributed EM algorithms for density estimation and clustering in sensor networks. IEEE Transactions on Signal Proc. **51** (2003) 2245–2253
2. Rabbat, M., Nowak, R.: Distributed optimization in sensor networks. In: Intl. Symposium on Information Proc. in Sensor Networks (IPSN), Berkeley, CA (2004)

3. Blatt, D., Hero, A.: Distributed maximum likelihood estimation in sensor networks. In: Intl. Conf. on Acoustics, Speech, and Signal Proc., Montreal, Canada (2004)

4. Liu, J., Chu, M., Liu, J., Reich, J., Zhao, F.: State-centric programming for sensor and actuator network systems. IEEE Pervasive Computing Magazine **2**(4) (2003)

5. Gerkey, B., Mataric, M.: A market-based formulation of sensor-actuator network coordination. In: AAAI Spring Symp. on Intel. Embed. and Dist. Sys. (2002)

6. Coates, M.: Evaluating causal relationships in wireless sensor/actuator networks. In: Intl. Conf. on Acoustics, Speech, and Signal Proc., Philadelphia, PA (2005)

7. Lemmon, M.D., Ling, Q., Sun, Y.: Overload management in sensor-actuator networks used for spatially-distributed control systems. In: Proceedings of the ACM Sensys Conference. (2003)

8. Kandel, E., Schwartz, J., Jessell, T.: Principles of Neural Science. Third edn. Appleton & Lange, Norwalk, CT (1991)

9. Neil, D.: Compensatory eye movements. In Sandeman, D., Atwood, H., eds.: The Biology of Crustacea, Neural Integration and Behavior. Academic Press, New York (1982) 133–163

10. Glantz, R., Nudelman, H., Waldrop, B.: Linear integration of convergent visual inputs in an oculomotor reflex pathway. J. of Neurophys. **52**(6) (1984) 1213–1225

11. Nowak, R., Mitra, U., Willett, R.: Estimating inhomogeneous fields using wireless sensor networks. IEEE J. on Selected Areas in Comm. **22**(6) (2004) 999–1006

12. Christensen, O.: An Introduction to Frames and Riesz Bases. Birkhauser, Boston, MA (2002)

13. Duffin, R., Schaeffer, A.: A class of nonharmonic Fourier series. Transactions of the American Mathematical Society **72**(2) (1952) 341–366

14. Daubechies, I.: Ten Lectures on Wavelets. Society for Industrial and Applied Mathematics, Philadelphia, PA (1992)

15. Bolcskei, H., Hlawatsch, F., Feichtinger, H.: Frame-theoretic analysis of oversampled filter banks. IEEE Transactions on Signal Proc. **46**(12) (1998) 3256–3268

16. Candès, E., Donoho, D.: New tight frames of curvelets and optimal representations of objects with piecewise C^2 singularities. Communications on Pure and Applied Mathematics **57**(2) (2004) 219–266

17. Strohmer, T., Heath Jr., R.: Grassmannian frames with applications to coding and communcations. Applied and Comp. Harmonic Analysis **14**(3) (2003) 257–275

18. Goyal, V., Kovačević, J., Kelner, J.: Quantized frame expansions with erasures. Applied and Computational Harmonic Analysis **10** (2001) 203–233

19. Gao, J., Harris, C., Gunn, S.: On a class of support vector kernels based on frames in function hilbert spaces. Neural Computation **13** (2001) 1975–1994

20. Balan, R., Casazza, P., Heil, C., Landau, Z.: Density, overcompleteness, and localization of frames, I. Theory. Preprint (2005)

Roadmap Query for Sensor Network Assisted Navigation in Dynamic Environments

Sangeeta Bhattacharya, Nuzhet Atay, Gazihan Alankus,
Chenyang Lu, O. Burchan Bayazit, and Gruia-Catalin Roman

Department of Computer Science and Engineering,
Washington University in St. Louis

Abstract. Mobile entity navigation in dynamic environments is an essential part of many mission critical applications like search and rescue
and fire fighting. The dynamism of the environment necessitates the mobile entity to constantly maintain a high degree of awareness of the changing environment. This criteria makes it difficult to achieve good navigation performance by using just on-board sensors and existing navigation
methods and motivates the use of wireless sensor networks (WSNs) to
aid navigation. In this paper, we present a novel approach that integrates
a roadmap based navigation algorithm with a novel WSN query protocol
called Roadmap Query (RQ). RQ enables collection of frequent, up-to-
date information about the surrounding environment, thus allowing the
mobile entity to make good navigation decisions. Simulation results under realistic fire scenarios show that in highly dynamic environments RQ
outperforms existing approaches in both navigation performance and
communication cost. We also present a mobile agent based implementation of RQ along with preliminary experimental results, on Mica2 motes.

1 Introduction

Mobile entity navigation is a crucial part of many mission critical applications
like fire fighting and search and rescue operations in disaster areas. These scenarios usually involve dynamic environments that make navigation dependent
on up-to-date knowledge of the changing environment. Moreover, information
about a large region around the mobile entity is required in order to achieve
good navigation performance. For example, in the case of a robot navigating a
region on fire, the robot would need real-time temperature information about
the surrounding areas in order to navigate the region without getting burnt.
Also, due to the highly dynamic and unpredictable nature of spreading fire,
temperature information of the surrounding areas would be needed frequently
for continuous awareness of the neighboring environment. On-board sensors have
a limited sensing range and hence cannot provide sufficient information required
to make good navigation decisions. Wireless sensor networks (WSNs), on the
other hand, present new opportunities to obtain frequent, up-to-date information about a large expanse of the surrounding area. Information obtained from
the WSN can be used by the mobile entity to make good navigation decisions,

P. Gibbons et al. (Eds.): DCOSS 2006, LNCS 4026, pp. 17–36, 2006.
© Springer-Verlag Berlin Heidelberg 2006

with reduced risks. Moreover, WSNs are easily deployable and are also economically feasible. Once deployed, a WSN can serve several mobile entities and can also be employed to coordinate the movement of multiple mobile entities.

The use of WSNs for navigation in dynamic environments presents important new challenges. Since frequent sensor data updates are required to maintain continuous awareness in dynamic environments, the data collection process can induce a heavy communication workload on the WSN, which usually has limited bandwidth and energy. The resulting network contention and congestion may cause excessive communication delay and loss of sensor data, which may significantly affect the safety and navigation performance of the mobile entity. Therefore, it is important to design efficient query protocols that can collect updated sensor data needed for safe navigation at minimum communication cost.

In this paper, we present a novel roadmap-based approach for navigation in dynamic environments. Our approach consists of two components; a roadmap-based navigation algorithm for the mobile entity and a distributed query protocol called *Roadmap Query (RQ)* for the WSN. The navigation algorithm uses a *roadmap* of the region, which is a virtual graph consisting of possible paths in the region, to search for a safe path to the goal. The path is selected based on roadmap edge weights derived from current sensor data that is collected from the WSN using RQ. RQ achieves communication cost savings by querying nodes only in the vicinity of the mobile entity, called *query area*, and by using a novel sampling strategy that queries only a few selected nodes lying along roadmap edges in the query area. The selective sampling strategy eliminates communication cost resulting from the collection of unnecessary and redundant data, while still enabling RQ to provide sufficient data needed for successful navigation in dynamic environments.

The main contributions of this paper are as follows. (1) We propose a new approach to mobile entity navigation that integrates roadmap based navigation algorithms with distributed query protocols; (2) We present Roadmap Query (RQ), a robust query protocol optimized for navigation in highly dynamic environments; (3) We provide a mobile agent based implementation of a sensor-network assisted navigation system on Mica2 motes; (4) We show through simulations that RQ achieves better navigation performance than existing protocols at only a small fraction of communication cost, in face of realistic fire scenarios and node failures.

2   Related Work

Several methods for robot navigation have been proposed in the past. These methods either assume a priori knowledge of the environment or use on-board sensors to avoid obstacles. A priori knowledge of the environment is not helpful in dynamic environments while on-board sensors have a limited sensing range and hence do not provide information about a sufficiently large region. Recent work in this area suggests integrating WSNs with mobile entities to enable navigation in dynamic environments. The proposed methods fall into two distinct categories.

The first category uses some form of global flooding initiated by the goal, the obstacle or the mobile entity itself. While this approach is effective in relatively static environments, it is unsuitable for dynamic environments since the need to constantly maintain a high degree of awareness of the changing environment (e.g., a spreading fire) would cause frequent flooding of the network. Thus, this approach may suffer from high communication cost and network contention, which would lead to poor navigation performance. It also wastes energy, thereby decreasing network lifetime. Protocols suggested in [1] and [2] fall into this category. Both protocols construct global navigational fields to guide the robot to the goal. In [1], the goal generates an attractive potiential field that pulls the robot towards the goal, while an obstacle generates a repulsive potential field that pushes the robot away from the obstacle. We will henceforth refer to this method as the Dartmouth Algorithm (DA). Unlike DA, the method in [2] uses value iteration to compute the magnitude of directional vectors that guide the robot to the goal. The approach presented in [3] addresses navigation of mobile sensor nodes, to increase coverage of event locations. In this approach, the goal (an event location) initially floods the network to locate a suitable mobile sensor node. Mobile sensor nodes respond to the flood by sending a response to the goal. The protocol then creates a navigation field around the path taken by the response, to draw the mobile node to the goal. Since the navigational field is only around a path that does not change until the goal changes, this approach cannot efficiently handle dynamic obstacles. Unlike the above approaches, the approach used in [4] assumes that a path already exists in the network, and uses controlled flooding to guide the robot to the start of the path, after which the robot follows the path. This approach is not applicable to dynamic environments where an initially safe path may quickly become unsafe due to changing conditions.

The second category of protocols do not use global flooding but instead use a local query strategy to achieve navigation. Our earlier work, presented in [5], which we will henceforth call Local Query (LQ), falls into this category. In LQ, the path from the start to the goal is built incrementally as the mobile entity traverses the region, by querying *all* nodes in the vicinity of the mobile entity. This approach avoids global flooding by making local decisions. While this method is more efficient than global flooding in a dynamic environment, it still wastes significant amount of energy and bandwidth by unnecessarily collecting information from *all* nodes in the query area. In contrast, RQ uses a selective sampling strategy to collect only necessary information, which is dependent on the environment and changes with it. As a result of this strategy, RQ achieves better navigation performance than LQ at only a small fraction of LQs communication cost (shown in Section 7). This feature makes RQ especially suitable to resource-constrained WSNs. Furthermore, LQ ignores the issue of sensor node failures. In contrast, RQ is designed to handle sensor node failures caused by dynamic obstacles (e.g., being burnt by fire). The robustness of RQ is crucial in such harsh environments where nodes can be easily destroyed. Another new contribution of this work is that unlike LQ, which was implemented in native code, RQ has been implemented using *mobile agents* that can dynamically reprogram nodes in the current query area as the mobile entity moves. An important advantage of our mobile-agent-based implementation is that it enables the adaptive deployment of navigation applications into pre-deployed WSNs with limited resources.

3   Problem Formulation

The navigation problem that we address in this paper is to find a *safe path* for a mobile entity through a sensor field from a start point p_s to a goal point p_g. We define a *safe path* to be a path that is clear of *dynamic obstacles*, i.e., obstacles whose location or shape changes with time (e.g., car, fire).

In this paper, we consider fire as the representative example for a dynamic obstacle. Thus, the temperature of the region traversed by the mobile entity is a function of time and is affected by the location and movement of fire. In this case, the problem can be restated as that of finding a safe path for a mobile entity, from start to goal, without the mobile entity getting burnt. The mobile entity is assumed to get burnt if the temperature at its location is higher than a threshold Δ_{burn}. A safe path is now redefined as one where the maximum temperature along the path taken by the mobile entity remains below the threshold Δ_T, while the mobile entity is on the path. Even though our solution is designed assuming fire as the dynamic obstacle, it can be generalized to other types of dynamic environments where safety is defined by changing sensory values (e.g., chemical spills, hazardous gas and air pollution).

We make the following assumptions in the paper: (i) Nodes are location aware. (ii) The mobile entity communicates with the WSN through an on-board gateway device (e.g., PDA) (iii) Nodes have a limited sensing range R_S. R_S is chosen such that if the temperature sensed by a node is below the threshold Δ_T, then the temperature at any point within the sensing range is below the threshold Δ_{burn}. Hence, edges with nodes having temperature above Δ_T are unsafe. The sensing range is thus dependent on the tunable parameter Δ_T. A lower Δ_T results in a longer sensing range.

4   Navigation Algorithm

Our navigation algorithm adapts the roadmap method that is commonly used for navigation in robotics, to make it more suitable for dynamic environments and for integration with WSNs. The roadmap method builds a roadmap of the region and uses it to find a path from the start to the goal. It only considers paths on the roadmap instead of all possible paths in the region and hence, has low computational complexity. Furthermore, it is particularly suitable for WSN assisted navigation, since it reduces the amount of sensor data that must be collected from the WSN by requiring information only along the roadmap edges.

Thus, our navigation algorithm first constructs a roadmap of the region and then incrementally finds safe sub-paths (consisting of roadmap edges) leading to

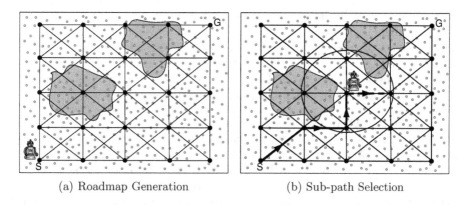

(a) Roadmap Generation   (b) Sub-path Selection

Fig. 1. Working of the Navigation Algorithm. The figures show the grid roadmap. Roadmap vertices and sensor nodes are depicted by black dots and gray dots, respectively. The start and goal are denoted by S and G respectively. The gray shaded regions denote fire. Figure (a) shows the generated roadmap. Figure (b) shows the sub-path selected (dotted arrow) within the query area (shown by the circle). The solid arrows show the path taken by the robot to reach its current location.

the goal. Sub-paths are selected based on edge weights that are repeatedly updated using temperature information obtained from the WSN, through RQ. The detailed working of our navigation algorithm is as follows. After constructing the roadmap, the mobile entity issues a query to obtain the maximum temperature along the roadmap edges lying within the query area. We assume a circular query area of *query radius* R_q, centered at the current location of the mobile entity $p_e(t)$. After issuing the query, the mobile entity waits for a time T_w to receive the query result which contains the maximum temperatures along the roadmap edges within the query area. At the end of the wait period T_w, the mobile entity computes the edge weights based on the temperature information obtained and finds a safe sub-path to the goal. If the mobile entity finds a safe sub-path, it starts moving along the sub-path. Otherwise, it reissues the query. This entire process is repeated every time the mobile entity reaches the end of a sub-path, until it safely reaches the goal. Note that the navigation algorithm handles the dynamics in the environment by generating the path incrementally, based on fresh information collected from the current query area.

Thus, the roadmap navigation algorithm has three stages, (i) roadmap generation, (ii) roadmap edge weight assignment and (iii) sub-path selection. The roadmap generation stage occurs at the start of the navigation process while the roadmap edge weight assignment and sub-path selection stages, occur repeatedly till the mobile entity reaches the goal safely. These stages are discussed next.

(i) Roadmap generation: We use a grid as the roadmap, as shown in Figure 1(a), where the grid points form the *roadmap vertices* and the edges form the *roadmap edges*. Note that the grid points are virtual points that are placed in space, without considering sensor locations. Traditional roadmap methods (e.g., Probabilistic Roadmap Methods [6]) randomly choose points in space and connect them to construct a roadmap. A benefit of using a grid is that roadmap information can be easily included in a query message without significantly increasing the message size (see Section 5.1). The grid size is a tunable parameter that is a tradeoff between communication cost and navigation performance.

(ii) Roadmap edge weight assignment: The roadmap edge weights are used to find a short, safe sub-path on the roadmap that leads to the goal, as the mobile entity traverses the roadmap. In order to balance path safety and path length, we use an edge weight function that is a weighted function of the normalized maximum edge temperature and the normalized edge length. The maximum edge temperature is provided by the RQ protocol that queries the WSN. The weight of an edge e is thus

$$W_e = \begin{cases} \alpha(\frac{\delta_e}{\Delta_M}) + (1-\alpha)(\frac{l_e}{L}) & \delta_e < \Delta_T \\ \infty & \delta_e \geq \Delta_T \end{cases} \tag{1}$$

where δ_e is the maximum temperature on e based on recent query results, l_e is the length of e, Δ_M is the maximum possible temperature, L is the maximum edge length among all roadmap edges E and $\alpha \leq 1$ is the weight given to the temperature field. The tunable parameter α determines the tradeoff between safety and path length. δ_e in equation 1 is obtained from the query result and is approximated as $\delta_e = max(\delta_s), s \in S$, where S is the set of nodes that cover edge e and δ_s is the temperature at a sensor $s \in S$. A sensor is said to *cover* an edge if the edge or part of the edge lies within its sensing circle. On the other hand, an edge is said to be covered if certain points on the edge are covered. The points on the edge that need to be covered are determined by the query protocol and will be discussed later. If an edge e is not covered by the nodes that respond to the query, then W_e is pessimistically set to ∞ so as to avoid traversing that edge.

Edge weights are timestamped and expire after a certain interval Δ_{exp}. Δ_{exp} should be chosen carefully, since a large Δ_{exp} will not account for the dynamism of the environment while a small Δ_{exp} may cause the mobile entity to oscillate. Thus, at the end of a query the edges within the query area have edge weights based on up-to-date temperature information obtained from the query result while the edges in the past m query areas have edge weights based on old temperature data, where m depends on Δ_{exp}. All other edges have weights based only on the edge length.

(iii) Sub-path selection: A sub-path consisting of edges lying within the query area is selected by running the Dijkstra's shortest path algorithm [7] on the roadmap. The result of the Dijkstra's algorithm is a path with the least weight from the mobile entity's location to the goal, at that instant. The sub-path consisting of edges within the query area, is extracted from this least-weight path. This stage is illustrated in Figure 1(b).

5   Roadmap Query

In this section, we present the RQ protocol. RQ collects updated temperature information from nodes covering the roadmap edges in a query area. It is issued by the navigation algorithm, every time the mobile entity reaches the end of a sub-path, until the mobile entity reaches the goal. In addition, to improve safety, it is also issued when the temperature at the mobile entity location rises above the threshold Δ_T. The temperature at the mobile entity location is obtained using an on-board sensor.

5.1   Basic Roadmap Query Protocol

RQ minimizes the communication workload on the network by reducing the number of nodes involved in the query process. This is achieved by optimizing the query protocol in accordance with the roadmap-based navigation algorithm. Since the navigation algorithm requires the maximum temperature only along the roadmap edges, the query message, is forwarded only along the roadmap edges lying within the query area. Moreover, due to the high density of sensor nodes, the query message is not forwarded by all nodes along an edge, but only by some selected nodes. These selected nodes form a backbone of nodes along the roadmap edges that fall within the query area and are called *backbone nodes*. RQ requires all backbone nodes to respond to the query. Nodes that hear the query message but are not on the backbone, respond to the query only if they satisfy a certain criteria and are called *non-backbone nodes*. The backbone and non-backbone nodes form a tree structure with the mobile entity as the root. The formed tree is used to aggregate and deliver the query results to the mobile entity. Thus, RQ reduces communication cost not only by reducing the number of nodes that forward the query message but also by reducing the number of nodes that respond to a query, within a query area.

In order to achieve communication cost reduction, RQ requires the queried nodes to have knowledge of the roadmap and to maintain 2-hop neighborhood information. Since we use a grid as the roadmap, the first requirement is easily met by including the location of the bottom left corner of the grid and the grid square size in the query message. Each queried node uses this information, to calculate the grid points and edges. The second requirement requires all nodes in the network to maintain 2-hop neighborhood information which may introduce some overhead. Neighborhood information is maintained through *hello messages* [8], which contain the ID of the sending node and the IDs of its 1-hop neighbors. Hello messages are broadcasted periodically by each node at an interval called the *hello period*. On receiving a hello message, the receiving node records the sending node as its neighbor and also stores the neighborhood information of the sending node. Each entry in the neighborhood table is associated with a timestamp that corresponds to the time the most recent hello message was received from that neighbor. The timestamp field is used to detect failed neighbors. We have described a simple neighborhood management technique but more sophisticated techniques [9] can also be used. Note that similar neighbor-

1. if *Query message* received
2.   Accept if in current query area.
3.   Set sending node as parent.
4.   Set h_i to hop count in msg plus 1.
5.   Apply forwarding rule to see if msg should be re-broadcasted.
      If yes, re-broadcast msg.
6.   Apply reply rule to see if query result should be sent. If yes,
      calculate time to send result and set timer *SendTimer* to time-
      out at the right time.
7. else if *Query reply* received
8.   if result not yet sent then store else discard.
9. else if *SendTimer* timed out
10.   Send aggregated *Query result* to parent.

Fig. 2. Roadmap Query (RQ) Algorithm

hood information is also required by other common services such as routing and power management.

RQ uses two rules to determine which nodes should forward the query message or respond to the query. We call the rule that determines if a node should forward the query message, as the *forwarding rule* and the rule that determines if a node should respond to a query, as the *reply rule*. The forwarding rule identifies backbone nodes while the reply rule identifies non-backbone nodes.

Forwarding Rule: By the forwarding rule, if a node receives a query message that is being propagated along edge $e = \overrightarrow{p_{e1}p_{e2}}$ where p_{e1} and p_{e2} are the endpoints of the edge, and the arrow denotes the direction of query message propagation, then, the node rebroadcasts the message only if it covers edge e and is the closest to p_{e2} among its neighbors that can also hear the same query message. A node knows if a neighbor can hear the same query message by checking its neighborhood table that contains 2-hop neighborhood information. Note that, by this method, only a few nodes along the edge, called backbone nodes, rebroadcast the query message. Thus, some nodes do not rebroadcast the query message, thinking that another node that is closer to the endpoint will rebroadcast the message. These nodes listen for a certain time interval to see if the query message is rebroadcasted. If the query message is not rebroadcasted within this time, these nodes rebroadcast the message. This method takes care of situations where the node selected to rebroadcast the query message does not receive the query message due to collision or other factors.

Reply Rule: The reply rule states that a node should send a query reply, if (i) it is a backbone node, or, (ii) its temperature is above Δ_T and it covers a roadmap edge that falls within the current query area. The first condition draws query results from the minimum number of nodes that entirely cover all roadmap edges within the query area. The second condition identifies non-backbone nodes and adapts the number of nodes responding to the query, according to the danger level. This condition enforces the safety of the path by drawing query results

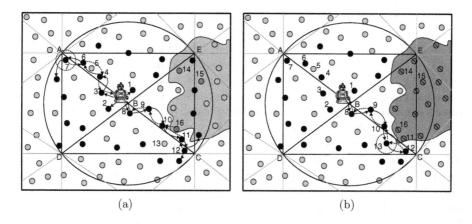

(a)   (b)

Fig. 3. Working of RQ protocol. The figures show the backbone (colored black) and non backbone nodes (colored dark gray) in a query area (solid black circle) centered at B. Roadmap edges within the query area are colored black. Figure (a) shows the query message (solid arrow) and query reply (dotted arrow) propagation along edges BA and BC. The shaded region represents a region with temperature above the threshold. Figure (b) shows a possible case where RQ fails. The shaded region represents very high temperature at which all nodes in the region (crossed out) have failed. In this case, even though the fire spreads across edge BC, the mobile entity considers BC safe since it hears from enough active nodes with temperatures below Δ_T, that cover BC.

from nodes if they sense a dynamic obstacle (e.g. fire) near roadmap edges lying within the query area.

Given these two rules, RQ works as follows. On receiving a query message, a node i that lies within the query area, sets the sending node j as its parent, if the link between the nodes is symmetric and sets its hop count h_i to $h_j + 1$ where h_j is the hop count of the sending node and is contained in the query message. The hop count is used to send the query results at a time that facilitates data aggregation. Node i then applies the forwarding rule to determine if it should rebroadcast the message. If it is required to rebroadcast the message, it rebroadcasts the message and then applies the reply rule to determine if it should respond to the query. If it needs to respond to the query, it calculates the time t_r at which the result needs to be sent and sets a timer to timeout at that time. When the timer times out, node i sends its parent an aggregated query result, deduced from its information and the information obtained from its children. If node i and its children are along the same edge, the MAX function is applied to the sensor readings. Otherwise, the results are just merged into one message.

The query reply time t_r is calculated such that it facilitates data aggregation and is set to $t_0 + \frac{h_{max} - h_i}{h_{max}} \times T_w$, where t_0 is the time at which the mobile entity sends the query request and h_{max} is a tunable parameter denoting the maximum possible hop count within a query area. Thus, a node waits for time interval $T_r = t_r - t_c$, where t_c is the time at which the node receives the query message, to receive query replies from its children. The RQ algorithm is shown in Figure 2 and is illustrated in Figure 3(a).

Figure 3(a) illustrates different aspects of the RQ protocol. (i) It shows the backbone (colored black) and non-backbone (colored dark gray) nodes selected by the RQ protocol in response to a query issued by a robot positioned at B. The resulting query area is shown by a solid circle centered at B. The roadmap edges lying within the query area are colored black. (ii) The figure illustrates the query message (solid arrow) propagation along edges BA and BC. The query message is propagated from node 2 to node 3, to node 4 and then to node 6 and finally to node 7 along edge BA. The query message propagation along edge BA stops at node 7 since it covers endpoint A, i.e. A lies within node 7's sensing range (shown by dotted circle centered at 7). Node 7 then propagates the message along the adjoining roadmap edges in the query area. Note that the query message is forwarded by node 4 and then by node 6 and not by node 5. This is because of the forwarding rule. When node 5 hears the query message from node 4, it sees that it has a neighbor, node 6, that is also node 4's neighbor (hence, it must have also heard the query message) and that is closer to endpoint A. Thus, by the forwarding rule it does not rebroadcast the query message. (iii) The figure illustrates the outcome of the reply rule when a portion of the query area (shaded region) has temperature above the threshold. By the reply rule, nodes 14, 15 and 16 in this area must reply to the query, since their temperatures are above the threshold and they cover a roadmap edge lying within the query area. These nodes are thus, non-backbone nodes. (iv) The query reply (dotted arrow) propagation along edges BA and BC is also shown. Note how a non-backbone node, node 16, becomes a leaf node, under parent node 10.

5.2   Extension to Handle Node Failures

Robustness to node failures is especially important in dynamic environments since nodes can be destroyed by harsh environments such as fire. The basic RQ protocol cannot handle certain situations arising due to node failures, as shown in Figure 3(b). In the figure, the shaded region depicts a spreading fire, that burns nodes in the region. Due to node failures, edges BE and EC are not covered by working nodes. Hence, the robot does not receive sufficient information about these edges and considers them unsafe. However, edge BC is completely covered since even though the fire burns node 11, node 13 (which is unaware of the nearby fire) takes its place in forwarding the query message, thus giving the robot the false impression that the edge is safe. If the robot were to choose to traverse edge BC, it would collide with the dynamic obstacle, the fire, and get burnt. This scenario shows the importance of fault-tolerance in dynamic environments. Therefore, we extend RQ to avoid such situations.

In order to make RQ fault-tolerant we include node failure information in the query results. The mobile entity, uses this information to avoid paths with failed nodes, assuming that node failures are due to destruction by fire. Node failure information is obtained, by requiring nodes to send a list of failed neighbors that cover roadmap edges in the current query area, along with their sensor reading. Also, the reply rule is modified slightly such that nodes now send a query reply if (i) they are backbone nodes, (ii) their temperatures are above a threshold and they cover a roadmap edge lying within the query area or (iii) they have failed neighbors that cover roadmap edges lying within the query area.

It can be seen that with these modifications RQ is successful in situations like the one depicted in Figure 3(b). This is because, by the modified reply rule, the robot is informed about the failed nodes 11 and 16 by either node 10, node 12 or node 13. Since node failure is assumed to be due to destruction by fire, the robot infers that the edge BC is not safe and does not traverse that edge. Thus, the modifications make RQ robust to situations where the fire destroys only some nodes along an edge leaving enough working nodes with temperatures below Δ_T to cover the edge, which would give the mobile entity the false impression that the edge is safe.

The modified RQ protocol depends on node failure information, which is easily obtainable. Since each node maintains a neighborhood table and receives periodic hello messages from its neighbors, a node knows if a neighbor has failed, if it hasn't heard from the neighbor in n hello periods. The choice of n has to be made carefully, since a lower value of n will result in more false positives while a higher value of n will result in delayed awareness of danger, thus leading to poor navigation performance. In our simulations, we set $n = 2$.

5.3 Analysis

In this section, we show that RQ successfully gathers the information required by the navigation algorithm within a query area, under the following two conditions.

The first condition is a sensing covered network. In a sensing covered network, every point in the region is covered by at least one sensor. Without this network property, it is impossible to guarantee that a roadmap edge is covered by any sensor at all. A sensing covered network is desirable, as it increases the mobile entity's awareness of the surroundings thus improving its navigation path.

The second condition is the double range property, by which, the communication range R_C of a node is at least twice the sensing range R_S of the node, i.e., $R_C \geq 2R_S$. The double range property guarantees network connectivity in a sensing covered network [10] and hence is a desirable property for such networks. Since the sensing range depends on the temperature threshold Δ_T, we can achieve the double range property by selecting an appropriate Δ_T.

Query message propagation in RQ, starts at a node s that receives the query message from the mobile entity and is closest to the mobile entity's location. From node s, the query message is forwarded along edges covered by s and then along edges that are connected to them, and so on. Message propagation from one edge to another occurs at nodes that cover the intersection point of two or more edges. Note, that only roadmap edges that *completely* lie within the query area, are considered per query. Since these edges lie completely within the query area, they form a connected subgraph. Given the above, we can prove that "*Given a sensing covered network with $R_C \geq 2R_S$, every node covering a roadmap edge* *lying completely within a query area receives the query message from the mobile entity, under RQ"*. The proof [11] is omitted due to space limitations.

This property of RQ is very useful in environments that do not cause node failures (e.g., chemical spill and air pollution). Environments like fire that cause node failures violate the sensing coverage condition, in which case RQ only provides best effort service. We note that it is extremely difficult to provide any guarantees in the presence of node failures. However, as shown in our simulations, RQ still provides sufficiently good performance in a number of realistic scenarios.

6   Implementation

We implemented the basic RQ protocol on Agilla [12, 13], a mobile agent middleware for the TinyOS [14] platform. A mobile agent based implementation enables RQ to be used in a pre-deployed WSN without requiring the RQ protocol to be pre-installed on the WSN. A WSN running some other application can be quickly re-utilized to run the RQ protocol by just injecting mobile agents containing the protocol into the network. The capability to flexibly reprogram a WSN for a different application is particularly important to WSNs that have limited storage and long operational lifetime [12, 13, 15]. For example, a WSN deployed in a building for temperature monitoring can be quickly re-programmed to run the RQ protocol in case of a fire emergency. The RQ protocol can then be used to guide people safely out of the building.

An Agilla application consists of one or more mobile agents that coordinate with each other, to achieve application-specific behavior. An agent is programmed using a high-level language supported by Agilla. Agilla provides primitives for an agent to move and clone itself from sensor node to sensor node while carrying its code and state, effectively reprogramming the network. New mobile agents can be injected onto a sensor node, thereby allowing new applications to be installed after the network has been deployed. To facilitate inter-agent coordination, Agilla maintains a local tuple space and neighbor list on each sensor node. Multiple agents can communicate and coordinate through local or remote access to tuple spaces. Prior experiences with Agilla have demonstrated that it can provide efficient and reliable services needed by highly dynamic applications such as fire tracking [13].

6.1   RQ Using Agents

In the agent based implementation of RQ, the mobile entity injects an *explorer agent* into the network that collects the edge weights and delivers them to the mobile entity. Once injected into the network, the explorer agent clones itself on nodes lying along the roadmap edges according to the forwarding rule. The reply rule is applied to determine the agents that need to respond to the query. The agent migration sets up a tree structure along the roadmap edges within the query area, which is used to collect the query result. Per node query results are aggregated such that a list of per-edge-maximum-temperatures is forwarded

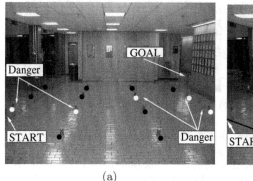 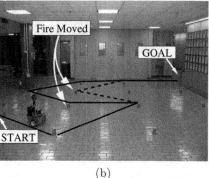

(a)   (b)

Fig. 4. (a) Experimental environment. (b) The robot avoids the initial path (dotted line) and follows a safer path (solid line) when the fire spreads.

along the tree branches to the mobile entity through remote tuple space operations. The mobile entity processes the query result and takes appropriate action as explained before.

6.2 Experiments

We used a Pioneer-3 DX robot by ActiveMedia [16], as the mobile entity in our experiments. The robot controller carried a mote as a communication interface to a WSN consisting of Mica2 motes. The WSN was arranged in a 4x4 grid, with a grid square length of 2 meters, as shown in figure 4(a). Each node was assigned an (x,y) coordinate based on its position in an euclidean co-ordinate system. In the figure, the node in the lower-left corner was assumed to be the origin of the co-ordinate system with coordinate (0m, 0m). The coordinate of the node in the upper-right corner is therefore (6m, 6m).

The goal of the robot, in the experiments, was to move from (0m, 1m) to (7m, 7m) while avoiding the fire. Experiments were conducted with two types of fire: (a) static fire, and (b) dynamic fire. In the static fire experiments, the temperatures of the motes were fixed throughout the experiment. Fire was simulated by assigning predefined high temperature values (70^oC) to motes located at (0m, 2m), (2m, 2m), (6m, 2m), (4m, 4m), and (6m, 4m) (motes with white dots in Figure 4(a)), and 30^oC to the remaining motes. Dynamic fire was simulated by assigning the same predefined values as in the static fire, but the values were changed during the experiment. More specifically, the temperature of the mote located at (2m, 4m) was increased while the temperature of the mote located at (2m, 2m) was decreased, thus simulating a fire spreading northwards.

Static fire. The path found in this scenario is shown as the dotted line in figure 4(b). As is seen, the robot successfully avoids dangerous places by staying close to motes with normal temperature.

Dynamic fire. The dynamic fire scenario shows the reaction of the robot when the fire changes location. In this case, the robot follows the same path as the

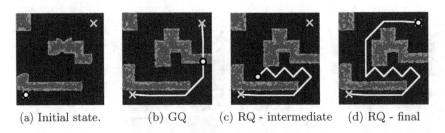

|(a) Initial state. | (b) GQ | (c) RQ - intermediate | (d) RQ - final |

Fig. 5. Path selected by GQ and RQ. (a) Initial state of the environment. The circle depicts the mobile entity; the cross at the bottom left corner marks the starting point; and the cross at the top right corner marks the goal. The blue region represents the safe region and has temperature below $60°C$. The red and green regions represent fire with temperature above $150°C$ and above $60°C$, respectively. (b) Mobile entity gets burnt on path selected by GQ. (c) Mobile entity incrementally builds path in RQ. (d) Mobile entity safely reaches the goal on path built by RQ.

static fire until it reaches (4m, 2m). At this point, the fire at (2m, 2m) moves to (2m, 4m). The robot then successfully finds a new path to avoid the fire. This scenario is illustrated in figure 4(b). The solid line shows the path followed by the robot, while the dotted line represents the initial path.

These experiments demonstrate that a robot can use our agent based implementation of RQ, to successfully find a safe path in the presence of dynamic obstacles. We further evaluate the performance of RQ and compare it with existing approaches through simulations under realistic fire scenarios.

7   Simulation Results

In this section, we present the results obtained from simulations in NS-2. We evaluate and compare RQ with existing protocols, using 9 different realistic fire scenarios, obtained using the NIST Fire Dynamics Simulator (FDS) [17]. In all the scenarios, the fire starts in different locations scattered over the region and then spreads over the region over time. This behavior presents two different environments, (1) which is very dynamic and occurs when the fire is still spreading and most of the region is still not on fire, and (2) which is less dynamic and occurs when a large part of the region is already on fire. We test the performance of the algorithms in both environments, by starting the mobile entity at two different times of 50s and 200s, after the fire starts spreading.

We evaluate the RQ protocol both with and without the extension for handling node failures to observe the difference in performance caused by the extension. We refer to the basic RQ protocol as B-RQ and the RQ protocol with the extension as Robust RQ (R-RQ). We also compare our protocol to other approaches like LQ [5], DA [1] and Global Query (GQ). LQ uses local flooding while DA and GQ use global flooding. LQ and DA were discussed earlier in the related works section (Section 2) and hence are not described here.

In GQ, the mobile entity broadcasts a query message, which is flooded throughout the entire network. On receiving the query message, the nodes respond with their location and temperature. These responses are aggregated and delivered to the mobile entity which uses the data to compute the edge weights of all roadmap edges in accordance with Equation 1 to obtain a complete path from the start to the goal. Since this method employs global data collection, it has a high communication cost. In addition, it also suffers from a long query latency, which significantly reduces a mobile entity's awareness of the region. This leads to situations where the mobile entity gets burnt while traversing a path that changes from being safe to being unsafe, due to lack of awareness of the changing environment. This situation was observed in a simulation run and is shown in figures 5(a) and 5(b). The white line in Figure 5(b) depicts the safe path that is initially computed by the mobile entity. As the mobile entity traverses the path, a part of the path is engulfed by fire. The mobile entity is unaware of this until it is very close to danger, at which point it stops moving and issues a query to find a safe path. However, due to the significant query latency, the fire spreads to the location of the mobile enitity and burns it, before it finds a safe path. The outcome of using the RQ protocol for the same scenario is shown in figures 5(c) and 5(d). Since the RQ protocol uses local queries with low query latency (due to low communication cost) it computes successive safe sub-paths that successfully lead it to the goal, around the regions on fire.

Each simulation was run with 900 nodes uniformly distributed in a $450m \times 450m$ area. The mobile entity's velocity was set to $3m/s$ and a 5×5 grid was used as the roadmap, with each grid square, measuring $90m \times 90m$. The communication range and bandwidth of the nodes were set to 45m and 40kbps, respectively. The sensing range of the nodes was obtained using the maximum temperature gradient δT at the border of a fire. Thus $R_S = \frac{\Delta_{burn} - \Delta_T}{\delta T}$. δT was found to be $4.5^oC/m$ from the simulation scenarios. Δ_T and Δ_{burn} were set to 60^oC and 150^oC, respectively, assuming a robot as the mobile entity. These settings result in $R_S = 20m$, which satisfies the double range property ($R_C \geq 2R_S$). The query radius (R_q) of B-RQ, R-RQ and LQ was set to 90m in all the simulations, since it was experimentally found to be optimal. Performance under different query radii is omitted due to space constraints but can be found at [11].

As mentioned in Section 4, an edge is considered covered if certain points on the edge are covered. The selection of the points differs with the query protocol. Since LQ and GQ query all nodes within a query area, any number of points (≥ 2) on an edge can be considered in these algorithms. Note that the query area in LQ is a circle (of certain radius) centered at the mobile entity location while the query area in GQ is the entire region. In our simulations of LQ and GQ, we considered coverage of five equidistant points on an edge to indicate edge coverage. However, since fewer nodes respond to queries in RQ, we considered coverage of only the endpoints of an edge to indicate edge coverage in RQ.

The wait time T_w, during which the mobile entity waits for the query results, reflects the query latency in LQ, GQ and RQ. Based on experimental results, it was set to a value that permitted at least 90% query results to be received

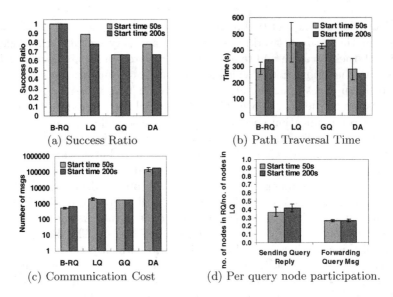

(a) Success Ratio

(b) Path Traversal Time

(c) Communication Cost

(d) Per query node participation.

Fig. 6. Performance comparison in the absence of node failures

by the mobile entity, in these protocols. RQ achieves a query latency that is approximately half of that of LQ, due to its low communication cost (Figure 6(c)). Correspondingly, T_w was set to 10s in RQ, 20s in LQ and 250s in GQ.

We use the following metrics to evaluate the performance of the different algorithms. (1) *Success Ratio*, defined as the ratio of the number of scenarios in which the mobile entity safely reaches the goal to the total number of scenarios. This is the most important metric for the application. (2) *Path Traversal Time*, defined as the average time taken by the mobile entity to reach the goal, over scenarios where the mobile entity successfully reaches the goal in all protocols being compared. The path traversal time includes the query latency (which depends on the performance of the query protocol) and the time that the mobile entity spends in navigation. (3) *Communication Cost*, defined as the average number of messages sent per scenario, over scenarios where the mobile entity successfully reaches the goal in all protocols being compared.

In the following subsections, in addition to the average results, we also present 90% confidence intervals. Confidence intervals are not provided for the case where the robot start time is 200s and there are no node failures, since there are only two scenarios where the mobile entity safely reaches the goal in all protocols being compared.

7.1   Performance in the Absence of Node Failures

Performance Comparison. In this sub-section, we compare the navigation performance and communication cost of GQ, LQ, B-RQ and DA, in the absence of node failures. Since R-RQ is designed specifically to handle node failures, we do not include it in this section.

Success Ratio : Figure 6(a) shows the success ratio of the different protocols at start times 50s and 200s. As seen in the figure, B-RQ performs better than all the other algorithms and enables the mobile entity to safely reach the goal in all tested scenarios. LQ does not perform as well, because unlike B-RQ, it queries all the nodes in a query area and hence incurs a long query latency, which slows down the mobile entity's progress towards the goal. As a result, the mobile entity sometimes gets caught up amidst the spreading fire with no safe path leading to the goal, or gets burnt while waiting for the query results.

The success ratio of GQ is the lowest. This is because GQ does not update the selected path to the goal based on the changing environmental state. At a start time of 50s, the mobile entity usually gets burnt due to lack of awareness of the fire encroaching the chosen path. On the other hand, at a start time of 200s, the mobile entity usually fails to obtain a safe path, because, by the time the mobile entity obtains the query results, which can take as long as 250s, most of the region is already engulfed by fire, disconnecting the start from the goal. The success ratio of DA is lower than that of B-RQ and LQ, because of high data loss due to contention caused by the high communication cost of DA (Figure 6(c)).

Path Traversal Time : Figure 6(b) shows the path traversal time obtained by the different protocols. We see that DA achieves the least path traversal time. This is because DA has very low query latency compared to the other algorithms, as it requires the mobile entity to query only nearby nodes. The path traversal time of B-RQ is comparable to that of DA. This is because, B-RQ also has a low query latency, since it queries only a few nodes per query. LQ and GQ, on the other hand, have longer path traversal times, since they query a large number of nodes and hence have long query latencies. Since all protocols achieve similar path lengths (shown in [11]), the difference in the path traversal times is dominated by the difference in query latencies.

Communication Cost : A comparison of the communication cost is presented in Figure 6(c). DA has an extremely high communication cost, since it requires all nodes in the network to maintain the potential fields, resulting in frequent flooding of the entire network caused by the spreading fire. In comparison, B-RQ has the least communication cost, since it queries only a few nodes that lie along the roadmap edges in a query area, per query. The extent by which RQ reduces the number of nodes that participate in a query, in comparison to LQ is shown in Figure 6(d). The figure shows that the number of nodes that forward the query message, per query in RQ, is only about 25% that of LQ and the number of nodes that send a query reply, per query in RQ, is only about 40% that of LQ. Thus, the total number of nodes that participate in a query in RQ is only about 40% that of LQ, on average. This significant reduction in the number of participating nodes per query is the reason behind the dramatic reduction in communication cost achieved by RQ. This saving is found to be 73% for a start time of 50s and 63% for a start time of 200s, from Figure 6(c). As a result of the reduced communication cost, RQ also has a lower query latency in comparison to LQ (almost 50% that of LQ), thus increasing its navigation performance, in terms of success ratio and path traversal time. This implies that greater the dynamism of the environment, the better will RQs performance be, in comparison to LQ. Another positive outcome of reducing the number of nodes participating per query is that when coupled with a power management protocol, it enables more nodes to sleep, thereby increasing network lifetime.

As expected, B-RQ also achieves huge communication cost savings over GQ. In particular, it achieves 70% and 62% lower communication cost for a start time of 50s and 200s, respectively. Note that, the large differences in communication cost between B-RQ, LQ and GQ, are not clearly visible in Figure 6(d), due to the logarithmic scale. The low communication cost of B-RQ is one of its main advantages, which enhances its navigation performance and also potentially increases network lifetime.

Overall, B-RQ achieves a higher success ratio and a significantly lower communication cost than all the other protocols as a result of its efficient forwarding and query reply rules. These results highlight the effectiveness of optimizing the query protocol in accordance with the navigation algorithm, in order to navigate successfully in dynamic environments.

7.2   Performance in the Presence of Node Failures

Since it is important to design robust protocols that can tolerate node failures caused by harsh environments, we now compare the performance of the algorithms in the presence of node failures. Nodes are assumed to fail at a temperature of 150^oC.

Performance Comparison. In this sub-section, we compare the navigation performance and communication cost of LQ, DA, B-RQ and R-RQ. GQ is not considered in these simulations, due to its poor performance in the earlier simulations.

Success Ratio: As shown in Figure 7(a) R-RQ effectively improves the success ratio of B-RQ when the mobile entity starts at 50s, at which time many new nodes start failing, due to the spreading fire. This is because R-RQ informs the mobile entity about failed nodes, thus warning the mobile entity about danger areas. On the other hand, R-RQ does not outperform B-RQ when the mobile entity starts at 200s, since by that time, the environment is relatively stable. In contrast, the performance of LQ and DA are affected significantly by node failures. R-RQ achieves upto 49% improvement in success ratio over LQ and up to 77% improvement in success ratio over DA. This demonstrates that R-RQ is particularly important in dynamic environments.

Communication Cost: Figure 7(b) shows the communication cost incurred by B-RQ, R-RQ and LQ. The commmunication cost of DA is not shown as it is significantly higher than the other protocols. As expected, the communication cost of LQ is much higher than that of B-RQ and R-RQ since it queries all the nodes in a query area. The communication cost of R-RQ is only slightly

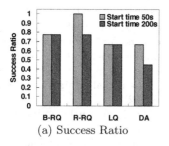
(a) Success Ratio

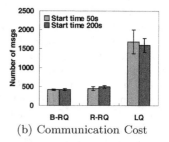
(b) Communication Cost

Fig. 7. Performance comparison in the presence of node failures

more than that of B-RQ since it requires nodes to respond if they have failed neighbors even if their temperatures are below the threshold. More specifically, R-RQ achieves 73% and 69% savings in communication cost over LQ, at a start time of 50s and 200s, respectively.

8   Conclusion

In summary, we propose a novel approach that integrates roadmap-based navigation with efficient query protocols for navigation in *dynamic* environments. We present the Roadmap Query (RQ) protocol that is specially optimized for collecting fresh data needed for navigation in the presence of dynamic obstacles and sensor node failures. We also present a mobile-agent based implementation of our navigation approach on a physical testbed consisting of Mica2 motes and a robot. Our simulation results demonstrate that RQ can significantly improve the success ratio of navigation while introducing minimum communication cost under realistic fire scenarios and node failures. Our results highlight the importance of joint optimization of navigation and WSN query protocols for efficient navigation in dynamic environments.

Acknowledgement

This work is funded in part by the NSF under an ITR grant CCR-0325529 and the ONR under MURI research contract N00014-02-1-0715.

References

1. Li, Q., Rosa, M.D., Rus, D.: Distributed algorithms for guiding navigation across a sensor network. (In: MobiCom'03)
2. Batalin, M.A., Sukhatme, G.S., Hatting, M.: Mobile robot navigation using a sensor network. (In: ICRA'04)
3. Verma, A., Sawant, H., Tan, J.: Selection and navigation of mobile sensor nodes using a sensor network. (In: PerCom'05)
4. Corke, P., Peterson, R., Rus, D.: Coordinating aerial robots and sensor networks for localization and navigation. (In: DARS'04)

5. Alankus, G., Atay, N., Lu, C., Bayazit, B.: Spatiotemporal query strategies for navigation in dynamic sensor network environments. (In: IROS'05)
6. Kavraki, L., Svestka, P., Latombe, J.C., Overmars, M.: Probabilistic roadmaps for path planning in high-dimensional configuration spaces. IEEE Trans. Robot. Automat. **12**(4) (1996) 566–580
7. Cormen, T.H., Leiserson, C.E., Rivest, R.L.: Introduction to algorithms. 6th edn. MIT Press and McGraw-Hill Book Company (1992)
8. Whitehouse, K., Sharp, C., Brewer, E., Culle, D.: Hood: a neighborhood abstraction for sensor networks. (In: MobiSys'04)
9. Woo, A., Tong, T., Culler, D.: Taming the underlying challenges of reliable multhop routing in sensor networks. (In: Sensys'03)
10. Xing, G., Wang, X., Zhang, Y., Lu, C., Pless, R., Gill, C.: Integrated coverage and connectivity configuration in wireless sensor networks. TOSN **1**(1) (2005) 36–72
11. Bhattacharya, S., Atay, N., Alankus, G., Lu, C., Roman, G.C., Bayazit, B.: Roadmap query for sensor network assisted navigation in dynamic environments. In: Technical Report WUCSE-05-41, Department of Computer Science and Engineering, Washington University in St. Louis. (2005)
12. Fok, C.L., Roman, G.C., Lu, C.: Rapid development and flexible deployment of adaptive wireless sensor network applications. (In: ICDCS'05)
13. Fok, C.L., Roman, G.C., Lu, C.: Mobile agent middleware for sensor networks: An application case study. (In: IPSN'05)
14. (TinyOS community forum) http://www.tinyos.net/.
15. Levis, P., Culler, D.: Mate: A tiny virtual machine for sensor networks. (In: ASPLOS X)
16. (Activmedia) http://www.activemedia.com.
17. McGrattan, K.: Fire dynamics simulator (version 4) technical reference guide. National Institute of Standards and Technology (2004)

Stabilizing Consensus in Mobile Networks

Dana Angluin, Michael J. Fischer, and Hong Jiang*

Department of Computer Science, Yale University

Abstract. Inspired by the characteristics of biologically-motivated systems consisting of autonomous agents, we define the notion of stabilizing consensus in fully decentralized and highly dynamic ad hoc systems. Stabilizing consensus requires non-faulty nodes to eventually agree on one of their inputs, but individual nodes do not necessarily know when agreement is reached. First we show that, similar to the original consensus problem in the synchronous model, there exist deterministic solutions to the stabilizing consensus problem tolerating crash faults. Similarly, stabilizing consensus can also be solved deterministically in presence of Byzantine faults with the assumption that $n > 3f$ where n is the number of nodes and f is the number of faulty nodes. Our main result is a Byzantine consensus protocol in a model in which the input to each node can change finitely many times during execution and eventually stabilizes. Finally we present an impossibility result for stabilizing consensus in systems of identical nodes.

1 Introduction

1.1 Fault-Tolerant Consensus

Coordination problems in distributed systems require nodes to agree on a common action. Lamport, Pease, and Shostak formulated this problem as the agreement problem [1, 2], which remains a fundamental problem in distributed computing. It is usually trivial to reach agreement in reliable systems. In practice, however, different components in a system don't always work correctly. Mission-critical control systems require agreement among non-faulty components even when some components are faulty. The problem was originally defined for Byzantine faults in which a faulty node in a network can behave arbitrarily. More benign are crash faults in which a faulty node stops all activity at a certain point in the execution but behaves correctly until then. Sometimes the recovery of crashed processes is also considered. Lamport, Shostak, and Pease [1, 2] gave a synchronous f-resilient solution for any f with authentication in the case of a complete communication graph and proved the impossibility result that consensus is not solvable without authentication unless the number of faulty processes is less than one-third of the total. Dolev [3] considered the Byzantine agreement problem in networks that are not completely connected. The first polynomial communication algorithm for Byzantine agreement was designed by

---

* Supported by NSF grant ITR-0331548.

P. Gibbons et al. (Eds.): DCOSS 2006, LNCS 4026, pp. 37–50, 2006.
© Springer-Verlag Berlin Heidelberg 2006

Dolev and Strong [4], whose work was subsequently improved by Dolev, Fischer, Fowler, Lynch, and Strong [5]. Fischer, Lynch, and Paterson [6] showed that in a fully asynchronous environment, there is no 1-resilient solution to the consensus problem, even for crash failure. A survey on fault-tolerant consensus by Fischer [7] provides an overview of early work on fault-tolerant distributed systems.

One of the reasons for the impossibility results for fault tolerance in the asynchronous model is that messages may be delayed arbitrarily as long as they are eventually delivered. Therefore, there is no way to distinguish between a crashed node and a slow node, or a lost message and a delayed message. Since distributed algorithms are expected to terminate, each non-faulty process is required to *commit* to an output value at some point in its execution when it knows the decision value that all non-faulty processes will agree on.

1.2   Motivation

In some persistent ad hoc networks, especially biologically-motivated systems, it is not important that each process be aware of the global status. For example, imagine the aggregation and migration of birds. During the initial gathering or a direction change, the movement of the birds is usually chaotic. Although each bird is not aware of the status of the whole flock, the flock eventually converges to a stable state in which all the birds head in roughly the same direction. Each bird adjusts its heading according to what it perceives but does not commit to a direction at any point, because it is possible that some other birds are still changing directions.

Vicsek *et al.* [8] described a compelling model of dynamics in order to investigate the emergence of self-ordered motion in systems of autonomous agents with biologically motivated interaction. Each agent's heading is updated from time to time according to a local rule. They demonstrated that all agents eventually move in the same direction despite the absence of centralized coordination and the changing neighbor set of each agent as the system evolves. Agents do not know accurately when the whole system converges, except for an estimation of expected convergence time when the agents communicate synchronously and certain topological properties are guaranteed as the network evolves. Jadbabaie, Lin, and Morse [9] provided a theoretical explanation for the above behavior and investigated other similarly inspired models. Agent failures were not considered in these two papers.

This model is related to some practical applications. For example, imagine the nodes as unmanned planes that cooperate with each other to determine some common behavioral parameters such as direction or speed. Some of the planes might be captured by enemy forces who inject malicious programs in order to disrupt the consistency of the system. It is desirable if system consistency could be maintained as long as not too many nodes are compromised.

We discuss asynchronous fault-tolerant consensus in a similar scenario. We relax the requirement of the original consensus problem that agents know when a decision is final. We also investigate persistent distributed systems that run for an extended period and whose inputs may change from time to time, as well as systems that have incomplete or evolving interaction graphs. We also prove that stabilizing Byzantine consensus cannot be solved in systems consisting of identical nodes.

1.3 Other Related Work

Because agreement is a fundamental problem in building distributed systems, and most practical systems are not synchronous, various methods have been used to circumvent the impossibility result in [6]. In many practical systems, nodes periodically send "I'm alive" messages (pings) to each other to detect possible crashes. In theoretical models, such techniques are captured by the abstract concept of failure detectors [10]. Generally, a failure detector is a module that provides information to processes about previous failures. Failure detectors differ in strength depending on whether they are always correct or whether they detect all failures [11]. Some failure detectors can be implemented in practical systems using timeouts.

The agreement problem can be solved in a randomized asynchronous model which allows each process to flip coins during execution. The problem statement is modified to require that the processes eventually terminate with probability 1. The first randomized solution for consensus was given by Ben-Or [12]. Rabin [13] and Feldman [14] produced more efficient algorithms.

The k-agreement problem [15] is a weakened problem statement that only requires all the decisions to be in a set of k values. Another weakened variation is the approximate agreement problem [16] which allows inputs, decisions, and messages to be real numbers and requires the difference between any two decision values to be within a small tolerance ϵ and that any decision value is within the range of input values.

Lamport invented the PAXOS algorithm [17] for a partially synchronous model of distributed systems. Asynchrony is considered timing failure. Other failures allowed are loss, duplication and reordering of messages, and crash failure of processes. Process recovery is also allowed. The PAXOS algorithm guarantees safety, meaning that in spite of timing, process, and link failures, and process recoveries, non-faulty nodes do not decide on inconsistent values. When the system stabilizes (no failure occurs, and a majority of the processes are active) for a sufficiently long time, termination can also be achieved.

Although mobile computing has been under active study for many years, research on fault-tolerance issues has been limited, and many desirable goals are yet to be achieved. The problems studied include ad hoc routing [18, 19], which allows nodes to exchange data despite the limited transmission range of wireless interfaces by routing messages through multiple network hops, broadcasting and multicasting [20], transaction control [21, 22, 23], group communication [24, 25], leader election [26], and mutual exclusion [27]. Angluin *et al.* [28] proposed self-stabilizing solutions to problems like leader election, ring orientation, token circulation, and spanning-tree construction in a model of pairwise interacting anonymous finite-state sensors under a global fairness condition. Basile, Killijian, and

Powell [29] gave a survey of fault tolerance in mobile wireless networks. The fault models considered in existing works are usually mobility, network partitioning, and sometimes crash failures. Byzantine node failures have generally not been considered.

2   Model and Definitions

We consider a network of n mobile nodes. Each has a unique ID $\in [1, n]$, an input port, and an output port to send outputs to an external observer. By assigning unique IDs to the nodes, we give each node the ability to distinguish between different nodes. Also, the IDs are assumed to be unforgeable in the sense that a faulty node cannot impersonate a non-faulty node in direct communications (i.e. messages not involving any intermediate nodes). Each node communicates with other nodes by sending out messages from time to time. A message may be received by other nodes that are close enough to the sender, or they may fail to be received by anyone due to a transmission error or because everyone is out of range. The sender of a message does not know whether a certain message is received by some node, nor does it know the identities of the nodes that do receive the message. We assume the fairness condition that if some node i sends messages infinitely often, every other node receives messages from i infinitely often. We remark that this model is weaker than the asynchronous model of distributed systems in [6] because we do not assume that the asynchronous communication channels are reliable. Therefore one cannot expect to solve the terminating consensus problem in our model in presence of faults.

In the literature, two kinds of failures (equivalently, faults) are usually considered for the consensus problem. The benign type of failure is crash failure: a faulty node may crash at any time. When a node crashes, it stops operating, otherwise it honestly follows the protocol. We assume that if a node crashes when sending a message, the incomplete message is discarded by the recipient. Byzantine failures are more severe. A Byzantine node may behave arbitrarily without the limit of computational power or memory usage to which a non-faulty node is constrained otherwise.

We define the notion of *stabilizing consensus*. Instead of requiring that each node commit to a final output at some point, we assume that each node has a current output that may change as the execution proceeds. In practical applications, the output could be interpreted as some parameter that reflects the behavior of each node. For example, in a flock of mobile nodes, the current output of each node could be its current speed or current direction.

As in the usual convention, a configuration includes all nodes' local states and the pending messages. A configuration C is said to be *output-stable* if in all possible executions starting from C, the output of each non-faulty node does not change. If every non-faulty node outputs x in an output-stable configuration C, we say the outputs stabilize to x in C.

Definition 1 (Stabilizing Consensus). *A protocol P solves the stabilizing consensus problem if all of the following requirements are satisfied:*

Stabilization. *The system eventually reaches an output-stable configuration.*

Validity. *If all nodes have the same input x, the outputs of all non-faulty nodes eventually stabilize to x.*[1]

Agreement. *In any reachable output-stable configuration, all non-faulty nodes have the same output.*

In the following sections, we consider both crash faults and Byzantine faults. We also discuss consensus in a scenario where each node receives an input that may change finitely many times and define *consensus with stabilizing inputs*. Finally we show that stabilizing consensus cannot be solved in a system consisting of identical nodes in presence of one Byzantine fault.

3   Stabilizing Consensus with Crash Faults

The following is a simple protocol that solves consensus in the presence of crash faults, assuming the inputs are non-negative integers. The protocol is based on the idea of the protocol for synchronous distributed systems from [30]. But notice that, in general, a node does not know when its output stabilizes, and an execution does not terminate.

For each node i, x_i is its local input (a non-negative integer), and y_i is its output.

– At the beginning, node i sets $y_i = x_i$.
– Whenever i is able to send a message, it sends y_i.
– Upon receiving message y, node i sets $y_i = \min(y_i, y)$.

The following theorem establishes the correctness of the protocol.

Theorem 1. *The above protocol solves stabilizing consensus in presence of f crash faults for any $f < n$, where n is the total number of nodes.*

Proof sketch. It is easy to see that the outputs will stabilize, because the output of each node can only decrease and cannot be negative. Also it is clear that the validity condition is satisfied because if all nodes have the same input value, all messages will contain this same value, so all nodes will output that value.

For the agreement condition, suppose for the sake of contradiction that two non-faulty nodes i and j stabilize to different outputs, y_i and y_j. Without loss of generality, we assume $y_i < y_j$. According to the fairness condition, eventually j will receive a value of y_i from i and set $y_j = \min(y_i, y_j) = y_i$. This contradicts the assumption that the output of j stabilizes to y_j.

---

[1] Some authors consider a stronger validity condition that requires agreement on x if x is the common input value of just the *non-faulty* nodes. This is equivalent to the validity condition presented here in the case of Byzantine faults since a Byzantine faulty node's behavior is not constrained by its actual input. The same is not true for crash faults.

4 Stabilizing Consensus with Byzantine faults

4.1 A Protocol for Fixed Inputs

In this section we give protocols tolerating Byzantine faults. We assume that when a node receives a message, it knows the identity of the sender.

We first consider a system where each node i receives a fixed local input x_i at the beginning. For simplicity, we assume $x_i \in \{0,1\}$. We give a protocol that tolerates f Byzantine faults, assuming $3f < n$ where n is the total number of nodes.

Initially, a node i estimates that every node has input 0. If i's input is 1, it always sends its input. It also sends an echo message for another node j if one or both of two events have occurred: i received j's input from j at some time in the past, or i received echo messages for j from sufficiently many (at least $f + 1$) nodes. When i receives an echo message for j from enough (at least $n - f$) nodes, it changes its estimation of j's input to 1. The protocol guarantees that eventually all non-faulty node have the same estimation of any node j's input, and if j is non-faulty, the estimation agrees with j's actual input. Each non-faulty node outputs 1 when it estimates enough (at least $2f + 1$) nodes have input 1, and it outputs 0 otherwise. A more detailed description follows.

The state of each non-faulty node i consists of the arrays $I_i[n]$, $E_i[n][n]$ and $M_i[n]$, in which all elements are initialized to 0.

- When node i is able to send a message, it sends a message including one or more of the following components:
  If $x_i = 1$, it sends (init, i).
  For all j such that $I_i[j] = 1$ or $\sum_{k=1}^{n} E_i[j][k] \geq f + 1$, i sends (echo, j).
- When i receives (init, j) from j, it sets $I_i[j] = 1$.
- When i receives (echo, k) from j, it set $E_i[k][j] = 1$, and if $\sum_{k=1}^{n} E_i[j][k] \geq n - f$, i sets $M_i[j] = 1$

Output: The current output of node i is 1 if $\sum_j M_i[j] \geq 2f + 1$, otherwise its output is 0.

It is easy to see that the outputs will stabilize, because each node outputs 0 initially and can flip its output to 1 at most once.

Correctness can be established by verifying the following claims.

Lemma 1. *If any non-faulty node i has 1 as input, eventually every non-faulty node j sets $M_j[i] = 1$.*

Eventually every node receives (init, i) from node i, and all non-faulty nodes will repeatedly send (echo, i). Therefore any non-faulty node j will receive (echo, i) from at least $n - f$ nodes and set $M_j[i] = 1$

Lemma 2. *If any non-faulty node i has 0 as input, $M_j[i]$ is always 0 for any non-faulty node j.*

In this case, no non-faulty node receives (init, i) from node i. Suppose node j is the first non-faulty node that sends (echo, i). It must have been triggered by receiving (echo, i) from $f + 1$ faulty nodes, which contradicts the assumption. A non-faulty node j never sends (echo, i) and receives (echo, i) from at most f faulty nodes, so it will never set $M_j[i] = 1$.

Lemma 3. *For any i, if $M_j[i]$ stabilizes to 1 in any non-faulty node j, $M_k[i]$ eventually stabilizes to 1 in any other non-faulty node k.*

If any non-faulty node j sets $M_j[i] = 1$, it must have received (echo, i) from at least $n - f$ nodes among which there are at least $f + 1$ non-faulty nodes. The messages (echo, i) sent by these $f + 1$ nodes are received by all non-faulty nodes, therefore all non-faulty nodes will send (echo, i) to each non-faulty node k so that it sets $M_k[i] = 1$.

Theorem 2. *The above protocol solves the stabilizing consensus problem.*

Given the above claims, it is easy to see that the protocol satisfies stabilization, validity, and agreement.

4.2 Stabilizing Inputs

We define a model of stabilizing inputs to a network protocol in which the input to each node may change finitely many times before it stabilizes to a final value. We are interested in solving the consensus problem corresponding to the final stabilized input assignment. This consistent input and output convention makes a solution suitable as middleware in constructing more complex systems. Here we define what consensus means in this model.

Definition 2 (Consensus with Stabilizing Inputs). *A protocol P solves consensus with stabilizing inputs if all of the following requirements are satisfied:*

Stabilization. *If the inputs to the non-faulty nodes stabilize, the system eventually reaches an output-stable configuration.*

Validity. *If all non-faulty nodes have the same stabilized input x, their outputs eventually stabilize to x.*

Agreement. *In any reachable output-stable configuration, all non-faulty nodes have the same output.*

Note that fixed inputs are a special case of stabilizing inputs.

The following protocol achieves consensus with stabilizing inputs and tolerates f Byzantine faults, assuming $3f < n$ where n is the total number of nodes. We give only a high-level description of the protocol. Our purpose here is to establish the possibility of a protocol rather than to present an optimal implementation. In our description, each node needs to keep track of messages received in the past. This intensive memory usage could be reduced by garbage-collecting data that is no longer useful in subsequent computation. We defer details of implementation and optimization to the full version of the paper.

The basic idea of the protocol is similar to the protocol for fixed inputs. However, the protocol for fixed inputs is biased in the sense that 0 and 1 are treated differently, whereas the following protocol is not. Many instances of a consensus protocol are run in parallel. When a node detects that its input has changed, it tries to restart the instance of the consensus protocol concerning its input. Each node determines its current output according to the $2f + 1$ most stable (least-frequently changed) estimated inputs.

Each non-faulty node i maintains two arrays $M_i[n]$ and $C_i[n]$. The elements of M_i are initialized to 0, and the elements of C_i are initialized to -1. It also has a counter c_i initially equal to 0. Let $x_i \in \{0,1\}$ denote the current reading of the input port. Node i also maintains a variable x_i' and initially sets $x_i' = x_i$.

- When i is able to send a message:
  1. If $x_i \neq x_i'$, set $x_i' = x_i$ and $c_i = c_i + 1$;
  2. Always send (init, i, x_i, c_i);
  3. For all j, x_j, and c_j, such that i has received (init, j, x_j, c_j) from j, or i has received (echo, j, x_j, c_j) from at least $f + 1$ different nodes, send (echo, j, x_j, c_j).
- When i receives (init, j, x_j, c_j) from j, if $c_j \leq C_i[j]$, the message is ignored, otherwise it records this message in its event log. If i receives contradicting init messages from the same node ((init, j, x_j, c_j) and (init, j, x_j', c_j) with $x_j \neq x_j'$), only the first message is recorded.
- When i receives (echo, j, x_j, c_j), if $c_j \leq C_i[j]$, the message is ignored, otherwise it records this message in its event log, and if the same message has been received from at least $n - f$ different nodes, i sets $M_i[j] = x_j$ and $C_i[j] = c_j$

Output:

- Define the *stable set* S_i to be a set of $2f + 1$ distinct integers in $[1 \ldots n]$ that minimizes $\sum_{j \in S_i} C_i[j]$. In case of ties, the set that minimizes $\sum_{x \in S_i} x$ is chosen.
- Node i outputs 1 if $\sum_{j \in S_i} M_i[j] \geq f + 1$, otherwise it outputs 0.

The variable c_i is a counter for node i to keep track of how many times its input has changed. Each node also uses the counter array C_i to keep track of the number of times the other nodes change their inputs. Because messages can be delivered out of order, and "echo" messages corresponding to inputs at different time can co-exist in the network, the counters also ensure that obsolete messages are ignored.

Lemma 4. *The invariant $C_i[j] \leq c_j$ holds in any real-time snapshot of the system for any non-faulty nodes i and j.*

Proof. Suppose at some point in real time, $C_i[j] = a$, $c_j = b$ and $a > b$. Then i must have received (echo, j, m, a) for some m from at least $n - f$ nodes. Therefore j must have sent (init, j, m, a), because at most f nodes send (echo, j, m, a) otherwise. This contradicts $a > b$. Because j would have set $c_j = a$ before sending (init, j, m, a), it must be true that $b \geq a$.

Lemma 5. *Let i and j be two non-faulty nodes. If i's input stabilizes to x, $M_j[i]$ eventually stabilizes to x.*

Proof. Suppose $M_j[i]$ stabilizes to $y \neq x$. Then j must have received (echo, i, y, a) for some a from at least $n - f$ nodes, so i must have sent (init, i, y, a) to at least $f + 1$ nodes. Since x is the final input of i, eventually i sends (init, i, x, b) for some b to all nodes. According to lemma 4 $a < b$. Suppose the time j receives (echo, i, y, a) from the $(n - f)^{\text{th}}$ node is t, and the time it receives (echo, i, x, b) from the $(n - f)^{\text{th}}$ node is t'. If $t < t'$, j will set $M_j[i] = x$. If $t' < t$, y is ignored by j, because at t the value of $C_j[i]$ can only be greater than or equal to b and $a < b$. Therefore $M_j[i]$ couldn't have stabilized to y.

Lemma 6. *If i and j are non-faulty nodes, for any k, if $M_i[k]$ stabilizes to x, $M_j[k]$ also stabilizes to x.*

Proof. Suppose $M_i[k]$ stabilizes to x, $M_j[k]$ stabilizes to y, and $x \neq y$. Let (echo, k, x, a) and (echo, k, y, b) be the corresponding messages received by i and j respectively when they assigned the final values to $M_i[k]$ and $M_j[k]$.

1. Without loss of generality, we assume $a > b$. Since i must have received (echo, k, x, a) from at least $n - f$ nodes, there must be at least $f + 1$ non-faulty nodes among them. All non-faulty nodes would receive (echo, k, x, a) from these $f + 1$ nodes, and therefore would send (echo, k, x, a) to all nodes they encounter. Thus j would also receive (echo, k, x, a) from at least $n - f$ nodes. Because $a > b$, $M_j[k]$ could not have stabilized to y.

2. If $a = b$, i receives (echo, k, x, a) from $n - f$ nodes, and j receives (echo, k, y, b) from $n - f$ nodes. This cannot happen, because $n > 3f$, and according to the protocol, a non-faulty node only sends one of the two messages but not both.

Therefore $M_i[k]$ and $M_j[k]$ cannot stabilize to different values for any k. This property guarantees that all non-faulty nodes will eventually agree on the stabilized entries of vector M.

Lemma 7. *Let i and j be any non-faulty nodes. For any k, if $C_i[k]$ stabilizes to c, $C_j[k]$ also stabilizes to c.*

Proof. Suppose $C_i[k]$ stabilizes to c_1, $C_j[k]$ stabilizes to $c_2 \neq c_1$. Without loss of generality, we assume $c_1 > c_2$. Let (echo, k, x, c_1) and (echo, k, y, c_2) be the corresponding messages received by i and j respectively when they assign the final values of $C_i[k]$ and $C_j[k]$. Since i must have received (echo, k, x, c_1) from at least $n - f$ nodes, there must be at least $f + 1$ non-faulty nodes among them. All non-faulty nodes would receive (echo, k, x, c_1) from these $f + 1$ nodes, and therefore would send (echo, k, x, c_1) to all nodes they encounter. j would also receive (echo, k, x, c_1) from at least $n - f$ nodes. Because $c_1 > c_2$, $C_j[k]$ could not have stabilized to c_2. This property guarantees that all non-faulty nodes will eventually agree on the stabilized entries of vector C.

Lemma 8. *In any execution of the above protocol, if the inputs to the non-faulty nodes stabilize, the outputs of the non-faulty nodes eventually stabilize.*

Proof. Let i be any non-faulty node. If x_i stabilizes, c_i also stabilizes, because they always change at the same time. According to lemmas 4 and 5, $M_j[i]$ and $C_j[i]$ also stabilize for any non-faulty j. According to lemma 6 and lemma 7, all non-faulty nodes will eventually agree on the stabilized entries of the arrays M and C (at least $2f + 1$ entries in each), which include entries corresponding to non-faulty nodes and entries corresponding to faulty nodes that stabilize at all. If some of the entries in the M arrays corresponding to faulty nodes do not stabilize, the corresponding entries in the C arrays of the non-faulty nodes will eventually be greater than the stabilized entries, because the entries of C arrays are non-decreasing. Only the $2f + 1$ nodes corresponding to the C entries with the smallest values affect the output, therefore the faulty nodes will eventually be ignored.

Theorem 3. *The above protocol solves consensus with stabilizing inputs.*

If all non-faulty nodes have $x \in \{0, 1\}$ as input, according to lemma 5, for any non-faulty node i at least $f + 1$ M_i entries corresponding to the stable set will be x, therefore all non-faulty nodes will output x, and the validity condition is satisfied. According to lemmas 6, 7, and 8, agreement and stabilization are also satisfied.

5   Impossibility of Stabilizing Byzantine Consensus Among Identical Nodes

In this section we give the impossibility result that stabilizing consensus cannot be solved in the presence of a single Byzantine fault in a network of nodes that are identical other than their inputs. We note that any subconfiguration of an output-stable configuration is also output-stable.

Theorem 4. *The stabilizing consensus problem cannot be solved in a set of identical nodes in the presence of one Byzantine fault.*

Proof. Assuming there is a protocol P that solves this problem, consider a system $C = C_0 \cup C_1(C_0 \neq \phi, C_1 \neq \phi)$, in which C_0 is the set of nodes with input 0, and C_1 is the set of nodes with input 1. There exists a finite execution E of P in C that reaches an output-stable configuration in which the outputs of all nodes have stabilized to the same value. Without loss of generality assume the common output value is 0. Consider another system $C' = \{a\} \cup C_1$, in which a is a Byzantine node, and C_1 is the same as in C. Node a runs a two-phase protocol. In phase one, when it is a's turn to send a message, it nondeterministically chooses whether to remain in phase one or move to phase two. If it remains in phase one, it chooses one of the messages sent by nodes in C_0 in the execution E and sends that message to the recipient. Upon entering phase two, a faithfully imitates a nondeterministically chosen non-faulty node i from C_0 starting from the state i is in at the end of the execution E. There exists an execution E' of P in C' that simulates E, in the sense that every time there is a message in E sent between a node in C_0 and a node in C_1, there is a corresponding message in E' sent between a and the node in C_1, and at the end of E', node a will faithfully simulate one node in C_0. Thus, the configuration of the system C' at the end of E' is a subconfiguration of the system C at the end of E, and will continue so at every subsequent time. Thus the outputs of the non-faulty nodes (those in C_1) will remain 0 no matter how execution proceeds from this point. This violates the validity condition, because the inputs of all non-faulty nodes in C' are 1.

The proof does not depend on the specific communication model and fairness assumption; therefore stabilizing Byzantine consensus is impossible even with the strong fairness condition and two-way interaction model in [28], and unbounded memory. Note that theorem 4 rules out not only deterministic solutions, but also randomized solutions[2], in the sense that for any candidate protocol P, there exists an $\epsilon_P > 0$, such that the probability of an execution failing to reach consensus is always greater than ϵ_P. ϵ_P is any constant less than the probability that C' successfully simulates C to the point when all non-faulty nodes are output-stable.

6   Discussion

6.1   Upper Bound on Faults

It was shown that in synchronous systems, the number of Byzantine nodes must be strictly less than one third of the total number of nodes for any solution to the agreement problem [31, 2]. This bound still holds for stabilizing consensus in our model. We omit the proof here, because the original proof in [31] does not rely on synchrony and can be adapted to our model easily.

6.2   General Graphs

In some applications the movement of each node is restricted to a certain region, therefore some nodes may not be able to receive messages from some other nodes. It was proven in [3] that the Byzantine agreement problem can be solved in an n-node synchronous network graph G, tolerating f faults, if and only if the $n > 3f$ bound holds and G is at least $(2f + 1)$-connected. This result can also be transferred to our model. Intuitively, since G is at least $(2f + 1)$-connected, there are at least $2f + 1$ disjoint paths between any two nodes. Let each node send each message through $2f + 1$ disjoint paths. Then the majority of the copies the recipient receives are sent via paths that do not contain faults. Thus, it is possible to implement reliable communication between any two nodes, and the

---

[2] A randomized solution would guarantee that consensus be reached with probability 1, assuming some probabilistic distribution of the nodes' coin flips and the choices of the scheduler.

above algorithms still work for such communication graphs with messages sent over multi-hop links.

In some systems, nodes are moving around, but the fairness condition does not hold, meaning that some nodes do not have infinitely many chances to receive messages from some other nodes. Some nodes only have chance to receive finitely many messages from some others, and some won't get close enough at all. In their self-stabilizing group membership protocol, Dolev, Schiller and Welch [25] used random walks of a mobile agent as a means of information dissemination. Similarly, one or more non-Byzantine message carriers could be used as a message-ferrying service to simulate our communication model. The message carriers do not have to be reliable as long as they successfully deliver messages infinitely often.

6.3 Communication Model and Message Overhead

In our model, there is no reliable way for two nodes i and j to make sure that a message sent from i is received by j and that both nodes are aware of the event. As a consequence, all non-faulty nodes send infinitely many messages since they never know when it is safe to stop sending.

If we augment the model with a stronger communication mechanism by which the sender of a message can learn whether the message is received and if so by whom, the protocol could be made eventually quiescent, that is, once the inputs have stabilized, each non-faulty node eventually stops sending messages. In our protocol, each node initially only needs to make sure every node has received its input. After that, it could enter a passive mode in which it listens to other nodes and only sends messages in response to messages received. When it knows that its response message has been received by all necessary recipients defined by the protocol, it can again become passive. Eventually all non-faulty nodes enter the passive mode and do not initiate new messages. According to our protocols, the faulty nodes that keep sending messages are eventually ignored, and the subsystem consisting of the non-faulty nodes becomes quiescent. Note that this does not mean the protocol terminates, because generally each node does not know whether other non-faulty nodes are passive, but a node can conserve energy by not sending unnecessary messages. This suggests that energy-efficient implementations of our protocols are possible when appropriate lower-level service is provided, either by a lower-level protocol or by additional devices.

7 Conclusions and Future Work

In this paper we defined and investigated fault-tolerant stabilizing consensus in a model inspired by natural phenomena. We considered crash faults and Byzantine faults in fully asynchronous and decentralized mobile networks, as well as systems with stabilizing inputs and systems with incomplete or evolving connectivity. The algorithms are useful in controlling distributed systems, such as sensor networks, that simulate certain biological behaviors. They are also useful as a middleware layer that provides service to higher-level protocols. One drawback of the algorithm for stabilizing inputs is that it involves unbounded counters, unless there is a bound on the maximum number of times the inputs could change. It is open whether there exists a protocol for this problem with bounded memory. In many practical ad hoc networks, the graph representing possible communications changes over time. It is open for future research whether stabilizing consensus can be solved in these systems without additional message carriers, possibly using authentication and a fault-tolerant ad-hoc routing protocol.

References

1. Lamport, L., Shostak, R., Pease, M.: The byzantine generals problem. In: Advances in Ultra-Dependable Distributed Systems, N. Suri, C. J. Walter, and M. M. Hugue (Eds.). IEEE Computer Society Press (1995)
2. Pease, M., Shostak, R., Lamport, L.: Reaching agreement in the presence of faults. Journal of the ACM **27** (1980) 228–234
3. Dolev, D.: The byzantine generals strike again. Journal of Algorithms **3**(1) (1982) 14–30
4. Dolev, D., Strong, H.R.: Polynomial algorithms for multiple processor agreement. In: Proceedings of the 14th annual ACM symposium on Theory of computing, San Francisco, California, United States (1982) 401–407
5. Dolev, D., Fischer, M.J., Fowler, R., Lynch, N.A., Strong, H.R.: An efficient algorithm for byzantine agreement without authentication. Information and Control **52**(3) (1982) 257–274
6. Fischer, M.J., Lynch, N.A., Paterson, M.S.: Impossibility of distributed consensus with one faulty process. Journal of the ACM **32**(2) (1985) 374–382
7. Fischer, M.J.: The consensus problem in unreliable distributed systems (a brief survey). Technical Report YALEU/DCS/TR-273, Yale University (1983)
8. Vicsek, T., Czirók, A., Ben-Jacob, E., Cohen, I., Shochet, O.: Novel Type of Phase Transition in a System of Self-Driven Particles. Physical Review Letters **75** (1995) 1226–1229
9. Jadbabaie, A., Lin, J., Morse, A.: Coordination of groups of mobile autonomous agents using nearest neighbor rules. IEEE Transactions on Automatic Control (2002)
10. Chandra, T.D., Toueg, S.: Unreliable failure detectors for reliable distributed systems. Journal of the ACM **43**(2) (1996) 225–267
11. Chandra, T.D., Hadzilacos, V., Toueg, S.: The weakest failure detector for solving consensus. Journal of the ACM **43**(4) (1996) 685–722
12. Ben-Or, M.: Another advantage of free choice: Completely asynchronous agreement protocols. In: Proceedings of the Second Annual ACM Synmposium on Principles of Distributed Computing, Montreal, Quebec, Canada (1983) 27–30
13. Rabin, M.O.: Randomized byzantine generals. In: 24th Annual Symposium on Foundations of Computer Science, IEEE, Los Alamitos, California, United States (1983) 403–409
14. Feldman, P.N.: Optimal Algorithms for Byzantine Agreement. PhD thesis, Massachusetts Institute of Technology (1988)
15. Chaudhuri, S.: More choices allow more faults: Set consensus problems in totally asynchronous systems. Information and Computation **105**(1) (1993) 132–158

16. Dolev, D., Lynch, N.A., Pinter, S.S., Stark, E.W., Weihl, W.E.: Reaching approximate agreement in the presence of faults. Journal of the ACM **33**(3) (1986) 499–516

17. Lamport, L.: The part-time parliament. ACM Transaction on Computer Systems **16**(2) (1998) 133–169

18. Beraldi, R., Baldoni, R. The Electrical Engineering Handbook Series. In: The handbook of ad hoc wireless networks. CRC Press, Inc. Boca Raton, FL, USA (2003) 127–148

19. Royer, E., Toh, C.: A review of current routing protocols for ad-hoc mobile wireless networks. IEEE Personal Communications (1999) 46–55

20. Williams, B., Camp, T.: Comparison of broadcasting techniques for mobile ad hoc networks. In: ACM International Symposium on Mobile Ad Hoc Networking and Computing. (2002) 194–205

21. Barbara, D.: Mobile computing and databases - a survey. Knowledge and Data Engineering **11**(1) (1999) 108–117

22. Bobineau, C., Pucheral, P., Abdallah, M.: A unilateral commit protocol for mobile and disconnected computing. In: 12th International Conference on Parallel and Distributed Computing Systems. (2000)

23. Pitoura, E., Bhargava, B.K.: Data consistency in intermittently connected distributed systems. Knowledge and Data Engineering **11**(6) (1999) 896–915

24. Briesemeister, L.: Group Membership and Communication in Highly Mobile Ad Hoc Networks. PhD thesis, School of Electrical Engineering and Computer Science, Technical University of Berlin, Germany (2001)

25. Dolev, S., Schiller, E., Welch, J.: Random walk for self-stabilizing group communication in ad-hoc networks. In: 21st Symposium on Reliable Distributed Systems. (2002)

26. Malpani, N., Welch, J.L., Vaidya, N.H.: Leader election algorithms for mobile ad hoc networks. In: Proc. Fourth International Workshop on Discrete Algorithms and Methods for Mobile Computing and Communications. (2000) 96–103

27. Walter, J.E., Welch, J.L., Vaidya, N.H.: A mutual exclusion algorithm for ad hoc mobile networks. Wireless Networks **7**(6) (2001) 585–600

28. Angluin, D., Aspnes, J., Fischer, M.J., Jiang, H.: Self-stabilizing population protocols. In: Ninth International Conference on Principles of Distributed Systems. (2005) 79–90

29. Basile, C., Killijian, M.O., Powell, D.: A survey of dependability issues in mobile wireless networks. Technical report, Laboratory for Analysis and Architecture of Systems, National Center for Scientific Research, Toulouse, France (2003)

30. Dolev, D., Strong, H.R.: Authenticated algorithms for byzantine agreement. SIAM Journal of Computing **12**(4) (1983) 656–666

31. Fischer, M.J., Lynch, N.A., Merritt, M.: Easy impossibility proofs for distributed consensus problems. Distributed Computing **1**(1) (1986) 26–39

When Birds Die:
Making Population Protocols Fault-Tolerant

Carole Delporte-Gallet[1], Hugues Fauconnier[1],
Rachid Guerraoui[2], and Eric Ruppert[3]

[1] Université Paris 7, France
[2] MIT, USA and EPFL, Switzerland
[3] York University, Canada

At vobis male sit, malae tenebrae
Orci, quae omnia bella devoratis:
tam bellum mihi passerem abstulistis. [6]

Abstract. In the population protocol model introduced by Angluin *et al.* [2], a collection of agents, which are modelled by finite state machines, move around unpredictably and have pairwise interactions. The ability of such systems to compute functions on a multiset of inputs that are initially distributed across all of the agents has been studied in the absence of failures. Here, we show that essentially the same set of functions can be computed in the presence of halting and transient failures, provided preconditions on the inputs are added so that the failures cannot immediately obscure enough of the inputs to change the outcome. We do this by giving a general-purpose transformation that makes any algorithm for the fault-free setting tolerant to failures.

1 Introduction

Consider an ad hoc mobile network in which each agent is a very simple component, such as a tiny sensor with very severe constraints on memory and power. Such systems have been envisioned, for example, in Berkeley's Smart Dust project [10]. An agent can communicate with other nearby agents through wireless communication. To make use of data collected by the agents of such a system, it is necessary to aggregate the data in some way [11, 13].

Angluin *et al.* [2] introduced the notion of a computation by a population protocol to model this situation. In their model, the computation is carried out by a collection of agents, each of which receives a piece of the input. These agents move around and information can be exchanged between two agents whenever they come into contact with each other. The goal is to ensure that every agent can eventually output the value that is to be computed (assuming a fairness condition on the sequence of interactions that occur). The agents are simple devices, and can be represented as finite state machines. The abstraction also makes absolutely minimal assumptions about the movement of the system's components. In particular, the algorithms designed for such systems cannot dictate

P. Gibbons et al. (Eds.): DCOSS 2006, LNCS 4026, pp. 51–66, 2006.
© Springer-Verlag Berlin Heidelberg 2006

the movement of the agents. Can interesting computations still be performed in such a model? Angluin *et al.* showed the answer is yes, assuming no agents fail. For example, protocols exist to compute parity, majority and constant-threshold functions, as well as boolean combinations of such functions.

A motivating example for their model was a flock of birds, in which each bird carries a monitoring device that measures the bird's body temperature. The devices can signal other devices within a small distance. They showed that this sensor network could be used, for example, to determine whether at least five birds in the flock have an elevated temperature, to trigger an alert indicating that there might be an illness sweeping across the flock. In this paper, we study what happens when some of those ill birds drop dead: Can interesting computations be done in the population protocol model in a way that tolerates failures?

If malicious failures can occur, it is very difficult to do anything useful in the model: a single Byzantine agent (in collusion with the adversarial scheduler of interactions) could move around the system, driving each agent into an arbitrary reachable state by having a sequence of interactions with it. Thus, we consider two types of less catastrophic failures. A *crash failure* causes an agent to cease interacting with other agents. A *transient failure* is a momentary failure that can arbitrarily corrupt the state of an agent. The agent continues executing its algorithm correctly after the transient failure occurs. Transient failures include, as a special case, sensing failures, which cause the input to an agent to be incorrect. This is because the input is part of the state of an agent and can therefore be corrupted by the transient failure. However, transient failures are more general, since they can affect the entire state of the agent. For example, they can corrupt any partial data that the agent has collected, as well as its "programme counter" which keeps track of what part of the algorithm it is executing. (Such general transient failures might be caused by electromagnetic interference from the environment during an interaction or by soft errors due to alpha particle strikes.) We shall assume that both crash failures and transient failures can occur in an execution and that we have a known upper bound on the number of failures of each type that should be tolerated.

Clearly, some functions that can be computed without failures will be impossible to compute in a model with failures. For example, if we consider the possibility of experiencing a single halting failure, a population will not be able to compute with certainty a threshold function that is 1 if at least five of the birds are ill and 0 otherwise. Consider an execution with exactly five ailing birds, one of which dies (along with its sensor) before the bird comes into contact with any other birds. The output should be 1, but this run cannot be distinguished by any live agent from a run where there are four feverish birds and the output should be 0. However, with at most one failure, we can still distinguish whether the number of ill birds is greater than five or less than five. We discuss two ways to formalize this. We can restrict the domain of the function to be computed, by adding a precondition that the number of ill birds will either be greater than five or less than five. Alternatively, we can say that the protocol will compute the result correctly when the number of sick birds is different from five, but may output either 0 or 1 in the case where exactly five birds are sick. We explore both approaches: the former in Sect. 5 and the latter in Sect. 6.

In short, we show that, for *any* function that can be computed by a population protocol in a failure-free environment, it is possible to design a population protocol that computes the function in a way that tolerates crash failures and transient failures, provided preconditions are added or incorrect responses are permitted for inputs that are very close to other inputs that have a different response, as described above for the example about birds.

As one might expect, we use replication to achieve fault-tolerance, but in a way that is different from traditional approaches. Given a protocol that computes a function in a failure-free environment, we run several copies of the protocol. Because of the severe limitation on the memory of each agent, we need a constant fraction of the agents to cooperate to simulate one instance of the original protocol; otherwise there would not be enough space to store the states of all of the simulated agents. We divide the agents into g groups of approximately equal size. Each group simulates one instance of the failure-free protocol by having each agent in the group simulate approximately g agents of the original protocol. The value of g is chosen to ensure that the output produced by the largest number of groups' simulations is correct.

2   Related Work

The population protocol model was introduced by Angluin, Aspnes, Diamadi, Fischer and Peralta [2]. They defined the concept of stable computation of a function in this model, focussing on stable computation of predicates, which are functions whose output is a binary value. They showed that the predicates computable in this model include all that can be expressed using Presburger arithmetic and that they are all included within the complexity class NL.

They also considered variants of the model where interactions are restricted. First, the interactions can be constrained by considering a particular communication graph, which has an edge between the nodes that represent two agents if those agents are permitted to come into contact with each other. Second, they considered a randomized version of the model, where interactions are chosen randomly and uniformly, and the output must be computed with high probability. In both cases, the power of the system is increased.

Angluin, Aspnes, Chan, Fischer, Jiang and Peralta [1] further studied the model with a non-complete communication graph. They described properties of the communication graph itself that can be computed by the agents in the system. For example, the system can determine whether it contains an odd cycle.

Angluin, Aspnes, Eisenstat and Ruppert [4] considered population protocols where the interactions between pairs of agents are one-way. Each interaction has a sender and a receiver, and the sender cannot discover any information about the receiver's state in such an interaction. Full or partial characterizations of the predicates that can be stably computed (with no failures) in several variants of this model were given.

The question of tolerating failures in the population protocol model was raised by Delporte-Gallet, Fauconnier and Guerraoui [7]. They described how an example protocol can be adapted to tolerate failures. However, their approach is not generally applicable to all population protocols.

The transient failures that we consider in this paper can corrupt the internal states of agents arbitrarily. We assume that the number of such failures is bounded. Research on self-stabilizing systems [8] assumes that *any* number of processes can have corrupted states, requiring that the system eventually return to a correct configuration. Angluin, Aspnes, Fischer and Jiang incorporated the notion of self-stabilization into the population protocol model [5]. They gave some self-stabilizing protocols for classical problems such as leader election and token passing. The types of problems they studied differ from the problems we discuss here. They concentrated on stably maintaining some property (*e.g.* having a unique leader, having a legal colouring of the communication graph), whereas we focus on computing functions of inputs initially distributed across the system. This makes it necessary for us to assume a bound on the number of transient failures, so that those inputs are not lost. Also, we are concerned with creating a general-purpose transformation that converts an arbitrary algorithm that works in the failure-free setting into a fault-tolerant algorithm.

The way we transform the specification of a problem for the failure-free population protocol model into a specification for the fault-tolerant model is, in spirit, analogous to the way such transformations have been done in traditional distributed systems. Consider for instance the seminal atomic commit problem from distributed databases [9]. In a failure-free distributed system, one would typically require a transaction to commit if and only if all servers vote "yes", *i.e.*, none detected a concurrency conflict. Such a specification is clearly impossible to implement (even in a synchronous system) if one server can fail: it is indeed impossible to distinguish an execution where all servers voted "yes" and one initially crashed, from an execution where this initially crashed server voted "no". It is thus typical to allow a transaction to sometimes abort even if all servers vote "yes" (and one of them fails or is suspected to have failed), or commit a transaction even if a minority of servers vote "no" (*e.g.*, in a replicated system).

Our approach to describing functions that can be computed in the failure-prone population protocol model is also related to the condition-based approach of Mostefaoui, Rajsbaum and Raynal [12]. They described exactly what sort of precondition must be placed on the possible inputs to the consensus problem in order for it to become solvable in an asynchronous system with f halting failures using shared read-write registers.

3   Population Protocols

Our formalization of the population protocol model is based on the work of Angluin *et al.* [2]. We present a version that assumes non-deterministic, two-way interactions can take place between any pair of agents, but also allows halting failures and transient failures. A halting failure causes an agent to cease functioning and play no further role in the execution. A transient failure corrupts the state of an agent, but the agent otherwise follows its algorithm correctly.

Each agent in the system is modelled as a finite state machine, and algorithms must be *uniform*: each finite state machine is "programmed" identically and the programming does not depend on the number of agents in the system. This makes the model strongly *anonymous*, since there is not enough space in the state to give each agent a unique identifier.

Let X be a finite input alphabet and Y be a finite output alphabet. Each agent is provided with an input drawn from X. Since agents are essentially interchangeable, an input to the system can be thought of as a multiset of elements from X. Let $\mathcal{X}$ be a set of all multisets of elements from X. Let $\mathcal{D} \subseteq \mathcal{X}$ be the set of all input multisets that can actually occur. In general, $\mathcal{D}$ may be a proper subset of $\mathcal{X}$, since there may be preconditions on what inputs are permitted. The goal of an algorithm is to compute a function $f : \mathcal{D} \to Y$. Each agent must eventually output the value of this function for the input multiset that was initially provided to the agents.

We now describe how to specify a population protocol. Let Q be the finite set of states that each agent may take. A population protocol is defined by an input assignment $i : X \to Q$, a transition function $\delta : Q \times Q \to \mathcal{P}(Q \times Q) - \{\emptyset\}$, and an output assignment $o : Q \to Y$. (The notation $\mathcal{P}(S)$ is used to denote the power set of S.) If two agents in states q_1 and q_2 encounter each other, they can change into states q_1' and q_2', where $(q_1', q_2') \in \delta(q_1, q_2)$. Without loss of generality, assume the transition function is symmetric: $\delta(q_1, q_2) = \delta(q_2, q_1)$. The protocol is called *deterministic* if $\delta(q_1, q_2)$ is a singleton set for all $q_1, q_2 \in Q$.

Let $I \in \mathcal{D}$ be an input for the system. An *execution* of the protocol on input I is an infinite sequence of configurations, $C_0, C_1, C_2, \ldots$, each of which is a multiset of states drawn from Q. The initial configuration C_0 is the multiset $\{i(x) : x \in I\}$. The configuration C_k must be obtainable from C_{k-1} by one of the following four types of transitions:

Ordinary transition: $C_k = C_{k-1} - \{q_1, q_2\} \cup \{q_1', q_2'\}$ where $\{q_1, q_2\} \subseteq C_{k-1}$ and $(q_1', q_2') \in \delta(q_1, q_2)$.
Halting failure: $C_k = C_{k-1} - \{q\}$.
Transient failure: $C_k = C_{k-1} - \{q\} \cup \{q'\}$.
Null step: $C_k = C_{k-1}$.

The output of an agent in state q is $o(q)$. We say that the execution *stably outputs* $v \in Y$ if every agent eventually outputs v and never changes its output thereafter. Formally, this means there is an i such that for all $j > i$, $o(q) = v$ for every $q \in C_j$.

If every sequence of interactions is considered to be a possible execution in the model, it would be possible to have isolated agents that never interact with one another. So the model must incorporate a fairness guarantee. Simply requiring that every pair of agents eventually meet is insufficiently strong for some interesting protocols, since the two agents might meet only at inopportune times, when their states prevent a particular kind of interaction from happening. So the research on population protocols has assumed a stronger fairness condition. In a *fair* execution, if a configuration C occurs infinitely often and a configuration C' can be reached from C by an ordinary transition, then C' occurs infinitely often. If, for example, we associate probabilities with different interactions, then an execution will be fair with probability 1. A protocol *stably computes* a function $f : \mathcal{D} \to Y$ if, for every input $I \in \mathcal{D}$, every fair execution on input I stably outputs $f(I)$.

4   The Simulation

In this section, we describe how any population protocol A that stably computes a function f in a failure-free setting can be adapted to run in a setting where a bounded number of crash and transient failures can occur. To do this, we construct an algorithm B that divides agents into groups and simulates, within each group, an execution of the original protocol A. We shall show in Sect. 5 that, if we add a precondition on the inputs, this simulation will correctly compute f. We first define the kind of precondition on the inputs that will be required.

Recall that X and Y are an input and output alphabet, $\mathcal{X}$ denotes the set of all multisets of elements from X, and $\mathcal{D} \subseteq \mathcal{X}$.

Definition 1. Let $a, b \in \mathbb{N}$. A function $f : \mathcal{X} \to Y$ is called (a, b)-*robust for $\mathcal{D}$* if, for any input multiset $I \in \mathcal{D}$ and any input $I' \in \mathcal{X}$ that can be formed from I by removing up to a elements and then adding up to b elements, $f(I) = f(I')$.

Example 2. Let $X = Y = \{0, 1\}$. Let f be the majority function: for any multiset S of 0's and 1's, $f(S) = 1$ if and only if S contains more 1's than 0's. Let $\mathcal{D}$ be the set of all input multisets where the number of 0's differs from the number of 1's by at least k. Then f is (a, b)-robust for $\mathcal{D}$ for any parameters a and b satisfying $a + b < k$. This is because, starting from any input multiset in $\mathcal{D}$, the number of input values that would have to be added and removed to change the output of f total at least k.

Let $f : \mathcal{X} \to Y$ be any function that can be stably computed by a population protocol in the failure-free environment. We shall show that if f is $(c + t, t)$-robust for $\mathcal{D}$, then f restricted to inputs from $\mathcal{D}$ can also be stably computed in an environment where up to c crash failures and up to t transient failures may occur.

Let A be a population protocol that stably computes f in the failure-free setting. The algorithm A is specified by the state set Q_A, input and output assignment functions i_A and o_A, and the transition function δ_A. Let $Q_{init} = \{i_A(x) : x \in X\}$. We shall build an algorithm B which simulates A in a way that tolerates up to c crash failures and t transient failures. We first describe the simulation. Its correctness is argued in Sect. 5.

The fault-tolerant algorithm B will divide agents up into g groups (where g is a constant to be chosen later), and simulate the original algorithm within each group. There will be roughly n/g agents in each group, where n is the number of agents in the system. (Recall that agents do not know the value of n.) Each of the agents that comprise a group will simulate up to $2g$ distinct agents of the original algorithm A. (For clarity, we shall hereafter refer to the agents of algorithm B as "agents", and the simulated agents of algorithm A as "threads".) No thread will be simulated by two agents in the same group (except as the result of a transient failure).

In B, each agent's state contains seven fields:

- *init* stores an initial value from Q_{init}, initialized to $i_A(x)$, where x is the input for the agent. (This field is never changed by the algorithm.)
- *joined* is a boolean variable that says whether the agent has joined a group yet. Initially, it is set to *false*.
- *group* stores a value from $\{1, 2, \ldots, g\}$, initially g, which will eventually be the name of the group this agent joins.
- *sum* will be used for a division subroutine and can take values in the range $\{0, \ldots, group - 1\}$, initially 1.
- *sim* stores a multiset of up to $2g$ elements from Q_A representing the states of the threads that the agent is simulating, initially $\emptyset$.
- *given*[1..g] stores an array of g boolean values, with each entry initially set to *false*. This will keep track of which groups contain a thread that has been given a copy of this agent's input value.
- *output*[1..g] stores an array of g values from Y, representing the output values from the simulations carried out by each of the g groups. It can be initialized arbitrarily.

Note that the state set of algorithm B has $|Q_{init}|g(g+1)\binom{2g+|Q_A|}{2g}2^g|Y|^g$ states, and this quantity is independent of n, the number of agents in the system, as required by the model. (The number of bits needed to represent an agent's state in the simulation is $O(g \log |Q|)$.)

The first phase of an agent's actions is devoted to assigning the agent to one of the g groups. This phase ends when the agent's *joined* field is changed to *true*. The second phase will be devoted to gathering input values from approximately g other agents and simulating, within each group, an execution of the original algorithm. We shall guarantee that each non-faulty agent's input value is eventually given to exactly one thread of exactly one agent in each group. Whenever two agents in the same group meet, they nondeterministically choose an interaction of two of their threads to simulate. In those groups that have no faulty agents, the simulation will be a faithful simulation of algorithm A, and the output of each thread within that group will eventually stabilize to the correct value. We shall choose g large enough so that agents will be able to recognize (and output) a value that is being produced by a group of agents that experienced no failures.

In phase 1, we first execute the division-by-g algorithm described by Angluin *et al.* [2] to split off, from the rest of the agents, group number g, which will contain approximately n/g agents. The remaining agents then execute a division-by-$(g-1)$ algorithm to split off group number $g-1$ (again of size roughly n/g). The remaining agents then divide by $g-2$, and so on. The *group* field of the state keeps track of which division is currently being worked on by the agent.

An agent is said to *join group* i when it sets its *joined* field to *true*, if its *group* field contains i at that time. Joining a group is an irreversible action for a non-faulty agent: once the *joined* variable is set to *true*, none of the fields *joined*, *group* or *sum* will ever change again.

To accomplish phase 1, if two agents whose *joined*, *group* and *sum* fields are $(false, i, s)$ and $(false, i, s')$ with $i > 2$ meet, they transition to $(false, i - 1, 1)$ and $(false, i, s + s')$ if $s + s' < i$ and to $(false, i, s + s' - i)$ and $(true, i, 0)$ if $s + s' \geq i$. We shall argue below that this has the effect of making about $1/i$ of the agents that set their group field to i eventually join group i: the *sum* field accumulates a count of agents who set their group field to i and when one count reaches i, an agent can join group i. When an agent whose group field is i has been counted (but does not join group i), it changes its group field to $i - 1$. When two agents whose *joined*, *group* and *sum* fields are both $(false, 2, 1)$, one transitions to $(true, 2, 0)$ and the other transitions to $(true, 1, 0)$. This has the effect of splitting the agents whose group field is set to 2; half of them join group 1 and half join group 2.

When an agent p joins group i, it sets its *sim* field to $\emptyset$ (if p's *given*[i] field is *true*) or to $\{init\}$ (if p's *given*[i] field is *false*). In the latter case, p also changes its *given*[i] field to *true*. If, at any time, an agent p_1 whose value of *given*[i] is *false* meets another agent p_2 that has joined group i and does not have a full *sim* field, p_2 adds p_1's *init* field to its *sim* field and p_1 sets *given*[i] to *true*. Interactions of this type will have the effect of creating, for each correct agent p (and possible some faulty ones), a thread inside the *sim* field of exactly one agent in group i initialized with the initial state that p would have in algorithm A.

Whenever two agents p_1 and p_2 that have joined the same group meet, the transition function non-deterministically chooses two elements q and q' from the union of the two agents' *sim* multisets (both elements could possibly be from the same agent's *sim* multiset) and changes the two states q and q' to a pair of states given by $\delta_A(q, q')$. If the union of the two *sim* multisets contains fewer than two elements, no state change occurs in either agent.

Whenever an agent p_1 meets an agent p_2 that has joined some group i and has a non-empty *sim* field, p_1 sets its *output*[i] field to $o_A(q)$, where q is the first element of p_2's *sim* field. The output assignment function for B is defined by taking the element that appears with the highest multiplicity in the field *output*.

Our simulation B is non-deterministic, even if the original protocol A is deterministic: when two agents in the same group meet, they non-deterministically choose which two threads should interact. However, it is not difficult to remove this non-determinism of B by making use of the non-determinism of the order in which interactions occur, using the technique described by Angluin *et al.* [1]. Each agent stores a "choice counter" which dictates which of the finite number of possible outcomes should result from an interaction. The counters are incremented by a circulating token. However, in our model, the token could be lost when a failure occurs. So instead, we can increment the choice counter of an agent when it encounters an agent in another group.

5 Correctness

Consider an infinite fair execution $C_0, C_1, C_2, \ldots$ of the simulation B on input multiset $I \in \mathcal{D}$. We first show that eventually about n/g agents join each group. Let

$$
\begin{aligned}
c_i = {}& \text{the number of crash failures of agents which have } group = i \\
& \text{immediately before the crash,} \\
t_i = {}& \text{the number of transient failures of agents which have } group = i \\
& \text{immediately before the failure,} \\
t'_i = {}& \text{the number of transient failures of agents which have } group = i \\
& \text{immediately after the failure,}
\end{aligned}
$$

$$x_g(j) = n,$$

$x_i(j) = $ the number of ordinary steps in $C_0, \ldots, C_j$ that caused an agent to set its $group$ field to i, for $i < g$,

$W_i(j) = \{p \in C_j : p.group = i \text{ and } p.joined = false\}$, and

$J_i(j) = \{p \in C_j : p.group = i \text{ and } p.joined = true\}$.

Note that $W_i(j)$ and $J_i(j)$ are multisets. They represent the agents in configuration C_j that are waiting to finish the division-by-i algorithm and those that have joined group i, respectively. Consider the sum $S_i(j) = i|J_i(j)| + \sum_{p \in W_i(j)} p.sum$. Initially, $S_i(0) = x_i(0)$. The only time an interaction between two agents changes this sum is when one agent sets its group field to i, which increases the value of S_i by 1. Thus, an ordinary step changes the values of S_i and x_i in the same way. If an agent's $group$ value is i when it crashes, the crash decreases the value of the sum by at most i. If an agent's $group$ value is i just before it experiences a transient failure, that failure can decrease the sum by at most i. If an agent's $group$ value is i just after it experiences a transient failure, that failure can increase the sum by at most i, since the process's sum field cannot exceed its $group$ field. Thus, we have

$$x_i(j) - i(c_i + t_i) \leq S_i(j) \leq x_i(j) + it'_i \tag{1}$$

If the interaction that causes the change from C_j to C_{j+1} happens because two agents in $W_i(j)$ meet, one agent is removed from $W_i(j)$ to form the set $W_i(j+1)$, and never returns to $W_i(j')$ for $j' > j$ (unless by a transient failure). So, eventually (*i.e.* for sufficiently large values of j), $W_i(j)$ will contain at most one element, so we shall have $0 \leq \sum_{p \in W_i(j)} p.sum \leq i$. So (1) implies that, eventually,

$$x_i(j) - ic_i - it_i - i \leq S_i(j) - i \leq S_i(j) - \sum_{p \in W_i(j)} p.sum = i|J_i(j)| \leq S_i(j) \leq x_i(j) + it'_i. \tag{2}$$

Dividing the bounds in (2) by i yields the following bounds on the size of $J_i(j)$.

$$x_i(j)/i - c_i - t_i - 1 \leq |J_i(j)| \leq x_i(j)/i + t'_i. \tag{3}$$

If an agent has set its *group* field to i (either by a legitimate interaction or by having a transient failure) before C_j, but did not subsequently change it to $i - 1$, then either it is still in $W_i(j)$ or $J_i(j)$, or it has failed. For sufficiently large j, $W_i(j)$ contains at most one agent, so for $i > 1$ we shall have

$$x_{i-1}(j) \geq x_i(j) + t'_i - |J_i(j)| - c_i - t_i - 1. \tag{4}$$

Combining (3) and (4) yields, for large j,

$$x_{i-1}(j) \geq x_i(j) + t'_i - (x_i(j)/i + t'_i) - c_i - t_i - 1 = x_i(j)\frac{i-1}{i} - c_i - t_i - 1. \tag{5}$$

Solving recurrence (5), using the boundary condition $x_g(j) = n$, gives us (for all i)

$$x_i(j) \geq ni/g - \sum_{\ell=i+1}^{g} (c_\ell + t_\ell + 1) \geq ni/g - c - t - g. \tag{6}$$

Finally, combining (3) and (6) yields

$$|J_i(j)| \geq \frac{n}{g} - 2c - 2t - g - 1 \geq \frac{n+t}{2g} \quad \text{(as long as } n \geq 2g(2c+2t+g+1)+t). \tag{7}$$

This means that there will eventually be at least $\frac{n+t}{2g}$ agents in each group. We call group i *correct* if $t'_i = t_i = c_i = 0$. Note that each agent's *sim* field is big enough to simulate $2g$ threads, so each correct group will be able to simulate enough threads to handle all n agents, plus t extra, bogus threads that could be generated by transient failures. (A transient failure could cause an agent that has already given an initial value to group i to give another initial value to group i.)

Let $sim_i(j)$ be the union of all the multisets that are stored in *sim* fields of states in $J_i(j)$. Consider the interactions that take place after C_j that set the $given[i]$ field of some agent to *true*. Let $future_i(j)$ be the multiset of the values in the *init* fields of those agents. These are the values that get added to the *sim* fields of agents in group i after C_j.

Lemma 3. *For each correct group i, $future_i(j)$ will be empty eventually (i.e. for sufficiently large j).*

Proof. There will eventually be at least $\frac{n+t}{2g}$ agents that join group i and each can hold $2g$ values in its *sim* field. At most $n + t$ values will be added to these fields (in total), so each agent whose $given[i]$ field is *false* will eventually either fail or meet an agent in group i that has enough room to take that agent's initial value. Eventually every agent's $given[i]$ field will become *true* and stay that way forever, so $future_i(j)$ will eventually become empty (and remain so forever). □

Lemma 4. *Let i be a correct group in the execution of B. There is a failure-free execution $D_0, D_1, \ldots$ of the population protocol A with input set $future_i(0)$ such that, for all j, $D_j = sim_i(j) \cup future_i(j)$.*

Proof. The only steps of B's execution that alter the multiset $sim_i(j) \cup future_i(j)$ are those involving interactions between two agents that have already joined group i and have at least two elements in total in their sim multisets. For each such step, two elements q_1 and q_2 in the sim multisets are changed to q_1' and q_2', where $(q_1', q_2') \in \delta(q_1, q_2)$. Thus, the corresponding step in the constructed execution of A is legal. All other steps of the constructed execution are null steps.

We must still show that the constructed execution is fair. Consider any configuration D that occurs infinitely often in the constructed execution. There is an infinite increasing sequence $j_1, j_2, \ldots$ such that $D = D_{j_1} = D_{j_2} = \cdots$. Let D' be a configuration that can be reached from D by some ordinary transition of A that changes two agents in states q_1 and q_2 to states q_1' and q_2'. We must show that D' occurs infinitely often in the constructed execution too.

Consider the sequence of steps $C_{j_1}, C_{j_2}, \ldots$. Since there are only a finite number of possible configurations, some configuration C must occur infinitely often in this sequence. By Lemma 3 the set $future_i(j)$ must be empty for all occurrences C, because it eventually becomes empty. So, in C, the union of the sim fields of agents in group i is equal to D, and therefore includes q_1 and q_2. Thus, there is an ordinary transition of the simulation B that changes q_1 and q_2 in those sim fields of C to q_1' and q_2' to form a new configuration C'. By the fairness property of the execution of B, C' must occur infinitely often. Note that the configuration of the constructed execution that corresponds to each of these occurrences of C' is equal to D'. So D' occurs infinitely often in the constructed execution. □

The following corollary follows immediately from the preceding lemma and the fact that A stably computes f.

Corollary 5. *Eventually, for every x in the sim field of any agent that has joined a correct group i, $o_A(x) = f(future_i(0))$.*

Now we show that the set $future_i(0)$ is sufficiently close to the input multiset I for correct groups.

Lemma 6. *For any correct group i, $future_i(0) = I \cup I^+ - I^-$ where $I^+, I^- \in \mathcal{X}$ and $|I^+| \leq t$ and $|I^-| \leq c + t$.*

Proof. Consider the multiset I^+ that contains, for each transient failure during the execution that leaves an agent in a state with the $given[i]$ field equal to $false$, the $init$ field of the agent immediately after it experiences the transient failure. This set contains at most t elements. Each of the values in $I \cup I^+$ can be given to a sim field of an agent in group i^* at most once, since doing so changes the $given[i^*]$ field of an agent from $false$ to $true$, and it remains $true$ until the agent experiences a transient failure. Thus, $future_i(0) \subseteq I \cup I^+$.

Furthermore, as argued in the proof of Lemma 3, every value in $I \cup I^+$ will eventually be transferred to the *sim* field of some agent in group i, unless the agent holding that value experiences a failure before the transfer occurs. So, at most $c + t$ of the elements of $I \cup I^+$ are not in $future_i(0)$. Let I^- be the set of those elements. Then $|I^-| \leq c + t$, and $future_i(0) = I \cup I^+ - I^-$.   □

Now, by choosing g appropriately, we can guarantee that the output produced by each agent in the simulation is the output produced by the simulated thread of some correct group, and this will be the correct output value.

Theorem 7. *If $f : \mathcal{X} \to Y$ is stably computable in an environment with no failures and f is $(c+t, t)$-robust for $\mathcal{D} \subseteq \mathcal{X}$, then $f : \mathcal{D} \to Y$ is stably computable in an environment with up to c crashes and t transient failures, provided $n \geq 2((|Y| + 2)(c + 2t) + 2)^2$.*

Proof. We use the simulation B described above, taking $g = |Y|(c + 2t) + 1$. The assumption that $n \geq 2((|Y| + 2)(c + 2t) + 2)^2$ guarantees that $n \geq 2g(2c + 2t + g + 1) + t$ for our choice of g, so the requirement for inequality (7) is satisfied.

Consider any execution of the simulation. By Corollary 5, there is some time after which every thread in every correct group i outputs $f(future_i(0))$. Also, there is a time after which no agent experiences a failure. After these two times have both passed, every agent will eventually meet an agent in each correct group i and store $f(future_i(0))$ in its local variable $output[i]$. Let C_j be the configuration of the execution of B when all of this has happened.

Let r be any agent. We shall show that, after C_j, r stably outputs a correct value. The most common value in r's $output[1..g]$ field occurs with multiplicity at least $c + 2t + 1$. Therefore, it is $output[i^*]$ for some *correct* group i^*, since at most $c + 2t$ groups can be incorrect. Therefore, the value that r outputs will be $f(future_{i^*}(0))$ for the correct group i^*.

Let $I' = future_{i^*}(0)$. By Lemma 6, $I' = I \cup I^+ - I^-$, where $|I^-| \leq c + t$ and $|I^+| \leq t$. By the robustness property of f, we have $f(I) = f(I')$. Thus, agent r stably outputs $f(I') = f(I)$, which is correct.   □

We have shown that $(c + t, t)$-robustness is sufficient to compute the function f in an environment with c crash failures and t transient failures. We now show that a weaker robustness condition is *necessary*.

Proposition 8. *Suppose that $f : \mathcal{D} \to Y$ can be stably computed by a population protocol in an environment with up to c crash failures. Then f can be extended to the domain $\mathcal{X}$ so that $f : \mathcal{X} \to Y$ is $(c, 0)$-robust for $\mathcal{D}$.*

Proof. Let y_0 be any element of Y. Let A be a population protocol that stably computes f. We extend f to all input multisets $I \in \mathcal{X}$ as follows: if A produces a stable output in some fair, failure-free execution E_I with input I, let $f(I)$ be that output value. Otherwise, define $f(I) = y_0$. Note that this is an extension of f since, for $I \in \mathcal{D}$, A stably computes f.

We now show that the extension of f is $(c, 0)$-robust. Let $I \in \mathcal{D}$ and let $I' = I - I^-$, where $I^- \in \mathcal{X}$ and $|I^-| \leq c$. We must show that $f(I') = f(I)$. Consider an execution of A on input I in which the agents with inputs from I^- immediately fail, and then the remaining agents execute $E_{I'}$. By the hypothesis of the proposition, this execution must stably output $f(I)$. But this execution was used to define $f(I')$, so $f(I') = f(I)$.   □

There is a gap between the $(c + t, t)$-robustness condition which is sufficient to compute a function in the presence of failures (Theorem 7) and the $(c, 0)$-robustness condition that is necessary (Proposition 8). Closing this gap remains an open question. However, for systems in which there are only crash failures (i.e. $t = 0$) the condition of $(c, 0)$-robustness is both necessary and sufficient.

6   Computing Multivalued Functions

We now generalize the model used for stably computing functions to cover the possibility that the output is not uniquely determined by the input multiset. As before, let X and Y be finite input and output alphabets, and let $\mathcal{X}$ be the set of all multisets of elements from X. Let $F : \mathcal{X} \to \mathcal{P}(Y) - \{\emptyset\}$ be a function, where $F(I)$ represents the set of legal outputs for the input multiset $I \in \mathcal{X}$. A population protocol is defined exactly as in Sect. 3. However, we have a weaker definition of stable computation for such multi-valued functions. We say that a protocol *stably computes* F if, in every fair execution on input I, there is a time after which every agent outputs only values in $F(I)$. Notice that the output of an individual agent may oscillate forever, but it eventually stabilizes in the sense that it eventually becomes a legal output and remains so forever. Furthermore, different processes may output different values. This definition of stable computation coincides with the original one in the case where $F(I)$ is a singleton set for all I.

This formulation of stable computation for multi-valued functions allows us to describe the performance of our simulation in a different way.

Theorem 9. *Let $c, t \geq 0$. Suppose $F : \mathcal{X} \to \mathcal{P}(Y) - \{\emptyset\}$ is a multivalued function that can be stably computed in an environment with no failures. Then the function $F_{c,t} : \mathcal{X} \to \mathcal{P}(Y) - \{\emptyset\}$ defined by $F_{c,t}(I) = \bigcup\limits_{\substack{|I^-| \leq c+t \\ |I^+| \leq c}} F(I \cup I^+ - I^-)$ is stably computable in an environment with up to c crash failures and t transient failures, provided $n \geq 2((|Y| + 2)(c + 2t) + 2)^2$.*

Proof (Sketch). We can run the simulation described in Sections 4 and 5, again taking $g = |Y|(c + 2t) + 1$. The proof is very similar to the proof of Theorem 7. Consider any execution on input I. It follows from Lemma 4 that the threads in each correct group i eventually stabilize to produce outputs in $F(future_i(0))$. Consider the portion of the execution after this has occurred and all failures have occurred, and then every agent has met some agent in each correct group.

Consider any moment after all of this has occurred. For any agent r, the most common value in r's *output* field at that time appears in its $output[i^*]$ field for some correct group i^*. Let $I' = future_{i^*}(0)$. By Lemma 6, $I' = I \cup I^+ - I^-$ where $|I^+| \leq t$ and $|I^-| \leq c+t$, so $F(I') \subseteq F_{c,t}(I)$. Thus the value that is output by r is in $F_{c,t}(I)$, as required.   □

Remark. It follows from this proof that, in an execution of the simulation on input I where $c' \leq c$ crash failures and $t' \leq t$ transient failures *actually occur*, the value produced as the output will be in $F_{c',t'}(I)$. In particular, if the execution happens to be failure-free, the value produced will be in $F(I)$.

Example 10. Suppose $X = \{1\}$ and $Y = \{0, 1, \ldots, 99\}$. Let $F(I) = \{|I|$ mod $100\}$. Then, $F_{1,2}(I) = \{F(I)-3, F(I)-2, F(I)-1, F(I), F(I)+1, F(I)+2\}$ (where addition is done modulo 100). Since F can be stably computed in the failure-free model [2], our simulation will stably compute $F_{1,2}$ in an environment that can have up to 1 crash and 2 transient failures. Thus, it is possible to count the number of agents modulo 100, even when failures can occur, if we are satisfied with an approximate answer.

7   Concluding Remarks

If the communication graph G, which specifies which pairs of agents can come into contact with each other, is not complete, our simulation technique can be applied in a straightforward way to compute any function that can be computed in the complete graph, provided G is $(c+1)$-connected so that c crashes cannot disconnect the graph. This can be done by having the two agents in each interaction non-deterministically choose whether to swap states, just as in the failure-free model [2].

Angluin, Aspnes and Eisenstat have recently shown that the only predicates that are stably computable in the failure-free population protocol model are those defined by semilinear sets of inputs [3]. This might make it possible to use a somewhat streamlined version of our simulation to compute all stably computable binary predicates in a fault-tolerant way. This is because the known protocols for computing semilinear predicates have a relatively simple form.

There are a number of directions for future work on fault-tolerant population protocols. One is to close the gap between the $(c+t, t)$-robustness condition that is sufficient to compute a function and the $(c, 0)$-robustness condition that is necessary. Angluin *et al.* describe another type of function computation in the population protocol model, where the output does not come from a finite alphabet [2]. Instead of producing the output at each agent, the output is distributed across the system, just as the input is distributed. As an example, the division-by-g algorithm that we use in our construction starts with n agents and outputs 1 at n/g of them, and 0 at all others. As is shown by our construction, we can at least approximate the result of the division algorithm in a failure-prone environment. It would be interesting to characterize the set of functions that can be computed in this sense, in a fault-tolerant way, if some limited inaccuracy in the outcome is permitted.

This paper was concerned with the fundamental computability question: is it possible to do computations in the presence of failures? Another issue to examine is how much complexity increases as a result of incorporating failures into the model. In our model, the powerful adversary can delay convergence to a stable output for an arbitrarily long time by isolating some agents from one another. Thus, to measure complexity, one would have to consider a weaker adversary. One measure would be the expected time to converge (after the last transient failure), given some probability distribution on the interactions.

Acknowledgements

We thank James Aspnes for helpful conversations. This research was funded by the Natural Sciences and Engineering Research Council of Canada, the ACI Fragile, and the Swiss National Science Foundation through NCCR-MICS.

References

1. DANA ANGLUIN, JAMES ASPNES, MELODY CHAN, MICHAEL J. FISCHER, HONG JIANG, AND RENÉ PERALTA. Stably computable properties of network graphs. In *Proc. International Conference on Distributed Computing in Sensor Systems*, volume 3560 of *LNCS*, pages 63–74, 2005.
2. DANA ANGLUIN, JAMES ASPNES, ZOË DIAMADI, MICHAEL J. FISCHER, AND RENÉ PERALTA. Computation in networks of passively mobile finite-state sensors. In *Proc. 23rd ACM Symposium on Principles of Distributed Computing*, pages 290–299, 2004. Expanded version to appear in *Distributed Computing*.
3. DANA ANGLUIN, JAMES ASPNES, AND DAVID EISENSTAT. Stably computable predicates are semilinear. In *Proc. 25th ACM Symposium on Principles of Distributed Computing*, July 2006. To appear.
4. DANA ANGLUIN, JAMES ASPNES, DAVID EISENSTAT, AND ERIC RUPPERT. On the power of anonymous one-way communication. In *Proc. 9th International Conference on Principles of Distributed Systems*, 2005.
5. DANA ANGLUIN, JAMES ASPNES, MICHAEL J. FISCHER, AND HONG JIANG. Self-stabilizing behavior in networks of nondeterministically interacting sensors. In *Proc. 9th International Conference on Principles of Distributed Systems*, 2005.
6. GAIUS VALERIUS CATULLUS. Carmen 3. In *Carmina*. "But curse upon you, cursed shades of Orcus, which devour all pretty things! Such a pretty sparrow you have taken away." (Transl. Francis Warre Cornish).
7. CAROLE DELPORTE-GALLET, HUGUES FAUCONNIER, AND RACHID GUERRAOUI. What dependability for networks of mobile sensors? In *Proc. First Workshop on Hot Topics in System Dependability*, 2005.
8. SHLOMI DOLEV. *Self-stabilization*. MIT Press, 2000.
9. VASSOS HADZILACOS. On the relationship between the atomic commitment and consensus problems. In *Proc. Workshop on Fault-Tolerant Distributed Computing*, pages 201–208, 1990.

10. J. M. KAHN, R. H. KATZ, AND K. S. J. PISTER. Next century challenges: Mobile networking for "smart dust". In *Proc. 5th ACM/IEEE International Conference on Mobile Computing and Networking*, pages 271–278, 1999.

11. SAMUEL MADDEN, MICHAEL J. FRANKLIN, JOSEPH M. HELLERSTEIN, AND WEI HONG. TAG: a tiny aggregation service for ad-hoc sensor networks. In *Proc. 5th Symposium on Operating Systems Design and Implementation*, pages 131–146, 2002.

12. ACHOUR MOSTEFAOUI, SERGIO RAJSBAUM, AND MICHEL RAYNAL. Conditions on input vectors for consensus solvability in asynchronous distributed systems. *Journal of the ACM*, 50(6), pages 922–954, 2003.

13. BOAZ PATT-SHAMIR. A note on efficient aggregate queries in sensor networks. In *Proc. 23rd ACM Symposium on Principles of Distributed Computing*, pages 283–289, 2004.

Stochastically Consistent Caching and Dynamic Duty Cycling for Erratic Sensor Sources*

Shanzhong Zhu, Wei Wang, and Chinya V. Ravishankar

Department of Computer Science and Engineering
University of California, Riverside, CA 92521
{szhu, wangw, ravi}@cs.ucr.edu

Abstract. We present a novel dynamic duty cycling scheme to maintain stochastic consistency for caches in sensor networks. To reduce transmissions, base stations often maintain caches for erratically changing sensor sources. Stochastic consistency guarantees the cache-source deviation is within a pre-specified bound with a certain confidence level. We model the erratic sources as Brownian motions, and adaptively *predict* the next cache update time based on the model. By piggybacking the next update time in each regular data packet, we can dynamically adjust the relaying nodes' duty cycles so that they are awake before the next update message arrives, and are sleeping otherwise. Through simulations, we show that our approach can achieve very high source-cache fidelity with low power consumption on many real-life sensor data. On average, our approach consumes 4-5 times less power than GAF [1], and achieves 50% longer network lifetime.

1 Introduction

Power-efficient sensor data acquisition has become important as large-scale sensor networks become increasingly practical. A framework for data acquisition in sensor networks was introduced in [2], and various power-efficient techniques have been proposed in [3, 4, 5, 6] for sensor data collection in multi-hop wireless environments. Typically, users present their queries to a base station (BS), which collects data appropriately and generates responses.

In this paper, we show how to combine two strategies for reducing sensors' power consumption: *base station caching* and *dynamic duty cycling*. These ideas have been applied independently, but little work exists on strategies for combining them effectively in sensor networks.

1.1 Caching to Reduce Data Transmissions

Caching is commonly used to reduce data transmissions, which dominate power consumption in sensor networks. Several models for caching have been explored in the literature. In the first such model, exemplified by [3, 2], queries explicitly specify the sampling rate for sensor data. Queries arriving at intermediate times

---

* This work was supported by a grant from Tata Consultancy Services, Inc.

© Springer-Verlag Berlin Heidelberg 2006

are handled using cached data at the BS. Source and relaying transmissions are scheduled to occur as required by the known sampling times. The sensors can also be put to sleep in between, saving even more power. Although simple, this model cannot offer any guarantees on the precision of the cached data, especially when the underlying source data change rapidly and unpredictably.

Another approach is represented by [7], in which sources continuously stream updates to a central server which handles a large number of aggregate queries registered by users. The server caches a copy of each source object. Sampling times are not presepecified, but each aggregate query is associated with a *precision requirement*, indicating the maximum error the user will tolerate. A *filter*, or error bound, is installed on each source, and only values exceeding the filter bounds will be sent to the server. Filter bounds are adaptively set to minimize transmission costs, while ensuring that the precision requirements are met.

Unfortunately, for all the reasons discussed in [8], this approach wastes power if directly used in sensor networks. In addition, since sampling intervals are not fixed, updates arrive unpredictably, so all relaying nodes must always have their radios on. It is well-known that sensors consume significant amount of power in the listening mode [9, 10]. For example, in MICA2, the power consumed in listensing/receiving mode (7mW) is very close to the power consumed in transmitting mode (10mW); while in MICAz, the power consumed in listening/receiving mode (19.7mW) is even higher than in transmitting mode (17mW) [11]. Thus, to conserve power in sensor networks, we must put sensors into sleep as often as possible, while still guaranteeing cache consistency requirements.

Erratic Data Sources and Stochastic Consistency. Power optimization is particularly challenging in sensor networks that monitor *erratic* data sources [12]. Erratic data are numerical data that change frequently and unpredictably, such as temperature, pressure, and humidity. It is hard to predict erratic data behavior, making it hard to ensure cache-source consistency.

Strict cache-source consistency is unrealistic in sensor networks, since power is limited and wireless channels are volatile. However, users are often willing to tolerate some error, as long as it remains within pre-specified bounds. *Stochastic consistency*, first introduced in [12], captures this idea, and guarantees the cache-source deviation is within an error bound with a certain confidence level.

For example, in a sensor network to monitor temperatures, a user may be satisfied with a value within $2°F$ of the true value, with confidence 90%. Therefore, the source sensor needs to update the cached copy at the BS only when the cache-source deviation is no longer within $2°F$ with confidence 90%. We address the issue of maintaining stochastic consistency with minimum power consumption in sensor networks.

1.2 Dynamic Duty Cycling to Reduce Power Consumption

Our goal is to let each sensor node dynamically adjust its duty cycle so that it is in sleep most of the time, but has its radio on whenever an update must be relayed to the BS. Lowering duty cycles is known to be an effective way to

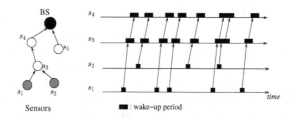

Fig. 1. Duty-cycling sensor nodes

extend the lifetime of the network [13, 14, 15, 1]. Our approach is illustrated in Figure 1, in which s_1 and s_2 are two sources, whose updates are delivered to the BS through s_3 and s_4. Both s_3 and s_4 turn on their radios only when an update packet is expected to arrive. The key challenge is how to let each relaying node estimate the arrival time of the next update, so that it can adjust its duty cycles accordingly.

Approaches such as GAF [1] and SPAN [15] try to maintain a routing backbone to ensure connectivity of the wireless ad hoc network, while allowing as many nodes as possible to sleep. At least one routing node is guaranteed to be within the transmission range of any node. GAF uses geographic location information (from GPS) to determine node equivalance for routing. SPAN uses a distributed randomized algorithm to maintain the backbone. Unfortunately, neither scheme exploits source data characteristics which may not require a connected backbone at all times.

In a sample network with 100 nodes uniformly distributed in a $1500m \times 300m$ region mentioned in [1], the resulting GAF routing backbone consists of 45 nodes. Hence, GAF always has 45 nodes listening, whether or not a message is active. In contrast, our approach captures source data characteristics, so that each source can *predict* update times, letting relaying nodes safely sleep till that time. Our approach will save more power especially under light source rates, since the relaying nodes are allowed to sleep more often than in GAF or SPAN (see Fig. 6).

We predict update times using the *Brownian motion* model [16], a stochastic model widely used to characterize randomly fluctuating data. Based on the user-provided consistency requirement and current data characteristics, the model adaptively determines the due time of the next update so that errors are bounded. The next update time is piggybacked on the current update message and delivered to the relaying nodes en route to the BS, which can safely turn off their radios and sleep before the arrival of the next update.

In our approach, each source delivers updates only at the times *predicted* by the Brownian motion model. In contrast, in approaches such as [7], updates are delivered at the times the source *detects* that the actual value has exceeded the error bound. The correctness of our approach is determined solely by how well our model matches future data behaviour under the stochastic consistency model. As shown by our extensive experiments (see Section 6), our method achieves high consistency (or *fidelity*) on various real-life sensor data, while saving a significant amount of power.

1.3   Our Contributions

We make several contributions in this paper. First, we experimentally verify that sensor data, such as temperature, humidity and ocean salinity, can be modeled as Brownian motions. This model has been successfully used in earlier work to model many other real-world erratic data sources [12, 17]. We confirm that model parameters, such as the *drift* and *diffusion* parameters, can capture the short-term linear trend and variance, respectively, with high confidence.

Next, we propose a dynamic duty cycling scheme based on the Brownian motion model, to allow nodes to turn off their radios frequently, while guaranteeing consistency requirements. A node will turn on its radio only when an update message is expected to arrive. In general, duty cycles are driven by the consistency requirements and source data characteristics.

Finally, we verify the correctness and efficiency of our approach with extensive simulations, which show that we can achieve high fidelity using far less power than GAF.

The rest of this paper is organized as follows: We review some related work in Section 2. Our system architecture and routing scheme are described in Section 3. In Section 4, we briefly introduce the Brownian motion model and perform experiments to verify its applicability on many sensor generated data. Our dynamic duty cycling scheme is presented in Section 5. The experimental results are presented in Section 6. Section 7 concludes our work.

2   Related Work

Various consistency models have been proposed to accommodate different requirements for cache freshness. For example, *quasi-caching* [18] allows the cached value to deviate from the source value in a controlled way (say, delay-bounded or error-bounded). *Probabilistic consistency* [19] guarantees that cached values are temporally consistent with the true value with a probability p. The concept of *stochastic consistency* was introduced in [12], and aims to provide an error-bounded cached copy with a given confidence. This model has been successfully used in pull-based replicated systems for erratic data streams [12]. We use this model in sensor environment.

2.1   Duty Cycling

Dynamic duty cycling is another technique widely used to achieve power efficiency in sensor networks. In GAF [1] and SPAN [15], nodes adaptively switch between sleeping and listening, while guaranteeing the existence of a capacity-preserving backbone routing network at any time. In GAF, each node used geographic location information (provided by GPS) to associate itself with a *virtual grid*. All the nodes in a virtual grid are equivalent for routing. *SPAN* is a distributed randomized algorithm, in which nodes can locally determine whether to sleep or stay awake in the backbone routing network, without knowledge of their geographic locations. Periodically, the set of routing nodes is changed to ensure even power dissipation. LEACH [20] aims to provide a cluster-based routing hierarchy where all sensor nodes are divided into clusters. A cluster head is elected to route data on behalf of the other nodes in each cluster. In our approach, duty cycles are driven by the source update rates, which are in turn governed by consistency requirements. In Section 6, we show that our approach lets sensors sleep more often, thus saving more power than GAF.

A periodic duty cycling scheme was introduced in *S-MAC* [21], in which nodes periodically switch between the listening and sleeping modes to conserve power. Neighbouring nodes exchange their listen/sleep schedules to synchronize their duty cycles. To deliver a packet, a sending node waits till the next hop node wakes up. However, significant latency will still be introduced since delays are accumulated along multiple-hop paths to the BS. A similar scheme was proposed in *STEM* [14], which assumes that two separate radios, a *wakeup* radio and a *data* radio, are available to each sensor node. To send a packet, the wakeup radio of the sending node polls the receiving node until it wakes up, and turns on its data radio. Again, data packets will experience significant delays because such delays at each hop will accumulate over the route. Both schemes are clearly not suitable in our situation, where updates must reach the BS as soon as possible, to ensure cache freshness. In TAG [3], the nodes along the aggregation tree are periodically synchronized with each other to relay and aggregate new sensor data. Since the source sampling rate is specified in the query, their synchronization scheme is much simpler than ours.

The success of our scheme relies on modeling the underlying data as Brownian motions. Applying probabilistic models to sensor data has been shown to be effective in conserving power while providing quality results [22,23,17]. Section 4 confirms earlier work that has shown that Brownian motions can model erratic data streams with high confidence [12,17].

2.2 Stochastic Consistency

Stochastic consistency [12] guarantees that the deviation between a cached value and the true value is within a pre-specified error bound ϵ with a confidence at least p. Let $v_i(t)$ and $c_i(t)$ be the source and the cached values, respectively, of object o_i at time t. The cache is stochastically consistent with source at t if

$$\Pr[\,|v_i(t) - c_i(t)| \le \epsilon\,] \ge p. \tag{1}$$

We must update the cached copies frequently enough to maintain stochastic consistency. On the other hand, to save cache/source communications, we must send the updates right before the confidence that cache-source deviation is within ϵ starts to drop below p. In Section 4, we discuss how to determine update times under this model.

3 Our System

Sensor networks typically consist of a BS with ample resources and a set of resource-limited sensor nodes communicating with the BS over multi-hop wire-

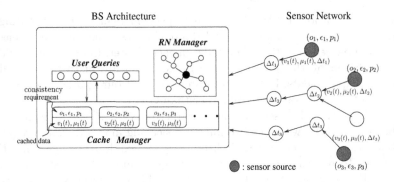

Fig. 2. The BS Architecture

less channels (see Figure 2). The BS serves as the destination for sensor data, and as the interface to user queries. It maintains caches to reduce communications and provide prompt responses. Our focus is on how to maintain cache-source consistency in a power-efficient way. Our caching system can support a broad spectrum of queries, ranging from monitoring single sensor's readings to aggregate queries as in [7,3].

Fig. 2 also shows an architectural schematic for the BS. The *cache manager* manages all cached objects. The object o_i represents a data source sensed by sensor s_i. Each object o_i is associated with a *consistency requirement* (ϵ_i, p_i) determined by user requirements. (Converting user requirements to object consistency requirements is an orthogonal concern we do not address. An example can be found in [7].)

An object's consistency requirement is also available at the corresponding source sensor, which determines when a cache update must be sampled and delivered. A cache update takes the form $(v_i(t), \mu_i(t))$, where $v_i(t)$ is the sampled value at time t, and $\mu_i(t)$ is the current *drift* parameter estimated at the source. The drift parameter is a Brownian motion parameter and represents the current linear trend of o_i. It helps to provide a more accurate cache value at the BS (see Section 5.1). The next update time Δt_u is adaptively evaluated under the stochastic consistency model, and included with each update so that each relaying node en route can sleep for time Δt_u. The BS responds to queries by retrieving the current values from the cache, calculating query results, and returning them to users.

The *RN manager* maintains a view of the *routing network (RN)*, which is a collection of routes through which sensors may reach the BS. Based on this information, the BS can determine a power-efficient route for each newly cached object source (see Section 3.1).

3.1   Routing

In principle, our dynamic duty cycling scheme is independent of the routing protocol, as long as routes are *persistent*, that is, the route for each source remains unchanged for a certain time. This property allows nodes on each route to obtain the wake-up time for the next update from each update message (see Section 5.3). Many ad hoc routing protocols for sensor networks generate persistent routes [4, 3, 24, 25]. In our work, we use a *energy-aware* routing scheme similar to [24] and [26], to avoid bottleneck nodes that would otherwise dissipate their power much faster than the others. First, we build a *routing network (RN)* through which sensor nodes can communicate with the BS. Typically, the RN includes good-quality wireless links to ensure reliable transmissions. Besides, as power is our major concern, it is desirable that each route in the RN be the shortest path from the source to the BS.

A common approach to building a RN is to assign a *level* number to each sensor node depending on its distance to the BS [3, 6, 26]. The BS is at level 0; Those nodes 1-hop away from the BS are at level 1, and so on. Initially, the BS broadcasts a query message containing its ID and level number. Upon receiving this message from its neighbours, each node determines its level and parents, and rebroadcasts the query message with its own ID and level number. After the query messages have flooded the entire network, a RN is formed where each node has one or more parents through which it can send packets towards the BS. Any path is the shortest one in the resulting RN. A more detailed description of constructing the RN can be found in [26]. The above algorithm must be run periodically to accommodate topology changes. To allow the BS maintain a view of the RN, each node must send a message containing its level number and parents to the BS.

After the RN is set up, the *RN manager* is responsible for determining a route for each newly cached object source. To balance power consumption, we choose routes with the maximum remaining power. A route's remaining power is defined as the minimum remaining power on its en-route nodes. Each node periodically determines its remaining power level and piggybacks the value on regular update messages destined for the BS. The RN manager periodically re-evaluates the remaining power on each route and chooses the one with the maximum power left.

4   Modeling Sensor Data

Sensor data are often numerical values, and change continuously. Modeling their behavior is central to our dynamic duty cycling scheme.

4.1   Standard and Drifting Brownian Motion Models

The Brownian motion model [27] is a continuous-time stochastic process widely used to characterize highly fluctuating data, and has been successfully used to model stock prices [28] and other erratic data sources [12] such as temperature and computer system loads. A *Standard Brownian motion (SBM)* $W(t)$ satisfies: 1) $W(0) = 0$; 2) $W(t) - W(s)$ is normally distributed with mean 0 and variance $t - s$ $(t \geq s)$; 3) $W(t) - W(s)$ is independent of $W(v) - W(u)$ if (s, t) and (u, v) are non-overlapping time intervals.

Table 1. Average *p*-values from the *W-S* test for various sensor traces and time intervals, confidence interval: 95%, all traces obtained from *TAO* Project

time interval	temp traces (Depth)		salinity traces (Long./Lat.)		humd traces (Long./Lat.)		slp traces (Long./Lat.)	
	36M	47M	5N/180W	2N/180W	0N/155W	2N/140W	2N/110W	2S/95W
10 min	75.21%	72.96%	77.44%	76.65%	80.14%	81.13%	76.54%	80.12%
15 min	72.38%	75.90%	76.26%	76.60%	79.47%	79.79%	75.88%	79.17%
20 min	73.45%	73.55%	76.10%	75.15%	76.45%	77.31%	75.45%	76.92%
30 min	71.47%	66.13%	75.00%	74.84%	72.60%	75.14%	74.38%	73.04%

A variant of the *SBM* is the *drifting Brownian motion (DBM)* $S(t)$, which includes a secular drift in the expectation of the process. The increment of $S(t)$ is modeled as:

$$\Delta S(t) = \mu(t)\Delta t + \sigma(t)\Delta W(t), \tag{2}$$

where $\mu(t)$ and $\sigma(t)$ are the *drift* and *diffusion* parameters for $S(t)$, respectively. $W(t)$ is a SBM process. The drift parameter models a secular upward or downward trend in the random data, while the diffusion parameter models the randomness of the data. Hence, the DBM is a combination of a predictable linear trend and a Brownian motion process. It is easy to see that the increment $\Delta S(t)$ also follows a Normal distribution: $\Delta S(t) \sim N(\mu(t)\Delta t, \sigma^2(t)\Delta t)$. In Section 4.2, we show that the DBM model is applicable to many real-life sensor data.

4.2 Verifying DBM on Sensor Data

In [12], we have already shown that real-life data sources such as stock traces, ocean temperatures, and system load data can be successfully modeled as DBM. In this work, we further verify that this model is appropriate for a wider variety of sensor genearted data, using the same methodology as in [12]. Since data increments are normally distributed in the DBM model, we will perform *normality tests* [29] on increments of sensed data. Various methods for normality testing have been proposed, and, as explained in [12], the *Wilk-Shapiro (W-S)* test [29] is the most appropriate in our case.

Our sensor data traces were taken from the *TAO* project [30] at the Pacific Marine Environmental Laboratory (PMEL). We tested four categories of data generated by ocean sensors: ocean temperature (*temp*), relative humidity (*humd*), salinity (*salt*), and sea level pressure (*slp*). Each category included data traces generated at different geographical locations (*Longitude/Latitude*) or ocean depths (*Depth*). Each data trace comprised about a year's data sampled at every 1 minute. Table 1 shows the results of the *W-S* test on these traces.

Each value in Table 1 represents the average *p*-value [16] evaluated on increment samples over a certain time interval. The *W-S* test calculates a test statistic for each data series. The *p*-value measures the probability that the test statistic will take on a value that is at least as extreme as the calculated value when the samples are normal. The *p*-value measures the probability that the tested data are drawn from a normal distribution. The significance level (α) of our test is set to 0.05. The larger the p-value, the stronger the confidence with which we may accept the samples as normal [29]. As shown in Table 1, The p-values for our data are far higher than α, indicating that we can believe the increments are normal with high confidence. For longer intervals, the p-value drops somewhat, suggesting the model may evolve in the long run.

The W-S test strongly supports our hypothesis that the sensor data are DBMs. On the other hand, since the model may evolve along with time, it is important to periodically estimate the model parameters $\mu(t)$ and $\sigma(t)$ to accurately characterize the underlying data. We will discuss how to estimate the two parameters in Section 5.2.

5   Dynamic Duty Cycling

We achieve power efficiency by operating sensors in low duty cycles, while guaranteeing that cache updates will arrive on time at the BS. The update intervals change dynamically due to the erratic nature of data sources. An approach to maintaining cache consistency similar to that of [7] would be for each source to constantly sample the underlying data. If it finds the sampled value deviates from the last update by more than ϵ, it forwards the value as a cache update. This approach forces relaying nodes to be awake all the time since update times are unpredictable. We need a more intelligent approach that can *predict* the due time of the next update, and let the relaying nodes sleep safely until that time.

Our approach models erratic data sources as Brownian motions, estimates the times when the cache-source deviation is expected to exceed ϵ, and schedules the next cache update at that time. When a source is ready to deliver an update, it also determines the time interval Δt_u until the next update, based on the DBM model. The source then sends the update along with Δt_u, so that each relaying node can obtain Δt_u. Since it knows that the next update from the source will arrive after time Δt_u, it can safely turn off its radio and sleep for time Δt_u. Our approach allows the relaying nodes to dynamically synchronize with each other and form a connected path whenever an update is ready to be sent.

We discuss how to adaptively derive Δt_u in Section 5.1. The drift and diffusion parameters must be estimated regularly from the underlying data. The issue of parameter estimation is discussed in Section 5.2. Our scheme to perform dynamic duty cycling is presented in Section 5.3. In Section 5.4, we analyze the power cost for our scheme.

5.1   Determining Cache Update Times

The drift and diffusion parameters characterize the current linear trend and randomness, respectively, of the sensor data. Each sensor source can use these parameters to adaptively determine Δt_u, the time till the next update. Let t_0 be the last time an update was delivered for object o_i, and $v_i(t)$ and $c_i(t)$ be the true and cached values of o_i at time t, respectively. Stochastic consistency requires the next update to be delivered before our confidence that the cache-source deviation is within ϵ drops below p. We must solve Δt_u from the following equation:

$$\Pr[\,|v_i(t_0 + \Delta t_u) - c_i(t_0 + \Delta t_u)| \leq \epsilon\,] = p. \tag{3}$$

Based on the DBM, we have $v_i(t_0 + \Delta t) = v_i(t_0) + \mu_i(t_0)\Delta t + \sigma_i(t_0)\Delta W(t)$, if $\mu_i(t_0)$ and $\sigma_i(t_0)$ are the drift and diffusion parameters estimated at time t_0, and $W(t)$ is the SBM (see Equation 2). Clearly, the expected value of $v_i(t_0 + \Delta t)$ is $v_i(t_0) + \mu_i(t_0)\Delta t$, which is also the best estimate the cache can make at time $t_0 + \Delta t$, given that the last cache update is $(v_i(t_0), \mu_i(t_0))$. Therefore, $c_i(t_0 + \Delta t) = v_i(t_0) + \mu_i(t_0)\Delta t$. We can easily derive that the cache-source deviation is normally distributed:

$$v_i(t_0 + \Delta t) - c_i(t_0 + \Delta t) \sim N(0, \sigma_i^2(t_0)\Delta t). \tag{4}$$

From Equations 3 and 4, we can obtain:

$$\Delta t_u = \frac{1}{2}\left(\frac{\epsilon}{\sigma_i(t_0)\,\mathrm{erf}^{-1}(p)}\right)^2, \tag{5}$$

where $\mathrm{erf}^{-1}(p)$ is the well-known *inverse error function* [31]. The detailed derivation of Equation 5 can be found in the Appendix of [32].

Δt_u must be recomputed on-line at sensor sources. Since $\mathrm{erf}^{-1}(p)$ can be precomputed and stored for the required p, computing Δt_u requires only some simple arithmetic operations, and is easily affordable for sensors.

5.2   Estimating Model Parameters

Obtaining accurate estimates for $\mu_i(t)$ and $\sigma_i(t)$ is critical to the success of our approach. According to the DBM model, increments follow the normal distribution $N(\mu_i(t)\Delta t, \sigma_i^2(t)\Delta t)$. Assuming both $\mu_i(t)$ and σ_i remain relatively constant over small time intervals, we may estimate $\mu_i(t)$ and $\sigma_i(t)$ by estimating the *mean* and *variance* of increment samples over a small time interval. The simplest unbiased estimators [16] of the mean and variance of a sample $\{x_1, \ldots, x_n\}$ are $\hat{x} = (\sum x_i)/n$ and $\hat{\sigma}^2 = \sum(x_i - \hat{x})^2/(n-1)$.

Let $\hat{\mu}_i(t)$ and $\hat{\sigma}_i(t)$ be the estimated values of $\mu_i(t)$ and $\sigma_i(t)$, respectively. Our estimation scheme works as follows: Let t_1 be the time of the next update. Starting at time $t_1 - \delta$, we sample the underlying data every h time units, where $h < \delta$. Thus, at time t_1, we collect n data samples: $v_i[1], v_i[2], \cdots, v_i[n]$, where $n = \delta/h + 1$. The obtained $n - 1$ increments $v_i[j + 1] - v_i[j]$ $(1 \leq j < n)$ are independent normal samples, and since δ is small, these samples are identically distributed. Thus, we calculate $\hat{\mu}_i(t_1)$ as follows:

$$\hat{\mu}_i(t_1) = \frac{(v_i[n] - v_i[1])}{\delta}. \tag{6}$$

We can also estimate obtain $\hat{\sigma}_i(t_1)$ from:

$$\hat{\sigma}_i^{\,2}(t_1) = \frac{1}{(n-2)h}\sum_{j=1}^{n-1}(v_i[j+1] - v_i[j] - \hat{\mu}_i(t_1))^2. \tag{7}$$

In typical sensors, such as those for light, temperature, or magnetic fields, the sampling time is on the order of $0.1ms$ [2].

A relatively small δ ensures accurate estimation of $\mu_i(t)$ and $\sigma_i(t)$, since these parameters remain constant during small intervals with high probability. On the other hand, a smaller h leads to more samples but may increase power consumption. A larger h saves power but may result in inaccurate estimates due to too few samples. Thus, we must choose both δ and h carefully to balance estimation accuracy and power consumption. In our experiments, we set $\delta = 10h$. Our results show that the obtained sample size is appropriate for our purpose.

5.3   Our Scheme

Each node can be in the *active* or *idle* state, depending on whether or not it is actively delivering/relaying update packets. Initially, all nodes are idle. During the RN setup phase, each node is assigned a *wakeup interval* t_w. It wakes up every t_w time units to check for pending caching requests from the BS. Upon receiving a request $R(s_i)$, an idle node switches to the active state, since it knows it will participate in relaying updates for source s_i. The choice of t_w must balance power consumption against response time (how long the BS must wait until receiving the first cache update). Larger t_w values let nodes sleep longer, but increase response times. We chose a moderate value for t_w in our experiments.

When the BS must query sensor s_i, it first consults the RN manager to find a route to s_i (see Section 3.1). It then sends the request $R(s_i)$ and the consistency requirement to s_i. If a node s_{j1} along the route finds the next node s_{j2} to be still asleep, s_{j1} will poll s_{j2} until it wakes up. Node s_{j2} records the sending node s_{j1}, so that it knows where to deliver s_i's updates. Each en-route node remains in listening mode until it receives the *first* update from s_i, and lets the Δt_u supplied by s_i drive its duty cycles after that point.

At the source s_i, Δt_u is evaluated on a regular basis according to Equation 5. A series of samples must be collected for parameter estimation before delivering the update message. Each update message contains the most recent sample $v(t_{update})$, the drift parameter $\mu(t_{update})$, and the next Δt_u.

Each node on the return route to the BS obtains Δt_u from the message containing the sensor update, and schedules to wake up at $t_{next} = t_{curr} + \Delta t_u - t_e$, where t_{curr} is the current system time, and t_e is a small time offset to accommodate variations in wireless transmission delays. In our simulations, we set $t_e = 10ms$ with moderate traffic in the network. Since a node may relay messages for several sources, it maintains a list to hold the future wakeup times. After relaying an update message, the node can safely turn off its radio and sleep until the next time entry in the list is due. Since the BS may make data requests while the node is asleep, the node must check for such requests to avoid poor response times. If the time interval until the next wakeup time entry is larger than t_w, the node must wake up at t_w to perform this check. More details on our scheme can be found in [32].

Every time a user requests o_i's value from the cache at the BS, the cache manager returns $v_i(t_l) + \mu_i(t_l)(t_{curr} - t_l)$, where t_{curr} is the current time, and t_l is the last time an update message was received.

5.4   Analysis of Power Consumption

Power is charged for communication, computation, and data sampling on sensor nodes. We ignore the power consumed by computation in our analysis since it is orders of magnitude lower than that by communication [2]. A sensor's radio may be in one of the following modes: *transmitting (T)*, *receiving (R)*, *idle(I)*, or *sleeping (S)*. In the idle mode, it listens to the wireless channel, waiting for incoming packets. In the sleeping mode, it turns off its radio, so the consumed power is negligible compared with other modes. Let the power consumed in transmitting, receiving, and idle mode be P_T, P_R, and P_I, respectively, and let P_S be the power consumed by sampling the underlying data. The total consumed power is simply $P = P_T + P_R + P_I + P_S$.

Our approach enables each node to remain in the sleeping mode most of the time and wake up only when an update is expected to arrive. Ideally, the idle time on each node is zero. Thus, $P = P_T + P_R + P_S$. Let P_{t_0} and P_{r_0} be the power consumed for transmitting and receiving one bit of data, respectively, and let P_{s_0} be the power consumed by sampling one piece of data from the environment (we assume each sensor has only one sensing module). Node s_i's power consumption rate at time t is

$$P_i(t) = \left(\sum_{s_j \in R_i} f_j(t) \right) M(P_{t_0} + P_{r_0}) + I(s_i)f_i(t)\left(MP_{t_0} + (\frac{\delta}{h} + 1)P_{s_0} \right), \quad (8)$$

where M is the size of the update message, R_i is the set of sources whose updates are relayed by s_i, $f_j(t)$ is the rate of updates generated by source s_j at time t, and $I(s_i)$ is an *indicator* function which is 1 if s_i is also a source, and 0 otherwise. Equation 8 suggests that a relaying node's power consumption is proportional to the aggregate amount of update traffic it relays, and a source's power consumption is also governed its updating frequency.

Combining Equations 5 and 8, we can further obtain:

$$P_i(t) = \left(\sum_{s_j \in R_i} (\beta_j \sigma_j^2(t)) \right) M(P_{t_0} + P_{r_0}) + I(s_i)\beta_i \sigma_i^2(t)\left(MP_{t_0} + (\frac{\delta}{h} + 1)P_{s_0} \right), \quad (9)$$

where $\beta_j = 2(\frac{\text{erf}^{-1}(p_j)}{\epsilon_j})^2$. In Equation 9, a relaying node's power consumption $P_i(t)$ is a function of its upstream sources' consistency requirements (ϵ_j and p_j) and their current data variance ($\sigma_j^2(t)$). Not surprisingly, the higher the variance, the greater the number of updates to be delivered to maintain a certain level of consistency, and the greater the power consumption. On the other hand, a more stringent consistency requirement (small ϵ_j and large p_j) also results in higher power consumption, which is illustrated in Fig. 6.

If s_i is a source, its power consumption is also governed by its own consistency requirement (ϵ_i and p_i) and data variance ($\sigma_i^2(t)$). Every time an update is required, additional δ/h samples must be collected for parameter estimation (see Equation 9). Since the power required for sampling (P_{s_0}) is very low for many typical sensors such as light, temperature, and accelerometer [2] (on the order of $0.1\,\mu J$), such sampling overhead is affordable.

6 Experiments

We conducted extensive experiments to demonstrate the correctness and efficiency of our dynamic duty cycling scheme using the ns-2 simulation package [33].

6.1 Simulation Setup

We uniformly deployed 100 sensor nodes in a $1500 \times 1500\,m^2$ region with the BS at the center. We chose UDP as the transport layer protocol and 802.11 [34] as the MAC layer protocol. The ns-2 simulator currently supports three propagation models, among which the *shadowing* model [35] is the most realistic and widely-used. This model has two parts: a path loss model, and a statistical model for the variation of reception at certain distances. We set the value of the path loss exponent as 2.0, and the value of the shadowing deviation as 4.0, representing a typical outdoor environment. We set the radio communication range to $250m$ and chose 0.95 as the rate of correct reception.

A subset of sensor nodes were chosen as sources. We used various categories of real-life sensor data as the source traces such as the ocean temperature traces (*temp*), the relative humidity traces (*humd*), and the ocean salinity traces (*salt*), all obtained from the *TAO* project [30] (see Section 4.2). Each source was associated with one data trace every time. In each experiment, we used 5 different traces from each category and demonstrated the average results. We purposely selected sources from the most distant nodes from the BS, since it is more challenging to maintain cache consistency for the distant sources. We chose a fixed payload size of 16 bytes for each update message including the data value, the drift parameter, and the Δt_u value.

To measure power, we adopted the power parameters from the *Chipcon* CC1000 RF transceiver [10], which is used as the radio module in MICA2 and MICA2DOT sensor models. When operated at $433MHz$, its receiving power is $22.2mW$, and the transmitting power is $31.2mW$, with the output power of $0dBm$. Each node was set to the same power level initially, and we measured the remaining power after the simulation ran for some time.

6.2 Measuring the Fidelity

We define the *fidelity* $f(o_i)$ as the percentage of time that the object o_i's source-cache deviation is within the error bound, that is,

$$f(o_i) = \frac{\text{the time cache-source error} \leq \epsilon}{\text{the total simulation time}}. \tag{10}$$

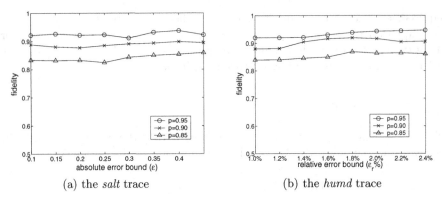

(a) the *salt* trace   (b) the *humd* trace

Fig. 3. The fidelity

Fidelity measures how well our scheme meets the consistency requirements. The higher the $f(o_i)$ value, the more confidence we have to achieve stochastic consistency. Ideally, the fidelity value must match the user-provided confidence probability p, indicating the drifting Brownian motion model is accurate in characterizing source data.

Fig. 3 shows the fidelity values for the *salt* and *humd* traces under absolute error bounds (ϵ) and relative error bounds ($\epsilon_r\%$), respectively. We randomly generated a topology of 100 nodes and picked a source s^* at the highest level. We associated different data traces to s^* one at a time, and measured its fidelity. Each data point in Fig. 3 represents an average fidelity value over five traces from the same category. To generate a certain amount of traffic in the network, we chose ten other sources in the topology, each associated with the same data trace, with a randomly chosen error bound.

Our scheme clearly achieves high fidelity for s^* under both categories of traces. The obtained fidelity value is very close to the corresponding confidence level p. For example, under the confidence level 90%, the average fidelity for the *salt* trace is 89.2%, while it is 89.5% for the *humd* trace. This is strong evidence for the accuracy of our DBM-based approach.

In Fig. 4, we show the impact of different network traffic loads on the fidelity. We used the same 100-node topology and chose a source s^*, associated with the *temp* trace with the error bound 0.1, at the highest level. We also chose a certain number of other nodes as sources, each associated with the same *temp* trace and the same error bound. We varied the number of sources from 10–65 and observed s^*'s fidelity. The amount of traffic in the network increases as the number of sources increases. Our scheme achieves a high and stable fidelity at confidence levels of 90% and 95%.

We also compared our scheme with GAF [1], an adaptive scheme that maintains a routing backbone in the wireless network, and puts other nodes to sleep as much as possible. We simulated GAF on the top of AODV [36]. With the same topology and input traces, GAF achieves the same fidelity as our scheme under light traffic loads, but much lower fidelity under heavy traffic loads.

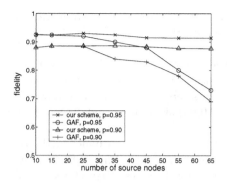

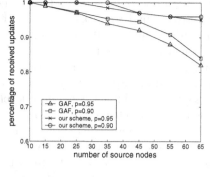

Fig. 4. Fidelity under various traffic loads (*temp* trace, $\epsilon = 0.1$)

Fig. 5. The percentage of received updates at BS (*temp* trace, $\epsilon = 0.1$)

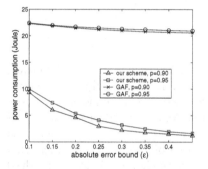

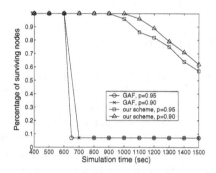

Fig. 6. The average power consumption per node (*temp* trace, 1000 *secs*)

Fig. 7. The fraction of surviving nodes (*temp* trace, $\epsilon = 0.1$, 100 nodes)

This behavior is explained by Fig. 5, which shows the percentage of s^*'s update packets received by the BS, under various traffic loads. Starting from 45 sources, the percentage of received packets begins to drop rapidly under GAF, while it remains stable under our scheme. Since GAF ensures that a connected routing backbone is always available, heavy traffic loads will lead to severe contention in the wireless channel. Our scheme, however, is more flexible in adjusting each node's duty cycle, causing less channel contention and increasing throughput under heavy traffic loads.

6.3 Power Consumption

In Fig. 6, we show the average power consumption per node in the 100-node sensor network, under our scheme and under GAF. Let E_{i_0} and E_{i_r} be node s_i's intial energy level and remaining energy after simulation, respectively. The average power consumption per node is $\frac{\sum_i (E_{i_0} - E_{i_r})}{100}$. To ensure that no node runs out of power during our simulation, we set a high initial energy level (100 J). We chose 15 sources located as far from the BS as possible, each associated with the same *temp* trace and the same consistency requirement. We increased the error bound from 0.1 to 0.45 and measured the power consumption for a simulation time of 1000 *secs*.

In general, our approach consumes far less power than GAF. As the error bound increases, the difference is more significant, since fewer updates are generated, and our approach allows the nodes to sleep more often. More power is consumed for a higher confidence level ($p = 0.95$) since more updates must be generated and delivered to the BS.

To compare sensor network lifetimes under our scheme and under GAF, we show the fraction of nodes surviving after a given simulation time in Fig. 7. The initial energy level was set to 15 J. For simulation time less than 600 *secs*, all the nodes survive under our scheme as well as under GAF. However, beyond 600 *secs*, the survivor fraction drops rapidly under GAF, while it still remains 100% under our scheme. With $p = 0.9$, we can achieve 67% longer network lifetime than GAF, while with $p = 0.95$, our lifetime is 50% longer.

7   Conclusions

We have proposed a novel approach to maintain stochastic consistency for erratic sensor sources. We achieve power efficiency by dynamically adjusting sensors' duty cycles. A node is guaranteed to be awake when an update message needs to be delivered/relayed, and asleep at other times. We model erratic sensor sources as drifting Brownian motions, and adpatively evaluate the model parameters at the sources. We have verified on various categories of real-life sensor traces that the DBM model faithfully captures the erratic data characteristics in the short term, and helps the source to adaptively evaluate when the next cache update is due, and notify the relaying nodes to wake up before this time.

Our scheme achieves high fidelity under the stochastic consistency model. Our fidelity is higher than that of GAF, which maintains a connected routing backbone and puts the other nodes to sleep, under heavy traffic loads and stringent consistency requirements, suggesting that we can attain higher throughput than GAF. Our approach also consumes significantly less power than GAF, since it is more flexible in adjusting each node's duty cycles, thus saving more power.

References

1. Y.Xu, J.Heidemann, D.Estrin: Geography-informed energy conservation for ad hoc networks. In: Proc. of the MobiCom Conf, Italy (2001)
2. S.Madden, M.J.Franklin, J.M.Hellerstein, W.Hong: The design of an aquisitional query processor for sensor networks. In: Proc. of the 2003 ACM SIGMOD Conf, San Diego (2003)
3. S.Madden, M.J.Franklin, J.M.Hellerstein, W.Hong: Tag: a tiny aggregation service for ad-hoc sensor networks. In: The 5th Symposium on OSDI. (2002)
4. C.Intanagonwiwat, R.Govindan, D.Estrin: Directed diffusion: A scalable and robust communication paradigm for sensor networks. In: Proc. of the ACM/IEEE MobiCom Conf. (2000)

5. Q.Han, S.Mehrotra, N.Venkatasubramanian: Energy efficient data collection in distributed sensor environments. In: Proc. of the 24th ICDCS Conf. (2004)

6. M.A.Sharaf, J.Beaver, A.Labrinidis, P.K.Chrysanthis: Tina: A scheme for temporal coherency-aware in-network aggregation. In: Proc. of the 3rd ACM MobiDE Workshop. (2003)

7. C.Olston, J.Jiang, J.Widom: Adaptive filters for continuous queries over distributed data streams. In: Proc. of the 2003 ACM SIGMOD, San Diego (2003)

8. A.Deligiannakis, Y.Kotidis, N.Roussopoulos: Hierarchical in-network data aggregation with quality guarantees. In: Proc. of the 9th EDBT, Greece (2004)

9. ASH Transceiver Designer's Guide, http://www.rfm.com, May, 2002.

10. Chipcon CC1000 RF Transceiver Datasheet, http://www.chipcon.com.

11. MPR/MIB Mote Sensor Hardware Users Manual, http://www.xbow.com.

12. S.Zhu, C.V.Ravishankar: Stochastic consistency, and scalable pull-based caching for erratic data sources. In: Proc. of the 2004 VLDB Conf, Toronto, Canada (2004)

13. F.Bennett, D.Clarke, J.B.Evans, A.Hopper, A.Jones, D.Leask: Piconet: Embedded mobile networking. In: IEEE Personal Communications Magazine. (1997)

14. C.Schurgers, V.Tsiatsis, S.Ganeriwal, M.Srivastava: Optimizing sensor networks in the energy-latency-density design space. In: IEEE Transactions on Mobile Computing. (2002) 1(1)

15. B.Chen, K.Jamieson, H.Balakrishnan, R.Morris: Span: An energy-efficient coordination algorithm for topology maintenance in ad hoc wireless networks. In: Proc. of the IEEE/ACM MobiCom Conf, Rome, Italy (2001)

16. S.Karlin, H.M.Taylor: A First Course in Stochastic Processes, 2nd Edition. Academic Press (1975)

17. S.Zhu, C.V.Ravishankar: A scalable approach to approximating aggregate queries over intermittent streams. In: Proc. of the 2004 SSDBM Conf, Santorini Island, Greece (2004)

18. R.Alonso, D.Barbara, H.Molina: Data caching issues in an information retrieval system. In: ACM Trans. Database Systems. (1990) 15(3)

19. H.Zou, N.Soparkar, F.Jahanian: Probabilistic data consistency for wide-area applications. In: Proc. of the 16th ICDE Conf. (2000)

20. W.Heinzelman, A.Chandrakasan, H.Balakrishnan: Energy-efficient communication protocols for wireless microsensor networks. In: Proc. of the Hawaii Intl. Conf on Systems Sciences. (2000)

21. W.Ye, J.Heidemann, D.Estrin: An energy-efficient mac protocol for wireless sensor networks. In: Proc. of the 21st InfoCom Conf, New York, NY (2002)

22. A.Deshpande, C.Guestrin, S.Madden, J.M.Hellerstein, W.Hong: Model-driven data acquisition in sensor networks. In: Proc. of the 30th VLDB. (2004)

23. G.Hartl, B.Li: infer: A baysian inference approach towards energy efficient data collection in dense sensor networks. In: Proc. of the 25th ICDCS Conf. (2005)

24. S.C.Huang, R.H.Jan: Energy-aware load balanced routing schemes for sensor networks. In: Proc. of the 10th Intl Conference on Parallel and Distributed Systems, Newport Beach, California (2004)

25. A.Woo, D.E.Culler: A transmission control scheme for media access in sensor networks. In: Proc. of the MobiCom Conf. (2001)

26. X.Hong, M.Gerla, W.Hanbiao, L.Clare: Load balanced, energy-aware communications for mars sensor networks. In: Proc. of the Aerospace Conf, vol 3. (2002)

27. A.M.Mood, F.A.Graybill, D.C.Boes: Introduction to the Theory of Statistics, 3rd Edition. McGraw-Hill (1974)

28. S.N.Neftci: An Introduction to the Mathematics of Financial Derivatives, 2nd Edition. Academic Press (2000)

29. H.C.Thode: Testing for Normality. Marcel Dekker, Inc. (2002)
30. The TAO Project, http://www.pmel.noaa.gov/tao/index.shtml.
31. http://mathworld.wolfram.com/InverseErf.html.
32. S.Zhu, W.Wang, C.V.Ravishankar: Stochastically Consistent Caching and Dynamic Duty Cycling for Erratic Sensor Sources, Technical Report, Univ. of California, Riverside (2005), http://www.cs.ucr.edu/~szhu/sensorcache.pdf.
33. The Network Simulator ns-2, http://www.isi.edu/nsnam/ns/.
34. Wireless LAN Medium Access Control (MAC) and Physical Layer (PHY) Specifications, IEEE 802.11 Standard, 1997 Edition.
35. T.S.Rappaport: Wireless Communications, Principles and Practice. Prentice Hall (1996)
36. C.E.Perkins, E.M.Royer: Ad hoc on-demand distance vector routing. In: Proc. of the 2nd IEEE Workshop on Mobile Computing Systems and Applications, New Orleans, LA (1999)

Distributed Model-Free Stochastic Optimization in Wireless Sensor Networks

Daniel Yagan and Chen-Khong Tham

Department of Electrical and Computer Engineering,
National University of Singapore
{daniel.yagan, eletck}@nus.edu.sg

Abstract. With the improvement in computer electronics in terms of process-ing, memory and communication capabilities, it has become possible to scatter tiny embedded devices such as sensor nodes to monitor physical phenomena with greater flexibility. A large number of sensor nodes, communicating over the wire-less medium, also allows information gathering with greater accuracy than cur-rent systems. This paper presents a new stochastic technique known as *Incremen-tal Simultaneous Perturbation Approximation (ISPA)* for performing optimization in wireless sensor networks. The proposed algorithm is based on a combination of gradient-based decentralized incremental (GBDI) optimization and *Simulta-neous Perturbation Stochastic Approximation* (SPSA) techniques. The former is based on *Incremental Sub-Gradient Optimization* (ISGO) techniques that allow the algorithm to be performed in a distributed and collaborative manner. The lat-ter component addresses the limitations of the GBDI component especially in real-world sensor networks. Specifically, the SPSA component is a *model-free* technique that finds the optimal solution without requiring a functional model such as an input-output relationship and a cost gradient. Simulation results show that the proposed ISPA approach not only achieves distributed optimization in a stochastic environment, but can also be implemented in a practical manner for resource-constrained devices.

1   Introduction

Optimization problems are central to the design, control and analysis of most engineer-ing systems. However, most of the common optimization techniques (i.e. linear pro-gramming, geometric programming, convex optimization) do not tackle a number of key issues especially when applied in actual real-world systems. For instance, a closed-form solution may not be available analytically due to the usually unknown dynam-ics of the system. It is thus evident that a mathematical framework which iteratively searches for the optimal solution is desired. Usually, the system designer may perform offline analysis and create rule-based heuristics. However, an unforeseen condition may have a significant impact on system performance. It is desirable that the algorithm be executed *online* or as the system interacts with the environment (i.e. perform *online optimization* and adjust to varying circumstances). Another issue is the notion of *noisy* data which may come from actual measured readings. The algorithm should handle inaccurate measurements as well as disruptive changes in the system. In addition, the

P. Gibbons et al. (Eds.): DCOSS 2006, LNCS 4026, pp. 85–100, 2006.
© Springer-Verlag Berlin Heidelberg 2006

implementation issues should also be dealt with in a practical and cost-effective manner. Extending these objectives to distributed systems with a number of processing devices poses a non-trivial problem in terms of achieving the globally optimal solution. Employing these optimization algorithms in resource-constrained devices can be considered a natural extension, but is definitely more challenging.

In this paper, we present an optimization technique that tries to achieve the above-mentioned objectives in sensor networks. In recent years, research on sensor networks has been increasing in momentum due to its enormous potential applications in real-time monitoring, tracking and estimation. It is known that wireless sensor networks provide the missing link between the digital and physical worlds. The major constraints in sensor networks are the limited energy, computation and communication resources. Although these constraints limit the data collection and communication capability of this type of network, in many applications, the main objective is not merely the common transfer of data from source to destination, but rather to *estimate* certain phenomena or functions of interest in the environment (i.e. average ambient temperature readings in a region or location of a moving object). The need for collaborative and distributed processing among the sensor nodes is thus evident in order to cope with the resource constraints.

We formulate estimation in real-time monitoring as a distributed optimization problem. This paper is organized as follows. Section 2 first describes a *gradient-based decentralized incremental* (GBDI) optimization scheme for the estimation of required system parameters [1]. This scheme is based on the *Incremental Sub-Gradient Optimization* (ISGO) algorithm [2], but applied in a decentralized manner. In Section 3, we highlight some weaknesses of the GBDI scheme especially in practical and real-world sensor network applications. We then present the optimization technique known as *Simultaneous Perturbation Stochastic Approximation* (SPSA) [3] in Section 4 for solving the limitations of the GBDI algorithm and for achieving the following objectives:

i. Robust estimation and the ability to handle noisy measurements;
ii. Stochastic objective cost functions which may be more applicable if the readings are noisy, and to capture other unknown system dynamics;
iii. An online and model-free approach;
iv. Local and global convergence in spite of non-convexity and non-determinism in the system; and
v. Practical and cost-effective distributed implementation.

Section 5 discusses how the SPSA algorithm can be easily applied in a *decentralized* manner. We propose a decentralized scheme known as the *Incremental Simultaneous Perturbation Approximation (ISPA)* algorithm that combines the GBDI and SPSA techniques. Section 6 presents simulation results to show the effectiveness of ISPA. Finally, we conclude and present possible research directions in Section 7.

2   Decentralized Incremental Optimization

Interesting prior work on decentralized incremental optimization were presented in [1, 4]. We first summarize their ideas as follows:

Consider the scenario of n sensor nodes uniformly distributed over a unit square area where each collects m measurements. Assuming that the objective is to compute or estimate the *average* value of all the measurements, the possible solutions are:

i. All sensors transmit to a central sink node which computes the average. This requires $O(mn)$ bits to be transmitted over $O(1)$ meter on the average.

ii. Each sensor computes a local average and transmit it to the sink node which then computes the global average. This requires only $O(n)$ bits over $O(1)$ meter.

iii. Find a path to the sink node, which passes through and visits each node once. The global average can be computed by a single accumulation process from the start node to the sink along the path where each node updates the estimate of the global average from its own local average. This requires $O(n)$ bits over only $O(n^{-1/2})$ meters.

The last approach may be more efficient than the first two approaches. It is known that radio communication has significant energy consumption and hence, the last approach also contributes to higher energy savings. It can also be shown that a similar procedure can be used to compute any average quantity.

The authors in [1] highlighted that computing averages can be considered as *minimizing* quadratic cost functions. Quadratic optimization problems are special since the solutions are linear functions of input data. In this case, a *single accumulation* procedure leads to a solution. It turns out that similar procedures exist for computing general types of estimates.

Specifically, many estimation criteria can be expressed in the following form:

$$f(\theta) = \frac{1}{n} \sum_{i=1}^{n} f_i(\theta) \tag{1}$$

where:
θ is the parameter of function to be estimated,
$f(\theta)$ is the cost function to be *minimized* which is expressed as the *average* of n local functions $\{f_i(\theta)\}_{i=1}^{n}$ from each sensor i.

For the average criterion,

$$f_i(\theta) = \frac{1}{m} \sum_{j=1}^{m} (x_{i,j} - \theta)^2 \tag{2}$$

and

$$f(\theta) = \frac{1}{mn} \sum_{i=1}^{n} \sum_{j=1}^{m} (x_{i,j} - \theta)^2 \tag{3}$$

where $x_{i,j}$ is the jth measurement at the ith sensor.

Computing the average of the sensor readings $x_{i,j}$ can thus be reformulated by finding θ that minimizes the cost function in (3). The decentralized incremental optimization algorithm in [1] operates by having an estimate of the parameter θ which is passed from node to node. Each node updates the parameter θ with respect to its local cost and gradient, and passes the updated value of θ to the next node.

For a quadratic cost function, such as in (3), a *single* pass or cycle among the nodes is enough to find the required estimate θ. Several cycles are needed for general cases of cost functions. The algorithm can be considered as an *incremental sub-gradient optimization* [2, 5] in which the number of cycles or passes among nodes are the number of iterations and can be analyzed theoretically. Essentially, the nodes collaborate to perform an *in-network decentralized* optimization for parameter estimation.

2.1 Incremental Sub-gradient Optimization

The basic concepts of incremental sub-gradient optimization are first reviewed in this sub-section. The concept of *sub-gradient* comes from the important property of the gradient of a convex differentiable function:

For a convex differentiable function $f : \Theta \to \mathbb{R}$, the following holds for the gradient of f at point θ_0 for all $\theta \in \Theta$:

$$f(\theta) \geq f(\theta_0) + (\theta - \theta_0)^T \nabla f(\theta_0) \tag{4}$$

In general, for a convex function f, a *sub-gradient* of f at θ_0, observing that f may not be differentiable at θ_0, is any direction g such that:

$$f(\theta) \geq f(\theta_0) + (\theta - \theta_0)^T g \tag{5}$$

The *sub-differential* of f at θ_0, denoted as $\partial f(\theta_0)$, is the set of all sub-gradients of f at θ_0. If f is differentiable at θ_0, then $\partial f(\theta_0) := \{\nabla f(\theta_0)\}$, which implies that the gradient of f at θ_0 is the only direction satisfying (4).

When considering the average measurement from n sensor nodes where each sensor collects m readings, the estimate parameter vector is defined as follows:

$$\hat{\theta} = \arg\min_{\theta \in \Theta} \frac{1}{n} \sum_{i=1}^{n} f_i(\{x_{i,j}\}_{j=1}^{m}, \theta) \tag{6}$$

where:

θ is the global parameter vector readings which describe the sensed phenomena (i.e. average temperature and light),
$\hat{\theta}$ is the estimate of the parameter vector θ,
$f_i(\theta) := f_i(\{x_{i,j}\}_{j=1}^{m}, \theta)$ is the convex cost function, which may not be differentiable, from the m readings of sensor i given parameter vector θ.

The formulation in (6) computes the *average*. In general cases, appropriate cost functions other than the average value can be formulated.

Gradient and sub-gradient methods (i.e. gradient descent) are commonly used for iteratively solving such optimization problems. Usually, these algorithms are performed in a *centralized* manner. For the sub-gradient descent approach for solving (6), the update equation is:

$$\hat{\theta}^{(k+1)} = \hat{\theta}^{(k)} - \alpha \sum_{i=1}^{n} g_{i,k} \tag{7}$$

where:

$\hat{\theta}^{(k)}$ is the estimate vector after k iterations,
$g_{i,k} \in \partial f_i(\hat{\theta}^{(k)})$ is a sub-gradient, where $\partial f_i(\hat{\theta}^{(k)})$ is the set of sub-gradients
α is a positive step size,
k is the iteration number.

At each iteration step, this centralized approach uses data from all n sensors. Since a distributed or decentralized approach is needed, the authors in [1] use the following *incremental* approach that divides the update equation in (7) into a cycle of n *sub-iterations*. We refer to their algorithm as a *gradient-based decentralized incremental* (GBDI) algorithm. Each sub-iteration tries to optimize a single component $f_i(\theta)$. The update equation after k cycles or iterations is as follows:

$$\hat{\theta}^{(k)} = \psi_n^{(k)} \qquad (8)$$

where $\psi_n^{(k)}$ is from the n sub-iterations:

$$\psi_i^{(k)} = \psi_{i-1}^{(k)} - \alpha g_{i,k}, \qquad i = 1, 2, ..., n \qquad (9)$$

where $g_{i,k} \in \partial f_i(\psi_{i-1}^{(k)})$ and $\psi_0^{(k)} = \psi_n^{(k-1)}$. Each sub-iteration can be mapped directly to the optimization of sensor i for its *local* cost function $f_i(\theta)$.

As mentioned in [1], diminishing step size in (9) is desirable so that the algorithm is guaranteed to converge to the optimal value. However, the rate of convergence generally becomes very slow as the step size gets smaller. In sensor networks, applications usually require deployment in a dynamic environment for the purpose of not only identifying (i.e. estimating), but also tracking the phenomena of interest. The authors in [1] have used a fixed step size to make the algorithm more adaptive to handle the non-stationary conditions in sensor networks. They have also stated that K cycles of the algorithm will produce an estimate $\hat{\theta}$ satisfying: $f(\hat{\theta}) \leq f(\theta^*) + O(K^{-1/2})$ where θ^* is the optimal solution. For a network of n nodes uniformly distributed over a unit square area and m measurements per node, the amount of communication required for the algorithm is approximately a factor of $K/(mn^{1/d})$ (i.e. d is the dimensionality of $\hat{\theta}$) less than the required number to transmit all measurement readings to a central location for processing.

3   Weaknesses of the GBDI Algorithm

We present two applications described in [1, 4] to identify the limitations of the GBDI algorithm.

3.1   Applications of the GBDI Algorithm

Robust Estimation. As sensor networks are deployed for monitoring, the need for a robust system is crucial especially in a hostile environment where some sensors may give *noisy* readings.

Suppose that each sensor collects a set of m pollution level measurements. The sample mean pollution level is obtained as: $R = \frac{1}{mn}\sum_{i,j} x_{i,j}$. If each measurement has a variance σ^2 and assuming independent and identically distributed samples, the variance of R is $\frac{\sigma^2}{mn}$. If a fraction of the sensors are damaged, say 10%, the variance of the readings will be $100\sigma^2$, i.e. the estimator variance of R increases by approximately 10 times. It is evident that the readings from the damaged sensors should be removed from the estimation process. If this is not possible, *robust* estimation techniques is the next logical step.

The GBDI algorithm can be used by replacing the classical least square loss function $\|x - \theta\|^2$ with a more general and robust loss function $\rho(x, \theta)$. The choice of $\rho(x, \theta)$ should give less weight to those samples that deviate greatly from the parameter θ. The cost function for the optimization can then be defined for this case as:

$$f_{robust}(\theta) = \frac{1}{mn}\sum_{i=1}^{n}\sum_{j=1}^{m}\rho(x_{i,j}, \theta) \tag{10}$$

where:

$$f_i(\theta) = \frac{1}{m}\sum_{j=1}^{m}\rho(x_{i,j}, \theta) \tag{11}$$

An example of $\rho(x, \theta)$ is the Huber loss function:

$$\rho_h(x; \theta) = \begin{cases} \|x - \theta\|^2/2 & \text{for } \|x - \theta\| \leq \gamma \\ \gamma\|x - \theta\| - \gamma^2/2 & \text{for } \|x - \theta\| > \gamma \end{cases} \tag{12}$$

The authors in [1] have showed that, by simply replacing the cost function with a more robust loss function, there is a significant improvement in the variance of the estimate of θ.

We observe that, as long as the new loss function $\rho(x, \theta)$ satisfies the conditions in (4) and (5), and its gradient can be computed *analytically*, it can be used as a substitute function in the GBDI algorithm. This is also application-specific and may include other considerations such as giving less importance to readings from damaged sensors in the loss function.

Energy-Based Source Localization. The GBDI algorithm can also be used for localization of energy-based sources such as acoustic sources.

Consider the case where a source emits energy isotropically and is positioned at an unknown position θ. The sensors are uniformly distributed in either a square or cube with side length $D \gg 1$ and each sensor knows its own location r_i for $i = 1, 2, ..., n$ relative to a fixed point. The task is to estimate θ from the nm sensor readings. The isotropic energy propagation model for the jth received signal strength at sensor i used is:

$$x_{i,j} = \frac{A}{\|\theta - r_i\|^\beta} + w_{i,j} \tag{13}$$

where:

A is a positive constant,
$\beta \geq 1$ is the attenuation characteristics of the medium,
$w_{i,j} \sim N(0, \sigma^2)$ are independent and identically distributed Gaussian samples with zero mean and variance σ^2, and $\|\theta - r_i\| > 1$, $\forall i$.

The maximum likelihood estimate of the source location is found by solving:

$$\hat{\theta} = \arg\min_{\theta} \frac{1}{mn} \sum_{i=1}^{n} \sum_{j=1}^{m} \left(x_{i,j} - \frac{A}{\|\theta - r_i\|^{\beta}} \right)^2 \tag{14}$$

where:

$$f_i(\theta) = \frac{1}{m} \sum_{j=1}^{m} \left(x_{i,j} - \frac{A}{\|\theta - r_i\|^{\beta}} \right)^2 \tag{15}$$

The gradient of $f_i(\theta)$ can be obtained as:

$$\nabla f_i(\theta) = \frac{2\beta A}{m \|\theta - r_i\|^{\beta+2}} \sum_{j=1}^{m} \left(x_{i,j} - \frac{A}{\|\theta - r_i\|^{\beta}} \right) (\theta - r_i) \tag{16}$$

3.2 Limitations of GBDI

The algorithm mainly relies on the cost function formulation, especially on the individual loss $f_i(\theta)$ such as in (11) and (15). Once an application-specific cost function is defined mathematically, performing the algorithm is straightforward as shown in (8) and (9).

It is apparent that the algorithm can be used to optimize other types of performance measures by defining an appropriate cost function. However, there are some limitations such as the following:

i. The cost function is dependent on an assumed model such as the energy isotropic propagation model in (13). This propagation model may not be accurate enough in actual real-world sensor networks. Fixing a model before obtaining the best estimate or optimal solution means that the algorithm may not be robust enough if the actual scenario departs from the assumed model. For instance, the sensor reading $x_{i,j}$ in (13) for sensor i assumes Gaussian independent and identically distributed readings which fail to capture the *correlation* among the m readings as well as the cross-correlation of other sensors' readings in the vicinity of sensor i. We may require a generic model-free approach to find the optimal solution.

ii. A new defined cost function may not be easy to differentiate analytically to obtain the gradient of $f_i(\theta)$ which is used in (9). It should be noted that the algorithm needs a *closed-form* expression for $\nabla f_i(\theta)$ beforehand at each sensor node. Another issue is the computational complexity of the gradient which is not viable in a tiny sensor node. Looking at (16), a sensor node needs to perform a number of complex calculations to obtain the gradient.

iii. The algorithm does not include the cost of transmitting the new estimate to the next node. This may play a role in minimizing the cost of energy consumption. It can be deduced that, if there is a large number of n sensors, performing the algorithm for a few cycles may even drain the energy of some sensors. If a number of sensor nodes fail along the path, the accuracy of the estimate suffers as well.

iv. Including other cost terms in the cost function may require careful consideration. As an example, the cost of transmission and battery or energy drain in an actual real-world scenario may be non-deterministic and unknown beforehand. The transmission energy obviously depends on a number of parameters such as the medium access control (MAC), routing, remaining battery and sleep-wake pattern. As mentioned earlier, fixing a model initially may not be advisable.

v. The validity and convergence of the algorithm uses the convexity and gradient $\nabla f_i(\theta)$ of the cost function. If the cost function becomes non-deterministic or stochastic and unknown, which is evidently more appropriate in a real-world sensor network, the GBDI algorithm may not apply and may suffer divergence. Therefore, thorough analysis is required to study the stochastic nature of the cost function for guaranteed convergence.

The next section presents a novel optimization technique that effectively addresses these limitations of the GBDI algorithm. Specifically, a model-free optimization approach is investigated with the following objectives:

i. Robust estimation and the ability to handle noisy measurements;
ii. Stochastic objective cost functions that include energy, communication and processing costs, which may be more applicable if the readings are noisy and to capture other unknown system dynamics;
iii. An online and model-free approach;
iv. Local and global convergence in spite of non-convexity and non-determinism in the system; and
v. Practical and cost-effective distributed implementation.

4   Online Model-Free and Robust Stochastic Optimization

Finding the best estimate parameter vector that satisfies (6) clearly falls under the general formulation of finding θ^* that solves $\min_{\theta \in R^p} L(\theta)$ where $L(\theta) : R^p \to R^1$ is a loss function over θ, a p-dimensional vector. We assume θ represents a vector of continuous parameters. In addition, we assume that the objective function $L(\theta)$ *cannot* be expressed as a closed-form expression. Finding the gradient with respect to θ is thus not possible. Hence, one can use an iterative approach that improves an initial guess θ_1 towards the optimal θ^*.

In classical deterministic optimization, it is assumed that one can obtain perfect information about the loss function and its derivatives if needed, and such information is used to improve the search direction in a deterministic manner. In practical scenarios, such an assumption may not be valid due to inevitable noise effects. However, deterministic methods, such as linear and non-linear programming, steepest descent, Newton-Raphson and conjugate gradient, provide a starting point for the analysis of non-deterministic or stochastic methods. Many techniques in both deterministic and stochastic optimization for continuous problems rely on the gradient vector of the loss function:

$$g(\theta) = \begin{bmatrix} \frac{\partial L}{\partial \theta_1} \\ \vdots \\ \frac{\partial L}{\partial \theta_p} \end{bmatrix} \qquad (17)$$

4.1   Gradient-Based Stochastic Algorithms

The most common *gradient-based* stochastic algorithm (i.e. stochastic gradient method) is the Robbins-Monro stochastic approximation (RMSA) which may be considered as a generalization of the following techniques: deterministic steepest descent, Newton-Raphson, neural network back-propagation and infinitesimal perturbation analysis-based optimization for discrete-event systems. RMSA relies on the direct measurements of the gradient $g(\theta)$ which yield an *unbiased estimate* of the gradient due to the presence of noise in the input data. RMSA has the form:

$$\hat{\theta}^{(k+1)} = \hat{\theta}^{(k)} - \alpha_k Y(\hat{\theta}^{(k)}) \tag{18}$$

where:

α_k is a non-negative step size,
$\hat{\theta}^{(k)}$ is the estimate vector after k iterations,
$Y(\hat{\theta}^{(k)})$ is the direct measurement of the gradient $g(\hat{\theta}^{(k)})$.

Obtaining $Y(\hat{\theta}^{(k)})$ in RMSA usually requires detailed knowledge of the functional relationship between the parameters being optimized and the cost function being minimized. Such a relationship can be difficult to obtain such as in the cases of non-linear feedback controller design and simulation-based optimization.

In areas such as recursive parameter estimation, there may be large computational savings in calculating or measuring the loss function itself, rather than the gradient. This is also one of the weaknesses mentioned in Section 3 for gradient-based methods. In contrast, *gradient-free* algorithms require only the measurement samples of the loss function which does not require the full input-output functional relationship for the gradient.

4.2   Gradient-Free Stochastic Algorithms

In these types of algorithms, no direct measurement of the gradient $g(\theta)$ is assumed. However, it is assumed that measurements of loss function $L(\theta)$ are available which may include added noise. The recursive update equation used is the general stochastic approximation update:

$$\hat{\theta}^{(k+1)} = \hat{\theta}^{(k)} - \alpha_k \hat{g}_k(\hat{\theta}^{(k)}) \tag{19}$$

where $\hat{g}_k(\hat{\theta}^{(k)})$ is the estimate of the gradient $g(\hat{\theta}^{(k)})$ at the iterate $\hat{\theta}^{(k)}$ based on loss function measurements.
Under appropriate conditions, (19) converges to θ^* in some stochastic sense usually *almost surely* (i.e. with probability 1) [3, 6]. The main component of (19) is the gradient estimate $\hat{g}_k(\hat{\theta}^{(k)})$. Two methods for calculating $\hat{g}_k(\hat{\theta}^{(k)})$ are presented here.

Let c_k be a small positive number, that decays as k gets larger. *One-sided* gradient approximations involve loss measurements $y(\hat{\theta}^{(k)})$ and $y(\hat{\theta}^{(k)} + perturbation)$, while *two-sided* gradient approximations involve two measurements of the form $y(\hat{\theta}^{(k)} \pm perturbation)$, where $y(\cdot) = L(\cdot) + noise$.

Finite Difference Stochastic Approximation (FDSA). Each component of $\hat{\theta}^{(k)}$ is perturbed one at a time, and the corresponding loss measurements $y(\cdot)$ are obtained. The ith component of estimate gradient vector $\hat{g}_k(\hat{\theta}^{(k)})$ for a two-sided finite-difference approximation is:

$$\hat{g}_{k,i}(\hat{\theta}^{(k)}) = \frac{y(\hat{\theta}^{(k)} + c_k e_i) - y(\hat{\theta}^{(k)} - c_k e_i)}{2c_k} \tag{20}$$

where e_i is a p-dimensional vector with one in the ith place and zeros elsewhere, for $i = 1, 2, ..., p$.

Simultaneous Perturbation Stochastic Approximation (SPSA). All elements of $\hat{\theta}^{(k)}$ are perturbed randomly together to obtain two measurements of $y(\cdot)$, but each component is formed from the ratio involving the individual components in the perturbation vector and the difference in the two corresponding elements:

$$\hat{g}_k(\hat{\theta}^{(k)}) = \left(\frac{y(\hat{\theta}^{(k)} + c_k \triangle_k) - y(\hat{\theta}^{(k)} - c_k \triangle_k)}{2c_k} \right) \begin{bmatrix} \triangle_{k,1}^{-1} \\ \triangle_{k,2}^{-1} \\ \vdots \\ \triangle_{k,p}^{-1} \end{bmatrix} \tag{21}$$

where $\triangle_k = \begin{bmatrix} \triangle_{k,1} & \triangle_{k,2} & \cdots & \triangle_{k,p} \end{bmatrix}^T$ is a p-dimensional random perturbation vector.

The perturbation vector satisfies conditions to be discussed in the Section 4.3. It can also be observed that the number of loss function measurements in each iteration for FDSA is $2p$, since each component of $\hat{g}_k(\hat{\theta}^{(k)})$ is obtained from two loss measurements (i.e. $y(\hat{\theta}^{(k)} + c_k e_i)$ and $y(\hat{\theta}^{(k)} - c_k e_i)$) that arise from the two-sided perturbation of the ith component of $\hat{\theta}^{(k)}$. In addition, each ith component is perturbed separately.

For SPSA, all the vector components of $\hat{\theta}^{(k)}$ are perturbed simultaneously (i.e. hence the name simultaneous), resulting in only two loss measurements $y(\hat{\theta}^{(k)} + c_k \triangle_k)$ and $y(\hat{\theta}^{(k)} - c_k \triangle_k)$, regardless of vector dimension p. This provides a potential advantage for SPSA to achieve larger computational savings to estimate θ. It should be noted that this advantage is only realized if the number of iterations required for effective convergence to θ^* does not increase in a way that cancels measurement savings per gradient approximation in each iteration. This is discussed further in the following subsection.

4.3 Simultaneous Perturbation Stochastic Approximation (SPSA) Theory

Sufficient conditions for the convergence of SPSA in the stochastic sense using a differential equation approach was presented in [3, 6]. Some of these conditions are summarized as follows:

i. The gain sequences a_k and c_k both go to zero at some specified rate;
ii. $L(\theta)$ is sufficiently smooth or several times differentiable near θ^*;
iii. The sequence $\{\triangle_{k,i}\}$ for $i = 1, ..., p$ are independent and symmetrically distributed about zero, with finite inverse moments $E\left(|\triangle_{k,i}|^{-1}\right)$ for all k, i. An example of such a distribution is the symmetric Bernoulli ± 1 distribution. Two common distributions that do not satisfy the conditions are the uniform and the normal.

It is also shown in [3, 6] that the probability distribution of an appropriately scaled $\hat{\theta}^{(k)}$ is approximately normal (with specified mean μ and covariance matrix Σ) for large k:

$$k^{\beta/2} \left(\hat{\theta}^{(k)} - \theta^* \right) \longrightarrow N(\mu, \Sigma) \qquad \text{as } k \to \infty \qquad (22)$$

where $\beta > 0$ depends on the gain sequences a_k and c_k,
μ depends on both the Hessian and third derivatives of $L(\theta)$ at θ^*,
Σ depends on the Hessian matrix at θ^*.

The asymptotic normality result can be used to study the relative efficiency of SPSA as compared with FDSA, which effectively justifies the use of SPSA. The efficiency depends on the shape of $L(\theta)$, values of the gain sequences a_k and c_k, distribution of the sequence $\{\triangle_{k,i}\}$ and noise measurement terms. It is shown in [6, 7] that in most practical problems, SPSA will be asymptotically more efficient than FDSA. In particular, by equating the asymptotic mean-squared error $E \left(\left\| \hat{\theta}^{(k)} - \theta^* \right\|^2 \right)$ in SPSA and FDSA, it is known that:

$$\frac{\text{Number of measurements of loss function in SPSA}}{\text{Number of measurements of loss function in FDSA}} \to \frac{1}{p} \qquad (23)$$

as the number of loss measurements in both procedures gets large. Equation (23) implies that the p-fold savings per iteration in gradient approximation as mentioned in Section 4.2 translates directly into a p-fold savings in the overall optimization process.

4.4   Extensions to the Basic SPSA Algorithm

The work in [8] has considered the problem of choosing the best distribution for the random perturbation vector $\triangle_k$. Based on asymptotic distribution results, it is shown that the optimal distribution for the components of $\triangle_k$ is symmetric Bernoulli. This simple distribution has also been effective in many finite-sample practical and simulation examples.

SPSA can also be used in feedback control problems where the loss function changes with time [9, 10]. A gradient smoothing idea (similar to *momentum* in neural networks literature) may help reduce noise effects and enhance convergence. Alternatively, it is possible to average several simultaneous perturbation gradient approximations at each iteration to reduce noise effects at the cost of additional function measurements [3].

An important extension of SPSA is the *global* minimization as discussed in [11, 12]. The first approach is based on a step-wise slowly decaying sequence c_k and possibly a_k, while the second approach is based on the principle of injected Monte Carlo noise in the right-hand side of the basic SPSA updating equation in (19). This latter approach is a common way of converting stochastic approximation to global optimizers.

The problem of constrained (equality and inequality) optimization using SPSA is presented in [13] using a projection approach. While the projection approach has a concise mathematical representation and may be simple to implement, it is often restricted in the types of constraints that can be handled. An alternative approach to constrained optimization is given in [14] that includes a penalty term in the loss function. In practical problems, constraints are only implicit in θ and the penalty function is well-suited to handle such cases.

5 SPSA for Distributed Model-Free Optimization

SPSA [3, 6, 9] is a gradient-free algorithm that has the following novel properties:

  i. Handles stochastic cost functions and shows robustness to noise in the loss function measurements;
 ii. An online and model-free approach that does not need the loss function gradient obtained from the explicit functional input-output relationship;
iii. Exhibits theoretical and experimental convergence for relative efficiency as compared with FDSA;
 iv. Local and global convergence in spite of non-convexity and non-determinism in the system;
  v. Handles constrained optimization; and
 vi. Practical and cost-effective implementation: SPSA needs only two loss measurements for each iteration, compared with FDSA that uses $2p$, where p is the vector dimension.

It can be easily seen that SPSA can be used for *estimating* the gradient for the decentralized incremental optimization method in [1, 4] for sensor networks.

Formally, by following the idea in Section 2.1, we divide the update equation of the *centralized* model-free SPSA algorithm in (19) into n sub-iterations, where n is the total number of sensors. The update equation after k cycles or iterations is:

$$\hat{\theta}^{(k)} = v_n^{(k)} \qquad (24)$$

where $v_n^{(k)}$ is from the n sub-iterations or sensors:

$$v_i^{(k)} = v_{i-1}^{(k)} - \alpha \hat{g}_{i,k}, \qquad i = 1, 2, ..., n \qquad (25)$$

$$\hat{g}_{i,k} = \left(\frac{y_i(v_{i-1}^{(k)} + c_k \triangle_k) - y_i(v_{i-1}^{(k)} - c_k \triangle_k)}{2c_k} \right) \begin{bmatrix} \triangle_{k,1}^{-1} \\ \triangle_{k,2}^{-1} \\ \vdots \\ \triangle_{k,p}^{-1} \end{bmatrix} \qquad (26)$$

where $v_0^{(k)} = v_n^{(k-1)}$ and y_i is the loss measurement of sensor i. Each sub-iteration is thus effectively mapped directly to the optimization of sensor i for its *local* measurement function y_i.

We propose the combination of the SPSA and GBDI components as the *Incremental Simultaneous Perturbation Approximation* (ISPA) algorithm. ISPA operates by having an estimate $\hat{\theta}$ that is passed from node to node, where each node updates the estimate parameter $\hat{\theta}$ with respect to its local loss measurement function and estimated gradient.

Essentially, ISPA addresses the limitations mentioned in Section 3. Other stochastic function terms can be included in the objective function to handle estimation of parameters in a cost-effective manner.

6   Simulations and Discussions

To verify the effectiveness of ISPA, we perform a simulation similar to the energy-based source localization discussed in Section 3.1. We compare ISPA with the model-based GBDI that was proposed in [1, 4].

We assume that 100 sensors are uniformly distributed in a 100×100 square area, where each sensor has 10 measurements. Each sensor i knows its current position r_i and the task is to estimate the location θ of the source in a decentralized manner.

For the model-based GBDI, we assume the isotropic energy propagation model in (13) to obtain the gradient in (16). For ISPA, as discussed in Section 4.2, a node needs to *measure* the loss function itself, since we assume that ISPA does not require the functional mapping found in (15), nor the gradient. However, for simulation purposes, we assume that the loss measurement for ISPA is defined as:

$$y_i(\theta) \equiv f_i(\theta) + noise = \frac{1}{m} \sum_{j=1}^{m} \left(x_{i,j} - \frac{A}{\|\theta - r_i\|^\beta} \right)^2 + z_i \qquad (27)$$

where $z_i \sim N(m_z, \sigma_z^2)$ are independent and identically distributed Gaussian samples with mean m_z and variance σ_z^2. For the simulation, we let $z_i \sim N(0, 100)$ and $w_{i,j} \sim N(0, 1)$.

We also use the symmetric Bernoulli ± 1 distribution for the random perturbation vector $\triangle_k$ and let $c_k = \alpha_k = 1/(k+1)$ as found in (19) and (21). We use the following parameter values: $A = 100$ and $\beta = 2$.

Figure 1 shows the loss function values for ISPA and GBDI. GBDI assumes perfect knowledge of the functional mapping and a closed-form expression for the gradient and thus, it can obtain the optimal limiting value of 0.00861. On the other hand, ISPA achieves close to the optimal performance despite its model-free and gradient-free properties. Figure 2 shows the *mean squared error* (MSE) for ISPA with respect to the mean

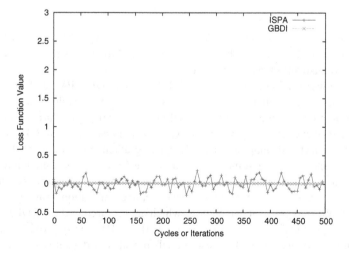

Fig. 1. Loss Function Comparison

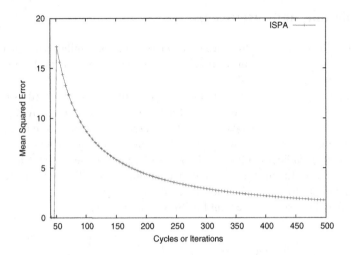

Fig. 2. Mean Squared Error (MSE) for ISPA

value of 0.00861 from GBDI obtained in Figure 1. Since the MSE decays with time, this indicates that ISPA converges to the optimal value.

We also study the case of time-varying measurements where, after a cycle of passes or iterations (i.e. the estimate $\hat{\theta}^{(k)}$ is passed among the n sensors in cycle k), we assume that the sensors have new measurement readings. The variation of readings may come from noisy data, instrumentation errors or node failures. For simulation purposes under GBDI, we use the same model defined earlier in (13) but the readings vary at every cycle. We also assume that with a small probability p_r, the sensor readings obey a similar isotropic energy propagation model as in (13):

$$x_{i,j} = \frac{A}{\|\theta - r_i\|^{\beta}} + \eta_{i,j} \tag{28}$$

where $\eta_{i,j} \sim N(0, 100)$ which is different from $w_{i,j} \sim N(0, 1)$. We let $p_r = 0.3$ in the simulation. For ISPA under the time-varying scenario, we assume the same model in (27), but we also vary the readings at every cycle or iteration.

Figure 3 shows the loss function values for ISPA and GBDI. ISPA clearly achieves better performance even when the sensor readings vary after each cycle or iteration. GBDI assumes perfect knowledge of the functional mapping and closed-form expression for the gradient. However, if the measurements vary due to noise effects, GBDI may not perform well due to its fixed mapping and model-based gradient computation. ISPA achieves better performance since it assumes a general stochastic cost or loss function without requiring the input-output relationship, and at the same time, it only requires possibly noisy loss function measurements. ISPA also achieves better computational savings by not performing complicated gradient calculations which are required in the GBDI algorithm.

In terms of practicality in actual sensor networks, ISPA clearly provides the advantage of computational savings. It is also model-free since it does not require the functional relationship and closed-form expression for the cost gradient.

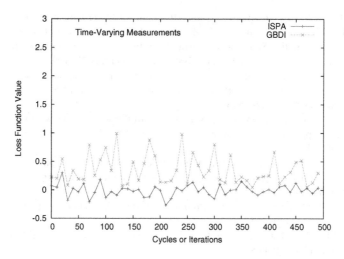

Fig. 3. Loss Function Comparison under Time-Varying Measurements

7   Conclusion

This paper presents a distributed incremental optimization technique for sensor networks which can be applied to parameter estimation and source localization tasks. The algorithm is able to cope with the stochastic nature of the environment through the use of stochastic cost functions without the need for complete knowledge of the functional relationship between the parameters being optimized and the cost function being minimized. It also handles the inevitable noise effects, as well as unknown system dynamics. The ISPA algorithm can be performed online and implemented in a practical and cost-effective manner. Furthermore, since ISPA is an SPSA-based algorithm, it exhibits local as well as global theoretical convergence properties. In future work, we plan to investigate the use of the proposed ISPA algorithm on actual distributed wireless sensor networks.

References

1. Rabbat, M., Nowak, R.: Distributed Optimization in Sensor Networks. In: Proc. Information Processing in Sensor Networks, Berkeley, CA, USA (2004)
2. Nedic, A., Bertsekas, D.: Incremental Subgradient Methods for Non-differentiable Optimization. Technical Report LIDS-P-2460, Massachusetts Institute of Technology, Cambridge, MA, USA (1999)
3. Spall, J.C.: Multivariate Stochastic Approximation Using a Simultaneous Perturbation Gradient Approximation. IEEE Trans. Automat. Contr. **37** (1992) 332–341
4. Rabbat, M., Nowak, R.: Quantized Incremental Algorithms for Distributed Optimization. IEEE J. Select. Areas Commun. **23**(4) (2005) 798–808
5. Nedic, A., Bertsekas, D.: Convergence rate of incremental subgradient algorithms. In Uryasev, S., Pardalos, P., eds.: Stochastic Optimization: Algorithms and Applications. Kluwer Academic Publishers (2000) 263–304

6. Spall, J.C.: A Stochastic Approximation Algorithm for Large-Dimensional Systems in the Kiefer-Wolfowitz Setting. In: Proc. IEEE Conf. on Decision and Control. (1988) 1544–1548

7. Chin, D.: Comparative Study of Stochastic Algorithms for System Optimization Based on Gradient Approximations. IEEE Trans. Syst., Man, Cybern. B **27** (1997) 244–249

8. Sadegh, P., Spall, J.: Optimal Random Perturbations for Multivariate Stochastic Approximation Using a Simultaneous Perturbation Gradient Approximation. In: Proc. American Control Conf. (1997) 3582–3586

9. Spall, J.C., Cristion, J.A.: A Neural Network Controller for Systems with Unmodeled Dynamics with Applications to Wastewater Treatment. IEEE Trans. Syst., Man, Cybern. B **27** (1997) 369–375

10. Spall, J.C., Cristion, J.A.: Nonlinear Adaptive Control Using Neural Networks: Estimation Based on a Smoothed Form of Simultaneous Perturbation Gradient Approximation. Stat. Sinica **4** (1994) 1–27

11. Chin, D.C.: A More Efficient Global Optimization Algorithm Based on Styblinski and Tang. Neural Networks **7** (1994) 573–574

12. Maryak, J.L., Chin, D.C.: Efficient Global Optimization Using SPSA. In: Proc. American Control Conf. (1999) 890–894

13. Sadegh, P.: Constrained Optimization via Stochastic Approximation with a Simultaneous Perturbation Gradient Approximation. Automatica **33** (1997) 889–892

14. Wang, I.J., Spall, J.: A Constrained Simultaneous Perturbation Stochastic Approximation Algorithm Based on Penalty Functions. In: Proc. American Control Conf. (1999) 393–399

Agimone: Middleware Support for Seamless Integration of Sensor and IP Networks

Gregory Hackmann, Chien-Liang Fok, Gruia-Catalin Roman, and Chenyang Lu

Department of Computer Science and Engineering,
Washington University in St. Louis, St. Louis MO 63130-4899, USA

Abstract. The scope of wireless sensor network (WSN) applications has traditionally been restricted by physical sensor coverage and limited computational power. Meanwhile, IP networks like the Internet offer tremendous connectivity and computing resources. This paper presents Agimone, a middleware layer that integrates sensor and IP networks as a uniform platform for flexible application deployment. This layer allows applications to be deployed on the WSN in the form of mobile agents which can autonomously discover and migrate to other WSNs, using a common IP backbone as a bridge. Agimone is the first system that allows mobile agents to migrate between sensor and IP networks. It facilitates data sharing between WSNs and the IP network through remote tuple space operations, allowing sensors to easily defer expensive computations to more-powerful devices. We demonstrate the expressiveness of Agimone's programming model by examining a prototype cargo-tracking application. We also provide an empirical evaluation that demonstrates the efficiency of Agimone using two WSNs consisting of Mica2 motes connected by an IP network.

1 Introduction

Wireless sensor networks (WSNs) consist of tiny sensors embedded within the environment. Many applications require that sensor nodes be deeply embedded in areas where they are difficult to physically access, such as scattered in forests, making it is impractical to physically gather the nodes in order to collect data or deploy new applications. This necessitates WSN systems in which the nodes operate for very long periods of time without physical access. Thus, data collection and application deployment is done over wireless networks. These long system lifetimes also mandate the flexibility to adapt to changing user requirements without completely reprogramming the sensors.

However, typical WSN platforms often lack sufficient support for flexible application deployment. For example, the TinyOS [1] operating system hardwires software components. Once deployed, application behavior can only be marginally tweaked by changing specific parameters defined prior to deployment. To complicate matters, the sensors' power consumption must be very low so that they can be deployed for months or even years without battery replacement. This requires that memory and other computational resources be scarce,

P. Gibbons et al. (Eds.): DCOSS 2006, LNCS 4026, pp. 101–118, 2006.
© Springer-Verlag Berlin Heidelberg 2006

and radio range and reliability be sacrificed [2]. These limitations impose severe restrictions on the complexity and scope of WSN applications.

Many of these restrictions can be eased by logically combining multiple, physically disconnected WSNs using a common IP network. For example, WSNs can be used for cargo tracking and monitoring by attaching sensors to individual cargo containers. However, containers are frequently too far apart to be covered by a single WSN, since they are housed in separate warehouses and eventually relocated by boat or rail. Thus, the sensors form multiple independent WSNs which are unable to directly communicate with each other. The utility of the cargo tracking application would greatly increase if the user could issue a query — such as searching the containers for a specific item — simultaneously to all of these containers, even though their WSNs are not physically connected.

PCs with attached WSN gateways, or embedded devices like Stargate [3], can act as gateways between the IP network and their respective WSNs. By coordinating these disjoint networks to act as one logical network, sophisticated WSN applications can be developed. This way, thousands of nodes located in clusters around the world can collaborate autonomously on a single task.

However, communication and coordination between these networks is a complex task, since WSNs are constantly forming and reshaping as the application evolves. Hence, WSN nodes must be able to determine the availability of other WSNs at run-time. Further, agent transactions across hosts should not be affected by temporal disconnections and other short-term communication failures. For these WSN applications to be useful to clients on the IP network, application developers must be able to channel data between devices on the IP network and nodes in the WSNs in a simple and straightforward manner.

Middleware aims to meet these needs, providing high-level programming constructs that greatly simplify WSN application development and increase utility. To address the limitations of existing WSN middleware systems, we have developed Agilla [4], a middleware for wireless sensors. Limone [5], a lightweight middleware for communication and coordination over IP networks, provides a similar programming model and benefits to devices ranging from PDAs to desktop computers. Both middleware use a mobile agent-based paradigm, where programs are composed of agents that can migrate across nodes.

Though these middleware offer similar programming models, they partition the application into two sets of distinct, incompatible APIs and data structures. This discrepancy is not limited to these two middleware platforms. WSN operating systems like TinyOS offer such different APIs and capabilities from general-purpose operating systems like Windows and Linux, that the need for two incompatible development platforms is inevitable. Traditionally, developers have been forced to manually develop a translation layer for each application that crossed middleware boundaries, a tedious and error-prone procedure.

The main contribution of this paper is providing a general-purpose model which WSN devices can use to exploit the vast computational resources, including other WSNs, found in IP networks such as the Internet. We have developed

Agimone, a thin and reusable integration layer between the Agilla and Limone middleware, which facilitates agent interactions that cross middleware boundaries. In Section 2, we discuss the shortcomings of the state-of-the-art and explain the motivation behind our general-purpose integration layer. Section 3 provides a brief overview of the programming models used by Agilla and Limone. Section 4 describes Agimone's architecture. Section 5 presents a cargo tracking application that highlights the capabilities and expressiveness of Agimone. A performance evaluation is provided in Section 6. We discuss related middleware systems in Section 7. Finally, we conclude in Section 8.

2 Problem Statement

As the number and size of WSN deployments increase, so does the capacity for sophisticated WSN applications. This potential remains largely untapped due to the difficulty in distributing and coordinating applications across WSN boundaries. In this section, we discuss how this potential can be realized using a middleware system that integrates IP networks and WSNs.

2.1 Cargo Tracking: A Motivating Application

Consider the problem of cargo tracking. 7 million cargo containers arrive annually into the United States, making it impossible to manually inspect every container. Instead, each shipping container can be equipped with a sensor, which will form a WSN with the other sensors and monitor the containers' contents. These sensors need to be accessed by many different types of users — such as customs agent, shipping companies, and customers — who have different and evolving requirements. It is impossible to predict all of these users' needs ahead-of-time, and so deploying a single monolithic application on each sensor is infeasible. Mobile agents are invaluable for this scenario. Each authorized user can deploy custom mobile agents to query the sensors on the containers.

However, the limited radio range of individual sensors forces WSNs to form physically-localized clusters. If we rely solely on the sensors' radios, users must interact individually with each of these clusters. This requirement is unreasonable and greatly limits the sophistication of WSN applications. Instead, the current state-of-the-art is to deploy base stations in each cluster. These base stations are connected together using a common IP network. This provides an infrastructure which WSN applications can exploit for inter-WSN interactions. It also provides a means for sensors to interact with clients on the IP network.

2.2 Challenges

Though these capabilities are essential, they are difficult to satisfy. The sensors that populate WSNs have vastly different capabilities from the devices connected to the IP network, preventing the deployment of a uniform software layer across all devices. Today, complex WSN applications consist of separate software support platforms for WSNs and the IP network. Application-specific software is

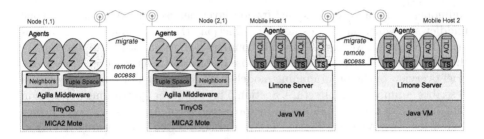

Fig. 1. The Agilla Architecture   **Fig. 2.** The Limone Architecture

used to pass messages and translate queries between these two classes of devices. However, writing this support layer requires programming experience with both types of devices. Also, this layer must often be modified and redeployed when application features or protocols change. This is unacceptable for applications which have a constantly-evolving set of capabilities, like cargo tracking.

In this work, we present a middleware platform that supports seamless integration of WSNs and IP networks into a uniform software platform. Our middleware's services facilitate the development of WSN applications which exploit the IP network as a resource for computation and communication. Mobile agents in a WSN are provided a list of all other WSNs attached to the same IP network. Agents can autonomously migrate over the IP network to any of the WSNs in this list. Finally, we provide a common data space where devices on the IP network and WSNs can share messages and data. These services offer application developers a straightforward yet powerful programming model for implementing complex WSN applications, like the cargo tracking application described above.

3   Background

This section provides a brief overview of the programming models offered by Agilla and Limone. More details on the implementation are available in [4] and [5].

3.1   Agilla

Agilla programs are mobile agents that coordinate through tuple spaces. Agilla's architecture is shown in Figure 1. Each agent is hosted on a virtual machine with dedicated instruction and data memory. An agent may execute special instructions that allow it to interact with the environment and move across nodes. Multiple agents can coexist on a node. Agilla provides agents with local data storage in the form of a heap and operand stack. Agilla agents use a stack-based architecture and are programmed in a bytecode language based on that of Maté [6], but tailored to the mobile agent paradigm. Like Maté, most Agilla instructions fit in a single byte. Agilla is available for Mica2, MicaZ, and Tyndall25 nodes and is distributed through TinyOS' source respository. See Agilla's website [7] for more details.

Agilla's tuple spaces offer a lightweight shared data space where the datum is a tuple that is accessed via pattern matching. This allows one agent to insert a tuple containing data (such as a sensor reading) and another to later retrieve it without the two knowing each other, thus achieving a high level of decoupling. Unlike messages passed over sockets, tuples placed in a tuple space survive temporal disconnections, which frequently occur due to node mobility or unreliable links. Tuple spaces offer many of the same programming benefits as shared data systems, but with far less message-passing.

Each sensor in the WSN has a single local tuple space. Data is stored in the form of fields; tuples containing one or more fields can be added to the tuple space using the **out** primitive. **rd** and **in** operations respectively remove and copy tuples from a tuple space; these operations are parameterized by *templates* that specify the forms of matching tuples. In Agilla, tuple and template fields contain one of a handful of well-known 16-bit data types (integer, string, sensor reading, etc.). Alternatively, a template's field may contain a type (e.g., **VALUE** or

```
1: pushn mrk      // string "mrk"
2: pushcl 15      // integer 15
3: pushc 2        // length of tuple
4: out            // out(<15, "mrk">)
```

Fig. 3. Agilla **out** Code Snippet

```
1: pushn mrk      // string "mrk"
2: pusht VALUE    // type VALUE (integer)
3: pushc 2        // length of template
4: rd             // rd(<VALUE, "mrk">)
```

Fig. 4. Agilla **rd** Code Snippet

STRING) rather than a specific value. This indicates that any value of the corresponding type is acceptable. The code snippets in Figures 3 and 4 demonstrate the **out** and **rd** operations, respectively.

The **rd** or **in** operations block until a matching tuple is available. Agents may also perform *probing* (non-blocking) tuple removals and copies using a different set of primitives. *Remote* tuple space primitives manipulate tuple spaces residing on remote sensors. Finally, Agilla offers a *reaction* mechanism, where a piece of code is executed when a specified type of tuple is placed in the local tuple space. These operations are described in further detail in [4].

Agilla agents may move or clone onto other hosts in the WSN using either *weak* or *strong* migration operations. Weak migrations include only the agent's code, so any computations must restart from the beginning on the new host. Strong migrations include computational state as well as code, so computations can resume after the agent is migrated. Because Agilla agents run on top of a virtual machine, agents can migrate between devices of different hardware architectures, provided that the radios are compatible.

3.2 Limone

Limone provides a similar agent-based programming model using tuple spaces for inter-agent communication. Its architecture is shown in Figure 2. Limone supports the same primitive tuple space operations as Agilla, as well as an analogous reaction mechanism. However, each Limone agent has its own dedicated tuple space, whereas (due to memory limitations) all Agilla agents on a single host share one tuple space. Limone also provides a pluggable device discovery mechanism, where each agent-specified *profile* is automatically propagated to other interested agents as new agents enter or leave the network.

Limone's tuple contents do not suffer from many of the restrictions imposed by their Agilla counterparts. Fields in Limone tuples are indexed by a user-specified name and can contain any Java data type of any size. Similarly, Limone templates are more flexible than Agilla templates. In addition to matching by name and type, Limone templates use *constraint functions* to provide a fine-grained way to specify matching values. For example, the constraint < "ID", Integer, GreaterThanConstraint(10)> matches fields named "ID" that contain an Integer greater than 10. Most constraints use either DefaultConstraintFunction (match any value of the correct type), or EquivalencyConstraintFunction (match only the specified value). Figures 5 and 6 provide code which demonstrate the syntax of Limone's **out** and **rd** operations.

```
ETuple t = new ETuple();
t.addField(new EField("ID", 15));
// Field <ID: 15>
t.addField(new EField("Flag",
      "mark"));
// Field <Flag: "mark">
getTS().out(t);
// out(<ID: 15, Flag: "mark">
```

Fig. 5. Limone **out** Code Snippet

```
ETemplate t = new ETemplate();
t.addConstraint(new EConstraint("ID",
      Integer.class, new DefaultConstraint-
      Function()));
// Match field ID containing any Integer
t.addConstraint(new EConstraint("Flag",
      String.class, new Equivalency-
      ConstraintFunction("mark")));
// Match field flag containing "mark"
ETuple tuple = getTS().rd(t);
// rd a tuple matching this template
```

Fig. 6. Limone **rd** Code Snippet

4   Architecture of Agimone

We have constructed the Agimone architecture (shown in Figure 7) which integrates the Agilla and Limone middleware platforms. Each WSN is associated with a base station such as a laptop or a Stargate. The WSNs are populated with Agilla agents which perform computations and collect sensor data. Inter-agent communication is facilitated by Agilla tuple spaces. Each WSN node hosts one Agilla tuple space, and up to three Agilla agents.

The IP network and WSNs are spanned by WSN gateways attached to these base stations: sensors can communicate with a nearby gateway wirelessly, while the base stations communicate with their attached gateways using a wired interface (e.g., UART or USB). The base stations communicate over the IP network using Limone. Communication in Limone is performed using tuple spaces; each Limone agent is provided with its own Limone tuple space.

WSNs discover each other using beacons where multicast routing is supported, or a centralized service directory elsewhere. We have implemented a simple Limone service registry that is suitable for a small number of agents.

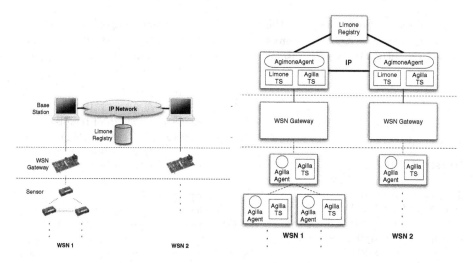

Fig. 7. Agimone Network Architecture   **Fig. 8.** Agimone System Components

However, it is not designed to scale for deployment on larger networks like the Internet. Since Limone's discovery mechanism is pluggable, applications that require greater scalability can use a more sophisticated protocol, like WSDL [8].

Agimone is populated with the following components, as shown in Figure 8:

- The AgimoneAgents are specific Limone agents which allow Agilla tuples and agents to traverse the IP network. These agents serve as the basis for the Agimone integration layer. Each base station hosts one AgimoneAgent.
- The Agilla and Limone tuple spaces, as described above.
- The Limone registry allows remote WSN discovery. Each application shares a single Limone registry.

In the rest of this section, we will describe Agimone's services in further detail.

4.1 WSN Discovery

Since new WSNs are formed and destroyed as the applications evolve, it is often necessary for agents in the WSNs to be aware of these changes at runtime. This is accomplished using a *WSN advertisement* scheme. Each base station's AgimoneAgent encapsulates information about the corresponding WSN in a WSN advertisement message. This advertisement describes the WSN's properties to Agilla agents. Since different applications may be interested in different properties of the WSNs, this advertisement is application-specific. For example, agents that comprise a cargo tracking application may be interested in knowing the location of each network. Thus, the WSN advertisements contain a 3-character string describing their locations, such as "dok" (dock) or "shp" (ship).

When a new WSN connects to the IP network, its corresponding AgimoneAgent beacons a well-known Limone registry with messages containing its WSN advertisement. The Limone registry forwards these advertisements to other Limone agents. Similarly, the registry notifies Limone agents when hosts leave the network. AgimoneAgents use these notifications to store up-to-date copies of all other WSN advertisements in their base station's Agilla tuple space.

```
1: pusht STRING
      // type STRING
2: pushc 1  // length of template
3: pushloc UART_X UART_Y
      // base station's location
4: rrdp      // rrdp(base station,
      // <STRING>)
```

```
1:      pusht STRING
            // type STRING
2:      pushc 1  // length of template
3:      pushloc UART_X UART_Y
            // base station's location
4:      rrdp      // rrdp(base station,
            // <STRING>)
5:      rjumpc OK
6:      halt     // if tuple not found, halt
7: OK   pushloc UART_X, UART_Y
            // base station's location
8:      smove    // migrate to base station
```

Fig. 9. WSN Discovery Code Snippet   **Fig. 10.** Migration Code Snippet

Agilla agents can access the base station's tuple space by performing remote tuple space operations with the special destination address (UART_X, UART_Y). Thus, they can select an appropriate WSN advertisement using a **rrdp** operation. The example code in Figure 9 places any available WSN advertisement containing a string on top of the Agilla agent's operand stack.

4.2   Migration Across WSNs

Using these advertisements, Agilla agents can select other WSNs and migrate to them with the assistance of the AgimoneAgent. This procedure is detailed in Figure 11. WSN advertisements are distributed in Steps 1 and 2, and placed in the base stations' Agilla tuple space in Step 3. The Agilla agent selects one of these WSN advertisements in Step 4 and places it on top of its operand stack.

Once the Agilla agent has an acceptable advertisement on its operand stack, it performs a strong migration to the WSN gateway, as shown in Step 5. Sample code to perform this operation is listed in Figure 10. This migration request is forwarded to the AgimoneAgent executing on the base station in Step 6. The AgimoneAgent extracts the destination WSN advertisement from the top of the agent's operand stack. It then encapsulates the Agilla agent into a Limone tuple of the form <Agent: *(encapsulated agent)*>. In Step 7, it places this tuple into the Limone tuple space of the destination network's AgimoneAgent.

On initialization, AgimoneAgent installs a reaction on its tuple space that notifies it of tuples in the form <Agent: Agilla Agent>. Thus, in Step 8, the AgimoneAgent on the destination base station is notified of the tuple's arrival. It extracts the agent from the tuple and injects it into the WSN gateway in Step 9. In Step 10, the agent migrates to the new WSN, where it resumes computation.

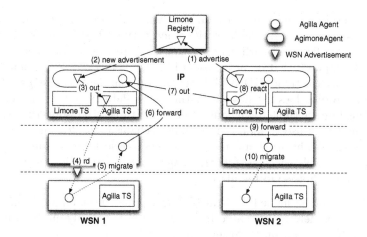

Fig. 11. Agilla Agent Migration Across Different WSNs

This process involves several transactions across WSNs and the IP network. However, this is transparent to the Agilla agent developer, who only invokes a single migration operation to the base station. Thus, developers can leverage the Limone network's infrastructure while still using the familiar Agilla APIs.

4.3   Cross-Middleware Interactions Via Tuple Spaces

So far, we have only considered the IP network as a way for distant WSNs to interact. However, it can also be used to support interactions between devices in a WSN and devices on the IP network. Because of the limited computational powers of wireless sensors, Agilla agents may wish to use devices on the IP network as a computational resource. Likewise, a Limone agent may wish to exploit the sensing resources of a remote WSN. Both goals can be achieved by giving Limone agents access to the Agilla tuple space that resides on each base station, providing both types of agents with a common data space for exchanging messages. However, directly exposing the Agilla tuple space API to Limone agents has some undesirable side effects. For example, though Limone agents can reside on any host in the IP network, they would only be able to interact with a WSN if they reside on a base station within its radio range.

Instead, the `AgimoneAgent` exposes each base station's Agilla tuple space to the Limone network by wrapping it in the Limone tuple space API. Other Limone agents communicate with Agilla agents by performing remote tuple-space operations on this Limone tuple space. The `AgimoneAgent` translates these operations to their Agilla equivalents and forwards them to the Agilla API. Hence, any tuples placed by Limone agents into this tuple space are available to Agilla agents in the corresponding WSN, and vice-versa. These Limone agents need not have a WSN gateway attached to their host to interact with the WSN, since an `AgimoneAgent` will communicate with the WSN on their behalf. Limone and Agilla agents interact with this shared tuple space using their respective

APIs. So, developers who are only familiar with one of these systems can still leverage resources made available by the other, without first learning a new API.

However, as discussed earlier in Section 3, there are restrictions on Agilla tuples and templates that do not exist in Limone. For example, a Limone agent may try to place the tuple <ID: 3.14, Flag: "mark"> in the `AgimoneAgent`'s tuple space. Since Agilla does not have a floating-point data type, there is no way to convert this Limone tuple to an equivalent Agilla tuple. To resolve this problem, the `AgimoneAgent` uses Limone's rejection mechanism to filter incoming tuple space operations. This mechanism allows agents to reject any remote operations issued on their tuple space. The `AgimoneAgent` places the following restrictions on all incoming tuples and templates:

- Fields cannot be named arbitrarily. Field names must impose a numerical order on the fields, as required by Agilla. That is, exactly one field must be named "1", exactly one field must be named "2", etc.
- Fields must contain Agilla data types.
- The only constraint functions are DefaultConstraintFunction (i.e., match by type) or EquivalencyConstraintFunction (i.e., match by exact value).

The `AgimoneAgent` will reject all non-conforming operations, since they have no Agilla equivalents. Conforming operations are converted to their Agilla counterparts and forwarded to the Agilla tuple space. The results are converted from Agilla tuples to Limone tuples (using the conventions specified above) and sent back to the request's originator.

4.4 Implementation Details

Agilla and Limone have been implemented and deployed on a wide variety of hardware. Agilla has two parts: a NesC-based portion that is installed on sensors, and a Java-based `AgentInjector` that is installed on base stations. Since storage is at a premium on many sensors, Agilla is necessarily compact: it consumes 49.66KB of flash ROM and 3.07KB of RAM. Agilla supports several different sensor architectures, including Mica2, MicaZ, and Tyndall25. For this paper, we used a CVS snapshot of Agilla 3.0, which can be downloaded from [9].

The Limone and Agimone packages are developed in Java according to the J2ME Personal Profile 1.0 [10] specification. This allows deployment on devices like PDAs and Stargates which cannot host full Java Standard Edition runtimes, as well as on desktop and laptop computers. Limone was designed for deployment on storage-constrained devices like PDAs: its bytecode distribution consumes only 132KB of storage space. Agimone is even more compact: it consumes 13KB of storage space. Agimone operates on any platform supported by Limone, which includes Windows Mobile, Windows XP, Linux, Solaris, and Mac OS X.

5 Case Study: Cargo Tracking

Using the architecture described in the previous section, we can implement a wide range of complex WSN applications. Cargo tracking is one such application that is well-suited for implementation using Agimone. As discussed in Section 2, cargo containers can be equipped with sensors that form WSNs in localized clusters. Many of these containers are located in remote warehouses and vehicles. So, users must be able to interact with these clusters without needing to be within the WSN's communication range. This can be achieved by connecting the WSNs' base stations using a common IP network, then deploying Agimone on them so that queries may traverse either network as needed.

In this section, we present a prototype application that uses mobile agents to track cargo. Our group had developed a similar application (demonstrated at SenSys '05 [11]) using a custom Limone agent to marshal messages between the sensor and IP networks. This custom agent had to be repeatedly modified and redeployed as our application's feature set evolved, greatly complicating development efforts. These difficulties motivated the creation of Agimone and a complete redesign of the application around it, resulting in much cleaner code overall and a simpler deployment process. Although Agimone was motivated by the cargo tracking application, we emphasize that Agimone is a *general purpose* middleware with a uniform programming model that can be used for a broad class of applications that need to integrate multiple WSNs and the IP network.

In the interest of space, we provide here a brief overview of two agents that are part of this application. More in-depth information about the application, including sample code, may be found in [12].

5.1   Watchdog Agents

Sensors attached to shipping containers can be equipped with various inexpensive sensor boards which can be used to detect attempted intrusions into the containers. As a demonstration of this potential, we have implemented two prototype agents that monitor the sensor's accelerometer and light readings, respectively. These agents loop, repeatedly reading the sensor until an unusual reading is detected. When this happens, an event is recorded in the local tuple space, and an alert tuple is placed in the base station's tuple space.

The `AgimoneAgent` on the base station automatically exposes these alert tuples to the Limone network. Remote Limone clients on the IP network can register reactions for these tuples. Limone automatically notifies these clients when any new alerts are generated. We can then do whatever processing we desire with these alerts (e.g., log it to disk and notify security personnel).

As a testament to Agilla's expressiveness, the watchdog agent that monitors the light sensor contains only 17 lines of code. The Limone client requires only 11 lines of code to automatically receive alerts and extract their contents. The Agilla agent and the Limone client were developed in only a few hours.

5.2   Intrusion Search Agent

A user, such as a shipping company or a port authority, may later want to search all the containers for any tampering recorded by the watchdog agents. Consider a scenario where containers are being moved between a ship and a loading dock, each of which has a corresponding WSN and base station. These base stations are connected by an IP link, e.g., Ethernet or 802.11b. Though users can search both WSNs simultaneously, a comprehensive search may be unnecessarily expensive. Ideally, the scope of such a search should be determined at runtime. For example, the user may know that containers on the ship are far more susceptible to tampering than the dock. So, the search for tampered containers should begin on the ship. If one of these containers has been tampered with, then the search should automatically expand to the dock, in order to determine the scope of the security breach.

We have developed a sample Agilla agent which consults WSN advertisements at runtime to locate WSNs and apply this searching policy. This involved adding only 23 lines of code to the previous Agilla agent. Owing to Agimone's flexibility, the Limone client used to monitor the watchdog agents' alerts required no modifications to support this new agent's alerts. Further, no additional support code had to be deployed to the base stations to support inter-WSN migrations.

6   Performance Evaluation

We evaluated our system by deploying it on two WSNs connected by an IP network. The WSNs are composed of Mica2 motes and are separated by using different radio channels. Each WSN has a single gateway attached to an IBM R40 laptop via a 115.2Kbps serial link. The laptops are connected via a 100Mbps wired Ethernet link. Since they are on the same subnet, discovery is performed using multicast beacons rather than a Limone registry. The laptops are configured with a 1.5GHz Intel Pentium M processor, 512MB of RAM, Windows XP and Java Standard Edition 5.0. Latencies are measured using Java's System.nanoTime() method, which uses the system's most accurate timer. This section presents micro-benchmarks examining the primitives that cross network boundaries. These benchmarks can be divided into three categories: tuple space operations, agent migration, and overall performance.

We have not compared the performance of Agimone to any other middleware systems. This is because to date no comparable systems exist: Agimone is currently the only middleware which supports the interaction of mobile agents across WSNs joined by an IP network. In this section, we focus on the cost of the inter-WSN operations supported by Agimone. The interested reader may consult [4] for a detailed discussion of Agilla's intra-WSN performance.

6.1   Tuple Space Operations

In the first set of benchmarks, we evaluate the cost of the tuple space operations **rinp**, **rrdp**, and **rout** across middleware boundaries. These operations may be performed by the `AgimoneAgent` on the tuple space belonging to the WSN gateway (PC-to-Mote), or by an Agilla agent on the base station's tuple space (Mote-to-PC). In the interest of brevity, we only provide here a brief overview of the benchmarks. The interested reader may find more technical details in [12].

Mote-To-PC. The first set of benchmarks determine the latency of an Agilla agent on the WSN gateway accessing `AgimoneAgent`'s Agilla tuple space. We created three benchmark agents, each of which performs one of the remote tuple space operations (**rinp**, **rrdp**, and **rout**) 100 times, over which the mean was calculated. Each benchmark was repeated 100 times. The operations have an average latency of 10 to 11 ms, as shown in shown in Figure 12.

Operation (Mote-to-PC)	latency (ms)
rinp	10.64 ± 0.15
rrdp	10.35 ± 0.06
rout	10.37 ± 0.07
Operation (PC-to-Mote)	latency (ms)
rinp	10.98 ± 0.17
rrdp	11.26 ± 0.19
rout	10.85 ± 0.07

Fig. 12. The Latency of Remote Tuple Space Operations

PC-To-Mote. The second set of benchmarks repeats the same operation in the opposite direction. In this case, since the latency can be directly measured, each experiment calculates the latency of one operation execution. Figure 12 shows the average results from 100 runs of each benchmark. The mean latency of PC-to-Mote tuple space operations is 10 to 11 ms.

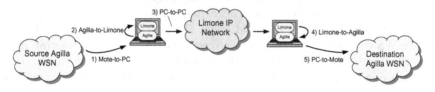

Fig. 13. The Five Stages of an Inter-WSN Agent Migration Operation

6.2   Agent Migration Operations

As discussed in Section 3.1, agent migrations enable agents located in one WSN to migrate across an IP network into another WSN. From an Agilla agent's perspective, an inter-WSN agent migration occurs by invoking a single operation. However, as discussed in Section 4, there many steps involved which are transparent to the agent. In this set of benchmarks, we identify five distinct stages involved in migrating a 36-byte agent across WSNs, as shown in shown in Figure 13, and measure the cost of each stage. Again, we refer the interested reader to [12] for more in-depth technical details.

The results of these benchmarks are shown in Figure 14. All benchmark results are presented as an average of 1000 runs. Note that stage 2 has a significant difference between mean and median latency. This difference is caused by sparse points with values orders of magnitude above the mean, which we suspect are caused by the process being interrupted by the OS or Java's garbage collector.

Stage 1: Mote-to-PC. Here, the agent moves from the source mote to the base station. We measured this procedure by deploying an agent which searches the `AgimoneAgent`'s tuple space for a WSN advertisement and then attempts to migrate to the base station. The mean latency of this stage is 36.12 ± 1.19ms.

Stage 2: Agilla-to-Limone. In this tage, the agent passes from the Agilla middleware on the base station to the Limone middleware. The cost of this operation should be negligible, since it only involves a few local method calls. This is borne out by our tests; the mean latency is 1.03 ± 0.16ms.

Stage 3: PC-to-PC. In this stage, the `AgimoneAgent` encapsulates the migrating agent into a Limone tuple and places it in the destination `AgimoneAgent`'s tuple space. We timed this stage by repeatedly migrating an agent between two base stations, then halving the round-trip time. This stage had a mean latency of 19.45 ± 0.26ms.

Stage	Mean Latency	Median Latency
1	36.12 ± 1.19ms	33.73ms
2	1.03 ± 0.16ms	$303.95\mu s$
3	19.45 ± 0.26ms	18.77ms
4	1.13 ± 0.16ms	$834.74\mu s$
5	28.16 ± 5.92ms	22.28ms

Fig. 14. The Latency of Each Agent Migration Stage (Average of 1000 Runs)

Stage 4: Limone-to-Agilla. In this stage, the `AgimoneAgent` extracts the encapsulated agent from the Limone tuple and passes it to Agilla's `AgentInjector`. Like stage 2, this only involves a few local method calls, so the latency should be negligible. We recorded the time between placing the tuple in the tuple space to passing the agent to the `AgentInjector`. The mean latency is 1.13 ± 0.16ms; as expected, this is negligible relative to other stages.

Stage 5: PC-to-Mote. In the final stage, the agent is injected into the destination WSN. Similarly to stage 1, we measured this latency by migrating an agent which immediately reads an advertisement tuple from the base station, and measuring the time between injection and receiving the tuple space request. The mean latency of this stage is 28.16 ± 5.92ms.

6.3   Overall Performance

The last set of benchmarks evaluate the latency of common sequences of operations. The In-and-Out benchmark measures the cost of migrating in and out of the same WSN. The End-to-End benchmark evaluates the cost of migrating from one WSN to a different WSN and back. These two benchmarks use the same 36-byte agent and are repeated 1000 times.

While Agimone simplifies programming and increases network flexibility, its use of virtual machines results in some overhead. We quantify this overhead by comparing the first two benchmarks above with native-code implementations. To isolate the cost of message-passing from execution, the native implementations exchange 36-byte data messages in place of 36-byte mobile agents.

In-and-Out. This benchmark injects an agent which migrates repeatedly between two WSNs, and measures the cost of moving the agent in and out of one WSN. When the agent is injected into the WSN, it immediately performs a **rrdp** to find the other WSN's advertisement, and then attempts to migrate to it. Thus, this benchmark measures the aggregate of the Mote-to-PC, PC-to-Mote, Limone-to-Agilla, and Agilla-to-Limone migration operations, and the Mote-to-PC tuple space operation. The results of this benchmark are shown in Figure 15.

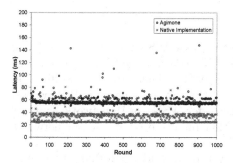

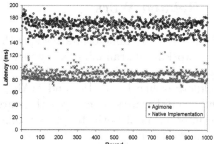

Fig. 15. The In-and-Out Agent Migration Latency

Fig. 16. The End-to-End Migration Latency

The mean In-and-Out latency is 62.18 ± 6.09ms, with a 55.76ms median. This is approximately the aggregate of the constituent stages (1, 2, 3, and 4).

The native implementation of In-and-Out is a Java application that sends a 36-byte query to the attached gateway sensor in two TinyOS packets; the sensor immediately sends 36 bytes of data back. The benchmark measures the time from sending the request to receiving the response. The native implementation has a mean latency of 30.09 ± 0.51ms, and a median latency of 26.29ms.

End-to-End. The End-to-End latency is measured by injecting the same agent and recording its round-trip time over the IP network. The results are shown in Figure 16. The mean round trip time is 179.19 ± 9.96ms, with a median of 167.96ms. This closely matches the sum of the various stages involved.

The native implementation of End-to-End adds to the In-And-Out benchmark by sending a 36-byte packet over the IP network to a remote base station after receiving a response from the WSN. The remote base station sends a 36-byte reply. The benchmark measures the time from querying the sensor node to receiving a response from the remote base station. The native implementation has an mean latency of 86.36 ± 2.15ms, and a median latency of 84.38ms.

The benchmarks presented in this section provide a general overview of Agimone's performance and overhead. All inter-network tuple space operations, regardless of direction, take about 10.5ms. A mobile agent takes about 85.9ms to migrate from one WSN to another. Of this, approximately 65ms is spent moving to and from the WSN and its base station, and 20ms is spent traversing the IP network. The latency of migrating into a WSN and back is about 60ms. Most of this time (>57ms) is spent on the serial link between the base station and WSN gateway. The actual transition from Agilla to Limone is less than 1ms in either direction. The overhead of Agimone compared to native code varies depending on the task. In the two operations presented, In-And-Out and End-to-End, there was a 32.09ms and 92.83ms increase in execution time relative to native code, respectively. Native code, however, is not nearly as flexible as mobile agents, and presumably requires more development time.

7   Related Work

There are many middleware systems that increase WSN flexibility by enabling in-network reprogramming. They include XNP [13], Deluge [14], Maté [6], SensorWare [15], Impala [16], and Smart Messages [17]. There are also coordination middleware like LIME [18], and MARS [19] that are designed for IP networks. These middleware systems are either targeted for WSNs, or IP networks, but not the integration of both. Recent systems that integrate IP and sensor networks are more closely related to Agimone.

The Hourglass [20] and Stream-based Overlay Networks (SBONs) [21] systems form an overlay network over the Internet out of servers connected to various WSNs. The system routes data streams generated within WSNs to applications on the Internet. The system also provides resource registration and discovery services to servers. Servers dynamically adapt to network conditions by installing stream operators like data filters and aggregators on the source, e.g., to reduce network congestion. Hourglass-SBON focuses on delivering data streams generated within WSNs to consumers on the Internet. Agimone, on the other hand, is a general-purpose middleware system that supports agent migration and coordination across WSNs and IP networks, as well as data sharing.

Tenet [22] provides a two-tiered architecture partitioned into resource-poor sensors and relatively powerful computers connected via an IP network. The higher-tier computers directly control sensors, which service them using well-established protocols. This moves much of the application development onto more-powerful computers, simplifying debugging. Unlike Agimone, Tenet's tasks cannot relocate autonomously or carry state across nodes. Therefore, Agimone provides a more flexible infrastructure for deploying adaptive applications. Also, Tenet uses message passing as its basic communication paradigm, which easily fails in the face of transient link failures. Agilla uses tuples for all inter-agent communication, which survive temporal communication failures.

SERUN [23] uses a three-level network architecture divided into inexpensive data-gathering sensors, data-processing *microservers*, and PC-class systems where end-users can issue queries. When a query is issued, a task is sent to a microserver that queries one or more sensors and processes the data according to the task's instructions. SERUN differs from Agimone in that it moves much of the application-specific code away from the low-power sensors and onto the microservers, and its tasks cannot autonomously migrate across microservers.

IrisNet [24] diverges from traditional WSNs by proposing an Internet-scale sensor network consisting of desktop PCs with low-cost sensors, e.g., web cams. IrisNet provides a query service for obtaining sensor data from anywhere on the Internet. Functionally, it is similar to TinyDB [25] in that it treats the network as a database. However, since IrisNet operates on relatively powerful machines rather than embedded sensors, it is best suited for applications where sensing capabilities are secondary to computational resources. In this sense, IrisNet is complementary to Agimone rather than an alternative.

8 Conclusion

In this paper, we have presented Agimone, a middleware system for integrating WSNs over the Internet and other IP networks. We have implemented an efficient layer that integrates Agilla and Limone, two existing mobile agent middleware platforms. By designing a cargo tracking application that uses Agimone, we have demonstrated how developers can easily take advantage of the functionality we provide. Our empirical performance data demonstrates the efficiency of our middleware on existing sensor and base station hardware. Though there is some runtime overhead associated with using mobile agents as compared to native code, the increase in developer productivity outweighs this performance penalty for all but the most time-critical of applications.

Acknowledgment

This research is supported by the Office of Naval Research under MURI research contract N00014-02-1-0715 and by the the NSF under NOSS contract CNS-0520220. Any opinions, findings, and conclusions expressed in this paper are those of the authors and do not necessarily represent the views of the research sponsors. We would also like to thank Boeing Corporation for their support on an earlier version of the cargo tracking application.

References

1. Hill, J., Szewczyk, R., Woo, A., Hollar, S., Culler, D., Pister, K.: System architecture directions for networked sensors. In: Architectural Support for Programming Languages and Operating Systems. (2000) 93–104
2. Zhao, J., Govindan, R.: Understanding packet delivery performance in dense wireless sensor networks. In: Proc. of the ACM SenSys. (2003)
3. (http://platformx.sourceforge.net/)
4. Fok, C.L., Roman, G.C., Lu, C.: Rapid development and flexible deployment of adaptive wireless sensor network applications. In: Proc. of the 24th International Conference on Distributed Computing Systems (ICDCS'05), IEEE (2005) 653–662
5. Fok, C.L., Roman, G.C., Hackmann, G.: A Lightweight Coordination Middleware for Mobile Computing. In DeNicola, R., Ferrari, G., Meredith, G., eds.: Proceedings of the 6th Internation Conference on Coordination Models and Languages (Coordination 2004). Number 2949 in Lecture Notes in Computer Science, Springer-Verlag (2004) 135–151
6. Levis, P., Culler, D.: Maté: a tiny virtual machine for sensor networks. In: ASPLOS-X: Proceedings of the 10th international conference on Architectural support for programming languages and operating systems, New York, NY, USA, ACM Press (2002) 85–95
7. (http://mobilab.wustl.edu/projects/agilla)
8. W3C-XML-Activity-On-XML-Protocols: W3c recommendation: Web services description language 1.1. http://www.w3.org/TR/wsdl (2003)
9. (http://mobilab.wustl.edu/projects/agilla/download/index.html)
10. (http://java.sun.com/products/personalprofile/index.jsp)

11. Hackmann, G., Fok, C.L., Roman, G.C., Lu, C., Zuver, C., English, K., Meier, J.: Demo abstract: Agile cargo tracking using mobile agents. In: Proceedings of the 3rd Annual Conference on Embedded Networked Sensor Systems (SenSys'05), ACM (2005) 303

12. Hackmann, G., Fok, C.L., Roman, G.C., Lu, C.: Agimone: Middleware support for seamless integration of sensor and ip networks. Technical Report WUCSE-05-56, Washington University in St. Louis Department of Computer Science and Engineering (2005)

13. ⟨http://www.tinyos.net/tinyos-1.x/doc/Xnp.pdf⟩

14. Hui, J., Culler, D.: The dynamic behavior of a data dissemination protocol for network programming at scale. In: Proceedings of the 2nd international conference on Embedded networked sensor systems, ACM Press (2004) 81–94

15. Boulis, A., Han, C.C., Srivastava, M.: Design and implementation of a framework for efficient and programmable sensor networks. In: Proc. of MobiSys, USENIX (2003) 187–200

16. Liu, T., Martonosi, M.: Impala: A middleware system for managing autonomic, parallel sensor systems. In: ACM SIGPLAN Symposium on Principles and Practice of Parallel Programming. (2003)

17. Kang, P., Borcea, C., Xu, G., Saxena, A., Kremer, U., Iftode, L.: Smart messages: A distributed computing platform for networks of embedded systems. Special Issue on Mobile and Pervasive Computing, The Computer Journal **47** (2004) 475–494

18. Picco, G., Murphy, A., Roman, G.C.: LIME: Linda meets mobility. In: Proc. of the 21st Int'l. Conf. on Software Engineering. (1999)

19. Cabri, G., Leonardi, L., Zambonelli, F.: MARS: A programmable coordination architecture for mobile agents. Internet Computing **4**(4) (2000) 26–35

20. Shneidman, J., Pietzuch, P., Ledlie, J., Roussopoulos, M., Seltzer, M., Welsh, M.: Hourglass: An Infrastructure for Connecting Sensor Networks and Applications. Technical Report TR-21-04, Harvard (2004)

21. Pietzuch, P., Ledlie, J., Shneidman, J., Roussopoulos, M., Welsh, M., , Seltzer, M.: Network-aware operator placement for stream-processing systems. In: Proc. of the 22nd International Conference on Data Engineering (ICDE'06, to appear). (2006)

22. Govindan, R., Kohler, E., Estrin, D., Bian, F., Chintalapudi, K., Gnawali, O., Rangwala, S., Gummadi, R., Stathopoulos, T.: Tenet: An architecture for tiered embedded networks. Technical Report CENS-TR-56, UCLA CENS (2005)

23. Liu, J., Cheong, E., Zhao, F.: Semantics-based optimization across uncoordinated tasks in networked embedded systems. Technical Report MSR-TR-2005-46, Microsoft Research, One Microsoft Way, Redmond, WA 98075 (2005)

24. Gibbons, P., Carp, B., Ke, Y., Nath, S., Seshan, S.: Irisnet: An architecture for a worldwide sensor web. IEEE Pervasive Computing (2003) 22–33

25. Madden, S., Franklin, M., Hellerstein, J., Hong, W.: The design of an acquisitional query processor for sensor networks. In: Proceedings of the 2003 ACM SIGMOD Int. Conf. on Management of Data. (2003) 491 – 502

Gappa: Gossip Based Multi-channel Reprogramming for Sensor Networks*

Limin Wang and Sandeep S. Kulkarni

Software Engineering and Network Systems Laboratory
Department of Computer Science and Engineering
Michigan State University
East Lansing MI 48824 USA
{wangliml, sandeep}@cse.msu.edu

Abstract. Reprogramming the sensor networks in place is an important and challenging problem. One way suggested for reprogramming is with the help of an UAV (Unmanned Ariel Vehicle). To reprogram a sensor network with the help of an UAV, one can either communicate the entire new program to one (or a few) sensor in the field, or let the UAV communicate parts of the code to a subset of sensor nodes on multiple channels at once. In the latter approach, the nodes need to communicate with each other to receive the remaining parts of the program.

In this paper, we propose a protocol for such gossip between nodes. To better utilize the multi-channel resources and reduce contention, our protocol provides a multi-channel sender selection algorithm. This algorithm attempts to ensure that in any neighborhood, at any time, there is at most one sensor transmitting on a given frequency. Moreover, our sender selection algorithm is greedy in that it tries to select the sender that is expected to have the most impact for each channel. Our protocol also conserves energy by putting the nodes that are unlikely to contribute or receive data shortly to "sleep" state. Through simulation, we show that our protocol is faster and more energy efficient than the existing reprogramming approaches that assume that the new program is initially located only on a small set of nodes.

1 Introduction

The problem of multihop reprogramming is necessitated by the facts that sensor networks consist of hundreds or thousands of sensor nodes and they are often deployed in remote or hostile environments (e.g., battle fields, forests). It is demanding and sometimes impossible to collect all the sensor nodes from the field for reprogramming. Therefore, it is necessary to reprogram sensor networks in place.

One way suggested for reprogramming is with the help of an UAV (Unmanned Ariel Vehicle). Specifically, in this approach, an UAV flies over the network and communicates the new code to the sensors. Reprogramming with the help of an UAV can be achieved in two ways. For one, the UAV could communicate the entire new program to one (respectively, subset or all) sensor in the field. (This approach is also similar to

---

* This work was partially sponsored by NSF CAREER CCR-0092724, DARPA Grant OSURS01-C-1901, ONR Grant N00014-01-1-0744, NSF equipment grant EIA-0130724, and a grant from Michigan State University.

P. Gibbons et al. (Eds.): DCOSS 2006, LNCS 4026, pp. 119–134, 2006.
© Springer-Verlag Berlin Heidelberg 2006

the case where UAV drops a sensor with the new program on the field). This sensor (respectively, set of sensors) then communicates the new program to the remaining sensors using approaches such as [1, 2, 3, 4, 5].

Another approach for reprogramming with the help of UAV is based on the observation that while sensors (e.g., Mica2/Mica2Dot, XSM, Telos) are limited to sending and receiving on one frequency at a time, the UAV is often more powerful and, hence, can communicate at multiple frequencies at once. Thus, the UAV can divide the program into multiple segments and transmit each segment on a different frequency. The sensors themselves choose one of these frequencies and receive the corresponding segment. Clearly, one advantage of this approach is that the contact time required for UAV is reduced. Moreover, by exploiting multi-channel resources, sensor nodes are able to split the network traffic among different channels, and hence, reduce contention. Finally, there is data redundancy, since every segment is associated with many sensor nodes. Compared to the type of communication that originates from one or a few seed nodes, this gossip based communication has the potential to enable higher concurrency and better utilization of channel capacity.

This paper focuses on the second approach. With this approach, the sensors then need to communicate the remaining segments with each other. We denote this problem as the gossip based multihop reprogramming of sensor networks. In particular, in this problem, each sensor is associated with one of the segments from the new program. (We also consider the case where only a subset of sensors have a segment, as long as each segment is with at least one sensor). We propose *Gappa*, a gossip based multihop reprogramming protocol, which utilizes multiple channels to rapidly and reliably reprograms all the sensors in the network. We implement *Gappa* on TinyOS platform [6] and evaluate it through simulations on TOSSIM [7].

Contributions of the paper

1. We propose a multi-channel sender selection algorithm, which tries to guarantee that on each channel, only one sender is selected to transmit in a neighborhood at a time. Moreover, the algorithm attempts to select the sender whose transmission is expected to have the most impact on each channel. To better utilize multi-channel resources, if a node loses in the sender selection for one channel, it will compete to transmit code on a different channel that is available. In this way, *Gappa* propagates code rapidly.

2. *Gappa* conserves energy by putting a sensor node to sleep if all the channels that it attempts to transmit code on are busy, and it is not interested in receiving the code segments that its neighbors are transmitting.

3. To enable gossip based communication, *Gappa* allows sensor nodes to receive segments that are out of order. There is less dependency on special nodes since every node that has a segment is a potential sender. This, combined with multi-channel usage and pipelining technique, leads to high concurrency in data exchange.

4. We implement *Gappa* in TinyOS platform, and evaluate its performance using TOSSIM. Through simulation, we compare with the state-of-art programming protocols, MNP [1] and Deluge [2], and show that *Gappa* reduces the reprogramming time and energy consumption significantly.

Organization of the paper. In Section 2, we present the gossip based multi-channel reprogramming protocol *Gappa*. In Section 3, we evaluate the performance of *Gappa* under different network settings. We also present the performance comparison with MNP and Deluge. We review related work in Section 4, and conclude in Section 5.

2 Protocol Description

We make no assumption about the underlying network topology and availability of global services, such as localization or time synchronization. We consider networks with stationary nodes. The sensor nodes are equipped with a single radio interface, thus can communicate on one channel at a time. But they can switch to different channels at run time.

The new program image that is to be deployed is divided into n segments (n is normally a small number from 1 to 20). Each segment has a fixed number of packets. We assume [1] that the number of available non-overlapping channels is at least $n + 1$. We select $n + 1$ non-overlapping channels, which are indexed from 0 to n. Without loss of generality, we define channel 0 as the *control channel* and channels 1 to n as the *data channels*. The control channel is used for transmitting the control messages (e.g., advertisements, requests), while the data channels are used for the actual data transmissions. Each data channel corresponds to one segment, i.e., segment k ($1 \leq k \leq n$) is always transmitted on channel k. The control channel is also the default channel, i.e., sensor nodes stay on channel 0 unless they are transmitting or receiving data packets of a certain segment.

Before we describe the algorithm in detail, we illustrate it using an example in Figure 1. The numbers marked on the sensor nodes represent the segments they have received. The edges represent communication links. We note several things from observing this simple network.

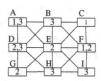

Fig. 1. Example sensor network

First, the nodes that have overlapping communication ranges cannot transmit the same segment simultaneously as it will cause significant collision on the shared data channel. For example, nodes A and B should not transmit segment 3 at the same time.

---

[1] Note that this condition holds for most sensor platforms that are popularly used. For example, a Mica2/Mica2Dot mote operating at 433MHz band can select separate channels with a minimum spacing of 150KHz [8]. Hence, in the 420-446MHz frequency range, the number of usable non-overlapping channels is 166. The basic principle in *Gappa* is to assign each segment a separate channel. Even in the case that the number available channels is not enough, we can assign multiple segments to one channel. However, this issue is outside the scope of this paper, as it is not required in current sensor networks.

Second, on each channel, the choice of the sender that transmits next is not uniform. If both nodes A and B want to transmit segment 3, B is a better choice than A, since more nodes in the neighborhood are expected to benefit from the transmission of node B.

Third, since sensor nodes can only communicate on one channel at a time, a node that has multiple segments must select one segment as the *preferred segment*, and will transmit this segment if it is selected as the sender. The choice of the preferred segment is decided by the status of its neighbor nodes. For example, node D might choose segment 2 as the preferred segment, because more nodes in its neighborhood request for segment 2 rather than segment 3. However, if D finds that E has decided to transmit segment 2, D cannot transmit segment 2 at the moment as it will cause collision with E on channel 2. In this case, transmitting segment 3 is a feasible alternative for node D.

Our algorithm consists of two parts: the control logic on the control channel and the operations on the data channels. We present these two parts in Section 2.1 and Section 2.2, respectively.

2.1   Operations on the Control Channel

Initially, all the sensor nodes are communicating on the control channel. Nodes perform two major tasks on the control channel: decide which nodes should switch to a data channel to perform data communication; and identify the nodes that are unlikely to contribute or receive data shortly and put them to sleep. The switching policy *tries to* guarantee that for each segment, at most one node in a neighborhood is selected to transmit the segment in the corresponding data channel. Moreover, it tries to select the sender that is expected to have the most impact. To achieve this goal, we extend the sender selection idea from MNP [1].

Multi-channel sender selection algorithm. In our algorithm, nodes perform sender selection using advertisements and requests. Each node maintains a sequence of <*SegID, ReqCtr*> pairs that indicate the segments it has received and the corresponding numbers of distinct requests (from different requesters) for those segments the node has received so far. *ReqCtrs* for all the segments are set to 0 when a node starts advertising. When a node receives a request that is *destined* to it from a "new" requester, it increments the *ReqCtrs* for the requested segments by one. Additionally, a node maintains a *preferred segment* ID, which is the segment that is requested by most number of nodes, i.e., has the highest *ReqCtr*. The preferred segment ID is set to 0 if the node has not received any request, and is recalculated whenever the *ReqCtrs* change. In the case that there are more than one segments have the highest *ReqCtr*, the preferred segment is randomly selected from these segments.

A node advertises the segments it has received, its preferred segment, as well as the *ReqCtrs* for all the received segments. Hence, an advertisement message includes the sequence of <*SegID, ReqCtr*> pairs, the preferred segment ID, and other information (program ID and size, source ID). When a node, say j, receives an advertisement message from a node, say k, if j needs any of the segments that are advertised, then it sends a request to k. The request message sent by j contains not only the IDs of the segments j expects to receive from k, but also k's preferred segment ID and the corresponding *ReqCtr* of k for that preferred segment (computed from the sequence of <*SegID, ReqCtr*>

pairs that k sent in the advertisement message). While the request is intended (*destined*) to k, it is sent as a broadcast message with k as one of the fields. Thus, when another node, say l, receives the request, l is aware of the fact that k is a potential sender. This allows us to count for the hidden terminal effect where l could not have received the advertisement message from k.

We note that a node sends a request to all senders that send the advertisement messages containing the code segments the node is interested in. This ensures that a node is aware of all the requesters who are likely to receive the code if it is chosen to transmit the code. Moreover, the sender selection is performed only among nodes that have the same preferred segment. (For example, if a node, say k, has preferred segment 3, and its neighbor l has preferred segment 1, they can transmit simultaneously on different channels (channel 3 and 1) without interrupting each other.) If k loses to l that has more requesters for the preferred segment, k takes the two actions in Figure 2.

---

1. Mask the preferred segment, and check to see if it has any segments that are not masked. If so, start advertising the remaining segments (and try to transmit on a different channel that is currently available). Otherwise, turn to idle stage.
2. Reset the *ReqCtrs* of all its segments to 0. Reset its preferred segment ID to 0. (This is due to the fact that some of the old requesters of this node may be receiving code from the node that wins the sender selection on a different channel.)

Fig. 2. Actions taken by a node that loses in the sender selection

A node on the control channel is in one of the two stages: advertise stage and idle stage. Initially, if a node has received a segment from an UAV, then it is in advertise stage; otherwise, it is in idle stage.

Tasks in advertise stage. A node S in advertise stage broadcasts an advertisement message every random interval (we use random interval to avoid message collision). It reacts to the requests, advertisements, and "SwitchChannel" messages as described next.

1. *Actions taken by an advertising node on receiving a request.* Every time S receives a request message, it checks to see if this message is *destined* to it. If the message is destined to it, and is from a "new" requester that S has not seen before, S increments the *ReqCtrs* of the requested segments by one, and recalculates its preferred segment ID. If the request message is destined to another node Q, Q has the same preferred segment ID as S and Q has more requesters, then S loses in the sender selection, and will take the actions in Figure 2.

2. *Actions taken by an advertising node on receiving an advertisement.* When S receives an advertisement message from a node Q, if S needs any of the segments Q advertises, S broadcasts a request message destined to Q after a short random interval (to avoid collision with other request messages sent to Q). As mentioned earlier, it puts the ID and *ReqCtr* of Q's preferred segment in the request message. In addition, S also checks to see if it loses to Q in the sender selection algorithm. If so, S takes the actions in Figure 2. Note that this sender selection procedure cannot cause deadlock, as the node with the highest *ReqCtr* of the preferred segment - with appropriate tie breaker on node ID - will succeed.

3. *Actions taken by an advertising node on receiving a "SwitchChannel" message.* If S receives a "SwitchChannel" message from a node Q, which indicates that Q is going to transmit its preferred segment in the corresponding data channel. If S is interested in receiving the segment from Q, then S switches to the data channel for that segment. Otherwise, if the segment that Q is going to transmit is the same as S's preferred segment, which means that S has lost the sender selection for this segment, then S takes the actions in Figure 2.

The advertise stage ends when a node has sent a given number of advertisements continuously (without resetting *ReqCtrs* or switching channels). At this point, if it has received one or more requests, it broadcasts "SwitchChannel" messages and switches to the data channel that is assigned to its preferred segment. Otherwise, it turns to idle stage.

Tasks in idle stage. A node in idle stage can choose to keep its radio on to listen to the channel, or turn its radio off to save energy. Thus, a node in idle stage is in one of the two states: listen state (with radio on) and sleep state (with radio off). The length of time t a node stays in idle stage, and whether it keeps its radio on (listen or sleep) when it is idle, are decided by the status of the node and its observation of neighbors. Specifically, a node maintains two boolean variables: *TendToReceive* and *TendToSend*, which indicates the node's intention to receive or transmit a segment. *TendToReceive* is set to *false* initially and when a node has successfully received a complete segment. When the node hears an advertisement, request, or "SwitchChannel" message, it checks to see if it needs any of the segments that are advertised (or requested) in the message. If so, the node sets its *TendToReceive* to *true*. *TendToSend* is set to *false* when a node starts advertising. When the node hears an advertisement or request message, if it finds that it has some segments that other nodes do not have (requested or not advertised), it sets its *TendToSend* to *true*.

When a node enters idle stage, if it has received the entire program, or its *TendToReceive* is *false*, then it goes to sleep state. Otherwise, it is in listen state. A node in sleep state does not sleep through the entire idle stage. Rather, it takes short naps (say, 4s), wakes up and checks the channel for a short amount of time (say, 0.5s) between naps. If messages received during this interval causes *TendToReceive* to become *true*, the node turns to listen state, and keeps its radio on for the rest of time (in idle stage).

The length of idle stage t is exponentially increased, starting with a minimum t_l to a maximum t_u. This allows us to dynamically adjust the rate of advertising: nodes advertise aggressively when reprogramming is actively in progress and advertise slowly to save energy when most nodes have received the code. t is reset to t_l in two situations. First, if a node switches to a data channel (to transmit or receive data), when it returns to the control channel, it will reset t to t_l. Second, if a node receives a request, or its *TendToSend* becomes *true* (i.e., it identifies *potential requesters*), it sets t to t_l.

Although a node in idle stage does not advertise or participate in sender selection, it still sends requests or switches channel when needed. When a node S is in listen state, it reacts to the advertisements and "SwitchChannel" messages as described next.

1. *Actions taken by a node in listen state on receiving an advertisement.* If S receives an advertisement message that advertises segments it is interested in, S broadcasts a request message destined to that advertising node.

2. *Actions taken by a node in listen state on receiving a "SwitchChannel" message.* If S receives a "SwitchChannel" message that contains the ID of the segment S is interested in receiving, S switches to the data channel for that segment, and as a result, the idle stage ends.

2.2   Operations on Data Channels

When a node S decides to become a sender, it broadcasts a "SwitchChannel" message for a few times, then switches to the data channel that is assigned to its preferred segment. When a neighbor node hears a "SwitchChannel" message from S, if it needs the segment, it will switch to the corresponding data channel, and turn to *download* state.

We note that although the sender selection algorithm attempts to keep only one active sender in a given neighborhood on each channel, it is possible to have multiple active senders due to time-varying link properties. Hence, when S enters the data channel, it listens to the radio for a short amount of time (say, 1s), which we call *pre-forward* state, before it starts forwarding data. If S hears any message when it is in pre-forward state, it realizes that another node is currently transmitting data on that data channel. Hence, it will mask the segment it is going to transmit (to ensure that it will not reenter this channel immediately), return to the control channel, and start advertising the remaining segments. In this case, those nodes that have followed S to this data channel (switched to this data channel after receiving S's "SwitchChannel" messages) are left on this channel. If the nodes are able to receive packets from the current sender, they will stay on this channel and receive data from the current sender. Otherwise, they will return to the control channel after a time out.

Gappa uses the a similar loss recovery mechanism as MNP. Each packet has a unique ID. Each node maintains a bitmap, which we call *MissingVector*, for each of the segments it is receiving. Each bit in a *MissingVector* corresponds to a packet. All bits are initially set to 1. When a node receives a packet in a segment for the first time, it stores that packet in EEPROM (external storage for motes), and sets the corresponding bit in the *MissingVector* for that segment to 0. In this way, we guarantee that each packet in a segment is written to EEPROM only once. Note that *Gappa* allows nodes to receive segments that are out of order. Thus, a node might have received several incomplete segments. It is necessary for a node to maintain bitmaps (*MissingVectors*) for all the segments it has not completely received. For simplicity, we assume that the *MissingVectors* are in memory. We note that the extension for storing them on EEPROM and loading only the *MissingVector* for the segment that is being received in memory is straightforward.

Each node also maintains a *ForwardVector*, which is a bitmap of the segment that it is going to transmit, and is an indicator of the packets the node needs to send. When the pre-forward stage times out, the sender node S turns to *forward* state, and broadcasts a "StartDownload" message several times. S includes its *ForwardVector*, which is initially set to 0 (i.e., all the bits are set to 0), in the "StartDownload" message. When a node hears a "StartDownload" message, it waits for a short random interval (to avoid collision with transmissions from other requesters), checks to see if the *ForwardVector* contained in the "StartDownload" message has already included all the packets it needs. If so, it keeps silent. Otherwise, it sends a "RequestPackets" message to S. The

"RequestPackets" message contains its loss information (*MissingVector*) for this segment. When S receives a "RequestPackets" message, it unions its *ForwardVector* with the *MissingVector* contained in the message. This updated *ForwardVector* is included in the "StartDownload" message that S sends next time. In this way, S's neighbor nodes are aware of the packets S is going to send, and hence, will not send requests repeatedly. We restrict the length of the segment to be no longer than 128 packets, so that the maximal size of *MissingVector* and *ForwardVector* is only 16 bytes, and thus fits into a radio packet.

After S has transmitted the "StartDownload" message for a few times, it starts transmitting the packets indicated in its *ForwardVector*. The download process ends when the receiver receives an "EndDownload" message from the sender. At this point, if the node has successfully received the whole segment, it includes the segment in the sequence it will use in future advertisements. When the download process ends, both the sender and the receivers return to the control channel, and restart advertising.

It is possible that the receiver never gets the "EndDownload" message. The reason can be the sender dies or returns to the control channel during pre-forward stage, or the "EndDownload" messages collide with other messages. To avoid being stuck in *download* state, a node in *download* state always sets a timer when it is waiting for the next packet. If the timer expires, it returns to the control channel.

As we mentioned earlier, a node masks a segment when it loses in the sender selection for that segment, or when it detects a busy data channel in *pre-forward* state. In both cases, the node cannot advertise or transmit this segment until the other node has finished transmitting the segment. Since this node does not know the exact time when the other node finishes transmitting, it keeps the segment masked for a certain amount of time. The mask bits are cleared when a node turns to idle stage or starts forwarding packets on a data channel.

2.3 The State Machine

In Figure 3, we show an overall picture of *Gappa*. *Gappa* operates as a state machine. Also, pseudo code of tasks in advertise stage on control channel can be found in [9].

3 Evaluation

We implemented *Gappa* on TinyOS platform, and evaluated it using TOSSIM [7]. TOSSIM is a discrete event simulator for TinyOS wireless sensor networks. In TOSSIM, the network is modelled as a directed graph. Each vertex in the graph is a sensor node. Each edge has a bit-error rate, representing the probability with which a bit can be corrupted if it is sent along this link.

Radio transmission in TOSSIM is simulated as follows. Each node maintains a variable *radio_active*, which is set to 0 at the initial state. Every time a node transmits a bit, it increments the *radio_active* values of all its neighbors. In a lossy model, the transmitted bit can be flipped if a bit error occurs. When a node finishes transmitting, it decrements the *radio_active* values of all its neighbors. A node hears a bit if its *radio_active* value is 1 or greater. Although TOSSIM only models radio transmission on

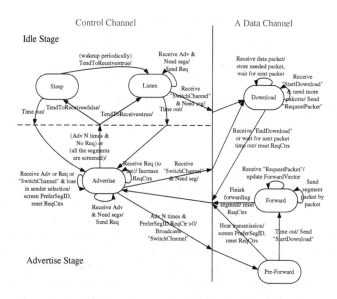

Fig. 3. *Gappa*: the state machine

a shared channel, we can make a few changes so that it also simulates radio transmission on multiple channels. Towards this end, each sensor node maintains an additional *frequency* variable, which is the channel number the node is currently communicating on. When a node transmits or stops transmitting, it only modifies (increments or decrements) the *radio_active* values of those neighbors that are on the same frequency. Moreover, when a node switches to a different channel (i.e., changes its *frequency* variable), its *radio_active* variable is reset to 0. Note that the switching channel action can be taken only when the radio transmission of the node on the current channel is completed.

We calculate the energy consumption by counting the operations performed during reprogramming. (Alternatively, we can also use PowerTossim [10] to evaluate power consumption. However, since each simulation lasts for tens of hours, the trace file generated during simulation, which is required by PowerTossim in order to compute the energy usage, becomes too large (of the order of several gigabytes) to process.) In Table 1 (from [11]), we list the costs of various operations on a Mica mote with a pair of AA batteries. We use Equation 1 to compute the energy consumption E (in joules), which is the product of charge Q (in coulombs, 1 nAh is the same as 0.0036 coulombs) and voltage V (in volts).

$$\begin{aligned}
E &= Q \cdot V \\
&= 0.0036 \cdot (20 \cdot n_{send} + 8 \cdot n_{receive} + 1250 \cdot t_{idle} \\
&\quad + 1.111 \cdot n_{read} + 83.333 \cdot n_{write}) \cdot 3 \\
&= 0.0108 \cdot (20 \cdot n_{send} + 8 \cdot n_{receive} + 1250 \cdot t_{idle} \\
&\quad + 2.222 \cdot n_{senddata} + 83.333 \cdot 2 \cdot n_{storedata})
\end{aligned} \tag{1}$$

Table 1. Charge required by various Mica operations

Operation	nAh
Transmitting a packet	20.000
Receiving a packet	8.000
Idle listening for 1 millisecond	1.250
EEPROM Read 16 Bytes	1.111
EEPROM Write 16 Bytes	83.333

In the above equation, n_{send} and $n_{receive}$ are the number of packets transmitted and received respectively during reprogramming, t_{idle} is a node's active radio time (in seconds), n_{read} and n_{write} are the number of reads and writes respectively executed by EEPROM. EEPROM is read and written in 16-byte blocks (lines). Hence, as each packet has 22 bytes data payload, each *data* packet transmitted involves 2 EEPROM reads, and each data packet stored corresponds to 2 EEPROM writes, i.e., $n_{read} = 2 \cdot n_{senddata}$, $n_{write} = 2 \cdot n_{storedata}$, where $n_{senddata}$ is the number of *data* packets transmitted, $n_{storedata}$ is the number of data packets that are stored in EEPROM.

Equation 1 shows that energy consumption is decided by idle listening time (t_{idle}), message transmissions (n_{send}, $n_{senddata}$) and receptions ($n_{receive}$), and the number of data packets stored ($n_{storedata}$). Among these, $n_{storedata}$ is decided by the size of program (divided in packets) to be transmitted, because our algorithm guarantees that each packet is written to EEPROM only once. Therefore, the key to reducing energy consumption is to reduce idle listening time and message transmissions and receptions. Among these, idle listening time (or active radio time) is the most important factor that affects the energy consumption.

In the current implementation, each segment has 128 data packets. t_l and t_u are set to 16 seconds and 512 seconds respectively. The simulations are performed in a grid topology. Due to the fact that the execution time of each simulation is of order of tens of hours, we do not provide confidence intervals. In Sections 3.1 and 3.2, we show how the algorithm performs under various network settings, specifically program sizes, network densities, network sizes. In these sections, we assume that initially all the sensors have received one segment (which is randomly decided) from an UAV. In Section 3.3, we consider the situation where only a subset of the sensors initially have a segment.

3.1 Varying Program Sizes and Network Densities

Experiment setup 1. (Dense network) . In the first set of experiments, we set the distance between every two neighbor nodes to 10 feet (a dense network). The simulations were performed in a 20x20 grid topology. We run *Gappa*, MNP and Deluge under the same network settings. For MNP and Deluge, we assume that initially only the base station, the node at the bottom-left corner, has the new program. For *Gappa*, every node has one segment (randomly chosen from 1 to n, suppose the program has n segments) initially. In Figure 4, we compare the completion time, average active radio time per node, and the average energy consumption per node of these three protocols, under different program sizes. We find that with the increase of program size, the completion time, average active radio time, and average energy consumption increase for all these

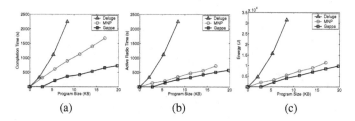

Fig. 4. Inter-node distance: 10 feet. (a) completion time (b) average active radio time per node (c) average energy consumption per node.

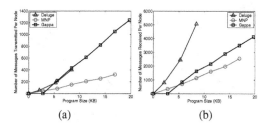

Fig. 5. Inter-node distance: 10 feet. (a) number of messages transmitted per node (b) number of messages received per node.

protocols. Among the three protocols, *Gappa* has the lowest completion time and energy consumption. To disseminate a program of the same size, *Gappa* saves 60-65% completion time and 22-33% energy compared to MNP, and saves 81-84% completion time and 86-88% energy compared to Deluge. In Figure 5, we show the average number of transmissions and receptions per node in there three protocols. We notice that the number of transmissions of *Gappa* is higher than the other two schemes. This is expected, as gossip based communication allows every node in the network to talk to each other, rather than require most nodes to listen to a few elected senders. Moreover, in *Gappa*, multiple senders are transmitting code in different data channels at the same time, while a receiver can only receive code in one data channel at a time. As a result, redundancy increases. Although the overall traffic increases, the traffic is diverted to different channels.

Experiment setup 2. (Sparse network). We repeated the same set of experiments for *Gappa*, MNP and Deluge on a sparse network, where the distance between two neighbor nodes is 15 feet. We show the completion time, the average active radio time per node, and the energy consumption per node of these three protocols in Figure 6. The results are similar. To reprogram the network with a program of the same size, *Gappa* saves 62-70% completion time and 17-26% energy compared to MNP, and saves 69-75% completion time and 73-81% energy compared to Deluge. In Figure 7, we show the average number of transmissions and receptions per node. *Gappa* has higher transmissions than the other two protocols.

We also compared *Gappa* with MNP and Deluge in the cases where the base station is placed in the center of the network and multiple base stations are used. We found that

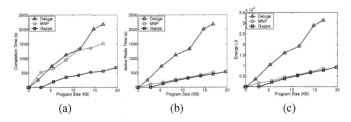

Fig. 6. Inter-node distance: 15 feet. (a) completion time (b) average active radio time per node (c) average energy consumption per node.

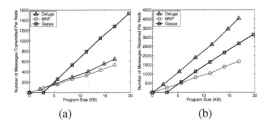

Fig. 7. Inter-node distance: 15 feet. (a) number of messages transmitted per node (b) number of messages received per node.

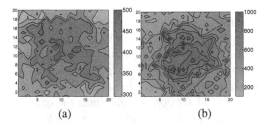

Fig. 8. Active radio time distribution of (a) *Gappa* and (b) MNP. Inter-node distance: 10 feet. Program size: 14KB.

Gappa still performs better than these two protocols in these situations. For reason of space, we refer a reader to [9] for the details of these experiment results.

Additional observations from Experiment setup 1-2. In addition to the average values, we also study the distributions of active radio time and radio communication. We consider the case where the program size is 5 segments (14.08KB, 640 packets). In Figure 8, we compare the active radio time distribution of *Gappa* and MNP (For Deluge, all the nodes's active radio time is the same as completion time). We note that the distribution of nodes' active radio time in *Gappa* is more even (ranges from 300-500s) than the distribution of nodes' active radio time in MNP (ranges from 200-1000s).

In Figure 9(a), we show the distribution of transmissions. Some nodes transmit more than others. These nodes are *distinct* senders, i.e., they are selected as senders many

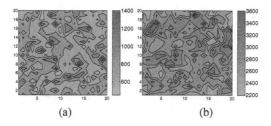

Fig. 9. *Gappa*: message transmissions and receptions. Inter-node distance: 10 feet. (a) transmissions (b) receptions.

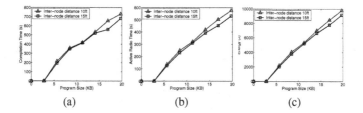

Fig. 10. *Gappa*: at inter-node distance 10 feet and 15 feet. (a) completion time (b) average active radio time per node (c) average energy consumption per node.

times. Note that these distinct senders are randomly distributed. In Figure 9(b), we show the reception distribution. We find that the distribution is even.

In Figure 10, we compare the performance of *Gappa* at different node densities. We note that *Gappa* performs well in both dense networks and sparse networks, although the performance in a sparse network is slightly better.

3.2 Varying Network Sizes

Experiment setup 3. In this section, we fix the inter-node distance to 10 feet and the program size to 5 segments (14.08KB, 640 packets), and conduct simulation on different network sizes (10x10, 15x15, 20x20 grid). In Figure 11, we show that the completion time, average active radio time per node, average energy consumption per node increase slightly when the network size increases. For example, the completion time for reprogramming a 15x15 network with a 14KB program is 512 seconds, while the completion time for reprogramming a 20x20 network with a program of the same size is only 531 seconds; although the number of nodes almost doubles, the completion time only increases 3.6%. This shows that *Gappa* scales well to large networks.

3.3 Varying Number of Seeds

Experiment setup 4. In this section, we study the situation where only a subset of nodes (seeds) have received a segment (randomly picked one) from an UAV initially. Each segment is received by at least one node. We conduct the simulation in a 20x20 network. The inter-node distance is set to 10 feet. The program size is 5 segments (14.08KB, 640 packets). We randomly select 5, 25, 50, 100, and 200 nodes as the seeds.

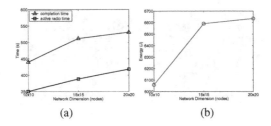

Fig. 11. *Gappa* at different network sizes. Inter-node distance: 10 feet. Program size: 14KB. (a) completion time and average active radio time per node (b) average energy consumption per node.

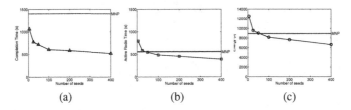

Fig. 12. *Gappa*: varying number of seeds. Inter-node distance: 10 feet. Program size: 14KB. (a) completion time (b) average active radio time per node (c) average energy consumption per node.

The results are shown in Figure 12. For comparison, we also draw the corresponding results of MNP. We note that even if the nodes that have received one segment from the UAV are only 1.25% (each segment with exactly one node), *Gappa* outperforms MNP in completion time. Additionally, from Figure 4 (a), it also outperforms Deluge.

4 Related Work

The existing work on delivering the entire program to all the sensors in the network includes TinyOS single-hop network reprogramming (XNP) [12] and multihop network reprogramming approaches, such as MOAP (Multihop Over-the-Air Programming) [3], Deluge [2], MNP [1], Infuse [4], and Sprinkler [5]. All these approaches assume that one (or a few) sensor has the entire new program initially, and communicate the new program to the remaining sensors in the network. By contrast, *Gappa* is designed for the scenario where some sensor nodes have received one segment on a selected channel from an UAV initially, and communicate with each other to receive the remaining segments.

The only work on multi-channel reprogramming we are aware of is [13], where the authors present preliminary experiment results (based on 25 nodes). In [13], the authors propose an algorithm, *Multi-Channel Deluge*, which divides nodes into *groups* based on node ID or geographically, and assigns a channel to each group. Similar to other existing reprogramming approaches, it also assumes that one or a few source nodes have the complete new program, and disseminate the new program to the entire network. In the algorithm proposed in [13], there are specially marked nodes in group 1 (the default group), which form a connected dominating set, so that all the nodes in the network can directly communicate to at least one node belonging to group 1. By contrast, in *Gappa*, all the nodes are equal. Hence, there is no dependency on special nodes.

5   Conclusion and Future Work

In this paper, we presented *Gappa*, a gossip based multihop reprogramming protocol for sensor networks, that is designed for the scenario where some sensor nodes receive one part of the new program from an UAV on a selected channel at the beginning, and communicate with each other so that all the nodes in the network receive the entire program after reprogramming. By exploiting multi-channel resources and pipelining technique, *Gappa* enables high concurrency on different channels and different locations, hence, propagates data rapidly. To reduce collisions on each channel, *Gappa* uses a multi-channel sender selection algorithm (based on the sender selection algorithm in MNP), which tries to guarantee that at any neighborhood, at most one sender transmits on one channel at a time. Among the competing senders on each channel, the multi-channel sender selection algorithm attempts to select the one whose transmission of the program on that channel is likely to have the most impact. In the case that a node loses the sender selection on one channel, it has the option to compete to transmit on another channel. If all the channels a node can transmit code on are busy, the node stops advertising for a certain amount of time. During that time, it can choose, based on its status, to wait to receive code with its radio on, or to turn off radio to save energy. In this way, *Gappa* reduces the active radio time of sensor nodes, hence, energy consumption, during reprogramming.

We evaluated *Gappa* through simulation on TOSSIM, and compared it with the other two state-of-art reprogramming protocols, MNP and Deluge, which both assume that initially only the base station(s) has the entire program. The simulation results show that under the same network settings, to propagate a program of the same size, *Gappa* saves up to 70% of completion time and up to 42% energy consumption compared to MNP, and saves up to 84% completion time and up to 88% energy consumption compared to Deluge. We also show that *Gappa* adapts well to different network densities and network sizes. Moreover, we note that *Gappa* distributes energy load more evenly. This is expected to help in maintaining a longer network lifetime.

We also considered the case where only a subset of nodes receive a segment of code initially. In the simulation, we study the worst cases where only 1.25% to 50% of nodes have received a part of the code. The simulations results show that even in these situations, *Gappa* still outperforms MNP and Deluge in completion time.

We note that *Gappa* works best in the case where the initial distribution of code segments is random. *Gappa* tries to reduce contention and enable high concurrency by transmitting different segments in their associated channels simultaneously. One possible extension of this work is to associate one segment to multiple channels. In this case, our multi-channel sender selection algorithm needs to be modified, so that a node chooses one channel from a list of available channels (randomly or follow a certain rule), and indicates its preferred channel it intends to transmit on in its advertisement messages. That way, two neighbor nodes that have the same segment can transmit simultaneously.

References

1. S. S. Kulkarni and L. Wang. MNP: Multihop network reprogramming service for sensor networks. *In Proceedings of the 25th International Conference on Distributed Computing Systems (ICDCS)*, pages 7–16, June 2005.
2. J. W. Hui and D. Culler. The dynamic behavior of a data dissemination protocol for network programming at scale. In *Proceedings of the second International Conference on Embedded Networked Sensor Systems (SenSys 2004)*, Baltimore, Maryland, 2004.
3. T. Stathopoulos, J. Heidemann, and D. Estrin. A remote code update mechanism for wireless sensor networks. Technical report, UCLA, 2003.
4. S. S. Kulkarni and M. Arumugam. Infuse: A TDMA based data dissemination protocol for sensor networks. *International Journal on Distributed Sensor Networks (IJDSN)*, March 2006.
5. V. Naik, A. Arora, P. Sinha, and H. Zhang. Sprinkler: A reliable and energy efficient data dissemination service for wireless embedded devices. *In Proceedings of the 26th IEEE Real-Time Systems Symposium*, December 2005.
6. TinyOS: A component-based OS for the networked sensor regime. `http://www.tinyos.net`.
7. P. Levis, N. Lee, M. Welsh, and D. Culler. Tossim: Accurate and scalable simulation of entire tinyos applications. In *Proceedings of the First ACM Conference on Embedded Networked Sensor Systems (SenSys 2003)*, Los Angeles, CA, November 2003.
8. *CC1000 Radio Stack Manual*, 2003. `http://www.tinyos.net/tinyos-1.x/doc/mica2radio/CC1000.html`.
9. L. Wang and S. S. Kulkarni. Gappa: Gossip based multi-channel reprogramming for sensor networks. Technical Report MSU-CSE-06-8, Department of Computer Science and Engineering, Michigan State University, Feburary 2006.
10. V. Shnayder, M. Hempstead, B. Chen, G. Allen, and M. Welsh. Simulating the power consumption of large-scale sensor network applications. *In Proceedings of ACM International Conference on Embedded Networked Sensor Systems (SenSys)*, November 2004.
11. A. Mainwaring, J. Polastre, R. Szewczyk, D. Culler, and J. Anderson. Wireless sensor networks for habitat monitoring. In *Proceedings of ACM International Workshop on Wireless Sensor Networks and Applications (WSNA'02)*, Atlanta, GA, September 2002.
12. Crossbow Technology, Inc. *Mote In-Network Programming User Reference Version 20030315*, 2003. http://webs.cs.berkeley.edu/tos/tinyos-1.x/doc/Xnp.pdf.
13. W. Xiao and D. Starobinski. Poster abstract: Exploiting multi-channel diversity to speed up over-the-air programming of wireless sensor networks. *In Proceedings of the Third ACM Conference on Embedded Networked Sensor Systems (SenSys) (Poster Session)*, November 2005.

The Virtual Pheromone Communication Primitive

Leo Szumel and John D. Owens

University of California at Davis,
Department of Electrical and Computer Engineering,
One Shields Ave, Davis, CA 95616, USA
{lpszumel, jowens}@ucdavis.edu

Abstract. We propose a generic communication primitive designed for sensor networks. Our primitive hides details of network communication while retaining sufficient programmer control over the communication behavior of an application; it is designed to ease the burden of writing application-specific communication protocols for efficient, long-lived, fault-tolerant, and scalable applications. While classical network communication methods expect high-reliability links, our primitive works well in highly unreliable environments without needing to detect and prune unreliable links. Our primitive resembles the chemical markers used by many biological systems to solve distributed problems (pheromones). We develop and analyze the performance of an implementation of this primitive called Virtual Pheromone (VP). We demonstrate that VP can attain performance comparable to classical methods for applications such as sleep scheduling, routing, flooding, and cluster formation.

1 Introduction

Most wireless sensor network (WSN) and ad-hoc networking applications demand *efficiency, long life, fault-tolerance*, and *scalability*. We refer to these as ELFS applications. The goal of this paper is to demonstrate that Virtual Pheromone (VP) is an effective tool for building ELFS applications.

It is commonly accepted that cross-layer design is necessary in order to achieve the levels of efficiency desired in most WSN and energy-constrained ad-hoc network applications [1, 2]. Eschewing the classical network layered abstraction model (OSI) enables advancements in energy efficiency. This *energy* efficiency comes at the cost of *programmer* efficiency, because integration and debugging of these applications can be very complex. It is desirable to have mechanisms that balance the utility of abstraction against the flexibility of cross-layer design. The sensor network field has entered an era of consolidation wherein point solutions are being generalized to create low-level abstractions that are useful across many applications; TinyOS 2.0 and UCLA's Tenet are examples [3, 4]. In short, some energy efficiency must be traded for generality and ease of programming if sensor networks are to become ubiquitous.

1.1 Motivation

Sensor networks bring computer and network technology in closer entanglement with the natural world than ever before. Assumptions made in classical computing no longer

P. Gibbons et al. (Eds.): DCOSS 2006, LNCS 4026, pp. 135–149, 2006.
© Springer-Verlag Berlin Heidelberg 2006

hold: communications are unreliable, nodes are unreliable, a deterministic mapping of the system state may be unattainable, and energy is a finite resource. The combination of a lossy and random environment means that the behavior of a sensor network cannot be perfectly determined; rather, we must be satisfied with specific behavioral qualities or bounds on expected performance.

Because ELFS are bound so tightly with the physical world, it stands to reason that natural communication mechanisms may provide insight into the design of artificial communication mechanisms appropriate for that environment. Many species use pheromones (scent signals) to communicate information and organize behavior in complex and challenging environments. For instance, ants successfully forage for food and build nests in spite of continually changing physical parameters—paths are blocked, hazards are presented, food sources come and go.

Natural systems are dynamic in the relationship between actors (or agents) and the environment in which they reside. In a data-collection wireless sensor network, the sources of dynamism are radio interaction and node failure. When a node is asleep, it cannot participate in multi-hop communication, and when packets are sent, they may collide with other transmitters' packets. In more complex scenarios, e.g. mobile or sensor-actuator networks, dynamism is increased along more parameters. We argue that pheromone-inspired communications can be useful in the simple case and are very desirable in the complex case.

1.2 Contributions

Because communication cost is expected to dominate sensor network energy costs, we have designed a communication primitive that is designed to approach the efficiency of applications that are specifically tailored to their tasks. Our goal is to trade a minimal amount of energy efficiency for increased programmer efficiency and code reuse. Moreover, the nature of our communication primitive leads to efficient, scalable application design, because every transmission is recognized as a broadcast. Efficiency is chiefly provided by a single-layer abstraction: simulated diffusion of virtual pheromone signals. This abstraction benefits from the presence of unpredictable links, rather than requiring that they are pruned out. Finally, by using a common primitive for all (or most) communication, a powerful optimization point is created such that enhancements to the primitive can benefit a large set of applications.

2 Overview

In the biological sense, a pheromone is a chemical marker. It can be deposited by an organism and detected by that and/or other organisms, e.g. for the direction of a male moth to a female [5]. A pheromone dissipates via an evaporative process; as a result the strength of the pheromone decays with increasing distance from the source in the spatial and temporal axes. Spatial decay maintains locality of communication; temporal decay reduces system complexity by ensuring that old information is purged from the system.

2.1   Interesting Properties

Why is a spatio-temporally decaying information process interesting in the context of sensor networks? Let us examine the properties of this information process that are in common with many sensor network applications. **Spatially-limited sharing** of information: sensor networks can only scale if internode information sharing is limited by some process. **Encoding of distance** to source: hop counts, for instance, are commonly used as a cost metric in routing algorithms. **Encoding of time** since deposition: the relevance of information is crucial in allowing nodes to collaborate with each other; there is a significant difference between a sensor reading from one minute ago and a reading from one month ago. **Superposition** of like pheromones: akin to aggregation by sum; allows reinforcement. **Implicit gradient** leading to source: gradients enable *efficient* and *scalable* routing algorithms because the routing information is stored in a distributed fashion.

The net result of these properties is that all information exchange is via constrained broadcast through a shared medium. This maps very well to the behavior of a radio communication system.

2.2   The VP Communication Primitive

VP also exhibits the five aforementioned properties, each of which serves a purpose in the design of an ELFS application. Efficiency is driven by low-cost transmission of information; VP requires no ACK signals and uses an efficient flooding technique (Section 3.1). Fault-tolerance requires a level of redundancy and robustness to errors; VP's inherent redundancy provides a tradeoff of reliability against cost, and because no explicit point-to-point communication is used, information transfer is highly tolerant of node failure (as long as the network remains unpartitioned). Application scalability requires highly localized communication and efficient coordination amongst neighboring nodes; VP pheromones propagate proportional to the strength of the initial deposit, which creates a user-selectable bound on distance. The encoding of distance and time in the pheromone strength, combined with superposition of like pheromones, allows neighboring nodes to coordinate without requiring expensive point-to-point protocols that are sensitive to faults.

2.3   Related work

The topic of communication methods in wireless networks is broad; we reference representative works from two important categories: localized communication/control primitives and applications using pheromone-like concepts.

In directed diffusion [6], information sinks (consumers) publish *interests* which are propagated through parts of the network; information sources (producers) publish named data objects which are routed along the interest gradient. Geolocation is assumed so that interests can be distributed locally. Directed diffusion is an application facilitating data transfer between sinks and sources and works best when a flow of information will pass from a source to one or more sinks. In contrast, VP is a lower-level primitive intended for many communication tasks (including diffusion-like routing; see Section 4.2).

RUGGED [7] is a routing protocol that utilizes *data gradients*, or "fingerprints," rather than the interest gradients created by directed diffusion. Fingerprints are disseminated by an environmental processes—not by the sensor network—and this creates a possible efficiency benefit. Simulated annealing techniques are used to overcome local minima or maxima in the sensor field, including regions in which the sensed level is zero. Fingerprint routing is targeted specifically at data-collection applications in which natural gradients exist in the phenomena to be measured.

Payton et al. explore using "virtual pheromones" [8] for robotic control. The pheromones are sent via infrared or other line-of-sight communication method; gradient descent necessarily indicates an unobstructed path that the robots may use. Characterization of the infrared source and its spatial decay is used as a distance estimator. Parunak et al. use a pheromone-inspired memory model [9] to assist in the coordination of distributed decision-making systems. Brooks et al. use biologically-, chemically-, and physically-inspired techniques (including one based on pheromones) in sensor network adaptation techniques [10]. They find these techniques to be very robust to errors while attaining satisfactory power consumption.

2.4 Qualitative Expectations

It is instructive to consider the expected behavior of VP, as compared to the classical primitive of point-to-point transmission, before delving into technical analysis. We expect VP to work well in dense networks since the signal can propagate many hops. We expect VP to perform less well when used, naively, to implement classical protocols that utilize handshaking, acknowledgements, or other point-to-point information exchanges. VP should be highly tolerant of node failure and spurious communication because it benefits from the redundancy of broadcast and can utilize unreliable links.

Furthermore, due to the large time constants involved in pheromone decay, we expect VP to be most useful in latency-insensitive applications. While fields can propagate quickly, the rate of accomplishing a specific action-reaction pair using pheromones is not likely to exceed a classical communication approach. For instance, any application that leverages the superposition principle of pheromone signaling will take some time to reach a steady state response.

We begin by discussing the parameters of a pheromone communication primitive, and the details our specific implementation, in Section 3. The metrics relevant to ELFS are presented in Section 4 and then applied to four application scenarios and analyzed in Section 4.2. Section 5 describes the future direction of this research, and Section 6 concludes the paper.

3 Design

We desire to implement a mechanism displaying the properties listed in Section 2.1: encoding of space and time, spatially limited flooding, superposition, and gradient generation. Energy efficiency constrains our implementation options. We analyze the design problem as the following set of sub-problems:

Programmer API. How should a program deposit and detect pheromones? What information should be provided, and in what form? Our goal is to minimize the amount of code required to handle communications, yet provide enough flexibility to make the primitive useful.

We believe deposition should encode, at a minimum, the following fields: TYPE, STRENGTH, and SOURCE. PAYLOAD (arbitrary data) should be an optional parameter. Reading a pheromone should provide at least the first two fields: TYPE and STRENGTH. SOURCE may be required for some applications. The arbitrary payload field could be used as a way to encode complex data, such as a non-integer data type, or an uniquifier.

Spatial Dissipation Model. The desired behavior is to have the strength of the field decay with increasing distance from the pheromone source. This encodes physical distance from the sender and limits the range over which data will propagate. Candidate dependent variables for this decay include hop count, RSSI, or geolocation.

Temporal Dissipation Model. The desired behavior is to have the strength of the field decay with increasing time since the deposition of the pheromone. Exponential decay is ideal because it favors recent information while allowing old information to persist at a low level. Half-life may be global, indepdenent for each pheromone, or chosen from a small set (e.g., {fast-decay, slow-decay}), depending on application needs.

Pheromone Encoding & Transmission Strategy. The transmission strategy of the pheromone information can have a large impact on the energy efficiency of the operation. If maximum propagation speed is desired, a packet must be sent for each deposition. If the constraint of propagation speed is relaxed, packet overhead may be amortized over several pheromone depositions.

3.1   VP: A Design Implementation

VP encodes TYPE, STRENGTH, SOURCE, and PAYLOAD in a table and provides two functions to store (and implicitly transmit) and retrieve pheromone signals: DEPOSIT and SMELL. Utility functions to support common uses of pheromones will reduce program size and be useful across applications. To facilitate the applications in Section 4.2, we have implemented a function to forward a packet along a pheromone gradient (FWDGRADIENT), and a function to return a list of distinct sources for a pheromone type (SMELLDISTINCT).

DEPOSIT(TYPE, STRENGTH, PAYLOAD): Deposit a pheromone identified by TYPE with strength equal to STRENGTH; PAYLOAD is optional. The subsystem decides if the information should be propagated to neighboring nodes according to the rules in Section 3.1.

SMELL(TYPE): Return the net strength of the pheromone(s) matching TYPE.

SMELLDISTINCT(TYPE): Return a list of each distinct pheromone detectable at this node. This bypasses superposition and includes the payload of each pheromone (if present).

FWDGRADIENT(PACKET, TYPE, DIRECTION): Forward a packet according to the gradient of pheromone TYPE; DIRECTION may be *uphill*, *downhill*, or *level*.

Table 1. A simple data structure for storing pheromones. The strength level associated with type a from source i is denoted l_{ai}. For example, $\text{SMELL}(a) = l_{ai} + l_{aj}$ and $\text{SMELL}(b) = l_{bk}$. Payload is labeled equivalently to strength.

type	source	strength	payload
t_a	s_i	l_{ai}	p_{ai}
t_a	s_j	l_{aj}	p_{aj}
t_b	s_k	l_{bk}	p_{bk}

Storage and Retrieval of Pheromone State. The internal data structure is a $4 \times N$ table (Table 1). To support superposition, each unique type/source pair may have its own entry in the table, but when the pheromone is read all rows with a common type are summed together. Payload is suppressed when using SMELL but may be read using SMELLDISTINCT.

Deposition and Propagation of Pheromones. When a pheromone is deposited with strength s, the table is first checked for a hit on (TYPE,SOURCE). There are three possible cases:

1. If there is no hit, the data is stored and then scheduled for transmission as a pheromone update *with strength* $(s-1)$.
2. If the hit has a strength less than s, the "no hit" action (1) is taken.
3. If the hit has a strength $\geq s$, the deposition is ignored.

This algorithm sets up a cone-shaped field *prior to to initiation of the decay process* (Section 3.1). The initial field has a slope of one pheromone unit per hop; as a result, distance is derived from radio hops.

Broadcast Suppression Technique. Pheromone updates attempt to limit redundant transmissions. A depositing node that wants to send an update moves through the phases listed in Table 2. While *observing*, the node snoops for and accumulates the count of concurrent depositions by neighboring nodes. In the *transmitting* phase, the snoop count is compared to a threshold, η. If this threshold is not exceeded, the node transmits the pheromone; in either case, it calculates the expected pheromone distribution time and waits for the propagation to complete. This delay equals the pheromone strength at this node, s_i, times the expected observation delay at each hop (including packet transmit time); this algorithm forms a schedule for initiation of the distributed decay process. The *decaying* phase lasts until the pheromone reaches a minimum threshold, ε (Section 3.1).

Time-Decay of Pheromones. All nodes concurrently run a decay process on their pheromone tables. The decay process is a discretized approximation of continuous exponential decay, updated every τ seconds: $s(t + \tau) = \alpha s(t)$. The process terminates when $s(t) < \varepsilon$, where α is constant, τ is the update interval, ε is the termination threshold, and $s(0)$ is the initial strength at this node as defined in Section 3.1.

Table 2. A pheromone deposition is a distributed process including three phases at every node i: observation (for suppression), transmission (or suppression) and a pause, and then decay. t_{tx} is the expected transmit time, s_i is pheromone strength at node i, and τ is a constant.

phase	duration
observing	$t_{obs} = [0, \tau]$
transmitting (or suppressing)	$(t_{obs} + t_{tx})s_i$
decaying	until pheromone is depleted

Table 3. memory requirements to store a pheromone (bytes). The *next decay time* is not transmitted. *System* variables are used to propagate and decay the signal, and include *user* variables, which are exposed by the API. The total footprint of each pheromone is between 7 and 14 bytes.

space	parameter	bytes
system	source ID	2–4
·	spatial metric	1–2
·	next decay time	1–2
· user	type	2–4
· ·	strength	1–2

VP maintains an event list large enough to keep one "next decay time" for each pheromone. For large scale systems with many pheromones, it would be more appropriate to have one global timer that decays all pheromones; this may require a smaller τ (Section 3.1).

Constants. We use the following constants in our implementation of VP: half-life=10 s, $\tau = 100$ ms, $\varepsilon = 0.1$, $t_{tx} = 2$ ms, and $\eta = 2$. t_{tx} is derived from a 250 kbps radio and 40-byte packet (1.28 ms to transmit, plus processing time). η is chosen to optimize efficiency (we tried values 1,2,3,4)—it must be tuned to the RF model (discussion in Section 5).

Memory Requirements. We divide the pheromone information into two categories: *user* and *system* (Table 3). Parameter precision may be adjusted to application requirements. Each pheromone requires 7–14 bytes of information (before compression) to transmit; a packet payload of 32 bytes can contain at least 2-4 pheromones.

4 Metrics and Evaluation of VP

Because VP is targeted at ELFS applications, we choose one metric for each ELFS quality:

Efficiency: the communication cost, in *packets sent to accomplish a task*. All else being equal, a task that requires less communication is more efficient.

Long-livedness: the lifetime of the network, in *seconds-until-dysfunction*. This is differentiated from efficiency because application requirements, such as coverage, depend on both node death distribution and the fault-tolerance of the algorithm. It is reasonable to expect, however, that an efficient algorithm will also be long-lived.

Fault-tolerance: the lifetime, as *long-livedness* is observed with these varying system parameters: *network density, random node failure probability, and partial network occlusion*.

Scalability: the *long-livedness* is observed when the network is scaled in *node placement density*, in nodes per unit area.

4.1   Experimental Methodology

Showcasing all possible applications of our primitive is beyond the scope of this paper; rather, we aim to discuss several applications and tasks that can be accomplished using our primitive and to provide evidence of satisfactory performance.

Our experiments are performed using a custom simulator called AHLPS. We also verify functionality with physical deployment on 25 Telos motes using a less complete implementation of VP. Whereas implementation on motes proves the functionality of the primitive, only simulation allows us to explore VP in the environment it was designed for: large scale networks.

We previously developed the Agent High-Level Pythonic Simulator (AHLPS) to allow simulation of agent behavior in large sensor networks [11]. A pheromone primitive was added to AHLPS to support the research in this paper. AHLPS uses the TOSSIM empirical radio link model [12] to simulate link quality[1] but does not simulate a PHY layer or complex MAC layer. As a result, radio contention is not modeled; this is acceptable in our simulations because the mean communication rate is low (generally less than a packet per second per node). AHLPS allows us to investigate, at a high level, the behavior of an agent-based program. Further verification is then performed in TOSSIM and/or on physical nodes using our agent framework [13].

4.2   Case Studies

We will now use results on the behavior of representative tasks and applications to verify that VP supports the requirements of ELFS: efficiency, long-livedness, fault-tolerance, and scalability. Any algorithm implemented using VP is referred to as "Pheromone."

One way to measure efficiency is to take the ratio of a cost (generally energy) and a benefit (generally network lifetime). Unfortunately, results of this measure are specific to each application. We desire to separate out an *indicator* of efficiency that will imply efficiency performance for many applications. Because most sensor networks communicate some shared information, we analyze the efficiency of distributing information using three methods: flooding, epidemic routing, and VP.

Fault-tolerance is also very application-specific because of the specificity of the term "fault." We attempt to pick two common fault scenarios that many networks expect to incur: a long-period disruption to a large section of the network, and intermittent

---

[1] While the TOSSIM model is based on empirical data ranging from 0 to 40 meters, we have scaled this data so that 1 distance unit under AHLPS is equivalent to 40 meters under TOSSIM.

dropouts of specific nodes. Because it is impossible to generalize all applications' susceptibility to these specific faults, we choose a basic measure: the ability to route information from one part of the network to another. This embodies two concepts: inter-node collaboration, and network connectivity; most sensor network applications require both.

Efficiency, fault-tolerance, and scalability are requisites for long-livedness, and analyzed separately. Our aim is to show that VP enables longevity extension in a way applicable to many applications. We chose to examine cover-constrained sleep scheduling because it can be used in many applications to extend lifetime. While not a proof of generality, such an example provides evidence for our argument.

Scalability is extremely important in many sensor network designs. Unfortunately, few truly scalable networks have been deployed in the field. We are developing a multi-hop clustering algorithm that we believe will be crucial in the scalable operation of very large networks, allowing collaboration and organization of large (but constrained) subsets of nodes. Multi-hop clustering has been mostly a footnote in the literature, although it does resemble the "Multiple Sink Network Design Problem" [14] posed by Oyman and Ersoy. Our goal is to show that our multi-hop clustering task scales perfectly using VP.

Efficiency. Flooding is the most basic form of dissemination. Epidemic algorithms, such as Trickle [15], perform efficiency optimizations using suppression. We compare the use of VP to disseminate a unit of information to a network of 1000 nodes using naive Flood and Trickle. Comparison is on the basis of packets sent, per node, per unit of information, with the independent variable of network density (Figure 1). The simulation is run for 20 seconds but statistics are collected at the earliest time for which *cover* (fraction of nodes having received the information unit) is 95% or higher.

Flood has a constant cost because all nodes participate in every flood event. Trickle benefits from increasing network density because more nodes can be suppressed; however, there is an unavoidable baseline cost in order to run the algorithm's periodic probes. Increasing the maximum probing interval would decrease this cost but also decrease the responsiveness of the system. Because each run is begun from scratch, we also plot "Trickle–adjusted," in which every node is given one "free" packet to perform the initial round of probes.

There are two reasons that Pheromone performs better than Trickle in this test. First, there is the aforementioned base cost. Second, Trickle's probe-response protocol means that, in the limit of extreme density, distributing a unit of information would require at least three transmissions: broadcast a new version, broadcast a suppressing data request, broadcast a data response. Pheromone would require only a minimum of two transmissions: broadcast of pheromone, and one suppressing re-broadcast.

The flooding coverage chart indicates that none of the algorithms perform satisfactorily below a density of 10. It is important to note that Trickle *guarantees* 100% cover in a connected network while Flood and Pheromone, with their probabalistic approach, attain 95-100% coverage. In addition, the lack of contention in the AHLPS radio model gives advantage to Flood and Pheromone's latency performance. Pheromone and Trickle serve different purposes, but it is clear that Pheromone is an efficient dissemination mechanism.

Memory and computational resources are not modeled in AHLPS. We note that all three algorithms are computationally simple, with Pheromone being the most complex

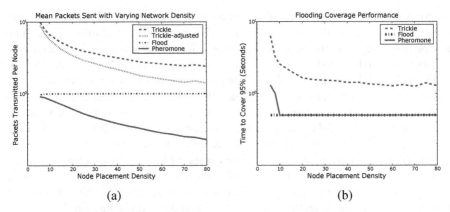

Fig. 1. Performance of Trickle, Flood, and Pheromone in information dissemination. Both Trickle and Pheromone benefit from increased network density. Pheromone is bounded above by both Trickle and Flood in terms of efficiency, while matching Flood's rapid speed.

because our current implementation uses a floating point multiplication (Section 3.1). The memory requirement for Pheromone is slightly greater due to the SOURCE parameter. (STRENGTH in Pheromone is balanced by HOPCOUNT in Flood and VERSION in Trickle.)

Long-Livedness. A well-referenced and effective sleep scheduling algorithm is PEAS [16]. We implemented PEAS under AHLPS as well as our own scheduler that uses VP. We use a metric from the PEAS paper: *time that 90% 3-cover is maintained*. *k-cover* is defined as the fraction of network area observed by at least k nodes. We use the same operating constants as Ye et al.: idle power 12 mW, sleep power 30 μW, transmit energy 600 μJ per packet, and receive energy 120 μJ per packet.

Pheromone has a single tunable parameter: THRESH. A single pheromone is sent as a beacon by all awakened nodes; if the perceived level at a given node is greater than THRESH, the node will go to sleep with probability proportional to the difference. PEAS has two parameters, λ, the mean rate at which nodes will wake up and probe for neighbors, and R_p, the probing range. We set $\lambda = 1$ Hz and $R_p = 1.0$ (the maximum radio range).

Sensing range is set to 2.0, or 2 times the maximum radio range. Since most good links are at a distance ≤ 0.2, the sensing range is significantly larger than the probing range. 3-cover failure for three or more seconds is considered a failure. We set THRESH = 1.0 to match PEAS' behavior (a node sleeps if any probe responses are heard). The fundamental difference between Pheromone and PEAS is that the former is probing (broadcast announcements) while the latter is polling (broadcast request, unicast response).

Figure 2 shows that While PEAS is very effective (a), it requires a high overhead in transmissions (b) because each broadcast probe is followed by multiple unicast responses. PEAS offers an additional benefit that Pheromone does not: the pursuit of λ probing rate. (Effectively, "λ" is fixed in Pheromone.) Pheromone offers performance comparable to PEAS at a much lower transmission overhead, indicating it would perform very well in applications with a higher transmit-to-idle cost ratio. Perhaps the biggest benefit of Pheromone is that it is easily integrated into a cross-layer algorithm design.

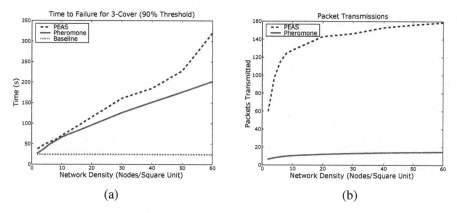

(a)   (b)

Fig. 2. (a) PEAS is highly effective at extending network lifetime. Pheromone is nearly as effective for lower densities but loses ground as density is scaled. (b) PEAS has a radio overhead nearly 10 times greater than Pheromone.

Fault-Tolerance. We wish to examine fault models sufficiently simple to avoid loss of generality in the results. We examine two models: *partial network occlusion* and *random network dropouts*. During an occlusion, a large portion of the network becomes unreachable (e.g. due to a signal jammer) between $t = [50, 150]$ s. During a dropout, nodes may become unreachable for a brief period (e.g., fast fading); the dropout rate is λ and the dropout interval is exponentially distributed with mean 1 s. Network size is 1000 nodes.

These fault models are applied to a partial implementation of DSR [17]. DSR is an on-demand point-to-point routing algorithm designed for ad-hoc networks. Because the DSR specification is complex, we implemented in AHLPS only the features involved in robust delivery from a single source to a single sink. While the DSR specification allows for asymmetric data/ACK paths, it does not specify a mechanism, so our implementation uses only symmetric paths.

We compare DSR to gradient ascent routing on a pheromone field. Packets are transmitted once per second and in the case of Pheromone, a pheromone is deposited at the sink once every 10 seconds. There are no tunable parameters for Pheromone. A significant difference is that Pheromone will follow multiple simultaneous paths when routing on a gradient (all uphill paths are followed), whereas DSR will discover parallel paths and then use just one at a time. Analysis of packet overhead (not presented here) indicates that Pheromone has a lower packet overhead for the data presented here.

Because AHLPS uses an empirical radio model, nodes have many neighbors at the fringe of connectivity and with correspondingly high loss rates. DSR does not perform well under this regime because it discovers, and attempts to use, faulty paths; it must explore many of these before finding a quality path. To alleviate this problem, we implemented an omniscient, zero-overhead, link-quality estimator (LQE) that permits only high-quality links. High quality is defined as an expected round-trip loss rate of at most 1%. Pheromone is permitted to run under both the empirical and LQE modes for comparison, but performs better without LQE because of the increased redundancy.

We see in Figure 3 that DSR outperforms Pheromone with the LQE feature (a) but is inferior without it. Random dropouts (b) affect DSR more than Pheromone because

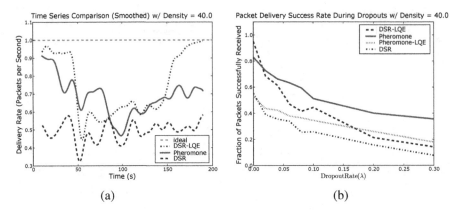

(a)   (b)

Fig. 3. DSR versus Pheromone gradient routing (a) during an occlusion event and (b) with varying dropout rate (λ). $\lambda = 0$ corresponds to no dropouts. Density is 40.0 nodes per unit area in both cases and LQE indicates an omniscient, zero-overhead, link-quality estimator. Pheromone does not require LQE, an important efficiency benefit, and is far more tolerant to random node dropouts. DSR, on the other hand, performs much better when LQE is available.

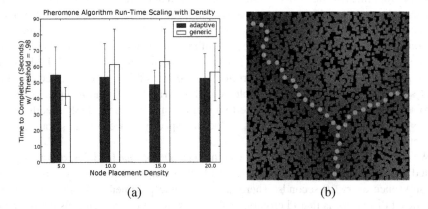

(a)   (b)

Fig. 4. (a) Clustering performance of two pheromone algorithms as a network of 4000 nodes is scaled in density. Time to complete cluster formation is independent of density. (b) A snapshot of cluster formation at $t = 19.0$ s using the Adaptive algorithm; lightness indicates pheromone strength (green dots added to emphasize cell edges).

DSR must explore new routes more often as λ is increased. Performance without LQE is important because in the real world LQE is not free, especially in mobile or time-fading environments.

Scalability. Our aim in this section is to demonstrate that a scalable application can be constructed from our primitive. We chose multi-hop clustering and implemented two versions using only VP for communication: *generic* and *adaptive*. Generic is based loosely on LEACH [18], in which a cluster head probability, p_{ch}, controls cluster head

Algorithm 1. The behavior of a cluster member. All nodes initially take this behavior and can change to cluster head (Algorithm 2) if no cluster head pheromones are present.

```
 1: scents ⇐ SmellDistinct('clustering_pheromone')
 2: if scents is empty then
 3:     WAIT for snoop interval
 4:     behavior ⇐ ClusterHead
 5: else
 6:     if two strongest scents are of equal strength then
 7:         my_membership ⇐ EdgeMember
 8:     else
 9:         my_membership ⇐ CellMember
10:     end if
11: end if
```

Algorithm 2. The behavior of a cluster head. Note that behavior can change to cluster member (Algorithm 1) if this cluster head loses an instant-runoff based on random pheromone IDs (Line 5). The runoff avoids mutual simultaneous annihilation of cluster heads.

```
 1: my_strength ⇐ 10 {10 hops}
 2: my_ID ⇐ random integer
 3: scents ⇐ SmellDistinct('clustering_pheromone')
 4: if detect one or more cluster heads in scents then
 5:     if my scent random ID < strongest scent's random ID then
 6:         behavior = ClusterMember
 7:     end if
 8: else
 9:     Deposit(type='clustering-pheromone', strength=my_strength, payload=my_ID)
10: end if
```

formation. Adaptive effectively adapts p_{ch} distributedly at each node. Both algorithms use a pheromone deposit to form clusters—the cluster head deposits a pheromone and nodes choose cluster membership based on observed pheromone strengths.

We consider the algorithm complete when the *mean cluster size* has stabilized within 98% of its final value. The scalability of both algorithms is apparent in Figure 4; density does not affect the time to completion. Both algorithms are scalable because of pheromones' suppression mechanism—in a network twice as dense, twice as many nodes will be suppressed during pheromone deposition.

Adaptive and Generic both stabilize at about $t = 40$ s and have similar mean transmission rates. Adaptive has a higher *peak* transmission rate, which may be a disadvantage in some applications. Given the similar performance of the two algorithms, we feel that Adaptive is superior because tuning of p_{ch} is not required. The Adaptive algorithm is presented in code listings 1.

5 Future Work

In future work it would be beneficial to examine the constants used in pheromone distribution and decay. Optimal η, the suppression constant, is likely to depend on the specific MAC and PHY. Higher η means a greater packet overhead and potential for collisions. Setting η too low will result in a lack of redundancy. In our experiments, one global pheromone decay half-life was sufficient. We would like to explore the benefit of allowing different half-life settings. Finally, local repair of the pheromone field (e.g. when a node is added or wakes up from sleep), as in that proposed by Han et al. [19], would drastically improve response time at a small overhead cost.

The VP API is designed around superposition but applications which utilize PAYLOAD need to read unique pheromones. Currently we uniquify the fields using PAYLOAD, but this can also be done by encoding a unique number into TYPE. The algorithms in our case studies could be further simplified by offering common operations, e.g.: "give me the n strongest pheromones matching this type", or "give me the payload of the strongest pheromone of this type." The design and selection of these operations is the topic of future work.

6 Conclusion

We have shown that VP addresses the needs of ELFS applications: efficiency, long-life, fault-tolerance, and scalability. We compare our Pheromone algorithms to existing, well-known point solutions for the following problems: dissemination, sleep scheduling, and routing. In addition, a novel scalable clustering application is examined. In every case, algorithms using VP attain comparable performance because they leverage the abundance of lossy links in the RF environment rather than trying to avoid them. Simultaneously, VP algorithms are simple to program and rely on a spartan API, which creates a powerful common optimization point.

The key contribution of VP is that it builds a simple conceptual interface to the radio that is consistent across a broad range of system parameters (e.g. density, node dropout rate). Applications built using VP can be fault-tolerant without having to re-implement custom error control mechanisms or having to rely on link quality estimators.

Our experience using VP is that applications must think about every communication as a broadcast, and this encourages the programmer to utilize that fact. The programs we developed are small (10–30 lines of Python code) and thus easier to understand. The use of parameter "tuning" can make optimization difficult; in our opinion it is imperative to design *adaptive* algorithms such as the clustering presented in Section 4.2.

References

1. Levis, P., Madden, S., Gay, D., Polastre, J., Szewczyk, R., Woo, A., Brewer, E., Culler, D.: The emergence of networking abstractions and techniques in TinyOS. In: Proceedings of the First USENIX/ACM Symposium on Networked Systems Design and Implementation. (2004) 1–14
2. Goldsmith, A., Wicker, S.B.: Design challenges for energy-constrained ad hoc wireless networks. IEEE wireless communications 9(4) (2002) 8–27

3. Levis, P., Gay, D., Handziski, V., Hauer, J.H., Greenstein, B., Turon, M., Hui, J., Klues, K., Sharp, C., Szewczyk, R., Polastre, J., Buonadonna, P., L.Nachman, G.Tolle, D.Culler, A.Wolisz: T2: A second generation OS for embedded sensor networks. Technical Report TKN-05-007, Telecommunication Networks Group, Technische Universität Berlin (2005)
4. Govindan, R., Kohler, E., Estrin, D., Bian, F., Chintalapudi, K., Gnawali, O., Rangwala, S., Gummadi, R., Stathopoulos, T.: Tenet: An architecture for tiered embedded networks. Technical Report 56, Center for Embedded Networked Sensing (2005)
5. Kuwana, Y., Shimoyama, I., Sayama, Y., Miura, H.: Synthesis of pheromone-oriented emergent behavior of a silkworm moth. In: Proceedings of the International Conference on Intelligent Robots and Systems. Volume 3. (1996) 1722–1729
6. Estrin, D., Govindan, R., Heidemann, J., Kumar, S.: Next century challenges: Scalable coordination in sensor networks. In: Proceedings of the ACM/IEEE International Conference on Mobile Computing and Networking, Seattle, Washington, USA, ACM (1999) 263–270
7. Faruque, J., Helmy, A.: RUGGED: routing on fingerprint gradients in sensor networks. In: The IEEE/ACS International Conference on Pervasive Services. (2004) 179–188
8. Payton, D., Daily, M., Estowski, R., Howard, M., Lee, C.: Pheromone robotics. In: Autonomous Robots. (2001) 319–324
9. Parunak, H.V.D., Brueckner, S.A., Matthews, R., Sauter, J.: Pheromone learning for self-organizing agents. IEEE Transactions on Systems, Man, and Cybernetics—Part A: Systems and Humans **35**(3) (2005)
10. Brooks, R., Pirretti, M., Zhu, M., Iyengar, S.: Distributed adaptation methods for wireless sensor networks. In: IEEE Global Telecommunications Conference. Volume 5. (2003) 2967–2971
11. Szumel, L.: The agent high-level pythonic simulator. (2006) Work in progress.
12. Levis, P., Lee, N., Welsh, M., Culler, D.: TOSSIM: Accurate and scalable simulation of entire TinyOS applications. In: Proceedings of the First ACM Conference on Embedded Networked Sensor Systems. (2003) 126–137
13. Szumel, L., LeBrun, J., Owens, J.D.: Towards a mobile agent framework for sensor networks. In: Second IEEE Workshop on Embedded Networked Sensors. (2005) 79–87
14. Oyman, E.I., Ersoy, C.: Multiple sink network design problem in large scale wireless sensor networks. In: Proceedings of the International Conference on Communications, Paris, France (2004)
15. Levis, P., Patel, N., Culler, D., Shenker, S.: Trickle: A self-regulating algorithm for code propagation and maintenance in wireless sensor networks. In: Proceedings of First Symposium on Networked Systems Design and Implementation, San Francisco, CA (2004)
16. Ye, F., Zhong, G., Lu, S., Zhang, L.: PEAS: A robust energy conserving protocol for long-lived sensor networks. In: 3rd International Conference on Distributed Computing Systems. (2003)
17. Johnson, D.B., Maltz, D.A.: Dynamic source routing in ad hoc wireless networks. In Imielinski, Korth, eds.: Mobile Computing. Volume 353. Kluwer Academic Publishers (1996)
18. Heinzelman, W., Chandrakasan, A., Balakrishnan, H.: Energy-efficient communication protocols for wireless microsensor networks. In: International Conference on System Sciences, Maui, HI (2000) 3005–3014
19. Han, K.H., Ko, Y.B., Kim, J.H.: A novel gradient approach for efficient data dissemination in wireless sensor networks. In: Proceedings of the International Conference on Vehicular Technology (VTC). (2004)

Logical Neighborhoods:
A Programming Abstraction
for Wireless Sensor Networks

Luca Mottola and Gian Pietro Picco

Dipartimento di Elettronica e Informazione, Politecnico di Milano, Italy
{mottola, picco}@elet.polimi.it

Abstract. Wireless sensor networks (WSNs) typically exploit a single base station for collecting data and coordinating activities. However, decentralized architectures are rapidly emerging, as witnessed by wireless sensor and actuator networks (WSANs), and in general by solutions involving multiple data sinks, heterogeneous nodes, and in-network coordination. These settings demand new programming abstractions to tame complexity without sacrificing efficiency. In this work we introduce the notion of *logical neighborhood*, which replaces the physical neighborhood provided by wireless broadcast with a higher-level, application-defined notion of proximity. The span of a logical neighborhood is specified declaratively based on the characteristics of nodes, along with requirements about communication costs. This paper presents the SPIDEY programming language for defining logical neighborhoods, and a routing strategy that efficiently supports the communication enabled by its programming constructs.

1 Introduction

Wireless sensor networks (WSNs) typically exploit a single base station for collecting data and coordinating activities. Habitat monitoring [1], a common example application, is paradigmatic in this respect, featuring a *single* base station collecting data from a high number of *homogeneous* nodes. Nevertheless, decentralized architectures are rapidly emerging where multiple base stations are employed, different applications run on the same hardware, or heterogeneous nodes are deployed. These approaches find their extreme realization in wireless sensor and actor networks (WSANs) [2], where nodes not only gather data from the environment, but are also capable of affecting it by performing a variety of actions. Applications range from localization to control systems in tunnels or buildings, interactive museums, and home automation [3].

In contrast with mainstream WSNs, characterized by a single application gathering and reporting data, these decentralized settings are composed of many collaborating tasks, each affecting only a portion of the system. For instance, a WSAN for building control and monitoring can be decomposed in at least three main tasks, i.e., structural monitoring, in-door environment monitoring, and response to extreme events such as fire or earthquakes [4]. To realize the latter functionality, the nodes controlling water sprinklers must monitor nearby temperature sensors and smoke detectors and take appropriate measures when and *where* needed. Therefore, the application logic now

P. Gibbons et al. (Eds.): DCOSS 2006, LNCS 4026, pp. 150–168, 2006.
© Springer-Verlag Berlin Heidelberg 2006

resides *in the network*: including a central base station in the control loop degrades system performance and reliability without any sensible advantage [2]. Dealing with this change of perspective demands new programming abstractions to tame complexity without sacrificing efficiency. Indeed, the developer is concerned not only with the application logic, but also with identifying the system portions to be involved and how to reach them. As no dedicated programming constructs exist for the latter task, the result is additional programming effort, increased complexity, and less reliable code.

This work tackles the aforementioned issues through the notion of *logical neighborhood*, an abstraction replacing the conventional notion of physical neighborhood—i.e., the set of nodes in the communication range of a given device—with a logical notion of proximity determined by applicative information. Logical neighborhoods are specified declaratively using the SPIDEY language, conceived to be a simple extension of existing WSN programming languages (e.g., nesC [5] in the case of TinyOS [6]). Using our enhanced communication API, a message can be broadcast to a logical neighborhood, instead of nodes within communication range. This way, application programmers still reason in terms of neighborhood relations and broadcast messages, but can now specify declaratively *which* nodes to consider as neighbors and, therefore, the span of communication. As such, our abstraction may foster a fresh look at existing mechanisms, algorithms, and programming models by replacing their conventional notion of physical neighborhood with our programmer-defined, logical one.

Clearly, our programming abstraction is ultimately of practical interest only in the presence of an appropriate and efficient routing mechanism supporting it. In principle, existing solutions can be exploited (e.g., [7]), but they exhibit various performance drawbacks, as they are based on different assumptions and scenarios. Therefore, in this paper we also present a novel routing protocol that is expressly devised to support our abstraction and leverages the kind of *localized interactions* [8, 9] characterizing the aforementioned decentralized scenarios. The evaluation included in this paper shows that indeed this routing protocol efficiently supports logical neighborhoods, therefore demonstrating the feasibility of our overall approach.

The rest of the paper is organized as follows. Section 2 describes the logical neighborhood abstraction and the SPIDEY language. Section 3 illustrates the novel routing strategy supporting our communication abstraction, while Section 4 evaluates its performance. Section 5 compares our approach against related work. Finally, Section 6 ends the paper with brief concluding remarks.

2   Programming Constructs for Logical Neighborhoods

The proposed abstraction revolves around only two concepts: *nodes* and *neighborhoods*.

A (logical) node is the application-level representation of a physical node, and defines which portion of its data and characteristics is made available by the programmer to the definition of any logical neighborhood. The definition of a logical node is encoded in a *node template*, which specifies the node's exported attributes. This is used to instantiate the (logical) node, by specifying the actual source of data. To make these concepts more concrete, Figure 1 (top) shows a SPIDEY code fragment that defines a template for a generic device and instantiates it by binding each template attribute to an expression of

```
node template Device
   static   Function
   static   Type
   static   Location
   dynamic Reading
   dynamic BatteryPower

create node ts from Device
   Function as "sensor"
   Type as "temperature"
   Location as "room1"
   Reading as getTempReading()
   BatteryPower as getBatteryPower()
```

```
neighborhood template HighTempSens(threshold)
   with Function = "sensor" and
        Type = "temperature" and
        Reading > threshold

create neighborhood hts100
   from HighTempSens(threshold : 100)
   max hops 2
   credits 30
```

Fig. 1. Sample node (top) and neighborhood (bottom) definition and instantiation

the target language, e.g., a constant or function. Template attributes can be **static** or **dynamic**. The former represent information assumed to be time-invariant (e.g., the type of measurement a sensor provides), while the latter represents information changing with time, (e.g., the current sensor reading). The decision about whether an attribute is static or dynamic depends on the deployment scenario. Making the distinction explicit may enable optimizations at the routing layer, as discussed in Section 3.

A (logical) neighborhood is the set of nodes satisfying a constraint on the nodes' attributes. As with nodes, the definition of neighborhoods is encoded in a template, which contains a predicate that essentially serves as the membership function determining whether a node belongs to the logical neighborhood. For instance, the neighborhood template HighTempSens at the bottom of Figure 1 is based on the Device template in the same figure, and selects nodes that host temperature sensors and are currently reading a value higher than a given threshold. As exemplified in the SPIDEY code fragment, a neighborhood template can be parameterized, with the actual parameter values provided by expressions of the target language upon neighborhood instantiation. Moreover, the instantiation of a neighborhood template specifies additional requirements about *where* and *how* the neighborhood is to be constructed and maintained.

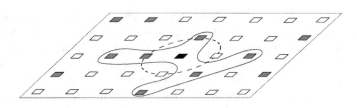

Fig. 2. A visualization of logical neighborhoods

For instance, Figure 1 specifies that the predicate defined in the `HighTempSens` template is evaluated only on nodes that are at a maximum of 2 hops away and by spending a maximum of 30 "credits". The latter is an application-defined measure of cost, further detailed next, which enables the programmer to retain some control over the resources being consumed during the distributed processing necessary to deliver messages to members of a logical neighborhood. A pictorial representation of the example, visualizing the logical neighborhood concept, is provided in Figure 2. There, the black node is the one defining the logical neighborhood, and its physical neighborhood (i.e., nodes lying in its direct communication range) is denoted by the dashed circle. The dark grey nodes are those satisfying the predicate in the neighborhood template in Figure 1 (bottom) when the threshold is set to 100°C. However, the nodes included in the actual neighborhood instance `hts100` are only those lying within 2 hops from the sending node, as specified through the **hops** clause during instantiation.

In essence, as graphically illustrated in Figure 3, templates define *what* data is relevant to the application, while the instantiation process constrains *how* this data should be made available by the underlying system. Separating the two perspectives has several beneficial effects. The same template can be "customized" through different instantiations. For instance, the very same template at the bottom of Figure 1 could be used to

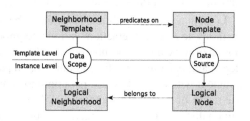

Fig. 3. Templates and their instantiation

specify a logical neighborhood with a different threshold or a different physical span. Moreover, this distinction naturally maps on an implementation that maintains a neighborhood by disseminating its template to be evaluated against the values exported by a node instance, and uses instead the additional constraints specified at instantiation time to direct the dissemination process.

SPIDEY provides additional simple and yet expressive constructs. Logical operators such as **and**, **or**, and **not** are provided to define complex predicates on node templates. Moreover, as logical neighborhoods essentially identify sets of nodes, it becomes natural to express a neighborhood as a composition with already existing ones, using conventional set operators such as union, intersection, subtraction, and inclusion. Finally, the SPIDEY language contains also features enabling the creation of *virtual nodes*, built by binding node attributes to aggregation functions operating on a logical neighborhood. Virtual nodes spare programmers from the burden of directly handling the communication needed to gather and aggregate data from the neighborhood members, and can be used recursively to create higher-level abstractions. More details can be found in [10]. The complete grammar of the SPIDEY language is shown in Appendix A.

Our language also provides the ability to control the cost involved in communicating towards a neighborhood, through the **credits** clause. Communication cost is defined in terms of the basic operation of sending a broadcast message to physical neighbors (the node's *sending cost*), and is measured in *credits*. The mapping between cost and credits is specified by the programmer on a per-node basis through a **use cost** construct, which delegates the computation of this mapping to an expression of the target language, e.g., a function. Therefore, the programmer can define a vast array of mappings, from a straightforward one where the sending cost is fixed, to sophisticated ones where it varies dynamically to adapt to context changes (e.g., low battery power). Moreover, different nodes can have different functions, e.g., yielding higher costs for tiny, battery-powered sensors, and lower costs for resource rich, externally-powered nodes. The overall number of credits necessary to communicate with the members of a logical neighborhood is evaluated as the sum of the costs incurred in by each node involved in routing, with each individual cost evaluated according to the function specified in the **use cost** declaration. Therefore, the ability to set the maximum amount of credits spent in communication in a logical neighborhood enables programmers to exploit different trade-offs between accuracy and costs. Neighborhoods endowed with many credits ensure a broader coverage but incur higher costs, while those with few credits may not reach all the specified nodes but limit resource consumption.

Logical neighborhoods must ultimately be used in conjunction with communication facilities, to enable interaction with the neighborhood members. On the other hand, the notion of logical neighborhood is essentially a scoping mechanism, and therefore is independent from the specific communication paradigm chosen. For instance, one could couple it with the tuple space paradigm to enable tuple sharing and access only within the realm of a logical neighborhood. In our current communication API we took the minimalist—and yet most general—approach of coupling logical neighborhoods with the standard broadcast-based message passing facility found in WSNs. As a result, our API includes simple `send` and `receive` operations mimicking those provided by the underlying operating system. For instance, our TinyOS implementation redefines the operations in the `GenericComm` module by extending the `send` operation with an additional parameter representing the logical neighborhood where a message must to be delivered, i.e., the scope of that particular message. Essentially, we are replacing the broadcast facility commonly made available by the operating system with one where message recipients are not determined by the physical communication range, rather by membership in a programmer-defined logical neighborhood. In addition, a `reply` primitive is also included to simplify communication from neighborhood members back to the message sender. To enable this degree of generality and flexibility, it is fundamental for our abstraction and API to be supported by efficient routing strategies. A description of our solution to the routing problem is described in the next section.

3   Routing for Logical Neighborhoods

The logical neighborhood abstraction is essentially independent of the underlying routing layer. Nevertheless, its characteristics cannot be easily accommodated by existing data-centric routing approaches. Indeed, these are usually conceived to solve the problem of data collection from a homogeneous nodes, thus focusing on how to collect efficiently the data from many sensors to a single node—the sink. In our approach the perspective is reversed: routing must efficiently transmit an application message from a single node (the sender) to those matching the neighborhood specification. Moreover, logical neighborhoods are a scoping mechanism, and therefore can be used in conjunction with several mechanisms other than data collection, e.g., to direct code updates only towards nodes with obsolete versions. As such, some of the techniques exploited by these protocols (e.g., route reinforcement based on data rates as in [7]) not only cannot be directly applied, but are actually complementary to ours. Moreover, our goal is to devise a protocol that captures the localized interactions that should characterize communication in decentralized, multi-sink WSNs and WSANs. This rules out solutions exploiting system-wide tree overlays as in TinyDB [11]. Finally, credit management is a distinctive feature of our approach that would anyway require appropriate integration.

Motivated by these considerations, this section describes a routing strategy designed to support efficiently and effectively the logical neighborhood abstraction. Our routing approach is *structure-less* (i.e., no overlay is explicitly maintained) and is based on the notion of *local search* [12]. Nodes advertise their *profile*, i.e., the list of attribute-value pairs specified by their template, and in doing so build a distributed state space containing information about the cost of reaching a node with given data. This information dissemination is localized and governed by the density of devices with similar profiles. Messages sent to a neighborhood contain its template, which determines a projection of the state space, i.e., the part to be considered for matching. In a nutshell, the message "navigates" towards members of a neighborhood by following paths along which the cost associated with a given neighborhood template is decreasing. The proposed routing approach is therefore composed of two parts: the *state space generation* and the *search algorithm*.

3.1   Building the State Space

In our scheme, whenever a new node is added to the system it broadcasts a PRO-FILEADV message containing the node identifier, a (logical) timestamp, the node's profile containing attributes and their values, and a cost field initialized to zero. The first two message fields are used to discriminate stale information, as the PROFILEADV message is periodically re-broadcast (possibly with different content) by the node. An example PROFILEADV is reported in Figure 4.

In addition, each node in the system stores a *State Space Descriptor* (SSD) containing a summary of the received PROFILEADV messages. An example is shown in Figure 5. The *Attribute* and *Value* fields store information previously received through a PROFILEADV message. For each entry, *Cost* contains the

Source	Timestamp	Node Profile		Cost
		Attribute	Value	
N54	72	Function	sensor	2
		Type	temperature	
		Location	room123	

Fig. 4. An example of PROFILEADV

minimum cost to reach a node with the corresponding information, and *Source* contains the identifier of such node. The *Links* field allows to store information more compactly, by retaining associations among entries instead of duplicating them in the SSD. In Figure 5 each entry is linked to the others as they all come from the same PROFILEADV advertised by node N8. *DecPath* and *IncPaths* contain routing information to direct the search process, as described in Section 3.2. Finally, each entry in an SSD is associated with a lease (not shown), whose expiration causes the removal of the entry not refreshed by a new PROFILEADV.

Upon receiving a PROFILEADV message, a node first updates the cost field in the message by adding its own sending cost, obtained by evaluating the expression in the

Id	Attribute	Value	Cost	Links	DecPath	IncPaths	Source
1	Function	sensor	5	2,3	N37	N98, N99	N8
2	Type	acoustic	5	1,3	N37	N98, N99	N8
3	Location	room123	5	1,2	N37	N98, N99	N8

Fig. 5. An example of *State Space Descriptor (SSD)*

use cost statement described in Section 2. Then, it compares each attribute-value pair in the message against the content of the local SSD. A modification (entry insertion or update) of the SSD is performed if an attribute-value pair: i) does not exist in the local SSD, or ii) it exists with a cost greater than the one in the message (after the local update above). The update (or insertion) of an SSD entry involves establishing the proper values in the *Links* field to keep track of the rest of the PROFILEADV message, updating the *DecPath* field with the identifier of the physical neighbor that sent the PROFILEADV, and setting the *Source* field to the identifier of the node whose information has been inserted in the PROFILEADV. For instance, assume the node storing the SSD in Figure 5 has a sending cost of 1, and receives the PROFILEADV in Figure 4. Its local SSD is then updated as described in Figure 6 (changes shown in bold). Note how the *Links* fields are updated so that only the minimum cost to reach an entry is kept, and yet the information about which entry came with which profile is not lost.

After a PROFILEADV has been processed locally, it is rebroadcast *only* if at least one SSD entry was inserted or updated, to propagate the state change. An example is shown in Figure 7(a). The PROFILEADV is rebroadcast as received, except for the updated *Cost* and *Source* fields. Interestingly, the propagation of PROFILEADV messages enables a node to determine if it lies, for some attribute-value pair, on a path where costs are increasing. This occurs when a PROFILEADV is overheard, through passive listening, with a cost greater than the corresponding pivot entry in the SSD. In this case, the identifier of the broadcasting node is inserted in the *IncPaths* field of the pivot entry.

Thus far, we assumed that PROFILEADV messages contain the whole node profile. Nevertheless, if some dynamic attribute changes frequently, there is a trade-off between the network load necessary to refresh the advertisements and the accuracy of the information being propagated. A straightforward alternative approach is to disseminate only part of the profile (e.g., static attributes) and perform additional matching at the receiver. These trade-offs are ultimately solved based on the characteristics of the deployment scenario, e.g., by considering information about the size of the logical neighborhood or the network density.

Finally, note how, as shown in Figure 7(a), profile advertisements do *not* flood the entire network, as a PROFILEADV is rebroadcast only upon an SSD update. Flooding occurs only for the first advertisement, or more generally when only one node contains

Id	Attribute	Value	Cost	Links	DecPath	IncPaths	Source
1	Function	sensor	**3**	**3,4**	**N77**	N98, N99	**N54**
2	Type	acoustic	5	1,3	N37	N98, N99	N8
3	Location	room123	**3**	**1,4**	**N77**	N98, N99	**N54**
4	**Type**	**temperature**	**3**	**1,3**	**N77**	-	**N54**

Fig. 6. The SSD of Figure 5 at a node with a sending cost of 1, after receiving the PROFILEADV message in Figure 4

a given attribute-value pair—a rather unusual case in the scenarios we target. Instead, for a given set of attribute-value pairs, the state space generation builds a set of non-overlapping *regions*, each containing a node with the considered information. Within a region, each node knows how to route a message addressed to a neighborhood template that includes attributes matching those of a node, along the routes stored in *DecPath*. Each region can be regarded as a "concavity" defined by costs in SSDs, with the target node at the bottom (cost to reach it is zero) and nodes with increasing costs around it. This is illustrated in Figure 7(b), where we show the SSDs after all the nodes performed at least one profile advertisement. Next we describe how this distributed state space is exploited for routing.

3.2   Finding the Members of a Logical Neighborhood

Local search procedures proceed step by step with subsequent *moves* exploring the state space [12]. At each step, a set of further local moves is available to proceed in the search process. Among them, some moves are accepted and generate further moves, while the remaining ones are simply discarded. In general, accepting moves depends on the heuristics one decides to employ given the particular problem tackled. In our case, a *move* is simply the sending of an application message containing the neighborhood template. Upon receiving a message, the move is accepted and further send operations are performed if the maximum number of hops, if any, has not been reached (as per the **hops** construct), and either i) the move proceeds along a decreasing path, or ii) enough unreserved credits are available on an exploring path. The notions of *decreasing path*, *exploring path* and *credit reservation* are at the core of our routing solution and are described next.

Decreasing paths. A path is *decreasing* if it gets the message closer to nodes whose profile matches the neighborhood template. To do so, message proceeds towards minima of the state space by traversing nodes that report an always smaller cost to reach a potential neighborhood member.

To determine decreasing paths, a node must identify the state space projection determined by a neighborhood template. To this end, the node finds in the local SSD the entry matching the neighborhood template with the greatest cost, if any. This entry is called *pivot*. If a pivot exists and is associated, via the SSD *Links* field, to a set of other entries matching the neighborhood template, the cost associated to the pivot represents the number of credits needed to reach the closest matching node via the path indicated by the *DecPath* field. For instance, imagine the application issues a send(m, n) operation through our enhanced communication API, to send the application message m to the neighborhood n, and assume n is defined to address all acoustic sensors. This neighborhood has its *pivot* in entry 2 of the SSD in Figure 6, and its predicate (Function = *sensor* and Type = *acoustic*) is matched via the link pointing from entry 2 to entry 1. Consequently, the node evaluates the cost to reach the closest acoustic sensor as 5 and forwards the message towards N37. Due to the state space generation process, messages following a decreasing path are certainly forwarded towards nodes matching the neighborhood template. Indeed, these paths simply follow the reverse paths previously setup by PROFILEADV messages originating from nodes whose profile contains information matching the neighborhood template. Additionally,

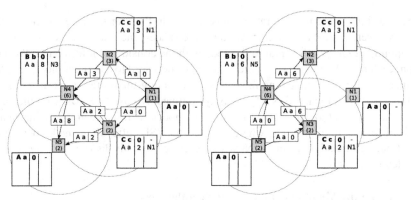

(a) Building the state space (time goes from left to right). Arrow labels denote sending of PROFILEADV messages, showing only the attribute-value (e.g., A a), and Cost fields. SSDs are shown with only attribute-value, Cost and DecPath fields. After N1 disseminated its profile, N5's PROFILEADV need *not* be propagated system-wide, but only where updates in SSDs are needed to make its presence known.

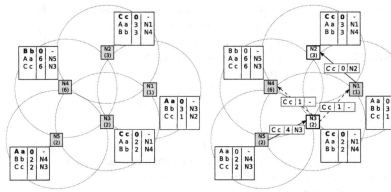

(b) After all the nodes performed at least one profile advertisement, the SSDs contain the costs to reach the closest node with a given attribute-value pair.

(c) A message navigating the state space: dashed lines represent exploring directions, solid lines denote decreasing paths. Arrow labels represent application messages showing only the unreserved credits and the intended recipient.

Fig. 7. Building and navigating the state space

note how the reply feature provided by our communication API can be implemented trivially by keeping track of the reverse path along which a message is received.

Exploring paths. If a message were to follow decreasing paths only, it would easily get trapped into local minima of the state space. To avoid this, we allow messages to be propagated also along *exploring paths*, i.e., directions where the cost to reach the closest node with a particular attribute-value pair is non-decreasing. Exploring paths include directions where the cost does not change (e.g., at the border between two regions) or where it increases. The latter are stored in the *IncPaths* SSD field, as discussed in Section 3.1.

Activating multiple exploring paths at each hop is ineffective, as it is likely to generate many routes that are shortly after rejoined. Therefore, exploration proceeds along a single increasing path, if available. Exploration on multiple paths, achieved through physical broadcast, is activated only when the message reaches a neighborhood member (i.e., a minima of the state space), or after the message has travelled for E hops, with E being a tunable protocol parameter. This design choice stems from the observation that increasing paths are key in enabling the message to "escape" local minima by directing it towards the boundary where a region confines with a different one, and a different decreasing path may become available.

Credit reservation. The instantiation of a neighborhood template may specify the credits to be spent for communicating with neighborhood members, as discussed in Section 2. To support this feature, the number of credits is appended by the sender to every application message sent to a given neighborhood. A node may decide to split these credits in two: one part *reserved* to be spent along decreasing paths and the other along exploring ones. The splitting occurs at the first node that identifies a decreasing path for the message being routed, and is effected by removing the reserved credits from the amount in the message, therefore effectively reserving the credits along the entire decreasing path. For instance, Figure 7(c) shows a message sent by N5 with 6 credits, targeting a neighborhood defined by a single predicate C = c. Neighborhood members are shown in white. As the pivot in N5's SSD reports a cost of 2 to reach the node N3 matching the predicate, the message is forwarded to N3 with 4 unreserved credits.

To deal with credit reservation, a node checks whether its identifier is inserted in the message by the sender node as the next hop along a decreasing path towards a matching node. If so, the node simply forwards the message to the next hop on the decreasing path (found in its SSD) without modifying the credit field, since the necessary credits have already been reserved by the first node on the decreasing path. Otherwise, if exploring paths are to be followed, the node "charges" the message for the number of credits associated to the node sending cost, as per the **use cost** declaration. The remaining (unreserved) credits are assigned to the exploring paths the local node decides to proceed on. Normally, all these credits are assigned to the single message forwarded along the increasing path. However, if multiple paths are explored in parallel through broadcast, according to the heuristics described above, the unreserved credits are divided by the number of neighbors before broadcasting the message. In Figure 7(c), N3 receives a message with 4 remaining credits. Since it is a neighborhood member, the message must be broadcast along all the available exploring paths. Therefore, N3 charges the message for its own sending cost (2) and divides the remaining credits by the number of its physical neighbors. This results in activating two exploring directions, each with a 1-credit budget.

4   Evaluation

This section reports about an evaluation of our routing protocol for logical neighborhoods. To this end, we implemented it on top of TinyOS [6] and evaluated it using the TOSSIM [13] simulator. Our goals were to verify that the protocol behaved as expected for what concerns the generation of the state space and the cost-aware routing of messages, and to characterize its performance. Clearly, this is key to assess the fea-

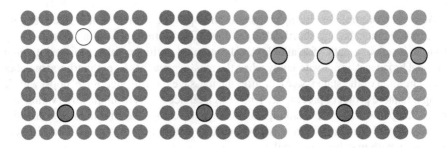

Fig. 8. State space generation. The first PROFILEADV message spreads throughout the system as no node disseminated its profile yet. Profiles advertised by other nodes propagate only until a smaller cost is encountered, partitioning space in regions centered on neighborhood members. Note how the white node does not receive the message in the first propagation—due to collisions—but eventually receives it in later retransmissions.

sibility of our approach and abstraction. The deployment scenario we simulated is a grid where each node can communicate with its four neighbors. This choice not only simplifies the interpretation of results by removing the bias induced by more unstructured scenarios, but also models well some of the settings we target, e.g., indoor WSN deployments [14].

Analyzing the Routing Behavior. Before characterizing the performance of our routing protocol, we analyze whether its behavior matches our design criteria. First, we verify separately the two basic mechanisms underlying our routing, i.e., the state space generation and its "navigation" by applicative messages addressed to a logical neighborhood. As for the former, the key property we want to verify is that the propagation of PROFILEADV messages is localized and partitions the system in non-overlapping regions, each with routing information towards a neighborhood member.

To simplify the analysis of results we developed a simple visualization tool that, given a simulation log and a neighborhood template, displays the propagation of PROFILEADV as well as applicative messages. Figure 8 shows a sample output of our tool where the logical neighborhood we consider selects three members (represented as circled nodes) based on their profiles, and the node sending cost is equal for all devices. The three snapshots correspond to the points in time when a given PROFILEADV, generated by one of the neighborhood members, has ceased to propagate. As it can be observed, the first PROFILEADV is propagated in the whole system, as no other profile information exists yet. However, when the second member propagates its profile, this is spread only until it reaches a node where the cost is less than the one in PROFILEADV. This process partitions the state space in two non-overlapping regions. Eventually, the system reaches a stable situation where the number of regions is equal to the number of neighborhood members, as shown in Figure 8—right.

For what concerns routing of applicative messages, Figure 9 shows the output of our visualization tool when a message is sent to the same neighborhood of Figure 8. The credits associated to the neighborhood are set as an over-approximation of the credits needed to reach the same three nodes along the shortest path. (More details about setting credits are reported later.) Note how the one in the picture is a worst-case scenario where the sender belongs to the same neighborhood the message is addressed to. In this situation, the message starts from a minimum of the state space, i.e., without any decreasing path. Therefore, the initial moves must be exploring ones, until a region different from the one where the message originated is reached. Despite this unfavorable initial situation, the message reaches all the intended recipients by alternating moves along decreasing paths with exploration steps.

The effectiveness of our mechanisms in reducing communication costs is unveiled when heterogeneous devices with different sending costs are deployed. Figure 10 shows a situation with a single neighborhood member and a message sender placed at the opposite corners of the grid, and where sending costs are assigned according to an integer approximation of a bi-dimensional Gaussian distribution. The figure shows the message dutifully steering away from the network center, where sending costs are higher, and striking a balance between the length of its route and the sending cost of the nodes on it. Thanks to the way our state space is generated through profile advertisements and SSD updates, this path is guaranteed to be, within a region, the one with the minimum cumulative sending cost.

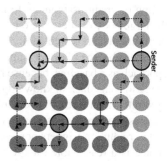

Fig. 9. An applicative message navigates the state space. Solid lines are decreasing paths, dashed lines are exploring paths.

Performance Characterization and Comparative Evaluation. Next, we wanted to study the performance of our protocol. Therefore, we defined a set of synthetic scenarios with a variable number of nodes placed 35 ft apart and with a communication range[1] of 50 ft. Each run lasted 1000 s—a value for which we verified all the measures exhibit a variance less than 1%. In dynamic scenarios, this approach provides more precise results than only averaging over multiple runs [15].

Each node is configured with a single (static) attribute whose value is randomly chosen from a predefined set $\mathcal{A}$ at system start-up. This profile is disseminated by PROFILEADV messages once every 15 s. A single sender node is placed in the center of the grid, generating applicative messages at the rate of 1 msg/s towards a single neighborhood defined with an equality predicate over the node profiles. In this setting, the number of receivers is determined by $|\mathcal{A}|$,

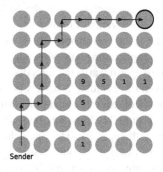

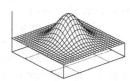

Fig. 10. A message navigating a state space where sending costs follow the distribution at the bottom

and in our case yields a number of neighborhood members of about 10% of the nodes in the system. The node sending cost is constant and identical throughout the system.

---

[1] We used the TinyOS' `LossyBuilder` to generate topology files with transmission error probabilities taken from real testbeds.

Credits are assigned by computing the average cost to reach each node in the system along the shortest path and weighing this value by the probability of the node being a receiver. Then, we increased this minimal value by about one third, to give each message some extra credits to spend on exploratory paths. This approach clearly overestimates the actual cost to reach a receiver, e.g., because it does not consider that two receivers may share part of the path from the sender. The definition of a model supporting fine-tuning of credit assignment to neighborhoods deserves further investigation based on the large body of literature on ad-hoc network density and random graph theory, and is our immediate research goal.

In the absence of directly applicable solutions to compare against, we chose a gossip approach as a baseline, because it is general enough to address the characteristics of our scenarios (e.g., lack of knowledge about the nature of applicative data) and yet generates less traffic than a straightforward flooding protocol. We set the protocol parameters so that gossip rebroadcasts a packet received for the first time with a probability $P = 0.75$, and our solution triggers new exploring directions once every $E = 4$ hops. This latter choice is a reasonable trade-off between generating too many redundant exploratory paths (E too small) and never activating exploratory paths within a region ($E > d$, with d the region diameter).

We based our evaluation on three metrics, namely i) the *message delivery ratio*, defined as the ratio between the messages received by neighborhood members and those that have actually been sent; ii) the *network overhead*, defined as the overall number of messages exchanged at the MAC layer, thus including PROFILEADV messages; and (iii) the average number of *nodes involved* in routing. This figure is further divided into the nodes processing a message at the MAC layer, and those processing one at the application layer. Message delivery is a measure of how effectively a protocol steers messages towards the intended recipients. On the other hand, in the absence of a precise model to evaluate a node's power consumption, ii) and iii) provide a sense of how a protocol exploits communication and computational resources, respectively.

Figure 11 illustrates our simulation results along the aforementioned metrics and w.r.t. the network size. Each chart is the average result of 5 different runs. As it is clear from the figures, our protocol outperforms gossip in all metrics. Message delivery is consistently higher than in gossip, and is even significantly less sensitive to an increase of the network size. As for network overhead, we provide additional insights by showing the results for our protocol with and without PROFILEADV advertisements, and by comparing against the ideal lower bound provided by routing along the minimum spanning tree rooted at the sender and connecting all neighborhood members (computed with a global knowledge of network topology). The chart evidences that we generate almost half of the overhead of gossip and yet deliver significantly more messages. The gap between the two is even more evident in the curve without the PROFILEADV messages, which essentially highlights how efficient is the pure routing mechanism, once the routing information is in place. This is particularly significant because the dissemination of PROFILEADV during state space generation is a fixed cost that is paid once and for all. In other words, adding another sender—regardless of the neighborhood it addresses—does *not* increment the overhead due to state space generation. In addition, the chart shows how the performance of our protocol in this case is closer to the ideal

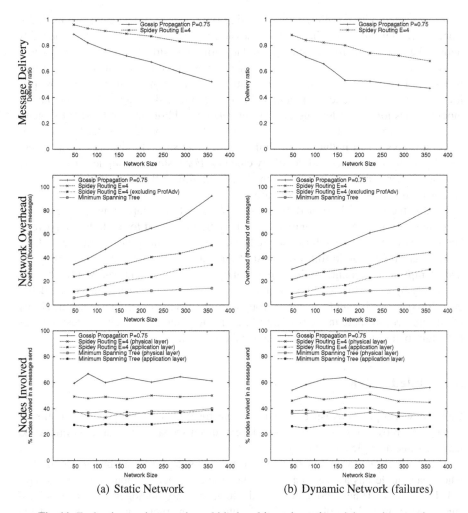

Fig. 11. Evaluation against gossip and ideal multicast, in static and dynamic scenarios

lower bound than to gossip. Finally, for what concerns the nodes involved in processing, Figure 11 shows that our performance at MAC layer is in between gossip and the minimum spanning tree, while at the application layer our routing requires only about half of the nodes exploited by gossip to process application messages and exhibits a performance closer to the minimum spanning tree. Therefore, our protocol is likely to provide a considerably longer network lifetime, although a precise characterization of the energy consumption is beyond the scope of this paper. This result is due to our guided exploration process, which privileges unicast messages (that, unlike broadcast, do not reach the application layers at all nodes in range), thus saving processing. In contrast, gossip explores the system in a completely "blind" way.

As shown in the right column of Figure 11, the evaluation was carried out also in a more dynamic scenario where 10% of the nodes are randomly turned off for 30 s and then reactivated without allowing any settling time in between. Clearly, we excluded from this random selection the intended message recipients, as this would irremediably impact the message delivery ratio. A similar setting has already been used in existing works on routing for WSNs (e.g. [7]) to simulate node failures or the addition of new nodes. As Figure 11 shows, our protocol still provides higher delivery than gossip at lower communication and computational costs, despite node failures. In particular, although nodes joining or leaving the system generate additional profile advertisements to change the shape of the state space, the network overhead remains far from the one of gossip. This result is due to the ability of the state space to change its shape very rapidly in response to network topology changes. For instance, a single PROFILEADV message dissemination among nodes in close proximity to the changing one is usually all it is needed to restore a stable situation.

Finally, the results illustrated in this section should be regarded as worst-case. Indeed, not only the credit assignment can likely be fine-tuned to waste less resources, but also our choice of neighborhood predicates (single disjuncts) is restrictive. Indeed, it forces each message to follow at most a single decreasing path at a time: neighborhood templates containing multiple elementary disjuncts instead can be routed more accurately by exploiting multiple decreasing paths, therefore further increasing delivery. Moreover, setting uniform costs throughout the system does not leverage the ability of our protocol to route in a cost-aware fashion. Nevertheless, we chose these settings to be fair to gossip, which does not provide these advanced capabilities.

5   Related Work

Only few works propose distributed abstractions for WSNs that support some notion of scoping. Moreover, unlike the strongly decentralized scenarios we target in this work, many assume a single data sink.

The work closer to ours is the neighborhood abstraction described in Hood [16], where each node has access to a local data structure where attributes of interest provided by (physical) neighbors are cached. However, only homogeneous nodes are assumed. Moreover, data collection is built into the constructs and therefore, as stated in Section 3, communication is expected to flow only according to a many-to-one paradigm. Finally, the current implementation considers only 1-hop neighbors and is mainly based on broadcasting all attributes and performing filtering on the receiver's side. Clearly, our framework is much more flexible as it provides a general application-defined neighborhood abstraction, which is decoupled from the application functionality and therefore can be used for purposes other than data collection (e.g., network reprogramming), as well as in conjunction with it to support efficiently heterogeneous scenarios.

The work on Abstract Regions [17], instead, proposes a model where <*key,value*> pairs are shared among the nodes belonging to a given *region*. The span of a region is based mainly on physical characteristics of the network (e.g., physical or hop-count distance between sensors), and its definition requires a dedicated implementation. Therefore, each region is somehow separated from others, and regions cannot be combined. This results in a much lower degree of orthogonality and flexibility with respect to our approach. Moreover, the concept of *tuning interface* provides access to a region's implementation, enabling the tweaking of low-level parameters (e.g., the number of retransmissions). Instead, our approach provides a higher-level, user-defined notion of cost that can be used to control resource consumption. In TinyDB [11], materialization points create views on a subset of the system. In this sense, common to our work is the effort in providing the application programmer with higher-level network abstractions. However, the approach is totally different, as TinyDB forces the programmer to a specific style of interaction (i.e., a data-centric model with SQL-like language) and targets scenarios where a single base station is responsible for co-ordinating all the application functionality. SpatialViews [18] is a programming language for mobile ad-hoc networks where *virtual networks* can be defined depending on the physical location of a node and the services it provides. Computation is distributed across nodes in a virtual network by migrating code from node to node. Common to our work is the notion of scoping virtual networks provides. However, SpatialViews targets devices much more capable than ours, focuses on migrating computation instead of supporting an enhanced communication facility as we do, and yet provides less general abstractions. Finally, in [19], the authors propose a language and algorithms supporting generic role assignment in WSNs with an approach that, in a sense, is dual to ours. In fact, their approach *imposes* certain characteristics on nodes in the system so that some specified requirements are met, while in our approach the notion of logical neighborhood *selects* nodes in the system based on their characteristics.

As for our routing protocol, we were influenced by Directed Diffusion [7] in using a soft-state approach based on periodic refresh for storing routes. However, our solution is radically different as it targets much more general scenarios. We do not assume data collection as the main communication functionality, and therefore we cannot rely on any knowledge about message content, required in Directed Diffusion for interpolation along failing paths. Similarly, we take into account an explicit notion of communication cost without relying on an application-defined notion of data rate. Moreover, an important difference is that our profile advertisements do not propagate to the whole network, unlike interests in Directed Diffusion. Finally, routing in Directed Diffusion is entirely determined by gradients, while we make the system more resilient to changes by allowing exploratory steps, whose use is nevertheless under the control of the programmer through the credit mechanism.

6   Conclusions and Future Work

This paper presented the SPIDEY language and a routing protocol supporting logical neighborhoods, a novel programming abstractions for WSNs. Logical neighborhoods capture sets of nodes with functionally related characteristics. SPIDEY constructs enable the programmer to specify neighborhoods declaratively, and yet control the trade-off between accuracy and resource consumption using an application-defined notion of cost. This latter information is used by our dedicated routing protocol, which supports efficiently our abstraction.

The benefits of our proposal impact two orthogonal aspects. First, developers can concentrate on the actual application goals while relying on logical neighborhoods as a way to logically partition the system and interact with it. We conjecture that applications built on top of our abstraction result in cleaner, simpler, and more reusable implementations. A qualitative and quantitative evaluation of the advantages our approach brings to the development task is currently being carried on. Second, our routing protocol achieves a longer system lifetime and a better resource utilization, by focusing only on the nodes that actually need to be involved.

In this paper, we coupled logical neighborhoods with the broadcast-based primitives typically provided by the operating system. As we pointed out, this choice simplifies the programmer's task, and opens up opportunities for adapting existing techniques by replacing physical with logical neighborhoods. Our future research goals involve the coupling of logical neighborhoods with different services (e.g., to support code deployment only in given portions of the system) as well as alternative communication paradigms. In particular, we plan to integrate logical neighborhoods with our tuple space middleware TINYLIME [20] supporting scenarios with multiple mobile sinks, to empower sinks with the ability to restrain data sharing to the desired set of nodes. Interestingly, this scenario is easily encompassed by our routing protocol, as routes are determined by the profiles of (static) sensors rather than the requests of (mobile) sinks. Finally, our immediate research goal is to devise an analytical model of our routing protocol, to provide the programmer with the ability to properly dimension the allocated credits based on the characteristics of the network, e.g., in terms of density and connectivity.

Acknowledgements. The work described in this paper is partially supported by the Italian Ministry of Education, University, and Research (MIUR) under the VICOM project, by the National Research Council (CNR) under the IS-MANET project, and by the European Union under the IST-004536 RUNES project.

References

1. Mainwaring, A., Culler, D., Polastre, J., Szewczyk, R., Anderson, J.: Wireless sensor networks for habitat monitoring. In: Proc. of the 1st ACM Int. Workshop on Wireless sensor networks and applications. (2002) 88–97
2. Akyildiz, I.F., Kasimoglu, I.H.: Wireless sensor and actor networks: Research challenges. Ad Hoc Networks Journal 2(4) (2004) 351–367
3. Petriu, E., Georganas, N., Petriu, D., Makrakis, D., Groza, V.: Sensor-based information appliances. IEEE Instrumentation and Measurement Mag. **3** (2000) 31–35
4. Dermibas, M.: Wireless sensor networks for monitoring of large public buildings (2005) Tech. Report, University of Buffalo. Available at www.cse.buffalo.edu/tech-reports/2005-26.pdf.
5. Gay, D., Levis, P., von Behren, R., Welsh, M., Brewer, E., Culler, D.: The nesC language: A holistic approach to networked embedded systems. In: Proc. of the ACM SIGPLAN Conf. on Programming Language Design and Implementation (PLDI'03). (2003) 1–11
6. Hill, J., Szewczyk, R., Woo, A., Hollar, S., Culler, D., Pister, K.: System architecture directions for networked sensors. In: ASPLOS-IX: Proc. of the 9^{nt} Int. Conf. on Architectural Support for Programming Languages and Operating Systems. (2000) 93–104

7. Intanagonwiwat, C., Govindan, R., Estrin, D., Heidemann, J., Silva, F.: Directed diffusion for wireless sensor networking. IEEE/ACM Trans. Networking **11**(1) (2003) 2–16

8. Estrin, D., Govindan, R., Heidemann, J., Kumar, S.: Next century challenges: scalable coordination in sensor networks. In: Proc. of the 5th Int. Conf. on Mobile computing and networking (MobiCom). (1999)

9. Qi, H., P.T. Kuruganti: The development of localized algorithms in wireless sensor networks. Sensors Journal **2**(7) (2002)

10. Mottola, L., Picco, G.: Programming Wireless Sensor Networks with Logical Neighborhoods. In: Proc. of the the 1st Int. Conf. on Integrated Internet Ad hoc and Sensor Networks (InterSense 2006), Nice (France) (2006) (Short paper). To appear. Available at `www.elet.polimi.it/upload/picco`.

11. S.R. Madden, M.J. Franklin, J.M. Hellerstein, Hong, W.: TinyDB: an acquisitional query processing system for sensor networks. ACM Trans. Database Syst. **30**(1) (2005) 122–173

12. L.A. Wosley: Integer Programming. Wiley (1998)

13. Levis, P., Lee, N., Welsh, M., Culler, D.: Tossim: accurate and scalable simulation of entire tinyos applications. In: Proc. of the 1st Int. Conf. on Embedded Networked Sensor Systems (SenSys). (2003) 126–137

14. Stoleru, R., J.A. Stankovic: Probability grid: A location estimation scheme for wireless sensor networks. In: Proc. of the 1st Int. Conf. on Sensor and Ad-Hoc Communication and Networks (SECON). (2004)

15. Yoon, J., Liu, M., Noble, B.: Sound mobility models. In: Proc. of ACM MobiCom. (2003) 205–216

16. Whitehouse, K., Sharp, C., Brewer, E., Culler, D.: Hood: a neighborhood abstraction for sensor networks. In: Proc. of the 2nd Int. Conf. on Mobile systems, applications, and services (MobiSys). (2004)

17. Welsh, M., Mainland, G.: Programming sensor networks using abstract regions. In: Proc. of the 1st USENIX-ACM Symp. on Networked Systems Design and Implementation (NSDI04). (2004)

18. Ni, Y., Kremer, U., Stere, A., Iftode, L.: Programming ad-hoc networks of mobile and resource-constrained devices. In: Proc. of the ACM SIGPLAN Conf. on Programming language design and implementation. (2005) 249–260

19. Frank, C., Römer, K.: Algorithms for generic role assignment in wireless sensor networks. In: Proc. of the 3rd ACM Conf. on Embedded Networked Sensor Systems (SenSys). (2005)

20. Curino, C., Giani, M., Giorgetta, M., Giusti, A., A.L. Murphy, G.P. Picco: TINYLIME: Bridging Mobile and Sensor Networks through Middleware. In: Proc. of the 3rd IEEE Int. Conf. on Pervasive Computing and Communications (PerCom). (2005) 61–72

A   SPIDEY Grammar

```
<node_template> ::= node template <node_templ_id>
                        ({static | dynamic} <field_name>)+

<node_instance> ::= create node <node_id> from <node_templ_id>
                    (<field_name> as {<target_lang_expr> |
                            <function_name>(<nhood_id>) every <time_period>})+

<nhood_template> ::= neighborhood template <nhood_templ_id>
                            [(<par_name>(,<par_name>)*)]
                        [with <node_predicates>]
                        [[{min | max}] cardinality <integer_value>]
                        [{union | intersect | minus | on}
                            <nhood_templ_id> [<par_bindings>]]*

<nhood_instance> ::= create neighborhood <nhood_id>[<par_bindings>]
                            from <nhood_templ_id>
                        [[{min | max}] hops <integer_value>]
                        [credits <numeric_value>]

<par_bindings> :: = (<par_name>:<target_lang_expr>
                            (,<par_name>:<target_lang_expr>)*)

<cost_function> ::= use cost <target_lang_expr>
```

Y-Threads: Supporting Concurrency
in Wireless Sensor Networks

Christopher Nitta, Raju Pandey, and Yann Ramin

Department of Computer Science
University of California, Davis
Davis, CA 95616
{nitta, pandey, ramin}@cs.ucdavis.edu

Abstract. Resource constrained systems often are programmed using an event-based model. Many applications do not lend themselves well to an event-based approach, but preemptive multithreading pre-allocates resources that cannot be used even while not in use by the owning thread. In this paper, we propose a hybrid approach called Y-Threads. Y-Threads provide separate small stacks for blocking portions of applications, while allowing for shared stacks for non-blocking computations. We have implemented Y-Threads on Mica and Telos wireless sensor network platforms. The results show that Y-Threads provide a preemptive multithreaded programming model with resource utilization closer to an event-based approach. In addition, relatively large memory buffers can be allocated for temporary use with less overhead than conventional dynamic memory allocation methods.

Keywords: Stack sharing, Multi-threading, Concurrency.

1 Introduction

Wireless Sensor Network (WSN) systems are inherently concurrent. Support for concurrency is needed in all layers of the WSN software stack. At the operating system level, hardware interrupts, low level I/O and sensor events are asynchronous. At the middleware level, specific services (such as time synchronization [1][2] and code distribution[3][4][5]) are highly concurrent in nature, and exist independently from other activities. For instance, most code distribution protocols have several concurrent activities: one may actively maintain a code distribution tree by periodically collecting neighbor information, while the other may cache and distribute code along the distribution tree. At the application level, programs may define both node-level (e.g., collect data) and group level activities (e.g., aggregate data), each occurring concurrently.

Concurrency exists not only at many different levels, but also in many different forms: rapid responses to specific events are easily represented using events and event handlers; concurrency among middleware services are better expressed using long running threads (LRT); concurrency and group concurrent activities are better defined using a combination of threads and atomic computations; and higher level system software abstractions (such as virtual machines and middleware) can be implemented

P. Gibbons et al. (Eds.): DCOSS 2006, LNCS 4026, pp. 169–184, 2006.
© Springer-Verlag Berlin Heidelberg 2006

easier using a threading mechanism. We have examined these concurrency models and believe there is a disconnect between the two main models.

WSN Operating Systems work with limited resources, RAM being the primary limitation. The limited RAM drives many embedded system designers to use an event-based programming model as in TinyOS[6] and SOS[7]. Though [8] shows that event-based and thread-based approaches can be interchanged, many applications do not lend themselves well to an event-based approach, especially those where true CPU concurrency is needed[9][10]. Often embedded system tasks run a cycle of work and waiting. Blocking is done at the highest level, computation is executed to completion and waiting occurs again. Preemptive multithreading pre-allocates RAM that cannot be used even while not in use by the owning thread. What is needed is a concurrency model that balances the programming needs through a preemptive threading model and at the same time meets the resource constraints of sensing devices.

We introduce a hybrid approach called Y-Threads. Y-Threads are preemptive multithreads with small thread stacks. The majority of work in Y-Threads is done by non-blocking routines that execute on a separate common stack. By separating the execution stacks of control and computational behavior, Y-Threads can support preemptive threading model with better memory utilization than preemptive multithreading alone.

We have implemented Y-Threads on several WSN platforms. Experimental results show that a Y-Thread version of a time synchronization application only increased energy consumption by 0.12% over the original purely multithreaded version. In addition the worst-case RAM requirement for the Y-Thread implementation was reduced by 16.5%. Experiments also show that Y-Thread implementations of a flash modification routine are more processing efficient than versions that dynamically allocate memory.

The rest of this paper is structured as follows. Section 2 discusses the motivation, programming model, and implementation of Y-Threads. Experimental test applications and results are described in Section 3. Section 4 discusses the existing concurrency models more in depth. Section 5 discusses the possibilities of future work on Y-Threads. We conclude in Section 6.

2   Y-Threads

Y-Threads are preemptive threads and are well suited to capture the reactive nature of many WSN programs. It is based on the insight that many WSN applications block, waiting for specific events to occur, that has motivated the development of Y-Threads. As events occur, they react by performing atomic computations, changing their state, and returning to the wait mode. Behavior of many such applications can be captured in terms of two sets of behavior: the first is a control behavior that is state-based and that guides the application through different state transitions as different events occur. The second are the different computational behaviors that occur during various state transitions.

For instance, consider the time synchronization code sample shown in **Fig. 1**. The control behavior is defined in the while loop: the application blocks while waiting for events (such as message arrival). Upon occurrences of these events, it performs specific actions (such as *processMsg*), and then goes back to wait for other events to occur.

We observe that the size of stack required to execute the control behavior is fairly small. By separating the execution stacks of control and computational behavior, we can support preemptive threading model and save on memory space. Y-threads implement this idea by providing support for both control and computational behaviors.

```
void timesync_ReceiveTask(){
  TimeSyncMsg msg;
  while(1) {
    recv_radio_msg(&msg);
    leds_greenToggle();
    if((state & STATE_PROCESSING) == 0 ){
      mess.arrivalTime = hal_sys_time_getTime32();
      processMsg(&msg);
} } }

void processMsg(TimeSyncMsg *msg){
  ...
}
```

Fig. 1. Time Synchronization code sample

2.1 Y-Threads Programming Model

Y-Threads provide the capabilities of preemptive multithreading with good utilization of limited RAM. Y-Threads are preemptive threads with small pre-allocated stacks. Y-Threads have the same semantics as general threads, but the majority of work is done by Run to Completion Routines (RCR) that execute on a separate common stack. Task stack sharing and a scheduling method for correct system operation are discussed in [11]. The difference in Y-Threads is that RCRs execute as the invoking thread and maintains the invoking threads priority, whereas [11] discusses run to completion flyweight tasks. The Y-Thread interface provides two primary functions, one to create the Y-Threads, and one to invoke the RCRs. **Fig. 2** shows the Y-Thread and RCR APIs. The APIs are discussed in further detail in Section 2.2.

```
typedef void (*oss_task_function_t)(void);
typedef void *oss_rcr_data_t;
typedef void (*oss_rcr_function_t)(oss_rcr_data_t);
oss_taskp_t oss_ythread_create(oss_task_function_t
task_func, uint16_t sz);
void oss_rcr_call(oss_rcr_function_t func,
oss_rcr_data_t data);
```

Fig. 2. Y-Thread and RCR API

Fig. 3 shows the memory map of two different threads each invoking an RCR. Task 1 has higher priority than Task 2 and waits until the semaphore is signaled. The arrow in each portion of **Fig. 3** signifies the current execution stack. In **a** of the figure Task1 is blocked, and Task2 is running. Task 2 invokes an RCR in **b,** which executes on a separate stack and then returns in **c.** Task1 is of higher priority, and is unblocked in **d** and invokes its RCR in **e.** The RCR returns and Task1 continues to execute on its stack in **f.** Notice the sharing of the common stack space for both RCRs. The stack sharing provides programming semantics of a much larger virtual stack for each task. The source in **Fig. 4** is an example that corresponds to the memory map in **Fig. 3.**

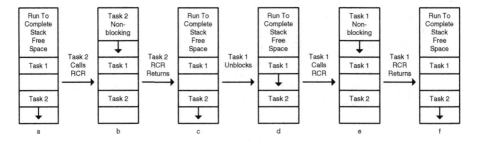

Fig. 3. Y-Threads Memory Usage

```
void Task1(){
  while(1){
    wait_sem(&sem1);
    oss_rcr_call(Task1Process,NULL);
} }

void Task2(){
  while(1){
    oss_sleep(100);
    oss_rcr_call(Task2Proccess,NULL);
    signal_sem(&sem1);
} }

void Task1Process(oss_rcr_data_t *data){
  ...//Do something
}

void Task2Process(oss_rcr_data_t *data){
  ...//Do something else
}
```

Fig. 4. Y-Thread Code Example

Y-Threads are similar to preemptive multithreads that spawn run to completion threads without the overhead of creating a run to completion thread. It may be possible to implement Y-Threads such that an RCR can be preempted by threads of higher priority since threads that are on the RCR stack cannot block. Currently Y-

Threads have not been implemented in this manner and such an implementation should rename RCR to Non-Blocking Routine (NBR).

Parameters can be passed to RCRs and results can be returned, since it is just a matter of copying data to and from the correct contexts. Since the invoking thread transfers execution control to the RCR during invocation, all data on the controlling thread's stack can be accessed by the RCR. The RCR signature is dependent upon the implementation of Y-Threads. The initial implementation of Y-Threads uses a single function to invoke the RCR. RCRs can invoke subroutines as long as none of the subroutines attempt to block execution. If an RCR invokes another RCR, it has the same semantics as invoking a subroutine since the invoking RCR is already executing on the RCR stack.

Y-Thread behavior can be emulated on any RTOS that has both preemptive threads and light weight or run to completion threads with extra overhead. Y-Threads are similar to OSEK[12] Extended tasks that spawn Basic tasks, in those systems that execute Basic tasks on a common stack. The exception is that they do not execute as Extended tasks and do not inherit their priority.

Y-Threads have better memory utilization than just pure preemptive multithreading since large amounts of memory can be automatically allocated and freed on the RCR stack. Preemptive multithreads must pre-allocate enough memory for the worst case stack utilization; therefore there is memory that is unused in each thread, but is also unusable by other threads. Y-Threads do not need to pre-allocate large amounts of memory for potential future use unlike purely preemptive multithreading. Y-Threads can also automatically allocate large amounts of memory on the RCR stack with less overhead than dynamic memory allocation. Y-Threads provide the advantages of preemptive multithreading without the disadvantage of high memory overhead.

2.2 Y-Thread Implementation

Y-Threads were first implemented on OS*[1] for the AVR ATMega128 using an invocation to a helper function *oss_rcr_call*. The signature of the RCRs is a single void pointer parameter with a void return type. The RCR data types and oss_rcr_call prototype can be seen in **Fig. 2**. Light Weight Threads (LWT) are implemented in OS* using an RCR wrapper. The calling convention of the ATMega128 under GCC passes most function parameters in registers making overhead of the *oss_rcr_call* helper function very low. The advantage of compiler support or *oss_rcr_call* in-lining is not likely to be as great for the ATMega128 as it may be for other architectures where parameters are primarily passed on the stack. Switching the control to the RCR stack was relatively straight forward in the ATMega128 since GCC uses registers and the frame pointer to access local variables and parameters.

OS knowledge is necessary for any Y-Thread implementation. OS knowledge is necessary since it must be known if the RCR stack is in use or not. Furthermore, if RCRs are implemented as "Non-Blocking Routines" as discussed in Section 2.1, the current top of the NBR stack must be stored during a context switch from a thread in the NBR. **Fig. 5** illustrates a higher priority thread preempting an NBR. In part a of the figure Task 2 of lower priority is executing a NBR. Task1 is unblocked by an ISR

---

[1] OS* is a light-weight synthesizable operating system for the Mica, Telos and Stargate families of sensor nodes currently under development by the SENSES group at UC Davis.

in part b and context switch from Task2's NBR to Task1 occurs. Task1 invokes an NBR in part c which executes on the common stack under Task2's ready NBR. Task 1's NBR would overwrite Task 2's NBR if the NBR stack top is not stored during the context switch. Due to the nature of the OS* scheduler, our current Y-Thread implementation does not need to store the top of the RCR stack.

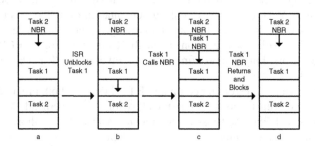

Fig. 5. NBR Preemption by higher priority thread

We also implemented Y-Threads for the TI MSP430. The RCR API is identical to that of the ATMega128 implementation. Unlike the ATMega128, the implementation of Y-Threads for MSP430 was complicated by GCC's use of the frame pointer. The difficulty arose in implementing a version of Y-Threads that operated properly under all optimization levels of GCC. One possible solution is to have implementations for different compiler options; this is similar to libraries that are developed for both banked and non-banked memory models.

3   Applications and Performance Evaluation

Since Y-Threads are preemptive multithreads with small stacks, any application that can be implemented using preemptive multithreads can also be implemented using Y-Threads. The applications evaluated in this paper show the advantage of a preemptive multithreading programming model, and the need for relatively large temporary memory allocations. All applications evaluated were compiled for the AVR ATMega128. All test data was either collected from AVRORA[2] or backed out of object dumps of the applications.

3.1   Run to Completion Routine vs. Subroutine and Light Weight Thread Invocation

The overhead associated with invoking an RCR compared to a normal subroutine is an important metric to evaluate. LWTs are run to completion threads on OS* and therefore are a possible alternative to RCRs for the execution of non-blocking code. A test program was written to evaluate the overhead associated with RCR, subroutine and LWT invocations. **Fig. 6** shows the overhead in instruction cycles for invoking

---

[2] AVRORA is an AVR simulator developed by UCLA Compilers Group and available at http://compilers.cs.ucla.edu/avrora/

each type of routine. LWTs must be posted to the scheduler prior to the thread switch; therefore the LWT thread switch is shown with and without the scheduler posting overhead. As expected the invocation overhead of the RCR is higher than the normal subroutine, but it is less than half as much as the LWT. The overhead of the RCR (177 instructions) is closer to that of a subroutine (94 instructions) than that of the LWT (462 instructions).

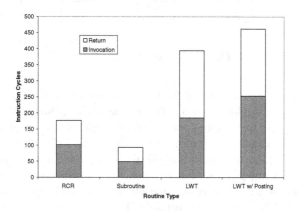

Fig. 6. RCR, Subroutine, and LWT Invocation Times

3.2 Dynamic Allocation vs. Y-Thread Automatic Allocation

Limited resource systems often need to allocate relatively large memory buffers for temporary use. Modifying flash is one such application that is necessary in many embedded systems. The nature of flash requires that entire pages, sometimes as large as 512B in size, be erased before reprogramming can occur. These resource constrained systems typically have between 1KB and 10KB of RAM, and therefore a flash page is relatively large compared to the entire system memory. **Fig. 7** illustrates the copy, erase, modify, write back cycle of modifying a flash page.

There are two methods to allocate relatively large amounts of memory for temporary use in a preemptive multithreading environment. Memory can be dynamically allocated through a function invocation such as *malloc* and then freed when not needed through an invocation of *free*. The other method is to automatically allocate the memory on the stack by invoking a function.

Dynamic memory allocation has the advantage that the allocation size is bound at run-time unlike automatic allocation that is bound at compile-time. The overhead of automatic allocation is constant and is typically lower than dynamic allocation. In a preemptive multithreading environment the drawback to automatic allocation is that the memory must be pre-allocated during the context allocation. The pre-allocated memory is unusable by other threads even when it is not in use by the thread. The advantage of Y-Threads is that they can automatically allocate memory in an RCR with less overhead than dynamic memory allocation. Further when a thread is not in an RCR the memory is available for other threads to use.

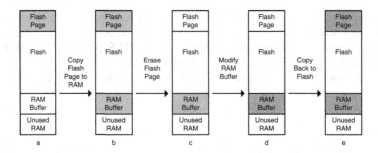

Fig. 7. Flash Page Modification

A flash modification application was implemented using four different methods. All test applications were implemented on OS* for the AVR ATMega128 and analyzed using AVRORA. The first method uses a small thread that dynamically allocates memory for the modification buffer via *malloc*. When the buffer is no longer needed it is freed. The second method utilizes a global buffer for the modification. The final two test applications utilize RCRs that automatically allocate the modification buffer on the stack. One of the methods uses the *oss_rcr_call* while the other inlines the *oss_rcr_call* to emulate compiler support for Y-Threads.

The flash modification function execution time is dependent upon the fragmentation of the heap for the *malloc* version. All other versions of the function were independent of heap fragmentation. **Fig. 8** shows the execution time of each of the function versions normalized to the RCR version. The *malloc* version is slightly slower than the RCR version in the best case and climbs to over 8% higher at only 16 fragments. The RCR inlined version and the global buffer version were 0.7% and 1.7% faster than the RCR version respectively. If Y-Threads were to be supported by the compiler or inlined through optimizations the RCR version would be approximately 1% slower than the global buffer version.

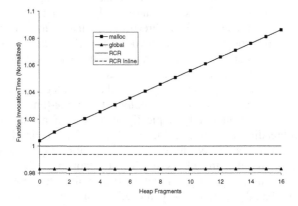

Fig. 8. Flash Modification Time vs. Heap Fragmentation

The *malloc* version while slightly slower than the RCR version in the best case performs worse as memory becomes fragmented. The non-determinism of the *malloc* version makes it undesirable for any application that has real-time requirements. Also the RCR version does not statically allocate memory that cannot be used like the global buffer version does.

3.3 Time Synchronization

Time synchronization is a common service that is often required in WSNs. The time synchronization code illustrated in **Fig. 1** was evaluated for performance. The *processMsg* subroutine was replaced with an RCR version for evaluation. **Fig. 9** shows the code for the RCR version of the application from **Fig. 1**. Notice that very few changes were required. The *processMsg* subroutine has floating point math that is relatively RAM intensive for the 8-bit AVR ATMega128 CPU. The simulation of the time synchronization routine was run on AVRORA for ten seconds.

```
void timesync_ReceiveTask(){
  TimeSyncMsg msg;
  while(1) {
    recv_radio_msg(&msg);
    leds_greenToggle();
    if((state & STATE_PROCESSING) == 0 ){
      mess.arrivalTime = hal_sys_time_getTime32();
      oss_rcr_call(processMsg, & msg);
} } }

void processMsg(oss_rcr_data_t *data){
  TimeSyncMsg *msg = (TimeSyncMsg *)data;
  ...}
```

Fig. 9. RCR Time Synchronization Code Sample

The energy consumption and the number of active CPU cycles were compared for the original and the RCR versions of the time synchronization algorithm. **Fig. 10** shows the results of the tests. The RCR version was active for 0.12% more instruction cycles than the non-RCR version, and consumed 0.02% more energy. The overhead of running the RCR version is insignificant in terms of energy usage.

RAM utilization was also compared for the both the RCR and original versions. **Fig. 11** illustrates the worst-case, thread stack, interrupt and RCR stack memory usage for both the original and RCR versions. The worst-case memory was calculated as the total memory required for maximum function invocations plus the maximum ISR requirements. The worst-case memory usage originally started at 278 bytes and was reduced to 232 bytes in the RCR version. Since the processMsg call was switched to an RCR call, the thread stack requirements were reduced from 142 to 58 bytes. The transition of processMsg to an RCR call alone should actually increase the total required RAM, but when interrupts are considered the worst-cast RAM utilization is actually reduced in the RCR version. In the original version LWTs execute on the scheduling stack, whereas the RCR version LWTs utilize the RCR stack. Sharing of the RCR stack is the main reason for the reduction in worst-case RAM utilization from that of the original version. As the number of Y-Threads increases, the worst-case RAM savings of RCR versions should increase because more RAM should be shared on the RCR stack.

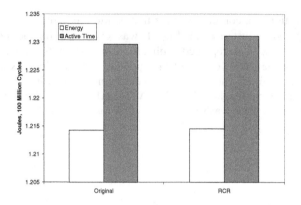

Fig. 10. Time Synchronization Energy and Active Time

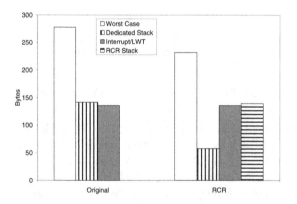

Fig. 11. Time Synchronization Memory Usage

3.4 "Delta" Based Code Distribution

The distribution of a new code image to an embedded system is difficult and often requires a significant amount of system resources. The method discussed in [13] of transmitting only deltas for system updates has much promise and could benefit from the use of Y-Threads. In order to minimize data transmission, deltas or differences between the new and existing image are transmitted.

The update application receives deltas and modifies flash pages. As discussed in [14], implementing state machines is often easier using preemptive multithreading than event based programming. The reception of delta packets can easily be done in a preemptive thread. Since the size of the deltas are relatively small compared to the size of a flash page, the delta packets can be stored on the thread stack.

A large buffer is required to construct the new flash pages from the deltas and the existing image. As discussed previously, the nature of flash requires that the entire page be erased prior to reprogramming. Prior to the implementation of Y-Threads, a statically allocated global buffer was used to implement the delta update application. A RCR can be used to construct the updated flash pages from the deltas and the existing image. The large buffer can be allocated on the RCR stack as discussed in the previous section. The delta update RCR constructs the new flash page in RAM, erases the page to be updated, and then reprograms the page.

4   Concurrency Model Discussion

The two main concurrency models are event-based microthreads, and preemptive multithreads. Event-based microthreads have very low overhead both in processing and memory utilization. Preemptive multithreaded systems allow for per thread state maintenance and thread blocking. The preemptive multithreaded programming model makes application development easier when true CPU concurrency is necessary.

Event-based microthreads and preemptive multithreading are not the only concurrency models that exist. Lazy threads and protothreads are two other pseudo-concurrency models. Both lazy threads and protothreads allow for blocking and are implemented using co-routine like mechanisms. Protothreads have been used successfully in Contiki[14][15]. All of the concurrency models must also deal with interrupts; therefore we discuss various implementation techniques for interrupt handlers.

4.1 Event-Based Microthreads

Event-based programming does not require a separate stack per execution context. Event handlers are typically implemented purely as function invocations making events efficient both in execution and memory utilization. Event-based systems often can be implemented in a high level language making the system easier to implement and more portable. **Fig. 12** shows the single stack utilization of event based threads.

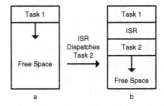

Fig. 12. Event-Based threading stack

The disadvantage of the event-based concurrency model is that events or "microthreads" must be run to completion; events are not allowed to block. The ease of system development can come at the cost of complicated application development for systems that require true CPU concurrency. A further disadvantage of microthreads is that state maintenance across event handler calls must be done via global variables [16]. The use of global variables to maintain state may limit the modularity that can be achieved especially if written in a language such as C.

4.2 Preemptive Multithreading

Preemptive threads require RAM for each execution context. Preemptive threads allow for blocking, and per thread state maintenance. State maintenance with true CPU concurrency is often easier to implement in a multithreaded environment. Often WSN applications need to implement communication protocols with similar timeouts and system states. The increased modularity and ease of application programmability comes at a cost.

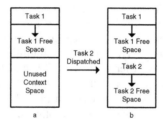

Fig. 13. Preemptive multithreading memory

Preemptive threads require assembly at the systems level for the context switch, complicating systems development. The context switch also has higher overhead than a function invocation. Preemptive threads must pre-allocate enough RAM to execute the deepest sequence of function invocations, and therefore hold an unused resource most of the time. Pre-allocation of memory is illustrated in **Fig. 13**. The pre-allocation of memory is the main drawback that has driven the use of event-based programming model in many embedded systems.

4.3 Lazy Threads

Lazy threads are a threading model that allows for parallel function invocations. The need for fast thread forking has driven the development of lazy threads. Lazy threads implementations attempt to allocate child threads on the parents stack in a "stacklet"[17]. If blocking of the child is necessary an entirely separate context must be allocated for the child and the child thread must be transferred. Unfortunately lazy threads are non-preemptive and therefore require cooperative multithreading. Implementations of lazy threads require compiler support to properly allow for continuations. No pointers to stack data are allowed since there is a possibility of context relocation, further limiting the flexibility of lazy threads.

4.4 Protothreads

Protothreads are pseudo threads that use a coroutine technique to implement blocking threads. Protothreads all execute on a single stack and are implemented entirely in C. The advantage of Protothreads is that they best utilize limited resources while providing a somewhat preemptive multithreading programming model. **Fig. 14** shows the coroutine thread intertwining on a single stack. There are a few limitations of Protothreads; a major limitation is that automatic variables are not saved across blocking waits. The automatic variable limitation is not intuitive, and therefore can lead to programming mistakes. Blocking waits may only occur in the Protothread function and cannot occur in subroutines called by the Protothread. The blocking wait subroutine limitation is not as severe as automatic variable limitation especially when considering that blocking is often done at the highest level.

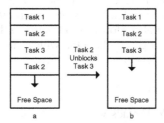

Fig. 14. Protothreads stack usage

4.5 Interrupts

Since interrupts can occur at any time, preemptive multithreads must accommodate for them by allocating extra space for the largest interrupt service routine (ISR). This extra overhead is also required for Y-Threads. A separate stack can be used for interrupt processing, which reduces the overhead required per thread to that required for context storage. The disadvantage of a separate interrupt stack is added processing overhead for the stack switch. A combination of separate and common stack processing for ISRs can be used to allow for good RAM usage while providing good performance for higher frequency ISRs. Having two types of ISRs reduces the ISR uniformity, making it a less ideal method.

Y-Threads already provide a method of processing on a common stack. This means that all ISRs can be written to execute on the current stack, and if the ISR needs extra space for processing it can invoke an RCR. The Y-Threading ISR programming paradigm is uniform, and provides the flexibility for either fast or large ISRs.

5 Future Work

The main areas of future work are related to development tools. Adding Y-Thread support to a language and the development of Y-Thread compatible libraries is a major area of future work. As stated in [18] threading must be part of the language for the best performance, and not just implemented in a library. This is true for Y-Threads as well. Another area of future work is to statically profile the Y-Thread software.

5.1 Language/Compiler Support

Language/Compiler support can improve the ease of programmability, and reduce the overhead of invoking Y-Thread RCRs. Currently Y-Thread RCRs are implemented using a helper function to call the RCR. The calling syntax is less than ideal since the function pointer and the data parameter must be passed to the *oss_rcr_call* helper function. Ideally the programmer would just invoke the RCR as if it were a normal function invocation, and the compiler would generate the stack switching wrapper necessary for the RCR call. If the language/compiler natively supported Y-Threads it would allow for RCRs with signatures other than a single void pointer with a void return type. Language/Compiler Support also would reduce the overhead of an RCR invocation. The data parameters which currently need to be copied twice would be copied directly to the RCR stack. The overhead of the *oss_rcr_call* function call would also be removed. Overall the added overhead of an RCR invocation could be reduced to checking if currently on the RCR stack, entering/exiting of an atomic state, and switching of the stack pointer. However, as stated in the Section 2, the compiler would need to have knowledge of the OS, which most likely would bind the language/compiler to a single RTOS.

5.2 Y-Threads and Libraries

The initial development of Y-Threads was driven by the observation that embedded systems often run a cycle of work and waiting. It was also observed that blocking is often done at the highest level. However not all blocking is done in this manner. If layers are built upon blocking calls then the dedicated Y-Thread stack may need to be larger than if blocking were to be done at a shallower location. The feasibility of using blocking libraries in Y-Threads must still be determined. Development of libraries using Y-Threads that block at higher levels is one possible option. It is possible that these libraries could be accessed using some form of message passing or shared memory, but work in this area is necessary to determine the practicality. The development of Y-Thread friendly libraries would be a logical extension of language/compiler support for Y-Threads.

5.3 Static Profiling

If a compiler existed that natively supported Y-Threads it could determine RCR stack requirements and if an RCR blocked. Determining the RCR stack requirements could improve memory utilization allowing for more RAM to be available for the heap, main stack and data space. If higher priority thread preemption is not allowed during an RCR invocation and there are no recursive function invocations or function pointer invocations within the RCR, then the memory requirements can be determined for the RCR stack. Either an error or a warning could be generated by the compiler if it detects that an RCR can invoke a blocking function since blocking is not allowed in an RCR. Statically determining if an RCR can invoke a blocking function or not only requires that no function pointer invocations exist within the RCR or its function invocations.

The RCR stack only needs to accommodate for the worst case stack utilization for all RCRs and any functions they invoke. If recursion exists within any of the RCRs then it will not be possible to determine the RCR stack requirements since the recursion depth is bound at runtime[9]. Function pointer invocations are bound at runtime and therefore would make a compile time analysis impossible. If higher priority thread preemption is allowed then the number of simultaneous RCR invocations cannot be bound at compile time and therefore the RCR stack requirements cannot be determined.

Determining if an RCR can invoke a blocking function, is a simple matter of following all possible function invocations from the RCR. This is similar to compilers detecting dead code.

6 Conclusions

Y-Threads can be implemented easily on systems that currently support preemptive multithreading as is the case with OS*. The advantage of Y-Threads is that they provide a preemptive multithreaded programming model with good memory utilization.

Automatic allocation of memory on the RCR stack can be done with less overhead than dynamic allocation. The RCR call overhead is constant making it more desirable than dynamic memory allocation for systems with real-time requirements. The RCR call and automatic allocation overhead is slightly higher than statically allocated global buffer implementation without the static allocation of the memory. The RCR memory is available for use by any thread, where static global allocation of memory can only be used by a single thread. This need for having memory available for other threads is shown in the time synchronization application that had less than 41% of dedicated memory for its threads.

The hybrid of the two concurrency models in Y-Threads has allowed for the best of both models. Future work could further reduce overhead with language support. Language support could also add the ability to detect blocking in RCRs and the RCR stack requirements for further optimization and application correctness checking. The use of Y-Threads in embedded systems specifically WSN could reduce the time of application development when compared to event-based approaches. Y-Threads also have the potential to increase the capabilities of preemptive multithreaded systems due to better resource utilization.

Acknowledgments. This work is supported in part by NSF grants CNS-0435531, CNS-0520269, and EIA-0224469. The authors would also like to thank Joel Koshy and the anonymous referees for their insightful comments on an earlier draft of this paper.

References

[1] M. Maroti, B. Kusy, G. Simon, A. Ledeczi, The Flooding Time Synchronization Protocol, Proceedings of the second international conference on Embedded networked sensor systems, 2004.
[2] J. Elson, Time Synchronization in Wireless Sensor Networks, PhD Dissertation, 2003.
[3] N. Reijers, K. Langendoen, Efficient code distribution in wireless sensor networks, Proceedings of the 2nd ACM international conference on Wireless sensor networks and applications, 2003.

[4] S. S. Kulkarni, L. Wang, MNP: Multihop Network Reprogramming Service for Sensor Networks, 25th IEEE International Conference on Distributed Computing Systems, 2005.

[5] J. Hui, D. Culler, The Dynamic Behavior of a Data Dissemination Protocol for Network Programming at Scale, Proceedings of the 2nd international conference on Embedded networked sensor systems, 2004.

[6] J. Hill, R. Szewczyk, A. Woo, System Architecture Directions for Network Sensors, Architectural Support for Programming Languages and Operating Systems, pages 93–104, 2000.

[7] C. Han, R. Kumar, R. Shea, A Dynamic Operating System for Sensor Nodes, Proceedings of the 3rd international conference on Mobile systems, applications, and services, 2005.

[8] H. Lauer, R. Needham, On the Duality of Operating System Structures, Proceedings of the Second International Symposium on Operating Systems, IRIA, 1978.

[9] R. Behren, J. Condit, E. Brewer, Why Events Are a Bad Idea (for high concurrency servers), 9th Workshop on Hot Topics in Operating Systems (HotOS IX), 2003.

[10] J. Ousterhout, Why Threads Are a Bad Idea (for most purposes), Invited talk given at USENIX Technical Conference, 1996.

[11] T. Baker, Stack-Based Scheduling of Realtime Processes, Journal of Real-Time Systems, 3, 1991.

[12] OSEK/VDX Operating System Version 2.2.3, available at http://osek-vdx.org/mirror/os223.pdf, 2005.

[13] J. Koshy, R. Pandey, Remote Incremental Linking for Energy-Efficient Reprogramming of Sensor Networks, Proceedings of the Second European Workshop on Wireless Sensor Networks, 2005.

[14] Dunkels, O. Schmidt, and T. Voigt, Using Protothreads for Sensor Node Programming, Proceedings of the REALWSN'05 Workshop on Real-World Wireless Sensor Networks, 2005.

[15] Dunkels, B. Gronval, T. Voigt, Contiki – a Lightweight and Flexible Operating System for Tiny Networked Sensors, Proceedings of the 29th Annual IEEE International Conference on Local Computer Networks, 2004.

[16] A. Adya, J. Howell, M. Theimer, W. J. Bolosky, J. R. Douceur, Cooperative Task Management without Manual Stack Management or, Event-driven Programming is Not the Opposite of Threaded Programming, Proceedings of the 2002 USENIX Annual Technical Conference, 2002.

[17] S. Goldstein, K. Schauser, E. Culler, Lazy Threads: Implementing a Fast Parallel Call, Journal of Parallel and Distributed Computing, 1996.

[18] H. Boehm, Threads Cannot Be Implemented as a Library, Proceedings of the 2005 ACM SIGPLAN conference on Programming Language Design and Implementation, 2005.

Comparative Analysis of Push-Pull Query Strategies for Wireless Sensor Networks⋆

Shyam Kapadia[1] and Bhaskar Krishnamachari[1,2]

[1] Department of Computer Science,
University of Southern California, Los Angeles 90089, USA
[2] Department of Electrical Engineering,
University of Southern California, Los Angeles, CA 90089, USA

Abstract. We present a comparative mathematical analysis of two important distinct approaches to hybrid push-pull querying in wireless sensor networks: structured hash-based data-centric storage (DCS) and the unstructured comb-needle (CN) rendezvous mechanism. Our analysis yields several interesting insights. For ALL-type queries pertaining to information about all events corresponding to a given attribute, we examine the conditions under which the two approaches outperform each other in terms of the average query and event rates. For the case of ANY-type queries where it is sufficient to obtain information from any one of the desired events for a given attribute, we propose and analyze a modified sequential comb-needle technique (SCN) to compare with DCS. We find that DCS generally performs better than CN/SCN for high query rates and low event rates, while CN/SCN perform better for high event rates. Surprisingly, for the cases of ALL-type aggregated queries and ANY-type queries, we find that there exist "magic number" event rate thresholds, independent of network size or query probability, which dictate the choice of querying protocol. While our analysis is based on a single-sink square-grid deployment, we believe the insights can be generalized to random deployments.

1   Introduction

The primary function of a sensor network is to enable information gathering. The simplest strategy is to have all sensors provide a continuous stream of all the data that they gather to a sink node. However, for many classes of applications where only a small subset of the collected information is likely to be useful to end-users, this simple approach can become very inefficient. Researchers have therefore advocated the use of data-centric techniques which allow for efficient in-network storage and retrieval of named data using queries [1]. A number of data-centric querying and routing techniques have been proposed and examined in recent years: directed diffusion [2], TAG/TinyDB [3], rumor routing [4], hash-based

---

⋆ This work has been supported in part by NSF grants numbered CNS-0435505 (NeTS NOSS), CNS-0347621 (CAREER), CCF-0430061, and CNS-0325875 (ITR).

P. Gibbons et al. (Eds.): DCOSS 2006, LNCS 4026, pp. 185–201, 2006.
© Springer-Verlag Berlin Heidelberg 2006

data centric storage [5], hybrid push-pull [8], comb-needles [9], ACQUIRE [10], and TTL-based expanding search [11, 12].

With the presence of an increasing number of choices of data-centric storage and querying techniques, it becomes of crucial importance to understand and quantify their performance (both in absolute terms and with respect to each other) with respect to key application, network, and environmental parameters. In particular, carefully developed mathematical models can provide deep practical design insights on protocol selection as well as protocol parameter optimization for different sensor network deployments.

There are several interesting prior studies on analytical modelling of query strategies [5], [6], [7], [8], [9], [10], [12]. The energy costs of data centric storage are compared with the two extremes of external storage and local storage in [5]. A hybrid push-pull query processing strategy is proposed and analyzed in [8]. Push and pull alternatives of directed diffusion are also analyzed in [7]. Shakkotai [6] has presented a comparison of the asymptotic performance of three random walk-based query strategies, showing that a push-pull rendezvous-based sticky search has the best success probability over time. The optimal parameter setting for the comb-needles approach is analyzed in [9]. The optimal replication level for queries disseminated using expanding ring searches is analyzed in [12]. A common thread through much of this literature on the analysis of query techniques is the argument that tunable hybrid push-pull strategies offer significant advantages. Our work builds on and complements these existing studies, as we aim to compare two distinct and important approaches to hybrid push-pull querying.

Following the nomenclature used to classify peer-to-peer networks, we can distinguish between two main categories of hybrid push-pull query strategies: structured and unstructured. The structured approach is exemplified by geographic hash table-based data centric storage technique [5]. The data from sources is placed at a location using the same hash that the sink uses to retrieve it. This significantly simplifies the query since the sink effectively "knows" exactly where to look for the stored information. The unstructured approach to push-pull querying is exemplified by the comb-needle approach [9]. In this approach, the absence of a hash implies that the sink does not have prior knowledge of the location of the information. In that case, the queries are disseminated in the form of a comb with horizontal teeth, while the sources send event notifications independently in the form of limited vertical needles in either direction. The inter-teeth spacing and needle size are chosen and optimized to ensure that sources and sinks can rendezvous with each other efficiently. To the best of our knowledge, these two distinct and important approaches to hybrid push-pull querying — the structured DCS and the unstructured comb-needle technique — have never been compared to each other. This is our objective in this paper.

We undertake a mathematical analysis comparing the expected total energy costs of both these approaches on a grid-based sensor deployment. Our modelling of these query strategies allows us to study the impact of several key parameters such as the size of the network, the event and query rates, the use of data aggregation (using summaries), as well as the type of queries. For a fair comparison, we carefully select optimized versions of each strategy. In particular, we allow the storage location to be chosen optimally for the hash based data-centric storage scheme, and we use optimized inter-tooth spacing for the comb-needle approach.

We consider two important types of one-shot queries in this paper. We refer to the first query type as an ALL-type query. These are global discovery-type queries, such as 'Give me the location of all the lions in the sensor deployed area?' or 'Return the locations that have temperature $\geq 60°$ F'. In this case, the desired information must be obtained from all nodes in the network with relevant event information. The second type of query, which we refer to as an ANY-type query, is a one-shot query where any event that has the information can reply to the querier. Examples of such queries are 'Give me any location where a lion has been spotted in the sensor deployed area?' or 'Give me any location where the measured temperature is greater than 60 F'. For the ANY-type queries, the entire network need not be necessarily covered by the combs in the comb-needle strategy. Based on this insight, we propose and analyze a modified sequential comb-needle querying scheme (see Section 2.3).

Our analysis yields a number of useful insights into the relative performance of structured and unstructured approaches to hybrid push-pull querying. In all cases, we find that the unstructured comb-needle approach outperforms the data centric storage strategy when the number of events per epoch is large, while the reverse is true for small number of events, particularly for higher query rates. A particularly surprising and strong finding of our analysis is that under the assumptions of our modelling (large square grid network with a single caching-enabled querying sink located at bottom left) for the cases of aggregated ALL-type queries as well as the ANY-type queries, there exist "magic numbers" dictating which approach should be used for a given application scenario. In particular, for ALL queries, we find that if the number of events per epoch is greater than about 40 (regardless of the query rate or size of the network), the comb-needle strategy always outperforms data centric storage. For ANY queries, when the number of events per epoch is less than 1.56 (regardless of the query rate or size of the network) the data centric storage strategy always outperforms the sequential comb-needle strategy. However, if the number of events is greater than 3.16 (regardless of the query rate or size of the network), the sequential comb-needle strategy always outperforms data-centric storage.

Even though the topology considered in this study is a square grid (more amenable to analysis), we believe that similar magic numbers will be obtained in case of a random deployment. This is because the behavior of the querying strategies considered in this study scales in a similar manner with the network topology-related parameters and is more critically affected by the application parameters such as average event and query rates. This remains a subject of future investigation.

The rest of the paper is organized as follows. In Section 2, we present a brief overview of the algorithms to be analyzed in our paper: data centric storage (DCS), the basic comb-needle (CN) algorithm, and sequential comb-needle (SCN) algorithm. We specify our modelling assumptions in Section 3. We derive and compare the costs of data centric storage and comb-needle strategies with and without summary aggregation for ALL-type queries in Section 4. Then we analyze and compare data centric storage with the sequential comb-needle algorithm in Section 5. Finally, we discuss our key findings, along with directions for future work in the concluding Section 6.

2   Overview of Algorithms

2.1   Data Centric Storage (DCS)

The data centric storage query dissemination strategy uses distributed hash tables to store the event data sensed by a particular node (see Figure 1(a)). All the events of a particular event type (i.e. event having similar attributes) are hashed to the same node location. The data is then transported from the various event nodes along the shortest path to the node at the chosen location. Assuming the presence of location information, the authors propose to use GPSR to perform the routing. Queries for an event are then directed along the shortest path to these named location, since the query nodes also use the same hash function. The query responses are sent on the reverse path along which the query is forwarded.

2.2   Comb Needle (CN)

In this query dissemination strategy, the event nodes send out the sensed information vertically up and down like a spike (needle) of a certain length (see Figure 1(b)). Let the length of the needle be denoted by s. The sink then sends out a query that traverses the network along a comb. The separation between the teeth of the comb is also s to ensure that at least one comb teeth hits each needle, so as to not miss out any event nodes. The information requested is then sent back to the sink along the shortest path. Note that this strategy is used when the average number of queries, Q, is less than or equal to the average number of events, E. When

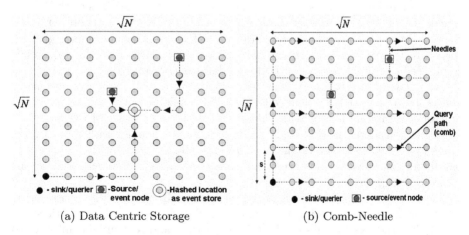

(a) Data Centric Storage   (b) Comb-Needle

Fig. 1. Illustration of the DCS and CN querying techniques

$Q > E$, a reverse comb-needle strategy is used where the query nodes form a needle and the event information is forwarded along a comb. Hence, the total cost for query dissemination in case of CN (C_{CN}) depends on the relationship between Q and E. However, in our case, since the sink is fixed and located at the left-bottom corner of the grid, we do not use the reverse comb needle scheme.

2.3 Sequential Comb-Needle (SCN)

The motivation behind introducing this query scheme is to efficiently resolve the ANY type queries in which case the query terminates as soon as the query hits the first event node of interest. In this way, the sequential comb needle scheme will always do better than the comb-needle scheme since it does not pay the extra cost incurred by the comb during query dissemination. In this scheme, similar to the comb-needle scheme, the event nodes form needles by spreading their information vertically to some nodes above and below them. The query originates from the sink and traverses the network as shown in Figure 2(a). Again, the size of the needles is denoted by s. Also, the distance between consecutive query horizontal traversals is s. The moment the query hits a node with the desired event information, the query path is truncated and the response is returned back to the sink.

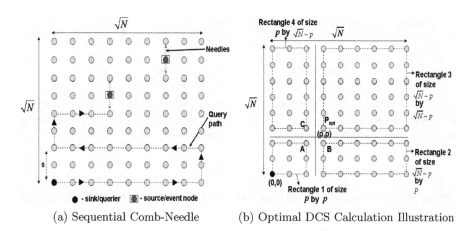

(a) Sequential Comb-Needle   (b) Optimal DCS Calculation Illustration

Fig. 2. Illustration of the SCN and the four-rectangle decomposition for calculating $d_2(p)$ - the average distance between all nodes and a storage point located at $P_{opt}(p,p)$, respectively

3   Model and Assumptions

We first present our modelling assumptions:

- We consider a $\sqrt{N} \times \sqrt{N}$ regular grid comprising N nodes. Each node has 4 neighboring nodes adjacent to it. Hence, the distances between the nodes are evaluated as Manhattan distances.

- Queries only occur at the sink node located at the left-bottom corner of the grid. This represents the interface of the sensor network to the outside world.
- We consider a time period, T, defined as an epoch. This represents the period of time when the event information stored by the nodes will be valid.
- Our analysis aims at optimizing the total expected energy cost incurred during each epoch. We use the total number of required unicast transmissions as the indicator of energy costs.
- Without loss of generality, we focus the analysis on queries and events for a single generic event attribute (i.e. event type). Events corresponding to this attribute are assumed to occur uniformly across the sensor network. Evaluation of complex queries comprising multiple attributes remains a topic of future research.
- We denote by E the average number of events that occur during epoch T.
- We denote by Q the expected number of queries that occur within epoch T. Since the event information does not change over an epoch, Q is always between 0 and 1 and represents the probability that a query is issued during that epoch.
- We assume the presence of a suitable MAC layer to handle collisions and contention.

4   Analysis of ALL-Type Queries

As mentioned earlier, ALL-type queries are of the type 'Give me all locations in the network where a lion was seen'. We first present the comparison of the data-centric storage and comb-needle scheme for such queries. We consider two cases: (a) When all the event information is sent to the sink (Without aggregation, i.e., with no summaries). (b) When only an aggregated summary of the event information is sent to the sink (With aggregation, i.e., using summaries).

4.1   Without Aggregation/Summaries

Cost of DCS with optimized hash location. Below, we calculate the average cost incurred in case of the DCS strategy in terms of the number of hops needed for query resolution in the epoch T.

There are 3 different query costs involved, C_{st} to store events, C_{qd} the query dissemination cost and C_{qr} the cost for the query response. Hence, we have the total cost in case of DCS given by,

$$C_{DCS} = C_{st} + C_{qd} + C_{qr} \qquad (1)$$

Since the position of the sink is fixed and known *a priori*, the DCS scheme can be optimized on the basis of the position of the hashed named location where all the event nodes send their data. Let $P_{opt}(x, y)$ denote the location of the node at that point. By symmetry, it is easy to see that P_{opt} will lie on the diagonal of the grid, otherwise, nodes on either side of the diagonal will have a larger distance to P_{opt}. Hence, they will pay more for transferring the event information to P_{opt}

as compared to the other nodes. Say, we have $x = y = p$. Let $d_1(p)$ denote the distance from the sink to P_{opt}, and $d_2(p)$ the average distance between any node on the grid to the node located at point P_{opt}.

Note that without summaries, all the event information has to be sent out in the reply to the sink. Hence, we have,

$$C_{DCS} = \min_p (d_2(p) \cdot E + d_1(p) \cdot Q + d_1(p) \cdot Q \cdot E)(\text{-Without summaries}) \quad (2)$$

We can determine the distance from the sink to P_{opt} trivially as

$$d_1(p) = x + y = 2p \quad (3)$$

The calculation of $d_2(p)$ is more involved. We can consider the grid as being divided into 4 rectangles as shown in Figure 2(b). The size of these rectangles is $p \times p$, $(\sqrt{N} - p) \times (\sqrt{N} - p)$, $p \times (\sqrt{N} - p)$ and $(\sqrt{N} - p) \times p$. The average distance from a node, located on a corner, to any node for a rectangle of size $X \times Y$ is given by Equation 38 (see Appendix), $\overline{D_{rect}} = \frac{X \cdot Y \cdot (X+Y-2)}{2 \cdot (X \cdot Y - 1)}$.

For rectangle 1, the average distance between the node on its right-top corner and the other nodes is given by $\frac{p^2}{p+1}$. Note that there are $p^2 - 1$ nodes in this rectangle other than the node at the right-top corner. Hence, the total distance between any node and the node on the right-top corner is given by,

$$D_1 = \frac{p^2}{p+1} \cdot (p^2 - 1) \quad (4)$$

Similarly, total distance for rectangle 3 is given by,

$$D_3 = \frac{(\sqrt{N} - p)^2}{\sqrt{N} - p + 1} \cdot ((\sqrt{N} - p)^2 - 1) \quad (5)$$

Since rectangles 2 and 4 are of the same size, we have their total distance given by,

$$D_2 = D_4 = 2 \cdot \frac{p(\sqrt{N} - p)}{2} \cdot \frac{\sqrt{N} - 2}{p(\sqrt{N} - p) - 1} \cdot (p(\sqrt{N} - p) - 1) \quad (6)$$

Also, note that the distance from A to P_{opt} is 2, while the distance from B and C to P_{opt} is 1. Hence, we need to add an additional $2p^2$ and $2p(\sqrt{N} - p)$ to account for the distances between all the points in rectangles 1, 2 and 4 to point P_{opt}.

From Equations (4), (5), and (6) we get,

$$d_2(p) = \frac{1}{N-1} \cdot \left[(\sqrt{N} - p)^2 \cdot (\sqrt{N} - p - 1) \right.$$
$$\left. + p \cdot (\sqrt{N} - p) \cdot (\sqrt{N} - 2) + p^2 \cdot (p - 1) + 2 \cdot p^2 + 2 \cdot p \cdot (\sqrt{N} - p) \right] \quad (7)$$

Simplifying the above expression we get,

$$d_2(p) = \frac{1}{N-1} \cdot \left[N \cdot \sqrt{N} - N - 2 \cdot N \cdot p + 2 \cdot \sqrt{N} \cdot p^2 + 2 \cdot \sqrt{N} \cdot p \right] \quad (8)$$

From Equation (2) we have,

$$C_{DCS} = \min_p (\frac{E}{N-1} \cdot \left[N \cdot \sqrt{N} - N - 2 \cdot N \cdot p \right.$$
$$\left. + 2 \cdot \sqrt{N} \cdot p^2 + 2 \cdot \sqrt{N} \cdot p \right] + 2 \cdot Q \cdot (E+1) \cdot p) \quad (9)$$

Using $\sqrt{N} + 1 = \sqrt{N} - 1 \approx \sqrt{N}$ and $N - 1 \approx N$, for large N, and simplifying the above equation we get,

$$C_{DCS} = \min_p ((\sqrt{N} - 2 \cdot p + \frac{2}{\sqrt{N}} \cdot p^2) \cdot E + 2 \cdot Q \cdot (E+1) \cdot p) \quad (10)$$

In order to determine the optimum value of p, we differentiate the above equation with respect to p and set it to 0. This yields the minimum value for C_{DCS} because the above expression is convex in p. Hence, we get the optimal value of p as,

$$p^* = \frac{\sqrt{N}(\sqrt{N}-1) \cdot E - Q \cdot (E+1) \cdot (N-1)}{2 \cdot E \cdot (\sqrt{N}+1)} \approx \frac{\sqrt{N}}{2 \cdot E}(E - Q \cdot (E+1)) \quad (11)$$

In the above expression, if $p^* \leq 0$, this implies that the event nodes should send all their information directly to the sink. This resembles the external storage scheme. In that case, the expression for C_{DCS} reduces to $\sqrt{N} \cdot E$. Hence, the cost for query dissemination then goes to 0. Also, the condition for which $p^* > 0$ is given by $Q < \frac{E}{E+1}$.

Putting the optimal value of p^* obtained from Equation (11) into Equation (10), we get the total cost for the DCS scheme (without summaries) as,

$$C_{DCS} = \begin{cases} \sqrt{N} \left[Q \cdot (1+E) - \frac{Q^2 \cdot (1+E)^2}{2 \cdot E} + \frac{E}{2} \right] & \text{if } Q < \frac{E}{E+1} \\ \sqrt{N} \cdot E & \text{Otherwise} \end{cases} \quad (12)$$

Cost of CN with optimized inter-tooth spacing. The derivation for the analysis of the comb-needle strategy is adapted from [9], however, here we use the exact expressions in case of the grid. First, we consider the case without summaries. As with DCS, there are 3 different costs involved, C_{needle} represents the needle costs for forwarding the event information to a subset of nodes, C_{comb} represents the query dissemination cost and C_{qr} represents the cost for the query response. Below, we present expressions for each of them.

$$C_{CN} = C_{needle} + C_{comb} + C_{qr} \quad (13)$$

Let s be the length of the needle formed by each node that senses an event. Then, the total needle cost is given by,

$$C_{needle} = s \cdot E \quad (14)$$

In the comb-needle strategy, the query is first sent out vertically upward from the sink and then fans out horizontally (see Figure 1(b)). The distance between consecutive horizontal fan outs is also s, also known as the teeth separation for the comb.

$$C_{comb} = (\sqrt{N} - 1) \cdot (1 + (\lceil \frac{\sqrt{N} - 1}{s} \rceil + 1)) \cdot Q \approx 2 \cdot \sqrt{N} \cdot Q + \frac{N \cdot Q}{s} \quad (15)$$

Note that the ceil is present because there is a horizontal fan out at $(0,0)$ and $(\sqrt{N} - 1, 0)$. Assuming, that each node where the comb tooth intersects with the needle, replies along the shortest path to the sink (see Appendix 6), we have the total query response cost given by,

$$C_{qr} = \frac{N}{\sqrt{N} + 1} \cdot E \cdot Q \approx \sqrt{N} \cdot E \cdot Q \quad (16)$$

Hence, the total cost for the comb needle strategy is given by,

$$C_{CN} = s \cdot E + 2 \cdot \sqrt{N} \cdot Q + \frac{N \cdot Q}{s} + \sqrt{N} \cdot E \cdot Q \quad (17)$$

Now we find the value of s that minimizes this total query cost. On solving we get, $s^* = \sqrt{N} \cdot \sqrt{\frac{Q}{E}}$

Hence, the total cost with the comb needle scheme without summaries is given by,

$$C_{CN} = \sqrt{N} \cdot (2 \cdot Q + 2 \cdot \sqrt{Q \cdot E} + E \cdot Q) \quad (18)$$

Comparison of DCS and CN. Figure 3(a) and 3(b) compare the normalized expected cost of querying (which is calculated as the total expected cost divided by the square-root of the number of nodes) with the DCS and CN strategies with respect to the two key parameters E and Q. Note that from Equations (12) and (18), we can see that the total expected cost of querying is proportional to $\sqrt{N}$ for both DCS and CN. We observe that CN outperforms DCS as the average number of events per epoch increases, while DCS outperforms CN when the per-epoch query probability increases.

Figure 4 shows the regions in the E-Q plane where DCS and CN outperform each other. This is generated by obtaining the zero-contour of the surface representing the difference in cost between DCS and SCN as a function of E and Q. We note that the equal-cost curve grows slowly with respect to E[1]. In particular, here, there is no threshold event rate beyond which CN is always better regardless of the query rate — we shall see later that this is not always the case.

4.2   With Aggregation/Summaries

Cost of DCS with optimized hash location. With summaries, all the event information can be compressed into a single packet and sent out to the sink,

---

[1] This can be shown rigorously in terms of the derivative of that curve, but we do not present that analysis here due to lack of space.

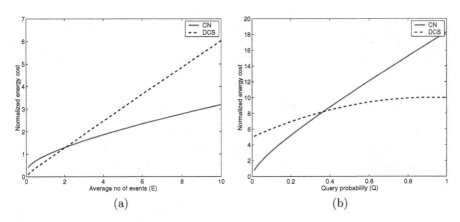

Fig. 3. Cost of CN and DCS for ALL-type queries, without summary aggregation, with respect to E (for Q = 0.1) and Q (for E = 10)

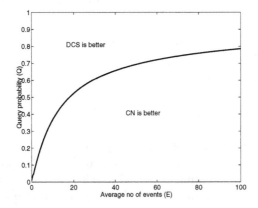

Fig. 4. Relative Performance of CN and DCS for ALL-type queries, without summary aggregation, with respect to event and query rates

hence, we have,

$$C_{DCS} = \min_{p} \left(d_2(p) \cdot E + d_1(p) \cdot Q + d_1(p) \cdot Q \right)(\text{-With summaries}) \quad (19)$$

Using a similar procedure to that used earlier for the case without summaries, since only the reply cost is different and everything else is the same, we obtain the total cost for DCS with summaries as,

$$C_{DCS} = \begin{cases} \sqrt{N} \left[2 \cdot Q - \frac{2 \cdot Q^2}{E} + \frac{E}{2} \right] & \text{if } Q < \frac{E}{2} \\ \sqrt{N} \cdot E & \text{Otherwise} \end{cases} \quad (20)$$

Cost of CN with optimized inter-tooth spacing. We now describe the CN cost with summaries. The query dissemination cost, C_{qd} and the needle cost

C_{needle} remain the same as was the case without summaries. For the reply cost, we note that reply from the various events can be aggregated on the way back to the sink. To account for this aggregation we approximate the reply cost to be the same as the cost for the comb i.e. the cost for query dissemination. This is because the events need only send their data horizontally toward the sink, the vertical path downward toward the sink will account for the aggregation. Hence, now we have the total cost for CN given by,

$$C_{CN} = C_{needle} + C_{comb} + C_{qr} = C_{needle} + 2 \cdot C_{comb} \qquad (21)$$

$$C_{CN} = s \cdot E + 4 \cdot Q \cdot (\sqrt{N} - 1) + 2 \cdot (\sqrt{N} - 1) \cdot (\lceil \frac{\sqrt{N} - 1}{s} \rceil) \cdot Q$$
$$\approx s \cdot E + 4 \cdot Q \cdot \sqrt{N} + 2 \cdot Ns \cdot Q \qquad (22)$$

Again, we solve for the optimum s to get, $s^* = \sqrt{N} \cdot \sqrt{\frac{2 \cdot Q}{E}}$

Hence, the total cost with the comb needle scheme with summaries is given by,

$$C_{CN} = 2 \cdot \sqrt{2 \cdot N \cdot Q \cdot E} + 4 \cdot \sqrt{N} \cdot Q \qquad (23)$$

Comparison of DCS and CN. Figures 5(a) and 5(b) compare the normalized expected cost of storage and querying with the DCS and CN strategies with respect to the two key parameters E and Q. We observe that even with summaries CN outperforms DCS as the average number of events per epoch increases, while DCS outperforms CN when the per-epoch query probability increases.

Figure 6 shows the regions in the E-Q plane where DCS and CN outperform each other. We can see that (unlike in the case without summaries) there exists an threshold Θ for the event rate beyond which CN is always better. This threshold can be derived analytically.

First, we can prove that when $Q \geq E/2$, $C_{DCS} = \sqrt{N}E$ is always smaller than C_{CN}, hence there is no solution for $C_{DCS} - C_{CN} = 0$ in this case. When $Q < E/2$, then we can write the expression for the equal-costs curve as follows:

$$\sqrt{N} \left[(2Q - \frac{2Q^2}{E} + \frac{E}{2}) - (2\sqrt{2QE} + 4Q) \right] = 0 \qquad (24)$$

As can be seen from the figure, the threshold event rate corresponds to the point when there is a query at every epoch. Setting $Q = 1$, and solving the above expression for E, we find that the threshold $\Theta \approx 39.78$. An important point to note is that this threshold is a "magic number" that is independent of the size of the network. It tells us a surprising design lesson: for a grid-based network where ALL-type queries are always injected from the bottom left corner, if there are more than 40 events on average in each epoch that must be aggregated in response to queries, then a comb-needle approach is preferable in terms of total energy cost as compared to a hash-based data-centric storage approach.

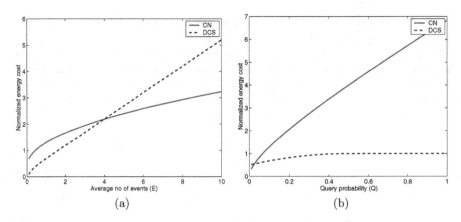

Fig. 5. Cost of CN and DCS for ALL-type queries, with summary aggregation, with respect to E (for Q = 0.1), and with respect to Q (for $E = 1$)

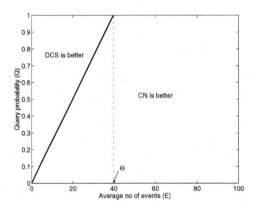

Fig. 6. Relative Performance of CN and DCS for ALL-type queries, with summary aggregation, with respect to event and query rates

5   Analysis of ANY-Type Queries

Recall that in case of ANY-type queries, the query need not visit every node in the network, it should be terminated as soon as it hits a node that has the desired information. Here, for such query types, we obtain the expressions for the data-centric storage scheme and the modified comb and needle scheme which we call the sequential comb-needle (SCN) scheme.

Cost of DCS. The cost for ANY-type queries remains the same as that obtained for ALL-type queries with summaries. This is because the data centric storage scheme stores all the information about a given event type at a named location. Hence, the reply to the ANY-type query can be considered similar to just returning the summary. Hence, the cost in case of DCS can be obtained from Equation 20.

Cost of SCN. We now derive the cost for the sequential comb-needles (SCN) approach. To determine the cost for the query transmission we need to obtain the average number of hops/transmissions till a node with the desired event information is hit. Since each event node replicates the data to s other nodes, and the separation between successive horizontal traversals along the query path is also s, the original grid with N nodes can be transformed to a new grid with $\frac{N}{s}$ nodes. The sequential comb-needle scheme then traverses this new grid as a chain of $\frac{N}{s}$ nodes. Denote $n = \frac{N}{s}$. Let X be a random variable that determines the number of hops till a event node is hit by the query. Note that, even in this compressed chain, $\frac{E \cdot s}{s} = E$ is the number of event nodes. Now, we have the cdf of X given by,

$$F_X(k) = P(X \leq k) = 1 - (\frac{n - k}{n})^E \qquad (25)$$

We can now obtain the pmf of X as,

$$p_X(k) = P(X \leq k) - P(X \leq k - 1) = (1 - \frac{k - 1}{n})^E - (1 - \frac{k}{n})^E \qquad (26)$$

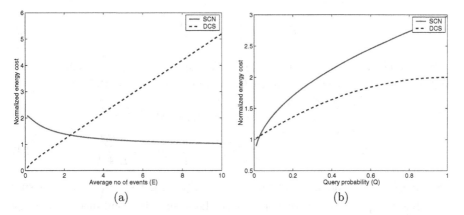

Fig. 7. Cost of SCN and DCS for ANY-type queries, with summary aggregation, with respect to E (for $Q = 0.1$), and with respect to Q (for $E = 2$)

Now the expected value of X can be obtained by using Equation 26 as follows,

$$E[X] = \sum_{k=1}^{n-1} k \cdot ((1 - \frac{k - 1}{n})^E - (1 - \frac{k}{n})^E) \qquad (27)$$

Let $f(k) = (1 - \frac{k}{n})^E$. Then we get,

$$E[X] = \sum_{k=1}^{n-1} k \cdot (f(k - 1) - f(k)) \qquad (28)$$

This summation can be opened up so that the consecutive terms can be grouped together to leave,

$$E[X] = \sum_{k=1}^{n-1} f(k) - n \cdot f(n) = \sum_{k=1}^{n-1} (1 - \frac{k}{n})^E \tag{29}$$

Note that $f(n) = 0$, hence, in the above expression by substituting $j = n - k$, we get,

$$E[X] = \sum_{j=1}^{n-1} (\frac{j}{n})^E = \frac{1}{n^E} \sum_{j=1}^{n-1} j^E \tag{30}$$

Approximating the summation by an integration, we get,

$$E[X] \approx \frac{1}{n^E} \cdot \frac{n^{E+1}}{E+1} = \frac{n}{E+1} = \frac{\frac{N}{s}}{E+1} \tag{31}$$

Note that E[X] just accounts for the number of horizontal steps taken by the SCN query path. We also need to account for the vertical steps that it takes. This can be approximated by determining the y-coordinate of the point where SCN hits the first event node. This is given by,

$$E[Y] = \frac{\frac{N}{s \cdot (E+1)}}{\sqrt{N}} \cdot s = \frac{\sqrt{N}}{E+1} \tag{32}$$

For simplicity, we assume that the query response path is the same as that taken by the query. Now, we can get the total cost in case of SCN as,

$$C_{SCN} = C_{needle} + C_{qd} + C_{qr} \tag{33}$$

$$C_{SCN} = s.E + \frac{\frac{N}{s}}{E+1} \cdot Q + \frac{\frac{N}{s}}{E+1} \cdot Q + 2 \cdot \frac{\sqrt{N}}{E+1} \tag{34}$$

Solving for the value of s that minimizes C_{SCN} we get $s^* = \sqrt{\frac{2 \cdot N \cdot Q}{E \cdot (E+1)}}$. Using this value, we get the total cost in case of the sequential comb-needle strategy as,

$$C_{SCN} = 2 \cdot \sqrt{\frac{2 \cdot N \cdot Q \cdot E}{E+1}} + 2 \cdot \frac{\sqrt{N}}{E+1} \tag{35}$$

Comparison of DCS and SCN. Figures 7(a) and 7(b) compare the normalized expected cost of storage and querying with the DCS and SCN strategies with respect to the two key parameters E and Q. We observe that SCN outperforms DCS as the average number of events per epoch increases, while DCS outperforms SCN when the per-epoch query probability increases.

Figure 8 shows the regions in the E-Q where DCS and SCN outperform each other. We can see that in this case, there are two significant thresholds for the event rate. Below a lower threshold Θ_{lower}, we find that DCS is always better (regardless of the query probability), and above an upper threshold Θ_{upper}, SCN is always better (regardless of the query probability). These "magic numbers" can be derived analytically.

First, similar to the analysis of the DCS and CN strategies with aggregated responses for the ALL-type queries, we can prove that when $Q \geq E/2$, $C_{DCS} = \sqrt{N}E$ is always smaller than C_{SCN}. When $Q < E/2$, then we can write the expression for the equal-costs curve as follows:

$$\sqrt{N}\left[(2Q - \frac{2Q^2}{E} + \frac{E}{2}) - (2\sqrt{\frac{2QE}{E+1}} + \frac{2}{E+1})\right] = 0 \qquad (36)$$

As can be seen from the figure, the threshold event rate corresponds to the point when there is a query every epoch. Setting $Q = 0$, and solving the above expression for E, we get the lower threshold $\Theta_{lower} \approx 1.56$. Setting $Q = 1$ and solving the above expression for E, we find that the threshold $\Theta_{upper} \approx 3.16$.

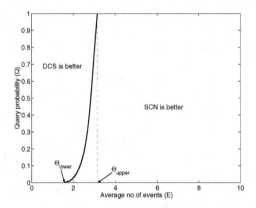

Fig. 8. Relative Performance of SCN and DCS for ANY-type queries with respect to event and query rates

6   Conclusions and Future Research Directions

We have presented a comparative analysis of two distinct and important approaches to hybrid push-pull querying in wireless sensor networks - the structured hash-based DCS, and the unstructured CN/SCN. We have examined their performance with respect to key environment, network, and application parameters including the event and query rates, network size, type of query, and the use of in-network aggregation.

We have found that the costs of DCS, CN, and SCN are all directly proportional to the square-root of the number of nodes in the network. Therefore, the relative performance of DCS versus CN/SCN is unaffected by network size. The exact shape of the relative best performance regions for the two approaches do change depending on the query type (ALL, ANY) and the use/non-use of summary aggregation; however, we find in all cases that the unstructured CN/SCN approach generally outperforms the DCS strategy when the number of events per epoch is large, while the reverse is true for small number of events, particularly for higher query rates. A possible explanation for this is that, relatively speaking, the query cost burden is reduced in structured strategies like DCS when compared with an unstructured strategy like CN/SCN because the use of hashing provides a predetermined location to pick up information about all events. But this comes at the expense of a higher cost burden in event notification since all events must be transmitted to a generally non-local hash location. Thus a hash-based push-pull scheme like DCS favors high query rates but low event rates, compared to a path-intersection based push-pull scheme like CN/SCN. It is possible that considering replication of event storage locations in DCS changes this tradeoff. This can be explored in future work.

Our analysis reveals the existence of event rate thresholds for aggregate ALL-type queries ($\Theta \approx 39.78$) as well for as ANY-type queries ($\Theta_{lower} \approx 1.56, \Theta_{upper} \approx 3.16$), that dictate which protocol should be used in a given application scenario regardless of the query probability. Moreover, we believe that these magic numbers will exist even in the case of a random deployment of sensor nodes. This remains a promising future research direction. We are currently implementing simulations and considering extending the analysis to further study the behavior of these strategies with a random deployment of sensor nodes.

Besides offering some concrete guidelines for practitioners, this study suggests a number of other interesting directions for future work. Our analysis can be extended to include other querying protocols enabling comparison of various proposed schemes under a common framework. These include extensions of the analysis taking into account different deployment topologies, different cost metrics (including other energy models, as well as delay), different types of queries (for example, complex queries involving multiple attributes) and allowing multiple querying sinks. The theoretical results we present should also be validated through experiments on a real application/test-bed.

Acknowledgements

We'd like to thank the members of the USC Autonomous Networks Research Group for their feedback on this paper. A special thanks to Joon Ahn, Sundeep Pattem, and Kiran Yedavalli for their technical input.

References

1. R. Govindan, "Data-Centric Storage in Sensor Networks, in Wireless Sensor Networks", (T. Znati, K. Sivalingam, C. S. Raghavendra Ed.), Kluwer Publishers, 2003.
2. C. Intanagonwiwat, R. Govindan, and D. Estrin, Directed Diffusion: A Scalable and Robust Communication Paradigm for Sensor Networks, In Proceedings of the Sixth Annual International Conference on Mobile Computing and Networks (MobiCOM), August 2000.

3. S. Madden, M.J. Franklin, J.M. Hellerstein, and W. Hong, TAG: a Tiny AGgregation Service for Ad-Hoc Sensor Networks, 5th Symposium on Operating System Design and Implementation (OSDI 2002), December 2002
4. D. Braginsky and D. Estrin, "Rumor Routing Algorithm For Sensor Networks", The First Workshop on Sensor Networks and Applications (WSNA'02), October 2002.
5. S. Shenker, S. Ratnasamy, B. Karp, R. Govindan, and D. Estrin, "Data-Centric Storage in Sensornets", ACM SIGCOMM, Computer Communications Review, Vol. 33, Num. 1, January 2003.
6. S. Shakkottai, Asymptotics of Query Strategies over a Sensor Network, INFO-COM'04, March 2004
7. B. Krishnamachari and J. Heidemann, "Application-Specific Modelling of Information Routing in Wireless Sensor Networks," Workshop on Multihop Wireless Networks (MWN'04) held in conjunction with the IEEE International Performance Computing and Communications Conference (IPCCC), April 2004.
8. N. Trigoni, Y. Yao, A. Demers, J. Gehrke, and R. Rajaraman. "Hybrid Push-Pull Query Processing for Sensor Networks", In Proceedings of the Workshop on Sensor Networks as part of the GI-Conference Informatik 2004. Berlin, Germany, September 2004.
9. X. Liu, Q. Huang, Y. Zhang, "Combs, Needles, Haystacks: Balancing Push and Pull for Discovery in Large-Scale Sensor Networks", ACM Sensys, November 2004
10. N. Sadagopan, B. Krishnamachari, and A. Helmy, "Active Query Forwarding in Sensor Networks (ACQUIRE)", *Ad Hoc Networks Journal-Elsevier Science*, Vol. 3, No. 1, pp. 91-113, January 2005.
11. N. Chang and M. Liu, "Revisiting the TTL-based Controlled Flooding Search: Optimality and Randomization", Proceedings of the Tenth Annual International Conference on Mobile Computing and Networks (ACM MobiCom), September, 2004, Philadelphia, PA.
12. B. Krishnamachari and J. Ahn, "Optimizing Data Replication for Expanding Ring-based Queries in Wireless Sensor Networks", *USC Computer Engineering Technical Report CENG-05-14*, October 2005.

Appendix

Average distance between a node located at the bottom-left corner and any other node within a $X \times Y$ rectangular grid.

This can be expressed by the following summation:

$$\overline{D_{rect}} = \frac{\sum_{i=0}^{X-1} \sum_{j=0}^{Y-1} (i+j)}{X \cdot Y - 1} \tag{37}$$

Evaluating the above expression, we get

$$\overline{D_{rect}} = \frac{X \cdot Y \cdot (X + Y - 2)}{2 \cdot (X \cdot Y - 1)} \tag{38}$$

Note that from this by setting $X = Y = \sqrt{N}$, we have the distance from the node at one corner to any point in the $\sqrt{N}$ by $\sqrt{N}$ square grid:

$$\overline{D_{square}} = \frac{N}{\sqrt{N} + 1} \tag{39}$$

Using Data Aggregation to Prevent Traffic Analysis in Wireless Sensor Networks

William Conner, Tarek Abdelzaher, and Klara Nahrstedt

Department of Computer Science, University of Illinois at Urbana-Champaign,
Urbana, Illinois, USA 61801-2302
{wconner, zaher, klara}@uiuc.edu

Abstract. When communication in sensor networks occurs over wireless links, confidential information about the communication patterns between sensor nodes could be leaked even when encryption is used to protect the actual contents of the messages. The communication patterns, which often reveal higher volumes of traffic near the sink, could allow an attacker to identify the vicinity of the sink node. With this information, an attacker could potentially disable the network by destroying the sink. In this paper, we present the *decoy sink protocol*, which protects the location of the sink in target tracking sensor network applications by forwarding data to a decoy sink for aggregation before the aggregated data is forwarded to the real sink from the decoy sink. Combining indirection and data aggregation in our protocol creates more traffic away from the sink and reduces the amount of traffic near the sink, which makes traffic analysis more difficult for attackers.

Keywords: Traffic analysis prevention, data aggregation, sensor networks.

1 Introduction

In recent years, many applications have been implemented on top of wireless sensor networks. The types of sensor network applications and their primary users vary greatly with applications ranging from habitat monitoring [1] used by scientists to shooter localization [2] used by soldiers. One particular class of sensor network applications that has received much attention from the sensor network research community is target tracking [3],[4],[5]. In target tracking applications, the sensors in the network collectively monitor the current position of one or more objects meeting some user-defined criteria (e.g., exceeding some amount of ferrous content as suggested in [6]). Target tracking can be used to monitor the locations of many types of objects, such as tanks in military scenarios [6] or animals in wildlife scenarios [7].

Although researchers developing target tracking applications have given much consideration to challenges, such as target detection [5] and handling multiple targets [4], very little attention has been given to the security of such systems. As Deng et al. observe, communication patterns in sensor networks are often very asymmetric with significantly more traffic appearing near the sink [8]. Many target tracking applications happen to exhibit this asymmetric communication pattern.

P. Gibbons et al. (Eds.): DCOSS 2006, LNCS 4026, pp. 202–217, 2006.
© Springer-Verlag Berlin Heidelberg 2006

In sensor networks, if the sink node and other nodes near the sink have higher amounts of traffic than nodes further away from the sink, then an opportunity exists for adversaries equipped with electronic listening devices to locate the sink through traffic analysis even if all the sensor nodes and sink are camouflaged in the background, as mentioned in [8]. As Deng et al. also observe in [16], such an attack does not depend on the adversary viewing the message contents. Therefore, symmetric encryption schemes, such as those described in [9],[10], and asymmetric encryption schemes, such as those described in [11],[12], will not protect the sink's location from being discovered by the adversary.

Assuming that the sink is the interface between the sensor network and the target tracking application's remote data storage and analysis backend, then an adversary can prevent target tracking by disabling the sink through physical destruction (e.g., detonating an explosive device in the sink's vicinity). Locating the sink allows an adversary to execute such attacks very efficiently since attacks can be focused on the sink and its immediate surrounding area. For example, consider the target tracking application depicted below in Figure 1. If an adversary determined the approximate location of the sink through traffic analysis, then it could physically damage the area surrounding the sink through some means, such as detonating an explosive device in the area shown by the shaded region in Figure 1. If the explosive destroys all the sensors in the shaded region including the sink, then the remote facility cannot acquire data from the sensor network.

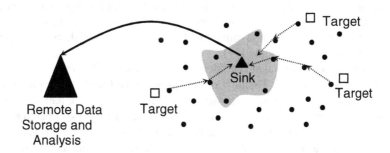

Fig. 1. Example application with remote storage and analysis

Preventing traffic analysis in sensor networks is a difficult problem. Since communication is over a wireless medium, adversaries can easily detect the existence of communication by using electronic listening devices. Although traditional traffic analysis prevention techniques, such as traffic padding described in [15], are useful in wired networks and some ad hoc networks, these techniques might consume too much energy in wireless sensor networks. Therefore, strategies that prevent traffic analysis in sensor networks must be designed for energy efficiency in addition to security.

In this paper, we present a novel approach for preventing traffic analysis in target tracking sensor network applications by combining the use of indirection and data aggregation. In our approach, the basic idea is that all sensors will first send their readings to some designated non-sink node (referred to as a *decoy sink node*) that will aggregate these readings into summary messages that will then be forwarded to the real sink node. Since the traffic pattern towards the decoy sink node will increase and the traffic pattern towards the real sink will decrease (due to aggregation), we are able to protect the location of the real sink from adversaries performing traffic analysis. We have extended our basic idea of using a single decoy sink node to our final solution where multiple decoy sink nodes are used to further increase the randomness of traffic patterns and provide robustness should an adversary attack a decoy sink node mistakenly believed to be the real sink. In the next section, we will discuss related work. In section 3, we will present our solution to preventing traffic analysis in target tracking sensor networks. Our performance evaluation follows in section 4. Finally, we will conclude in the last section.

2 Related Work

Much work has previously been done on preventing traffic analysis in traditional wired networks for applications such as electronic mail and Web browsing. In such networks, the major concern has typically been the prevention of eavesdroppers from determining the endpoints of a communication by protecting the identities of the source and destination. Onion routing, for example, provides anonymous communication by sending data from a source through a series of onion routers before the data ultimately reaches its destination [13]. Each message has several layers of encryption, where each layer corresponds to exactly one onion router along the path. Each layer is decrypted by the corresponding onion router to determine the next hop in the path for forwarding. Freenet, which is a peer-to-peer system that protects the anonymity of both content publishers and content downloaders, is another example [14].

Some previous work has also been done on preventing traffic analysis in wireless sensor network applications. Phantom routing protects the location of source sensors, which detect and report events, by having source sensors first route data in a random direction toward a phantom source and then route the data from that phantom source to the sink through flooding [17]. Phantom routing assumes that an adversary is initially positioned at the sink with a listening device. Phantom routing intends to protect the source sensors' locations by creating the illusion that sensor data is coming from several different directions (i.e., phantom sources). This prevents the adversary from backtracking to the real source location because a steady stream of data from a stationary source to the sink is not available. Our approach differs from phantom routing in several ways. First, we are concerned with protecting the sink node rather than source nodes. Also, our solution relies on data aggregation to protect the sink's location rather than depending solely on indirection. Unlike phantom routing, our approach does not assume that adversaries have a limited view of the sensor network.

Deng et al. also present algorithms for protecting the location of the sink in a wireless sensor network in order to prevent physical attacks made possible by locating the sink through traffic analysis [16]. Their approach is based on four techniques. The first two techniques introduce variation in the multi-hop path taken from a sensor node to the sink by having sensors forward data to a randomly chosen parent from among multiple parents and also taking random walks along the way. The third technique generates fake packets at nodes forwarding real data with a certain probability and these fake packets are then routed along random fake paths to non-existent sinks. The last technique introduces several random areas of high communication activity by creating random fake paths with a higher probability towards areas that have forwarded fake packets in the past. Unlike the algorithms presented in [16], our approach does not rely on creating additional fake packets to obscure the communication patterns toward the sink. Rather than adding fake traffic to hide the sink's location, we have chosen to first aggregate data away from the sink before forwarding the summarized data to the sink in an effort to conceal its location.

3 Decoy Sinks

The decoy sink protocol, which is presented in more detail later in this section, is a novel approach to traffic analysis prevention in wireless sensor network target tracking applications. Our solution combines indirection, which has previously been used in traffic analysis prevention [13],[14],[16],[17], and data aggregation, which is often used in sensor networks to save energy [18],[19]. We use this combination to protect the location of the sink. To the best of our knowledge, we are the first to use in-network data aggregation specifically for the purpose of preventing traffic analysis. Before describing the decoy sink protocol in depth, we will first discuss the various assumptions in our sensor network model, application model, and attack model.

3.1 Sensor Network Model

The sensor network model that we assume in the decoy sink protocol is quite simple. Many simplifying assumptions were made so that we could focus on our specific research problem of preventing traffic analysis rather than addressing the details of other open sensor network problems outside the scope of this paper.

In our model, the network consists of a large number of nodes equipped with both sensors for target detection and radios for communication. The specific type of sensor used is intentionally left unspecified since that choice is not constrained by our protocol. We also do not assume that the radios are capable of employing spread spectrum techniques to interfere with the ability of the attacker to detect communications with its listening device. Although spread spectrum techniques can be used in sensor networks consisting of MicaZ motes, our protocol is still useful for sensor networks consisting of older generations of motes, such as Mica2 [25],[26]. Sensors are also assumed to know their own locations through either manual configuration or a localization algorithm, such as in [20],[21],[22]. In our model, sensors are assumed to have energy constraints, so energy is a major concern in our protocol.

Our sensor network model assumes that communication between the sensors and the sink is over a multi-hop wireless network since our networks might consist of hundreds or even thousands of sensor devices. Due to the loss characteristics of multi-hop wireless sensor networks, as discussed in [23], we assume that there will be some probability p of packet loss at each hop.

3.2 Application Model

The type of application running on top of our sensor network is a simple target tracking application where we assume that sensors sample the environment once per sampling period in which they either detect a target or do not detect a target. If one or more sensors detect a target, then those sensors that detect the target will collaboratively apply some aggregation function to all of their location information and send a summarized report back to the sink over possibly multiple wireless hops. The reports are sent along a shortest path routing tree back to the sink. If a sensor does not detect a target, then no reports are created and sent from that particular node (however, that node might forward reports originating at other nodes).

The specific type of targets considered and detection measurements used in our application model is intentionally left unspecified since our decoy sink protocol is general enough to be useful in many different target tracking application scenarios. For specific target detection techniques, the reader should consider one of the target tracking applications presented in [3],[4],[5],[6]. Figure 2(a) below shows an example target tracking application where nodes send messages identifying their location back to the sink if they detect the target.

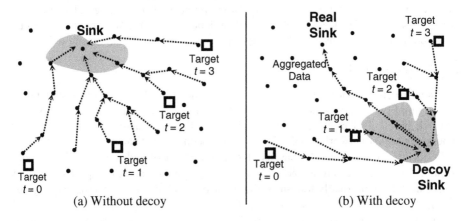

(a) Without decoy   (b) With decoy

Fig. 2. Example target tracking application

3.3 Attack Model

When discussing the decoy sink protocol in the next section, we assume that all messages are encrypted and that adversaries are not capable of breaking the cryptosystems used for encryption. Therefore, adversaries cannot determine the contents of messages. However, we also assume that adversaries have a global view of the sensor network's communication patterns over long periods of time. For example, an adversary might sparsely deploy several low-end listening devices capable of counting the number of transmissions overheard. Upon collecting these devices after some period of time and analyzing their data, the adversary can then determine the areas of highest communication (i.e., the best candidate areas for physical destruction).

Although it would be costly for the adversary, tightly synchronized high-end listening devices that communicate with each other might be able to collectively figure out the direction packets are traveling using timing correlation as described in [16]. However, we do not assume tightly synchronized high-end listening devices that communicate with one another to perform traffic analysis, because we assume the adversary wants to execute an inexpensive and efficient attack on the sink area that does not consume a lot of resources. If a dense field of high-end listening devices were required, then it might be more efficient for the adversary to destroy the entire field rather than attempting to locate the sink for a focused attack (e.g., detonating several explosive devices might be cheaper than deploying several high-end listening devices).

Phantom routing and the GSAT test used by Deng et al. to evaluate their solution both assume that adversaries have a view of traffic over a limited surrounding area within the sensor network with a single attacker gradually moving (based on local decisions) towards areas of higher traffic until they reach the source sensors or the sink, respectively [16],[17]. Our attack model places fewer restrictions on the adversary than these two assumptions since we assume that adversaries can have a global view of the number of transmissions received by each sensor. Our attack model makes assumptions closer to the other evaluation criteria presented by Deng et al. that measures the overall randomness in the entire sensor network's traffic patterns [16].

3.4 Decoy Sink Protocol

The decoy sink protocol that we propose is straightforward. Rather than forwarding data directly to the sink along the shortest path, sensors will forward their readings to an intermediate node other than the sink, which we refer to as the *decoy sink*. Before reaching the decoy sink, the data is first aggregated locally in the vicinity of the detected target. This aggregation is similar to the aggregation already used in many target tracking applications, such as [6]. We will refer to this sort of aggregation as *local aggregation*. Although other aggregation functions could be used, the local aggregation function that we consider is that every node that detects the target will broadcast its estimated distance from the target to its neighboring sensors and only the closest node to the target will actually forward its data to the decoy sink. The decoy sink will then perform additional aggregation on the aggregated data received from the sensors. Any further aggregation done by the decoy sink will be referred to as *remote aggregation*. After performing remote aggregation, the decoy sink will finally forward the summarized data to the real sink. While local aggregation is done to save energy, remote aggregation is done specifically to prevent traffic analysis by reducing the amount of data headed to the real sink.

In-network Data Aggregation. Sending a stream of readings to a decoy sink before forwarding them to the real sink creates a high traffic area near the decoy sink but it does not necessarily conceal the high traffic area near the real sink. To ensure that the amount of traffic near the real sink is reduced, we have the decoy sink perform remote aggregation on the readings before sending a stream of summarized data back to the real sink. Since the real sink is now receiving fewer messages from the decoy sink than it would otherwise receive directly from the sensors, the amount of traffic headed towards the real sink is significantly reduced. The contrast between Figure 2(b) and 2(a) illustrates the approach of the decoy sink protocol.

Multiple Decoy Sinks. One problem that could potentially arise when the decoy sink protocol is used is that attackers can use characteristics of the protocol to their advantage. One potential exploit would be for an attacker to physically destroy the decoy sink, which can be located by finding the highest traffic area. This traffic analysis attack is similar to the attack used on the real sink and might occur if an adversary mistakenly believes that the decoy sink is the real sink. Since all of the sensor data goes through the decoy sink, the sensor network target tracking application is now effectively disabled if the decoy sink is destroyed.

Another potential attack that uses characteristics of the decoy sink protocol is that an attacker could infer the location of the real sink by searching for areas that do not have the highest amount of traffic since traffic headed towards the real sink is significantly reduced due to in-network aggregation. Ideally, we want to avoid large variations in the traffic volume among different areas within the sensor network.

In order to protect against the above attacks, we propose deploying multiple decoy sinks at random locations in the sensor network to provide robustness against attacks on the decoy sink location and to more evenly distribute traffic in the sensor network. These decoy sink nodes can be chosen randomly and configured offline prior to deployment. Each decoy sink node will have a unique decoy identifier in addition to its unique node identifier (both of these identifiers are assigned offline prior to deployment). During initialization of the application, each decoy sink node will flood a decoy setup message containing its unique decoy identifier in order for the other sensor nodes to set up their routing information to the different decoy sink nodes. These decoy setup messages are similar to the message that the real sink floods during initialization to set up a shortest path routing tree with itself as the root.

Based on the decoy setup messages received, each sensor node will build a decoy routing table identifying the node along the next hop towards each individual decoy sink node. The decoy routing table entries are ordered by decoy identifier (from lowest to highest) and each sensor node will send all readings to a particular decoy sink for a certain time period T before switching to the next decoy entry in a round-robin fashion. We assume that the sensor nodes have their clocks synchronized using a clock synchronization algorithm, such as described in [27]. We also assume that each sensor uses the same time period T between decoy sink transitions and begins at the same decoy entry in their respective routing tables (e.g., at time t_0, each node will start with the lowest decoy identifier).

Figure 3 provides an example of a decoy routing table at some sensor node x when there are four decoys (assume that t is this sensor node's current clock time in seconds). This sensor node will forward any locally aggregated readings to the currently active decoy sink based on its decoy routing table. In this example, assuming that decoys are switched every T seconds where $T = 10$, then x will send its reading taken at time $t = 65$ to the decoy sink with decoy identifier 2 via the node with node identifier 604. Of course, other intermediate nodes along the path between node x and decoy 2 might need to forward the reading. After x specifies some decoy sink

Decoy Id	Next Hop Node Id	Decoy Active When?
0	47	if $0 \leq (t \bmod 4T) < T$
1	389	if $T \leq (t \bmod 4T) < 2T$
2	604	if $2T \leq (t \bmod 4T) < 3T$
3	105	if $3T \leq (t \bmod 4T) < 4T$

Fig. 3. Example decoy routing table

identifier y as the destination for its message, all of the other sensor nodes along the path must forward the message to y regardless of the time when they receive the message.

If a decoy sink area is destroyed, then the only data lost would be data sent during the time period when the decoy sink in that area was supposed to be active. When the other decoy sinks that have not been destroyed are active, the sensor network would be able to resume tracking the target. Multiple decoy sinks add robustness against physical attacks on the decoy sink location. If an attacker has to destroy several decoy sinks, then the attack becomes expensive, which contradicts the attacker's original goal of executing an efficient, focused attack. Another benefit of multiple decoy sinks is that traffic is more evenly distributed throughout the sensor network, which provides better sink location protection.

4   Performance Evaluation

The decoy sink protocol was evaluated through simulation of a wireless sensor network target tracking application using the JProwler discrete event simulation tool from Vanderbilt University [24]. Some of the simulation tool's code was modified in order to implement the customizations necessary for our simulation. Each simulation of the application consisted of 1000 sensor nodes placed randomly in a 300 meter × 300 meter field. The total simulated time for each simulation was 60 minutes. The radio range of each sensor was 30 meters with $p = 0.05$ where p is the probability of packet loss at each hop. To deal with packet losses, link-level retransmissions with passive acknowledgements, as briefly described in [23], were used. We set the maximum number of retransmission attempts to three.

In addition to the sensor nodes, a node representing the real sink was placed in the northwest corner of the field. Targets would appear at a random position along one of the following borders of the field: north, south, east, or west. Targets would then move at 4 meters per second across the field headed in one of the following directions: north to south, south to north, east to west, or west to east. As soon as one target finished crossing the field, another target would appear in a new random position. The same sensor network topology and target events were used for each simulation run to ensure a fair comparison. Upon detecting the target, the sensors sampling at a rate of 4 Hz would perform local aggregation as described in section 3.4 and send their data over multiple hops either to the real sink or the decoy sink depending on whether or not the decoy sink protocol was used in that particular simulation run. In the simulations, a sensor would detect the target if the target was within a range of 10 meters.

In each simulation, zero or more decoy sink nodes would be randomly chosen from the 1000 total sensor nodes with at most one active decoy sink node at any given time. Therefore, the decoy's current location was at possibly one or more different random positions throughout each simulation run when the decoy sink protocol was used. In our simulations, a new decoy sink node would become active in a new random location every T seconds as described in section 3.4. In each simulation, T was set to the total simulation time in seconds divided by the total number of decoy sink nodes.

The remote aggregation function used at the currently active decoy sink node was to buffer the readings it has received for the past r seconds, average those readings, send the average value to the real sink, and then clear the buffer. This procedure was done repeatedly at each currently active decoy sink. With the hope of evenly distributing the traffic, we tried to choose r in such a way that the real sink would receive approximately the same number of readings that each decoy sink received. Assuming that there are d decoy sinks with all sensor nodes sampling at rate s and that each decoy is active for roughly the same amount of time overall while targets are present, then each decoy sink receives approximate s / d messages per second on average. Therefore, we chose r such that $r = d / s$. This value of r would allow the real sink to also receive approximately s / d messages per second on average. Please refer to Table 1 to check the different values of r for the different numbers of decoy sinks used in our simulations.[1]

Table 1. Simulation parameters

Number of decoy sinks d	Remote aggregation period r	Sampling rate s
1	1 sec	4 Hz
4	1 sec	4 Hz
8	2 sec	4 Hz
20	5 sec	4 Hz

The specific metrics used to evaluate the decoy sink protocol were the following: protection of the sink's location, and overhead in terms of delay and energy. The results are discussed in more detail in the following subsections.

4.1 Sink Location Protection

The first measure of how well the real sink's location is protected from adversaries is checking how the number of transmissions received by each sensor node relates to the number of hops away from the real sink for that particular sensor node. The following figures illustrate how the number of transmissions received by each

---

[1] One exception to our rule for choosing an appropriate r was the special case when $d = 1$ since we would have to send every message even when remote aggregation was used (i.e., no remote aggregation would occur). For that case, we set $r = 1$, which is the same r used with four decoy sinks.

sensor node varies as the number of hops away from the real sink increases. The different graphs in the figures represent different numbers of randomly placed decoy sinks used during that simulation run and whether or not remote aggregation was used.

As shown in the comparison of Figure 4 with Figure 5, using the decoy sink protocol with one decoy sink creates a high traffic area away from the real sink. In Figure 5, when remote aggregation is not used, there is still a high traffic area present near the real sink. In Figure 5, when remote aggregation is used, the traffic headed towards the real sink is significantly reduced. However, there is a relatively high volume of traffic near the single decoy sink when remote aggregation is used and the decoy sink's location is not protected very well since multiple decoy sinks are not used.

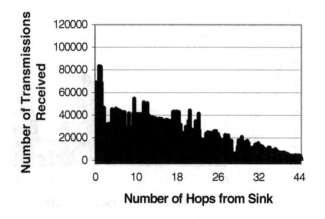

Fig. 4. No decoy sinks used

The benefits of combining multiple decoy sinks with remote aggregation become clearer in Figure 6, which shows that using remote aggregation can conceal the amount of traffic headed towards the real sink considerably while also concealing the locations of the various decoy sinks. When remote aggregation is not used with multiple decoy sinks, traffic near the real sink is relatively higher than other areas because the real sink is still receiving a steady stream of all the locally aggregated readings while the decoy sinks are only receiving a fraction of those readings. Therefore, simply using multiple decoy sinks without remote aggregation does not provide enough protection for the real sink's location. As shown in Figure 6, remote aggregation reduces the amount of traffic headed towards the real sink. Figure 6 also shows that using a larger number of decoy sinks tends to more evenly distribute the number of transmissions received at each node. However, this improvement in traffic analysis prevention might reduce the performance of the target tracking application since fewer readings are sent to the real sink due to the larger remote aggregation

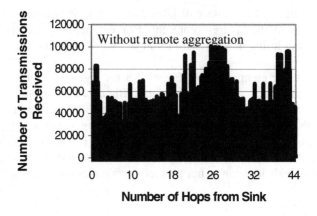

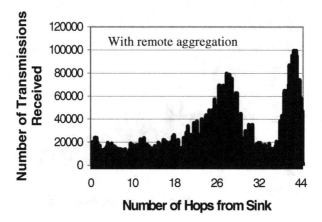

Fig. 5. One decoy sink used

periods that result from using more decoy sink nodes (please see Table 1). The effect of the decoy sink protocol on the performance of the target tracking application really depends on the characteristics of the application (e.g., speed of the targets). A tradeoff exists between protecting the real sink's location and tracking the target's position with more frequent readings.

Figure 7 uses the standard deviation of the number of transmissions received at each node to quantify the variability in the number of transmissions received at each node. When only one decoy sink is used in our simulations, the spread in the number of transmissions received at each node is quite large because the traffic is not more evenly distributed, as would be the case when multiple decoy sinks are used. As Figure 7 shows, increasing the number of decoy sinks when remote aggregation is used considerably decreases the spread in the number of transmissions received at each node. A lower standard deviation indicates a more even distribution of traffic, which makes traffic analysis more difficult.

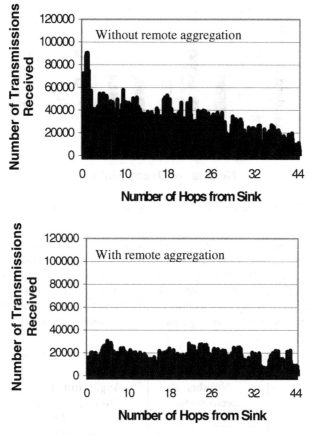

Fig. 6. Eight decoy sinks used

4.2 Delay and Energy Overhead

Since the decoy sink protocol involves sending data to an intermediate destination (i.e., the decoy sink) before the summarized data is forwarded to the ultimate destination (i.e., the real sink), we can expect some additional message overhead with respect to delay and energy. The expectation of lower delay and lower energy when the decoy sink protocol is not used is due to the fact that sensor readings are forwarded directly to the real sink without an intermediate destination along the way. Table 2 below illustrates the additional delay, which is measured in the average number of hops per message. Additional message delay along the path from the sensors detecting the target to the real sink is introduced when we send data to a decoy sink before forwarding it on to the real sink.

As expected, the average number of message hops increases since all data must first be forwarded to the decoy sink as an intermediate step before being forwarded to the real sink. However, one interesting result from Table 2 is that the average number of hops per message can be brought down considerably in our simulations if we

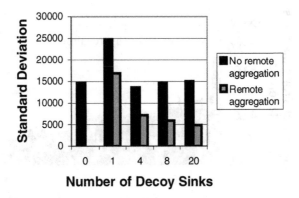

Fig. 7. Measuring spread in the number of transmissions received at each node

increase the number of decoy sinks used. In the case when only one decoy sink is used, the average number of hops per message is probably greater since that particular decoy sink happens to be far away from the real sink in our simulations. By using a larger number of random decoy sink nodes over time, we can balance long message hop distances due to decoy sink locations further away from the real sink with shorter message hop distances due to decoy sink locations closer to the real sink.

Table 2. Average number of hops per message (remote aggregation was used)

Number of decoy sinks	Avg. number of hops
0	22.61
1	58.95
4	35.94
8	37.80
20	35.29

An interesting result also appears in Table 3, which shows that the total amount of energy consumed using the decoy sink protocol is not necessarily that much greater than the amount of energy consumed without the decoy sink protocol. In Table 3, N denotes the number of decoy sinks in a particular simulation run. Of course, the energy costs depend heavily on how much we aggregate the data at the decoy sink nodes. For example, if we use eight decoy sink nodes, then the total number of messages transmitted is actually less than the case when no decoy sinks are used at all. The lower energy costs in this case is due to the more aggressive remote aggregation used when we have a larger number of decoy sink nodes (please refer to Table 1). In most of the simulations, locally aggregated message transmissions and link-level retransmissions were the dominant energy costs. These two types of messages must be sent regardless of whether or not the decoy sink protocol is used.

Table 3. Energy consumed (remote aggregation was used)

N	Number of Messages Sent/Forwarded for Each Message Type (LA = locally aggregated, B = broadcast announcing proximity to target, R = retransmission, RA = remotely aggregated, DS = decoy setup)					
	LA	B	R	RA	DS	Total
0	328235	46656	108698	0	0	483589
1	249679	48946	131623	147971	1002	579221
4	250844	50284	116003	69007	4008	490146
8	241075	48263	102498	39093	8016	438945
20	203221	48622	86392	16133	20040	374408

5   Conclusion

The decoy sink protocol presented in this paper combines the idea of indirection (a common technique for preventing traffic analysis in traditional wired and sensor networks) with in-network data aggregation (a technique typically applied to reducing power consumption in sensor networks) in an effort to prevent traffic analysis in wireless sensor networks.   The attack considered in this paper was adversaries attempting to locate and destroy the sink node in target tracking sensor networks. The basic decoy sink protocol operates by having sensor nodes send locally aggregated readings to a decoy sink node that will remotely aggregate the data before forwarding it to the real sink. Aggregation reduces the amount of traffic headed towards the real sink making traffic analysis more difficult. Using multiple decoy sink nodes is an extension to the basic protocol that adds robustness and more evenly distributes the traffic in the network. Our protocol was evaluated through simulation.

Acknowledgments. We thank the anonymous reviewers for their helpful comments on how to improve the final draft of this paper. This work was supported by the AT&T Labs Fellowship Program and Lucent Gift Fund. Any opinions, findings, and conclusions are those of the authors and do not necessarily reflect the views of the above agencies.

References

1. Szewczyk, R., Mainwaring, A., Polastre, J., Anderson, J., Culler, D.: An analysis of a large scale habitat monitoring application. In: 2nd ACM Conference on Embedded Networked Sensor Systems (2004)
2. Simon, G., Maroti, M., Ledeczi, A., Balogh, G., Kusy, B., Nadas, A., Pap, G., Sallai, J., Frampton, K.: Sensor network-based countersniper system. In: 2nd ACM Conference on Embedded Networked Sensor Systems (2004)
3. Aslam, J., Butler, Z., Constantin, F., Crespi, V., Cybenko, G., Rus, D.: Tracking a moving object with a binary sensor network. In: 1st ACM Conference on Embedded Networked Sensor Systems (2003)

4. Hwang, I., Balakrishnan, H., Roy, K., Shin, J., Guibas, L., Tomlin, C.: Multiple-target tracking and identity management. In: 2nd IEEE International Conference on Sensors (2003)
5. Gu, L., Jia, D., Vicaire, P., Yan, T., Luo, L., Tirumala, A., Cao, Q., He, T., Stankovic, J., Abdelzaher, T., Krogh, B.: Lightweight detection and classification for wireless sensor networks in realistic environments. In: 3rd ACM Conference on Embedded Networked Sensor Systems (2005)
6. Abdelzaher, T., Blum, B., Cao, Q., Chen, Y., Evans, D., George, J., George, S., Gu, L., He, T., Krishnamurthy, S., Luo, L., Son, S., Stankovic, J., Stoleru, R., Wood, A.: EnviroTrack: towards an environmental computing paradigm for distributed sensor networks. In: 24th International Conference on Distributed Computing Systems (2004)
7. Juang, P., Oki, H., Wang, Y., Martonosi, M., Peh, L.-S., Rubenstein, D.: Energy-efficient computing for wildlife tracking: design tradeoffs and early experiences with Zebranet. In: 2nd International Conference on Mobile Systems, Applications, and Services (2004)
8. Deng, J., Han, R., and Mishra, S.: Intrusion tolerance and anti-traffic analysis strategies for wireless sensor networks. In: The International Conference on Dependable Systems and Networks (2004)
9. Karlof, C., Sastry, N., Wagner,D.: TinySec: a link layer security architecture for wireless sensor networks. In: 2nd ACM Conference on Embedded Networked Sensor Systems (2004)
10. Perrig, A., Szewczyk, R., Wen, V., Culler, D., Tygar, J.: SPINS: security protocols for sensor networks. In: 7th International Conference on Mobile Computing and Networking (2001)
11. Watro, R., Kong, D., Cuti, S., Gardiner, C., Lynn, C., Kruus, P.: TinyPK: securing sensor networks with public key technology. In: 2nd ACM Workshop on Security of Ad Hoc and Sensor Networks (2004)
12. Gupta, V., Millard, M., Fung, S., Zhu, Y., Gura, N., Eberle, H., Chang Shantz, S.: Sizzle: a standards-based end-to-end security architecture for the embedded Internet. In: 3rd IEEE International Conference on Pervasive Computing and Communications (2005)
13. Reed, M., Syverson, P., Goldschlag, D.: Anonymous connections and onion routing. IEEE Journal on Selected Areas in Communications (1998)
14. Clarke, I., Sandberg, O., Wiley, B., Hong, T.: Freenet: a distributed anonymous information storage and retrieval system. In: International Workshop on Design Issues in Anonymity and Unobservability (2000)
15. Fu, X., Graham, B., Bettati, R., Zhao, W.: On effectiveness of link padding for statistical traffic analysis attacks. In: 23rd International Conference on Distributed Computing Systems (2003)
16. Deng, J., Han, R., Mishra, S.: Countermeasures against traffic analysis attacks in wireless sensor networks. In: 1st IEEE/CreateNet International Conference on Security and Privacy for Emerging Areas in Communication Networks (2005)
17. Ozturk, C., Zhang, Y., Trappe, W.: Source-location privacy in energy-constrained sensor network routing. In: 2nd ACM Workshop on Security of Ad Hoc and Sensor Networks (2004)
18. Madden, S., Franklin, M., Hellerstein, J., Hong, W.: TAG: a tiny aggregation service for ad-hoc sensor networks. In: 5th Symposium on Operating Systems Design and Implementation (2002)
19. Madden, S., Franklin, M., Hellerstein, J., Hong, W.: TinyDB: an acquisitional query processing system for sensor networks. ACM Transactions on Database Systems, vol.30, no.1 (2005)

20. He, T., Huang, C., Blum, B., Stankovic, J., Abdelzaher, T.: Range-free localization schemes for large scale sensor networks. In: 9th International Conference on Mobile Computing and Networking (2003)
21. Chan, H., Luk, M., Perrig, A.: Using clustering information for sensor network localization. In: International Conference on Distributed Computing in Sensor Systems (2005)
22. Moore, D., Leonard, J., Rus, D., Teller, S.: Robust distributed network localization with noisy range measurements. In: 2nd ACM Conference on Embedded Networked Sensor Systems (2004)
23. Kim, S., Fonseca, R., Culler, D.: Reliable transfer on wireless sensor networks. In: 1st IEEE International Conference on Sensor and Ad Hoc Communications and Networks (2004)
24. Institute for Software Integrated Systems at Vanderbilt University. JProwler: (http://www.isis.vanderbilt.edu/projects/nest/jprowler/)
25. Crossbow. MICA2 Datasheet: (http://www.xbow.com/Products/Product_pdf_files/Wireless_pdf/MICA2_Datasheet.pdf)
26. Crossbow. MICAz Datasheet: (http://www.xbow.com/Products/Product_pdf_files/Wireless_pdf/MICAz_Datasheet.pdf)
27. Maroti, M., Kusy, B., Simon, G., Ledeczi, A.: The flooding time synchronization protocol. In: 2nd ACM Conference on Embedded Networked Sensor Systems (2004)

Efficient and Robust Data Dissemination
Using Limited Extra Network Knowledge[*]

Ioannis Chatzigiannakis[1], Athanasios Kinalis[1,2], and Sotiris Nikoletseas[1,2]

[1] Computer Technology Institute, P.O. Box 1382, 26500 Patras, Greece
{ichatz, kinalis, nikole}@cti.gr
[2] Department of Computer Engineering and Informatics,
University of Patras, 26500 Patras, Greece

Abstract. We propose a new data dissemination protocol for wireless sensor networks, that basically pulls some additional knowledge about the network in order to subsequently improve data forwarding towards the sink. This extra information is still local, limited and obtained in a distributed manner. This extra knowledge is acquired by only a small fraction of sensors thus the extra energy cost only marginally affects the overall protocol efficiency. The new protocol has low latency and manages to propagate data successfully even in the case of low densities. Furthermore, we study in detail the effect of failures and show that our protocol is very robust. In particular, we implement and evaluate the protocol using large scale simulation, showing that it significantly outperforms well known relevant solutions in the state of the art.

1 Introduction

In this paper we study the problem of data propagation in wireless sensor networks. We propose a new protocol which is simple, local and uses limited extra knowledge of the network that is obtained in a distributed manner. The protocol uses *local information* regarding the surrounding actual network conditions, acquired by appropriately varying the range of wireless communication, and then *plans* a path of pairwise adjacent sensor devices that are used in the *forwarding* (i.e. propagation) of data towards the sink. Neighboring devices decide individually on whether to participate in propagation of events. The demand-driven sequence of *plan & forward* phases aims at better performance, compared to typical fixed transmission range data propagation, most needed in some frequently occurring situations like the case of low local densities of faulty sensor devices where fixed range protocols may trap in backtracking actions when no devices towards the sink are found; our protocol, by increasing the transmission range, may find such devices and avoid extensive backtracking.

This role-based approach where a limited number of devices do the high cost planning, while the rest operate in a low cost state, leads to systems that have increased

---

[*] This work has been partially supported by the IST Programme of the European Union under contract number IST-2005-15964 (AEOLUS), the Programme PYTHAGORAS under the European Social Fund (ESF) and Operational Program for Educational and Vocational Training II (EPEAEK II) and the Programme PENED of GSRT under contract number 03ED568.

P. Gibbons et al. (Eds.): DCOSS 2006, LNCS 4026, pp. 218–233, 2006.
© Springer-Verlag Berlin Heidelberg 2006

energy efficiency and high fault-tolerance, since the planning phases allow to optimize propagation paths and bypass obstacles (where no sensors are available) or faulty sensors (e.g. due to physical damage, power failure). We show that the cost of *forward planning* is amortized by the low energy short-range optimized hop-by-hop transmissions performed by our protocol; this selective spending of energy increases the lifetime of the network and the total number of events successfully reported to the control center.

The basic idea of our approach is to trade-off the cost of obtaining a certain amount of limited extra knowledge with the performance gains achieved using this additional knowledge in the subsequent propagation of data. By obtaining this extra knowledge about the network conditions (e.g. energy actually available at sensors, distance to the sink, faults etc.) at a somewhat global level, several performance measures (such as energy dissipation, latency, fault-tolerance) can be improved.

We implement and evaluate our protocol using simulation, showing that it significantly outperforms existing, well established relevant solutions in the state of the art. In particular, we demonstrate the above performance properties by comparing the new protocol to the well known **Directed Diffusion** paradigm [19] for information dissemination in wireless sensor networks using several important efficiency measures with a focus on energy dissipation, success rate and delivery delay. The extensive simulations that we present here, highlight the behavior of the internal mechanisms of the new framework and give useful insight on the fine-tuning of the various network parameters. The findings indeed demonstrate that our protocol achieves significant improvements in energy efficiency, higher success rates in faulty networks of low densities, and manages to disseminate data to their destination faster.

Related Work and Comparison. Local optimization protocols (like the Local Target Protocol, [13]) evolve in a greedy fashion trying to make optimal choices based on network knowledge within the typical fixed transmission range of the sensor currently possessing data under propagation. Such protocols tend to be more suitable in dense networks, with rather "uniform" conditions, i.e. where local samples of the network tend to be representative of it as a whole. Our protocol instead performs optimizations at a more global (yet limited) level, taking advantage of the extra knowledge obtained. Several protocols in the state of the art (most notably **Directed Diffusion**, [19]) try to maintain and update some global structure, such as a set of paths towards the sink to pull down data. Such approaches perform well in networks of low dynamics but their efficiency may drop in networks with many frequent changes and failures. Our protocol tries to become aware of the current, actual network conditions and accordingly optimize; however, this is done at a relatively local level in order to avoid collected knowledge becoming obsolete in the case of high dynamics. Furthermore, no structure or hierarchy are maintained by our protocol; once network information is obtained and optimized paths are chosen, data propagation happens in a hop by hop manner.

Our multi-hop approach is also in contrast to clustering protocols such as **LEACH** [17]. In such protocols, sensors self-organize themselves into clusters; in each cluster, only a single cluster head transmits directly to the sink, while the rest of the sensors propagate data to their cluster head. Such protocols perform well in small area networks of low event generation rate; however in larger networks of high event generation rates, transmissions happen at large distances and rotation of cluster heads may be too slow to avoid their energy depletion. Probabilistic forwarding schemes (like PFR, [9]) perform redundant optimized multi-path transmissions to combine energy efficiency and fault-tolerance. Such protocols, although well suitable in sparse networks, tend to spend a lot of energy in the case of high densities. For a survey of data propagation protocols, see [4]. Also, efficient protocols for fundamental problems in optical smart dust networks are proposed in [15], while routing communications methodologies are given in [1].

In [3], the computational complexity of the localization problem is studied for the first time, proving that it is NP-hard in sparse networks. We note that, also in view of this result, we avoid to solve a localization procedure since the network we study may be sparse; instead we obtain some implicit locality-related measurements (such as distances) at a local level. In fact, we neither assume an a-priori sense of orientation, since we only progressively build such a knowledge. Clearly, we can assume an explicit sense of orientation mechanism. For example, [16] presents self-stabilizing procedures for broadcast, flooding and sense of direction in wireless sensor networks. Such a sense of direction protocol can be used by our protocol in order to obtain local orientation references (i.e. for the sensors to know a general direction towards the sink).

Main Findings. Our extensive performance evaluation indicates that the amount of local information (on the surrounding actual network conditions) that is available to the protocol plays a crucial role in the overall performance of the network. This extra knowledge does not need to be accessible by all the devices of the network; allowing access to only a small group of devices suffices to considerably improve the overall performance. Interestingly, the additional energy spent by this small group (to obtain the extra knowledge) does not affect the overall energy dissipation, which is dominated by the high number of short-range transmissions employed during the data dissemination. Our protocol uses a simple collision-resolution mechanism to improve the network performance in cases of dense deployment of sensor devices and/or when no suitable underlaying MAC protocol is available. By carefully adjusting the protocol parameters we can trade-off latency with collision resolution. In fact, a limited increase to the delivery delays may lead to dramatic reductions to the total number of dropped packets.

In order to acquire a more complete view on the performance of our protocol, we conduct a comparative study with Directed Diffusion, a representative global structure based approach. The extensive experiments that we conducted, highlight the advantages of our approach that achieves significant improvements in energy efficiency, higher success rates in networks of low densities, and manages to disseminate data to their destination faster. We move beyond the typical study of networks with no failures, and investigate the performance of the two protocols in the presence of permanent node failures. We show that even under harsh conditions where more than 50% percent of the network becomes inoperable, out protocol still outperforms Directed Diffusion in all fundamental performance metrics (success rate, energy dissipation and delivery delay).

An early version of some of the ideas of our work have appeared in [12], a *brief announcement* in the 17th ACM Symposium on Parallelism in Algorithms and Architectures (SPAA 2005).

2   A Simple Model for Sensor Networks

We abstract the technological specifications of existing wireless sensor systems [14, 18]. Each node in our model is a fully-autonomous computing and communication device, is equipped with a set of monitors (e.g. sensors for temperature, humidity etc.) and is characterized mainly by its available power supply (battery) and the energy cost of computation and transmission of data. The communication equipment broadcasts messages to nearby devices within range $\mathcal{R}$ *that can vary* (i.e. the transmission power can be set at appropriate levels). Following [2, 10, 17, 20], for the case of transmitting and receiving a message we assume the following simple model where the radio dissipates E_{elec} to run the transmitter and receiver circuitry and ϵ_{amp} for the transmit amplifier to achieve acceptable SNR (signal to noise ratio). We also assume an r^2 energy consumption due to channel transmission at distance r. Thus to transmit a k-bit message at distance r in our model, the radio expends $E_T(k, r) = E_{elec} \cdot k + \epsilon_{amp} \cdot k \cdot r^2$ and to receive this message, the radio expends $E_R(k, r) = E_{elec} \cdot k$.

We consider a simple sensor network for remote surveillance of a region. In practice, such a network might consist of several hundreds or thousands of sensor devices deployed within that region. Let n be the total number of sensor devices, that are present in an area of size A. In some cases, the devices may be deployed in a regular fashion (e.g. a lattice, or a linear array) within that region. More generally, however, communication and networking protocols cannot assume structured sensor fields. Here, we assume that the sensor devices do not move and that the setting is two-dimensional.

A *user* of this remote surveillance system, which we call the sink $\mathcal{S}$, may contact the sensor devices in order to acquire information regarding the environmental conditions. In this sense, the user injects sensing tasks in the network, i.e. by broadcasting messages with a task description; the system can support a variety of task types [5]. Those sensor devices that match the task description report to $\mathcal{S}$ using hop-by-hop wireless communication and routing mechanisms described in Sec. 3.

The networks that we consider in our model are prone to failures due to the following reasons: (i) the components that make up a sensor device are of low-cost and also of low-reliability, (ii) the area of deployment may be harsh and unfriendly (e.g. terrain with water puddles, animals that run over the sensors) thus many operational failures may occur. We here model such situations by introducing the *failure rate* $\mathcal{F}$: the number of sensor devices that permanently fail to function per unit of time. For each unit of time, $\mathcal{F}$ failures occur at randomly chosen nodes, instantly, and no further computation and/or communication can be performed by these failed nodes.

3   Our Data Dissemination Protocol

The basic idea of our approach is that dissemination of information towards $\mathcal{S}$ is carried out within the wireless sensor network *using a series of interchanging phases*: (i) the *listening phase* (the device is sensing the environment and passively listening for messages), (ii) the *planning phase* (the device is preparing to propagate data to the sink S) and (iii) the *forwarding phase* (the device is participating in data propagation).

Given a particular environmental event that is sensed by a device p, and a surveillance (sensing) task that is set by S, a new message $\mathcal{M}$ is generated by p. Our goal is to use a limited (by β, a protocol parameter, that can be set by the implementor, described below) number of *long range* transmissions to collect information regarding neighboring nodes and then *plan* a series of *short range, low power* transmissions between nearby particles, based on certain optimization criteria, so that data is propagated to S. This *plan & forward* procedure provides (i) *high fault-tolerance* as long range transmissions allow to select paths that bypass obstacles (where no sensors are available) or faulty sensors (that have been disabled e.g. due to power failure), (ii) *increased energy-efficiency* because of the long range optimization performed and also as short-range hop-by-hop transmissions can effectively overcome some of the signal propagation effects in long-distance transmissions and (iii) *enhanced security* as the low energy transmissions protect from undesired discovery of the data propagation operation.

In our protocol, each sensor uses two small-sized data structures: (i) the *neighbors cache* that stores a small set of information about the active neighboring devices and (ii) the *path cache*, a list of node IDs that keeps track of the last path used to propagate data to S. The size of the neighbors cache is based on the density of the network while the path cache is bounded by the protocol parameter β. These structures are maintained during the listening phase and are extensively used during the planning phase.

Initialization Phase. We assume that there exists an initialization phase of the network during which all devices invalidate their local caches and execute an underlying localization protocol l. Since, in our model, the sensor devices cannot move, this phase is executed only once and does not impose any further overheads to the execution of the network. The protocol l is used by the sensors so that they can be able to estimate their distances within a certain accuracy factor, that depends on current technology advance and the actual protocol l. Let $d(i, j)$ be the *Euclidean distance of sensor devices i, j and $d_{es}(i, j)$ be the estimation of this distance measured by sensor devices*. Note that d_{es} is not necessarily an exact value but rather an estimate of the real distance d; we however assume that measured distances are analogous to real ones. Protocol l may operate without any common sense of orientation or any *geolocation abilities*, obviously, assuming special hardware equipment (e.g. smart antennae or GPS) makes this task even easier. Such a protocol is presented in [21] (and is compatible with our model of Sec. 2) that assigns *fictitious virtual coordinates* to all the devices of the network. In [22] the authors propose a greedy geometric routing protocol using pseudo (or virtual) coordinates, i.e. vectors composed by the hop distances from a node to a set of designated nodes (the anchors) in the network. Such a protocol can be used in our approach.

The Listening Phase. In this phase, sensor devices stay idle by passively listening for nearby devices that transmit announcements (see *planning phase* below) or that respond to the announcements, until (i) a new event is sensed that matches the *interests* given by S or (ii) a *message* $\mathcal{M}$ is received from another device.

When a device p listens an announcement from p', it first checks if $d_{es}(p, p') \leq \mathcal{R}_{close}$, a constant set by the protocol implementor (see Fig. 2). If this is true, it adds a new path $\mathcal{P}' = \{p'\}$ in the *path cache*. Then, it individually decides (based on *local criteria*) whether to respond to this announcement or not. The incentive here is to allow each device to control the energy consumption by ignoring some low-priority tasks, or deciding not to join a forwarding path when the network is dense and many neighboring devices have already joined. This decision can be based on a mechanism that considers various criteria regarding the conditions of the device (e.g. available energy, current load levels, etc.), the local conditions of the network (e.g. average neighborhood energy, local density, etc.) and even global conditions imposed by the network controller (e.g. operation-rule: all devices must join to increase success rate). In [12], we propose a mechanism that allows the devices to react locally on environment and context changes by using *a set of rules that are based on response thresholds* that relate individual-level plasticity with network-level resiliency, motivated by the nature-inspired method for dividing labor, a metaphor of social insect behavior for solving problems [6]. We plan to include this mechanism in extended versions of our protocol.

The device continues to passively listen to any device p'' that responded to the announcement of p', and if $d_{es}(p'', \mathcal{S}) < d_{es}(p, \mathcal{S})$ (i.e. p'' is closer to $\mathcal{S}$ than p), it adds p'' in the *neighbors cache*. This passive listening allows the devices to update their cache. In some sense, devices take advantage of any *long range* announcements conducted by nearby devices that undergo the *planning phase* to better understand the surrounding network conditions, and essentially reduce their own (future) needs to discover the neighborhood.

We here note that it is not necessary for the devices to constantly listen the radio channel, a very energy-consuming process. Our protocol can be combined with a lower-layer power conservation scheme like the one proposed in [11] or easily extended by incorporating sleep-awake schemes into the listening phase as done in [20]. Based on the performance evaluation presented in Sec. 5, our protocol operates well even for sparse networks, or otherwise, sensor networks implementing aggressive sleep-awake strategies.

The Planning Phase. A sensor enters the planning phase when data needs to be propagated. During this phase, p first examines the *path cache*. If the cache contains a valid path $\mathcal{P}$ of intermediate devices, it concludes and proceeds to the forwarding phase. If no such path exists, then p examines the *neighbors cache* in order to construct a new path $\mathcal{P}$ that will be used to forward $\mathcal{M}$ towards $\mathcal{S}$.

If the *neighbors cache* is outdated or empty, or because the cache contains very limited data regarding neighboring devices preventing p from constructing a sufficiently long path, p tries to discover all neighboring devices. Given a transmission range $\mathcal{R}$, p performs a *high power* data transmission with range $\beta \cdot \mathcal{R}$ (β is a protocol parameter) to announce its interest to disseminate data.

Remark that during an announcement, it is possible that p will not manage to discover all the neighboring devices because of message collisions occurring due to the concurrent responses of the nearby nodes. In order to tackle this problem, we implement a simple random backoff scheme during which p, after making the announcement, waits for a predetermined amount of time t_s. The nearby devices delay their response by a random period t_r, where $0 \leq t_r < t_s$. Of course, this mechanism can be avoided if the MAC protocol can properly handle collisions. In [8] distributed, contention-free self-organizing MAC protocols which do not assume a global time reference are proposed.

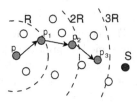

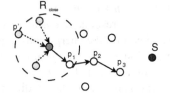

Fig. 1. Transmission example   **Fig. 2.** Nearby sensor devices react to *Announcements*

A distributed, local approach like in [8] can be assumed to run in combination with our protocol at a lower level to resolve medium access conflicts.

Still, it is possible that the device cannot detect any neighbor (e.g. because of low density, or high rate of failures, etc.). In this case the protocol has reached a *Dead-end* situation [21]. A possible way to overcome it is by repeating the transmission of the announcement (in case some devices have decided to participate, some previous inactive devices are now awake), or by using a *Range Variation* operation, similar to the one presented in [2] where the sensor device modifies its transmission range $\mathcal{R}$ according to a change-function, or even by using a *Backtracking* mechanism, similar to the one presented in [13].

Given that p has acquired enough information about the surrounding network condition, it selects a path $\mathcal{P}$ such that $\mathcal{M}$ is delivered to another sensor device p'' that is closer to $\mathcal{S}$ than p. This selection can be optimized in several ways, e.g. by selecting the particle with the higher available energy resources, the particle that has the lowest message load, or even the particle that has the most up-to-date cache. Clearly, the length of the path $\mathcal{P}$ is characterized by the locality of the information kept in the cache of p. If the knowledge about the neighboring devices is limited, path $\mathcal{P}$ will be short.

As soon as p'' is selected, p separates all the neighboring nodes, for which it has information in the cache, in β sublists ($\mathcal{L}_1, \mathcal{L}_2, \ldots, \mathcal{L}_\beta$) in a way such that $\forall p_j \in \mathcal{L}_i$: $(i-1) \cdot \mathcal{R} < d_{es}(p, p_j) \leq i \cdot \mathcal{R}$. Then p chooses one sensor device from each sublist L_i ($i \in \{1, \beta - 1\}$) so that the path $\mathcal{P} = p_1, p_2, \ldots, p_{\beta-1}, p''$ is defined. We here consider an optimization criteria for selecting one device from each sublist that is based on the relative distances of the nodes.

The sensor device executes a preparation procedure during which a bipartite multi-stage graph of β stages is generated based on $\mathcal{L}$. Each stage i of the graph $G(V, E)$ contains vertices that correspond to the sensor devices of $\mathcal{L}_i$ and the edges of G are between vertices of consecutive stages. Weights are assigned to the edges of G to reflect the estimated physical distance of the sensor devices that correspond to the adjoining vertices. Then, based on G, the protocol calculates the *shortest path* joining p and p''. The intuition for using a bipartite multi-stage graph is to reduce the total number of edges $m = |V|$ and therefore reduce the complexity of the *shortest path* operation given certain processing power limitations.

Note that it is possible that the operation of splitting $\mathcal{L}$ in β sublists ($\mathcal{L}_1, \mathcal{L}_2, \ldots, \mathcal{L}_\beta$) may result in some sublists $\mathcal{L}_i$ being empty, probably due to low-density of sensor devices on the particular sector of the transmission radius. In this case the protocol will produce a path $\mathcal{P}$ of length $l < \beta$. Certainly, there might exist other strategies for selecting path $\mathcal{P}$ that emphasize other aspects (such as available energy, distance from the $\mathcal{S}$) and/or may also include randomization techniques. We are currently working on such alternative choices.

As soon as the decision on such a path $\mathcal{P}$ is made, the protocol enters the *forwarding phase* by transmitting $(\mathcal{M}, \mathcal{P})$ to the first sensor device in $\mathcal{P}$ (i.e. in the example of Fig. 1, p_1). Then, every device p_j that receives $(\mathcal{M}, \mathcal{P})$ forwards $(\mathcal{M}, \mathcal{P} - \{p_j\})$ to p_{j+1}. When device p'' (i.e. in the example of Fig. 1, p_3) receives $(\mathcal{M}, \{\cdot\})$ the *forwarding phase concludes* and the protocol enters a new planning phase. Now p'' is responsible to further disseminate $\mathcal{M}$ towards $\mathcal{S}$.

The Forwarding Phase. In the forwarding phase, given a message of type $(\mathcal{M}, \mathcal{P})$, the sensor device does the following:

$\mathcal{P}$ *is not empty.* The message contains information about a path of sensor devices. If the path $\mathcal{P}' = \mathcal{P} - \{p_j\}$ is not empty, $\mathcal{P}'$ is added in the *path cache* and p sends $(\mathcal{M}, \mathcal{P}')$ to p_1 and sends a *success* message to `sender(`$\mathcal{M}$`)` (i.e. to the device it originally received the information from); in case p generated $\mathcal{M}$, no *success* message is sent. Otherwise, if $\mathcal{P}'$ is empty, the protocol enters the *planning phase*.

$\mathcal{P}$ *is empty.* The message contains no information about the path of sensor devices to use in order to propagate $\mathcal{M}$ towards $\mathcal{S}$. The protocol enters *the planning phase*.

After transmitting the packet, p will wait for p_1 to send a success message in order to ensure that $\mathcal{M}$ was received properly and the dissemination continues as planned. If p_1 does not respond within a predefined period of time, p assumes that the transmission fails and retransmits the packet to p_1. This process is repeated until either p responds with a success message or until the maximum number of retries has been reached (a protocol parameter). In this case, p decides that p_1 is no longer active, updates its cache (by removing p_1 and $\mathcal{P}$) and enters the planning phase.

4   Performance Metrics

In this work we wish to evaluate the performance of our protocol based on the following three fundamental metrics: the *success rate*, the *energy dissipation* and the *delivery delay*. These performance metrics characterize the ability of the protocol to coordinate the sensor devices so that all messages regarding the realization of environmental phenomena are transferred to $\mathcal{S}$, in an energy efficient way and with minimum delay. The importance of each of the above metrics depends on the nature of the application since there are inherent trade-offs between success rate, energy and latency.

In the previous sections we described the basic design properties of our protocol considering the existence of a single monitoring task and the dissemination of messages related to this task. In reality the system will have to handle multiple concurrent task initiations and in extend the diffusion mechanisms will be used for a large number of messages. Let K be the total number of crucial events $(\mathcal{E}_1, \mathcal{E}_2, \ldots, \mathcal{E}_K)$ that need to be reported to a particular $\mathcal{S}$ in the area and let us consider that a data dissemination protocol manages to report k number of these events.

Definition 1 (Success Rate). $\mathbb{P}_s$, is the fraction of the number of events *successfully* propagated to the sink over the *total number of events*, i.e. $\mathbb{P}_s = \frac{k}{K}$.

Definition 2 (Total Energy Dissipated). $E_{tot} = \sum_{i=1}^{n} \left(E_i^{init} - E_i \right)$, where E_i^{init} is the initial energy of sensor device i and E_i the available energy of device i at the end of the system operation.

Definition 3 (Delivery Delay). Let $\mathcal{D}$ be the *total period of time* that elapsed since the realization of a crucial event $\mathcal{E}$ until it was finally delivered to the sink $\mathcal{S}$.

Furthermore, we consider three protocol specific metrics that measure (i) the **Total Number of Announcements,** (ii) the **Average Path Length** and (iii) the **Number of Collisions** in terms of dropped packets. These performance metrics characterize the network management overhead imposed by the protocol stack (a combination of data dissemination and collision-resolution protocols) given a set of monitoring tasks. These metrics provide useful insights on the effect of the various network and protocol parameters to the overall performance of the system.

Based on the energy cost model used (see Sec. 2), the energy consumption for the transmission of a message is related to the distance that it is required to travel and its size in bits. Since announces sent during the planning phases are very short (e.g. a small, constant number of k_a bits), the extra energy spent due to increasing the transmission range is not increased a lot; in any case, this extra energy is worth spending since the additional knowledge obtained allows for much better path selection. Also, the propagation of the actual data packets during the forwarding phase (which may be longer than the short announces, i.e. k_i bits) is still performed in a multi-hop way, thus energy spent in each hop is in the order of $\mathcal{R}^2$.

In fact, based on the energy cost model, the total energy spent in each sequence of plan & forward phases is proportional to $k_a \cdot (\beta \cdot \mathcal{R})^2 + \beta \cdot k_i \cdot \mathcal{R}^2$. If $\beta < \frac{k_i}{k_a}$ then the energy cost of the announcements is smaller than the actual data propagation energy cost, i.e. the total energy is $\beta \cdot k_i' \cdot \mathcal{R}^2$, where k_i' is constant, similar to other, common multi-hop approaches.

Assuming that the sensor devices are random uniformly distributed on the area, the density can be calculated according to [7] as $\mu(\mathcal{R}) = \frac{(n \pi \mathcal{R}^2)}{A}$. Basically, $\mu(\mathcal{R})$ gives the number of sensor devices within the transmission radius of each device in region $\mathcal{A}$. Therefore, when $\mathcal{R}$ is set to $\beta\mathcal{R}$, the effective density $(\mu(\mathcal{R}))$ of the devices becomes β^2 times larger, which leads to many more devices responding to the announcement. To scan the same area and number of devices, a much greater number of short range transmissions (at least β^2) would be needed. Of course, the increased number of concurrent responses to the long-range announcement will potentially result in a high number of collisions, random back-off (or other collision resolution mechanisms) are more efficient in our case (as shown by the low latency achieved and the high success rate) since devices found during a single scan can be better coordinated with respect to many nearby announcements of shorter range.

5   Performance Evaluation

In this section we present a comparative evaluation study of our protocol (which we hereafter call the CKN protocol) with the well established Directed Diffusion (which we hereafter call DD) paradigm for information dissemination in wireless sensor networks [19]. We implement our protocol at the same level of the network stack with DD and use a higher layer sensing application that injects sensing tasks to the sensor network. A number of events is generated, corresponding to the sensing tasks, for propagation to the sink S. The experimental evaluation is conducted based on the commonly used Network Simulator (**ns-2** version 2.26), that provides a quite detailed implementation of the physical and MAC layers and allows detailed measurements of many variables (such as the energy dissipation) in simulations of wireless networks.

The sensor network is considered as a rectangular area of size $\mathcal{A} = 500m \times 500m$, where a number of $n \in [200, 400]$ sensor devices are randomly distributed in the area. We set $\mathcal{R} = 50m$, $\mathcal{R}_{close} = 50m$ and position S at point $(0, 0)$. Based on these settings, the corresponding $\mu(\mathcal{R})$ is in the range of $[6.28, 12.57]$. In [19] the experimental study conducted also considered networks of different sizes and number of sensor devices but with an almost fixed density $\mu(\mathcal{R}) \approx 9.817$. The energy available to the sensor devices was set to high levels ($E_i^{init} = 20J$), to create good enough initial conditions (in terms of available energy) where all the events can be delivered. This setting allows us to compare the energy dissipation of the protocols in a fair way. We implement the energy cost model of Sec. 4 in **ns-2**, and set the exact values of ϵ_{trans}, ϵ_{recv} and E_{idle} to match as close as possible the specifications of the mica mote platform [14].

We start the evaluation of our protocol by investigating the impact of the various parameters on the performance of the network. In the first set of experiments we examine the impact of parameter β, i.e. the parameter that controls the transmission range for the announcements and in extend the length of the path generated during the planning phase. Figure 3 depicts the six efficiency metrics discussed in Sec. 4. In this first set, we fix the search time to $t_s = 75ms$ and inject a set of sensing tasks that generate 2 events/sec. Each event is being sensed by a randomly chosen sensor device and in our simulation 1000 events are generated. The simulation duration is calculated according to the event rate and is long enough to allow all messages to be generated. Another 15 seconds of simulation time are added to allow the arrival of delayed messages.

The results depicted in Figure 3 demonstrate the ability to improve the success rate of our protocol in the cases of sparse deployment of sensor devices by adjusting β. For all network densities considered, setting $\beta = 2$ suffices to achieve a 100% success rate. In terms of energy consumption, as it was discussed in Sec. 4, the cost of the long range transmission of announcement messages is indeed amortized by the high number of short-range transmission of messages regardless of the density of the network and β. However, we are pleased to report that the latency of the network is improved when β increases. This is related to the fact that the devices transmit fewer announcements as β increases, thus reducing the overall delay as the devices spend less time waiting for the nearby devices to respond to their announcements. Therefore, although the long-range transmission of announcements lead to a higher number of collisions, this does not critically affect the performance of the network while the protocol manages to devise "good" paths based on the additional information acquired.

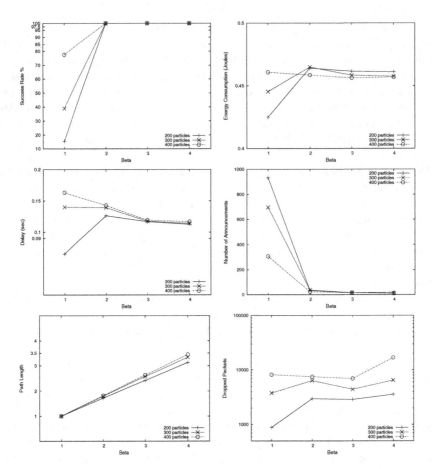

Fig. 3. Success Rate ($\mathbb{P}_s$), Energy Dissipation (E_{tot}), Delay ($\mathcal{D}$), Average Number of Announcements, Average Path Length and Average Number of Dropped packets for different values of $\beta \in [1, 4]$, various number of devices ($n \in [200, 400]$) and fixed search time ($t_s = 75ms$)

In the second set of experiments (see Figure 4) we examine the impact of parameter t_s, that is the time period that a device waits after making an announcement so that nearby devices can respond; we refer to t_s as the search time. The central idea for adjusting t_s is to allow the responses to spread over a longer period of time and in this way increase the effectiveness of the collision resolution protocol. Of course, by increasing t_s, the delivery delay of the network is also affected. However, as shown in Fig. 4, the overall degradation of the latency is limited while the number of dropped packets (i.e. the number of collisions) is dramatically reduced (notice that the figure is in logarithmic scale). This allows the devices to collect more information regarding the neighboring devices and thus devise longer paths. Interestingly, even if, by increasing t_s, the need to make an announcement is reduced, the overall energy dissipation seem to remain fixed, implying that the overall energy consumption is dominated by the short-range transmissions of info messages rather than by the long-range transmissions of announcements.

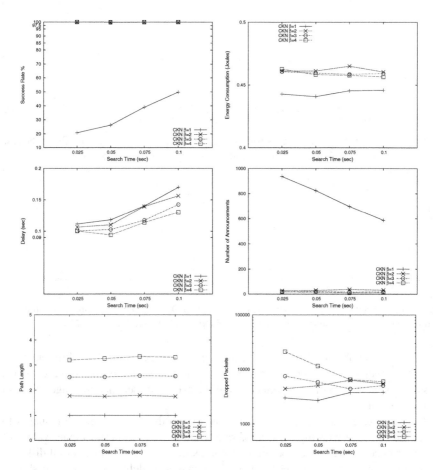

Fig. 4. Success Rate ($\mathbb{P}_s$), Energy Dissipation (E_{tot}), Delay ($\mathcal{D}$), Average Number of Announcements, Average Path Length and Average Number of Dropped packets for different values of $\beta \in [1,4]$, different search times $t_s \in [25, 100]ms$ and fixed number of devices ($n = 300$)

We now proceed with the comparative study of our protocol (**CKN**) with **DD**. In order to highlight the differences between the two different approaches, we first evaluate the two protocols in a "controlled" environment. In this set of experiments, since our protocol essentially variates the transmission range (based on β) in order to make long-range announcements, to make the comparison fair, when the network executes **DD**, we set the transmission range of the devices to $\beta \cdot \mathcal{R}$. Note however that unlike our protocol, **DD** does not vary the transmission range through out the execution of the network. For these experiments we consider only the three efficiency metrics discussed in Sec. 4.

In the first set of experiments (see Fig. 5), we measure the performance of the protocols when only 1 message needs to be disseminated to $\mathcal{S}$. This message is generated by the device positioned at $(500, 500)$, i.e. the device that has the greatest distance from $\mathcal{S}$. In the second set of experiments (see Fig. 6), we measure the performance of the two protocols when each device generates 1 message that needs to be disseminated to $\mathcal{S}$,

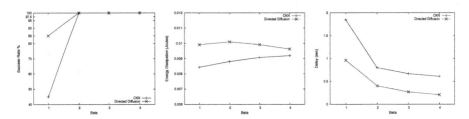

Fig. 5. Success Rate ($\mathbb{P}_s$), Energy Dissipation (E_{tot}) and Delay ($\mathcal{D}$) of CKN and DD in the case of 1 event (at point $500, 500$), for $\beta \in [1, 4]$, fixed $n = 300$ and fixed $t_s = 75ms$

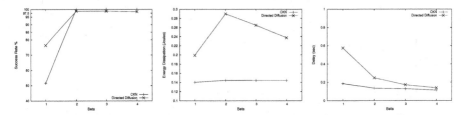

Fig. 6. Success Rate ($\mathbb{P}_s$), Energy Dissipation (E_{tot}) and Delay ($\mathcal{D}$) of CKN and DD in the case of 300 events (1 per each sensor), for $\beta \in [1, 4]$, fixed $n = 300$ and fixed $t_s = 75ms$

i.e. the two protocols must disseminate a total of n messages. These two different cases allow us to investigate the performance of the two protocols in the extreme case when a message is generated far away from $\mathcal{S}$ and in the average case when messages are sent from all possible positions.

The two different sets of experiments show that in the worst case, DD manages to deliver the message with higher success rate when $\beta = 1$, while for higher values of β, both protocols always succeed. Furthermore, still in the worst case, for all different values of β considered, it is seems that DD manages to deliver messages in shorter time, but at a higher energy cost. For the average case (see Fig. 6), again DD manages to deliver more messages than our protocol when $\beta = 1$, however, in terms of energy consumption and delivery delays, it is clear that our protocol significantly outperforms DD. These experiments also indicate the impact of β on the success rate and latency of our protocol, while the overall energy dissipation is affected at a very limited way that almost suggests that it is independent of β.

In the set of experiments shown in Fig. 7a we evaluate the two protocols for the general case, i.e. when we assume a set of sensing tasks that generate 2 events/sec and each event is being sensed by a randomly chosen sensor device. In this setting we generate a total of 1000 events and the simulation duration is calculated according to the event rate and is long enough to allow all messages to be generated. Another 15 seconds of simulation time are added to allow the arrival of delayed messages. In contrast to the previous two sets of experiments (see Fig. 5 and Fig. 6), in this set of experiments we used the original implementation of DD, as described in [19], i.e. the transmission range is always set to $\mathcal{R}$ regardless of β.

This experiment clearly shows the superiority of our approach in all three efficiency metrics considered here. By setting the parameter $\beta = 2$, our protocol achieves a 100%

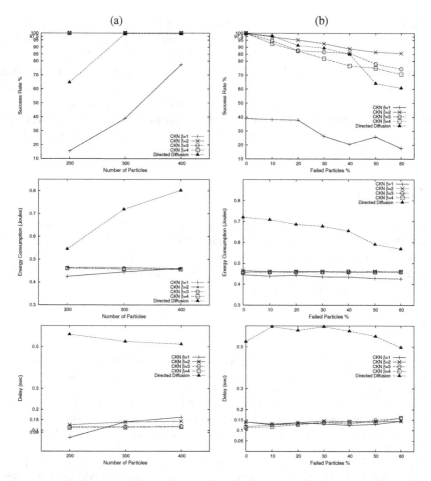

Fig. 7. (a) On the left, Success Rate ($\mathbb{P}_s$), Energy Dissipation (E_{tot}) and Delay ($\mathcal{D}$) of CKN and DD in the case of 1000 event, for $n \in [200, 400]$, $\beta \in [1, 4]$ and fixed $t_s = 75ms$. (b) On the right, Success Rate ($\mathbb{P}_s$), Energy Dissipation (E_{tot}) and Delay ($\mathcal{D}$) of CKN and DD in the case of 1000 event, when $0 \ldots 60\%$ of the $n = 300$ nodes fail ($\mathcal{F} \in [0, 0.36]$ failures/sec), $\beta \in [1, 4]$ and $t_s = 75ms$.

success rate, and delivers all messages to $\mathcal{S}$ with significantly lower delays and by spending fewer energy than DD.

In our last set of experiments we investigate the (more realistic) scenario where stopping failures occur at the sensing devices. In contrast to the previous settings where the operation of nodes was guaranteed, in this set we examine the behavior of the protocols under the presence of node failures. We use the failure rate $\mathcal{F}$ (defined in Sec. 2) to control the harshness of the environment and evaluate the fault-tolerance achieved by our protocol. We deploy $n = 300$ devices and set $\mathcal{F} = \frac{\alpha \cdot n}{T_{sim}}$, where $\alpha \in [0.1, 0.6]$ is a parameter that controls the fraction of nodes that fail and $T_{sim} = 500sec$ the total simulation time. Essentially, we allow the $10 \ldots 60\%$ of the nodes to fail during the simulation period. Based on the results shown in Fig. 7b, we observe that the failure rate mainly affects the success rate of the protocols while the energy consumption and delivery delay seems to be unaffected. However, in contrast to the case of low densities, β does not control the performance of the protocol; although for $\beta = 2$ the performance improves, further increases lead to reduced efficiency. This is explained by the fact that higher β lead to longer path lengths, and thus increased probabilities for a node failure to damage a path. In such cases the protocol is forced to make additional announcements and discover new paths that lead to slightly increased delivery delays and energy consumption. Although DD seems to follow a similar pattern of behaviour, our protocol achieves higher fault-tolerance for all cases of failure rates considered.

6   Future Work

We wish to extend our protocol by introducing adaptive mechanisms that will allow sensor devices to self-adjust the various parameters (e.g. β, t_s) in terms of the actual network conditions. We plan to compare the performance of our protocol with other, existing protocols and also using different network shapes, various distributions used to drop the sensors in the area of interest.

References

1. Alvarez, C., Diaz, J., Petit, J., Rolim, J., Serna, M.: Efficient and reliable high level communication in randomly deployed wireless sensor networks. In: 3nd International Mobility and Wireless Access Workshop (MOBIWAC 2004). (2004)
2. Antoniou, T., Boukerche, A., Chatzigiannakis, I., Mylonas, G., Nikoletseas, S.: A new energy efficient and fault-tolerant protocol for data propagation in smart dust networks using varying transmission range. In: 37th Annual Simulation Symposium (ANSS 2004). (2004) 43–52 IEEE Press.
3. Aspnes, J., Goldberg, D., Yang, Y.: On the computational complexity of sensor network localization. In: 1st International Workshop on Algorithmic Aspects of Wireless Sensor Networks (ALGOSENSORS 2004), Springer-Verlag (2004) 32–44 Lecture Notes in Computer Science, LNCS 3121.
4. Boukerche, A., Nikoletseas, S.: Protocols for Data Propagation in Wireless Sensor Networks: A Survey. In: Wireless Communications Systems and Networks. Kluwer Academic Publishers (2004) 23–51
5. Boukerche, A., Pazzi, R., Araujo, R.: A supporting protocol to periodic, event-driven and query-based application scenarios for critical conditions surveillance. In: 1st International Workshop on Algorithmic Aspects of Wireless Sensor Networks (ALGOSENSORS 2004), Springer-Verlag (2004) 137–146 Lecture Notes in Computer Science, LNCS 3121.
6. Bonabeau, E., Dorigo, M., Theraulaz, G.: Swarm Intelligence: From Natural to Artificial Systems. Oxford University Press (1999) A Volume in the Santa Fe Institute Studies in the Sciences of Complexity.
7. Bulusu, N., Estrin, D., Girod, L., Heidemann, J.: Scalable coordination for wireless sensor networks: self-configuring localization systems. In: International Symposium on Communication Theory and Applications (ISCTA 2001). (2001)
8. Busch, C., Magdon-Ismail, M., Sinrikaya, F., Yener, B.: Contention-free MAC protocols for wireless sensor networks. In: 18th International Workshop on Distributed Algorithms (DISC 2004), Springer-Verlag (2004) 245–259 Lecture Notes in Computer Science, LNCS 3274.

9. Chatzigiannakis, I., Dimitriou, T., Nikoletseas, S., Spirakis, P.: A probabilistic forwarding protocol for efficient data propagation in sensor networks. In: 5th European Wireless Conference on Mobile and Wireless Systems beyond 3G (EW 2004). (2004) 344–350. Also, in the Journal of Ad-Hoc Networks (2005)

10. Chatzigiannakis, I., Kinalis, A., Nikoletseas, S.: Wireless sensor networks protocols for efficient collision avoidance in multi-path data propagation. In: ACM Workshop on Performance Evaluation of Wireless Ad Hoc, Sensor, and Ubiquitous Networks (PE-WASUN 2004). (2004) 8–16 Also, in Performance Evaluation Journal: Special issue on PE-WASUN04 (to appear)

11. Chatzigiannakis, I., Kinalis, A., Nikoletseas, S.: An adaptive power conservation scheme for heterogeneous wireless sensor networks with node redeployment. In: 17th Annual Symposium on Parallel Algorithms and Architectures (SPAA 2005). (2005) 96–105. Also, in the Theory of Computing Systems Journal (TOCS): Special Issue on SPAA05 (to appear)

12. Chatzigiannakis, I., Nikoletseas, S.: A forward planning situated protocol for data propagation in wireless sensor networks based on swarm intelligence techniques. In: 17th Annual Symposium on Parallel Algorithms and Architectures (SPAA 2005). (2005) 214

13. Chatzigiannakis, I., Nikoletseas, S., Spirakis, P.: Efficient and robust protocols for local detection and propagation in smart dust networks. Journal of Mobile Networks and Applications **10**(1) (2005) 133–149 Special Issue on Algorithmic Solutions for Wireless, Mobile, Ad Hoc and Sensor Networks.

14. Crossbow technology inc., MICA motes http://www.xbow.com/Products/productsdetails.aspx?sid=71.

15. Diaz, J., Petit, J., Serna, M.: Evaluation of basic protocols for optical smart dust networks. IEEE Transactions Mobile Networks **2**(3) (2003) 186–196

16. Dolev, S., Herman, T., Lahiani, L.: Polygonal broadcast, secret maturity and the firing sensors. In: 3rd International Conference on Fun with Algorithms. (2004) 41–52

17. Heinzelman, W.R., Chandrakasan, A., Balakrishnan, H.: Energy-efficient communication protocol for wireless microsensor networks. In: 33rd IEEE Hawaii International Conference on System Sciences (HICSS 2000). (2000)

18. Hollar, S.: COTS Dust. Msc thesis, Engineering-Mechanical Engineering, University of California, Berkeley, USA (2000)

19. Intanagonwiwat, C., Govindan, R., Estrin, D.: Directed diffusion: A scalable and robust communication paradigm for sensor networks. In: 6th ACM/IEEE Annual International Conference on Mobile Computing (MOBICOM 2000). (2000) 56–67

20. Nikoletseas, S., Chatzigiannakis, I., Euthimiou, H., Kinalis, A., Antoniou, T., Mylonas, G.: Energy efficient protocols for sensing multiple events in smart dust networks. In: 37th Annual Simulation Symposium (ANSS 2004). (2004) 15–24 IEEE Press.

21. Rao, A., Ratnasamy, S., Papadimitriou, C., Shenker, S., Stoica, I.: Geographic routing without location information. In: 9th ACM/IEEE Annual International Conference on Mobile Computing (MOBICOM 2003), San Diego, CA (2003) 96–108

22. Wattenhofer, M., Wattenhofer, R., Widmayer, P.: Geometric routing without geometry. In: 12th Colloquia on Structural Information and Communication Complexity (SIROCCO 2005), Springer-Verlag (2005) 307–322 Lecture Notes in Computer Science, LNCS 3499.

Distance-Sensitive Information Brokerage in Sensor Networks

Stefan Funke[1], Leonidas J. Guibas[1], An Nguyen[1], and Yusu Wang[2]

[1] Department of Computer Science
Stanford University, Stanford, CA 94305
[2] Department of Computer Science and Engineering
The Ohio State University, Columbus, OH 43210

Abstract. In a sensor network information from multiple nodes must usually be aggregated in order to accomplish a certain task. A natural way to view this information gathering is in terms of interactions between nodes that are *producers* of information, e.g., those that have collected data, detected events, etc., and nodes that are *consumers* of information, i.e., nodes that seek data or events of certain types. Our overall goal in this paper is to construct efficient schemes allowing consumer and producer nodes to discover each other so that the desired information can be delivered quickly to those who seek it. Here, efficiency means both limiting the redundancy of where producer information is stored, as well as bounding the consumer query times. We introduce the notion of *distance-sensitive information brokerage* and provide schemes for efficiently bringing together information producers and consumers at a cost proportional to the separation between them — even though neither the consumers nor the producers know about each other beforehand.

Our brokerage scheme is generic and can be implemented on top of several hierarchical routing schemes that have been proposed in the past, provided that they are augmented with certain key sideway links. For such augmented hierarchical routing schemes we provide a rigorous theoretical performance analysis, which further allows us to prove worst case query times and storage requirements for our information brokerage scheme. Experimental results demonstrate that the practical performance of the proposed approaches far exceeds their theoretical (worst-case) bounds. The presented algorithms rely purely on the topology of the communication graph of the sensor network and do not require any geographic location information.

1 Introduction

Early sensor networks were primarily data acquisition systems, where the information collected by the sensor nodes was to be aggregated and routed to a central base station. Newer generations of sensor networks, however, act more as peer-to-peer systems, where arbitrary nodes in the network may wish to collect information about measurements and events elsewhere in the network. Furthermore, the information needed may be quite specific, with only a very small

P. Gibbons et al. (Eds.): DCOSS 2006, LNCS 4026, pp. 234–251, 2006.
© Springer-Verlag Berlin Heidelberg 2006

amount of sensor data being relevant for any particular query. This peer-to-peer view is necessitated as sensor networks expand to serve multiple geographically dispersed users, as more powerful mobile nodes move through a static sensor network and use it as a communication backbone to issue queries and collect data, or even to process complex queries, where sensor nodes may find it necessary to issue sub-queries themselves. The basic problem this creates is that of *information brokerage*: how *producers* of information, e.g., nodes that have collected data, detected events, etc., and *consumers* of information, i.e., nodes that seek data or events of certain types, can find out about each other and exchange the desired information.

Information brokerage is closely coupled with node naming and routing: even if we know the exact location of the information we want in the network, we still need to discover a good path for its retrieval. As sensor networks scale to larger sizes, the issue of *information locality* becomes more important. It is natural to expect that a consumer will be more interested in data collected near its current location. This is because such data can be accessed at lower communication cost/delay, and because in almost all sensor network applications local information has higher value and relevance to the task at hand than remote information. The main problem studied in this paper is what we call *distance-sensitive information brokerage* — information brokerage where the cost for a consumer node to discover a certain piece of information is proportional to the network distance to where that information was collected. We want to have this property of *distance sensitivity* even though neither the consumer node nor the producer node involved in the information exchange know directly anything about the location of the other node in the network. Current common information brokerage schemes *do not* have this property. For example, directed diffusion [14] performs flooding and thus in a 2-D network will visit $O(d^2)$ nodes to reach a distance d from a consumer (sink), while geographic hash tables (GHT) [24] may hash the information quite far away from a nearby producer-consumer pair.

Our goal in this paper is to develop a unified framework for routing and information brokerage which can provide provable guarantees for both good (almost-optimal) routing paths and distance-sensitive information brokerage. Note that in all producer-consumer brokerage schemes there is a trade-off between the time and space effort for information diffusion when producer nodes record new data and have new detections, vs. the query time of consumer nodes to discover this information. In this work we aim at minimizing the amount of work/storage that producers have to invest so that they can be discovered within the consumers' budget, that is, to be *distance-sensitive*. We focus on the static case typical for sensor networks where nodes do not move during the lifetime of the network, though links may fail or nodes may die. We also assume that, at any one time, only a small fraction of the network nodes have information to be made available to the network (such as sightings/measurements of rather exceptional events).

Related work. Hierarchies for addressing and routing within networks have been used for a long time and form the basis of the standard TCP/IP protocols. The basic idea is to define a tree-like hierarchy of node clusters, based on the inter-node distances in the network. This tree structure is then used to derive unique addresses for all nodes, based on which local routing schemes can be developed. Many previous hierarchical approaches have designated nodes as *gateways* to route between clusters [1, 22], causing unbalanced node traffic, as well as making routing sensitive to gateway node failures. Furthermore, since hierarchies partition the network, there can be nearby nodes that end up in different clusters even at the top level, causing the routing paths to be possibly much longer than the true shortest paths, violating distance sensitivity. Solutions have been proposed for this difficulty by introducing cross-branch links in the clustering hierarchy by Tsuchiya [28] and others, and empirical studies have established their effectiveness. For the case where the underlying communication graph has constant doubling dimension [11], it has been shown very recently in [12] (although in a different context) that paths of bounded dilation can be achieved. Chan *et al.* [3] present a different hierarchical framework to achieve slightly better results, but their construction is based on a probabilistic argument and a non-trivial derandomization thereof.

Since sensor networks are embedded in the physical world, several routing schemes attempt to exploit naming and routing using this host space to gain efficiency and avoid expensive preprocessing. For example, geographic routing schemes [2, 15, 16] name nodes by their geographic coordinates and provide local rules for forwarding messages towards their target. Such schemes may have problems in the presence of holes, in which case packets might get stuck in local minima of the distance function to the target. Protocols like GPSR ([15]) or the one proposed by Bose et. al. [2] have been developed to alleviate these problems, and indeed they can guarantee delivery of the packets by performing a perimeter routing step around network holes, after an appropriate planarization of the network graph. Still, the paths might be considerably worse than the true shortest paths. In particular, Kuhn et. al. in [17] show that if the shortest path has distance d, *any* geographic routing algorithm might produce a path of length $\Omega(d^2)$. Furthermore, in many situations, it is challenging to obtain geographic coordinates. Various approaches have been developed for cases in which either a few nodes [26] or no nodes [6, 21, 23] are aware of the geographic positions. However, a robust routing scheme with proven guaranteed performance for sensor networks with arbitrary underlying topology is still missing.

At the same time, in the past few years, sensor networks have started to serve more as information processing mechanisms instead of simply as data collection tools [7, 14, 18, 25, 29], requiring more sophisticated operations, such as data aggregation and range queries. From this point of view, the location of the sensor that owns information becomes less important than the information itself. This explains the rise of various *data-centric* information storage and retrieval schemes [27]. A representative is the Geographic Hashing Table (GHT) approach [24], where each type σ of information (like the measured occurrence of substance A) is mapped to a specific node v by using a geographic hash function which depends only on the information type σ and is commonly known to every node in the network. Upon detection of A, a producer sends a message to node v, indicating its possession of some data of type σ. Any consumer can then obtain the data by first visiting v to find out who owns it and then retrieving it from the owner (many variations are possible). This node v, however, might be far away from both the producer and the consumer even when the producer and the consumer are very close in the network. The problem can be partly alleviated by the GLS [19] approach (originally proposed for providing location services on mobile nodes), where a producer performs an information diffusion process by sending a message to a *list* of server nodes determined by its location and the data type. A consumer can then retrieve this data in time proportional to the distance to the producer *in the underlying hierarchy*, which in the case of GLS is a positional quad-tree. But since — as in the case of hierarchy-based routing strategies — nearby nodes might be far away in the hierarchy, the consumer still does not experience distance-sensitive query times. Furthermore this assumes the availability of geographic location information and an auxiliary ID server structure that has to be precomputed to allow for information broker-age[1]. Very recently, an approach [5] combining Geographic Hash Tables with landmark-based routing via the GLIDER [6] scheme has been proposed. There are also several other approaches focusing on data aggregation, multi-resolutional storage, or database-like queries [10, 8, 20]. They all suffer from the above two problems, however.

Closest to our approach is the work of [4], where the authors developed a location/address lookup service called L^+ for mobile nodes in a landmark-based routing scenario aiming at distance sensitivity in the queries, that is, the time required to lookup the address of a target node should be proportional to its distance. Their construction could also be extended to an information brokerage scheme (though they did not do so). Their simulation results show the effectiveness of the approach, even in a mobile scenario, but no rigorous theoretical analysis was conducted, which this paper aims to provide.

Our contributions. In this paper we analyze augmented hierarchical decomposition schemes for routing and information brokerage built on only the topology of the communication graph. While such routing hierarchies have been around for many years and a related theoretical analysis has been very recently found in a different context [12], this paper is the first to present a *unified generic* framework for both routing and information brokerage, and this framework as well as the accompanying analysis can be applied to various implementations of hierarchical decompositions (such as those in [1, 9, 22]).

Our framework can be applied to networks with arbitrary communication graphs and guarantees distance-sensitivity of both routing as well as information brokerage. If the shortest path metric of the communication graph has bounded doubling dimension [11], we can guarantee that the routing tables that

---

[1] The GLS paper did not address the information brokerage application, though the ID service presented there, which provides a mapping of unique node IDs to geographic coordinates, can be seen as a special case of information brokerage.

need to be installed at every node have size only $O(\log n)$ bits. A metric space has *bounded doubling dimension* if any ball with radius R in the metric space can be covered by a constant number of balls with radius $R/2$. In practice, sensor networks frequently experience low-level link and node volatility. We show through simulations that our routing protocol is robust: it performs gracefully against the failure of a small fraction of network nodes, due to the absence of any backbone or hub structure.

Our information brokerage scheme is built on top of the presented routing structure at no extra overhead. To our knowledge, this is the first approach that works for arbitrary communication graphs and has theoretical performance guarantees in terms of the effort on both the producer and consumer sides. In particular, after the producer stores references to its data at a small number ($O(\log n)$ in case of metrics of bounded doubling dimension) of nodes (in a multi-resolution manner), any consumer can retrieve it in a distance-sensitive way. Furthermore, by visiting only $O(d)$ nodes, the consumer can collect all occurrences of a particular type of data within a neighborhood of radius d. This kind of *range query*[2] can be useful for implementing efficient in-network processing and data aggregation. We are not aware of any other scheme that efficiently supports this type of query. Our information brokerage scheme inherits both load-balanced information diffusion and robustness against node failures from the employed routing scheme. All these results are backed by an evaluation in simulation which indicates that the practical performance of the analyzed schemes is significantly better than their theoretical guarantees.

We would like to emphasize again that even though a hierarchy is used in our approach, nodes high up in the hierarchy do not get any additional load comparing to nodes in lower levels. During routing, nodes in the hierarchy act as landmarks toward which packets are forwarded, but they only stay as landmarks when the packets are still far away from them, so a breakdown of a node in some sense *does not* hinder its function as a routing waypoint. Just like any other nodes, the failure of a node in high level of a hierarchy only has a local effect on the routing and information brokerage capabilities of the sensor network.

Outline. In Section 2, we introduce the notion of a hierarchical decomposition as a generic hierarchical framework for organizing a sensor network, and then describe our routing scheme and an information brokerage system under this framework. The performance of our framework is experimentally evaluated in Section 3 by extensive simulations. Finally, we conclude and discuss possible extensions to our scheme in Section 4.

2   Distance-Sensitive Routing and Information Brokerage

In this section, we first introduce the notion of a *hierarchical decomposition* (HD) constructed on an arbitrary sensor network in which geographic coordinates of

---

[2] We remark that we use the term *range query* differently here from work like [8, 20], where 'range' refers to a range in data space and not in the space of sensor locations.

the nodes may not be available. As it becomes clear later, a HD captures all the necessary properties we need for routing and information brokerage. We then show how to use a HD of the communication graph for efficient routing and for distance sensitive information brokerage.

2.1 Hierarchical Decompositions of Graphs

In the following we consider an undirected, weighted, connected graph of n nodes with node distances induced by the shortest path metric. We call a tree H of height h a hierarchical decomposition (HD) of S if

- each node $c \in H$ is associated with a set of nodes $S_c \subseteq S$ (*cluster*),
- for the root $r \in H$ (which is at level $h-1$) we have $S_r = S$,
- all leaves of H have the same level 0,
- for any node $c \in H$ at level k, we have that the cluster S_c associated with c has diameter less than $\alpha \cdot 2^k$ for some constant α,
- if $c \in H$ has children $c_1 \ldots c_l$, we have $S_c = \biguplus S_{c_i}$

In case of an unweighted graph (i.e. edge weights are all 1), the diameter of a connected n-node graph is at most n, and one can construct a hierarchical decomposition of height at most $h = 1 + \lceil \log n \rceil \leq 2 + \log n$. The following discussion focuses on this case for simplicity. We also remark that the constraint of all leaves being at the same level 0 is not mandatory and could be removed. We make this assumption for simplicity of the presentation and also because the hierarchical decomposition we use in our experiments has this property.

There are different ways to construct a HD. In our implementation we use a specific HD based on the discrete center hierarchy (DCH) from [9], which is also similar to the hierarchy in [12]. It can be built efficiently and distributedly on an arbitrary sensor network given only its connectivity graph (details omitted for lack of space).

2.2 Routing Using Hierarchical Decompositions

First we provide a routing scheme based on a hierarchical decomposition H of a communication graph. Important properties which we want to ensure are:

- **scalability:** the routing information that an individual node of the network has to store should be small compared to the network size; the load on the nodes should be distributed in a balanced fashion (so we disallow dedicated hub nodes or backbones).
- **efficiency:** the path generated by our routing scheme between nodes v and w should be only a constant factor worse than the optimal shortest path in the communication network.
- **robustness:** the impact of nodes or links failing should be limited and local. In particular, packets that get temporarily stuck due to node or link failures should be able to recover using local rules.

The scheme we provide has all these three properties.

An addressing scheme. Let $\Delta_{\max}$ be the maximum degree of H. We number the children $c_1, \ldots c_o$ with $o \leq \Delta_{\max}$ of a node c arbitrarily, and define the following addresses for the nodes of the tree:

- the root r has as address the h-dimensional vector $f(r) := (1, 0, 0, \ldots, 0)$ (remember h is the height of H)
- a node $c' \in H$ at level k which is child c_i of parent c is assigned the address $f(c') := f(c) + i \cdot e_{h-k}$, where e_{h-k} is the h-dimensional unit vector with a 1 at the $(h-k)$-th position

Essentially this constructs an IP-type address for each node of the tree and hence also for each cluster in the hierarchical decomposition. The entries in the vector are bounded by the maximum number of children $\Delta_{\max}$.

Connecting levels and efficient routing. Let us now extend the addressing scheme to an efficient routing protocol. For that we need the notion of *neighboring clusters* of a node:

Definition 1. *A cluster L at level k is called a* neighboring cluster *of a node v if there exists a node $q \in L$ such that $d(v, q) \leq \alpha \cdot 2^{k+1}$.*

Note that while the number of neighboring cluster of a given node in each level may be larger than the number $\Delta_{\max}$ in the decomposition tree, in 'well-behaved' decompositions, we expect this number to be small. In particular, when the hop distance between the nodes is a metric with bounded doubling dimension [11] and a DCH is used as a hierarchical decomposition, the maximum number λ_{max} of neighboring clusters of a node in any given level is a constant, and a node has at most a total of $O(h) = O(\log n)$ neighboring clusters. From now on, we assume this is the case (so $\Delta_{\max}$ and $\lambda_{\max}$ are constants) to simplify our presentation.

We let each node in the sensor network store its distances to all of its neighboring cluster. The routing of a message to a node w from v can then be done as follow. Node v first looks up its neighboring clusters and find the smallest cluster L containing w. v then locally forwards the message for w to *any* node closer to L than v.

We claim that the length of the path when the message arrives at the smallest cluster containing w is at most a constant times longer than the optimal, shortest path from v to w in the communication graph. If our hierarchical decomposition has singleton clusters associated with the leaves, this implies a complete path from v to w which is a constant factor approximation of the shortest path. In particular, we have the following result (proof omitted for lack of space).

Lemma 1. *Let v, w be two nodes in the network. Then the path generated by the above routing scheme from v to the cluster of the lowest level containing w has length at most $4 \cdot d_{vw}$ where d_{vw} denotes the shortest path distance between v and w in the communication graph.*

Since there are typically many close-to-shortest paths from a given node to a cluster, this routing scheme has a *natural robustness* against node or link failures. And even if none of the immediate neighbors is closer to the target cluster, inspecting a slightly larger *local* neighborhood most of the time results in a successful forwarding of the message towards the target cluster.

If every bit counts. Assuming that the maximum number of neighboring clusters λ_{max} and the maximum number of children Δ_{max} in the HD is a constant, each node has to store address and distance of $O(h)$ clusters. In case of a communication graph with unit edge weights and singleton clusters at the leaves, $h = \Omega(\log n)$ and hence each node has to store $\Omega(\log^2 n)$ bits: the address of a cluster has $h = \Omega(\log n)$ components and the bit-complexity of the distance value stored for a neighboring cluster might be $\Omega(\log n)$ as well.

If every bit of space is relevant, we can do still better. Let us first consider a more efficient way to store the addresses of neighboring clusters of a node v. The key observation here is the trivial fact that if at level $k - 1$ some cluster L is a neighbor of v, then in level k its parent $p(L)$ is also a neighbor. That means if a node has already stored the address of $p(L)$ it can store the neighbor L at level $k - 1$ at additional cost of only $\log \Delta_{max}$ bits. Hence for constant $\lambda_{max}, \Delta_{max}$, the addresses of all neighbors at all levels can be stored using $O(\log n)$ bits. The need for $\Omega(\log n)$ bits per distance value per neighbor can easily be reduced to a constant by just remembering *one edge* to an adjacent node in the communication graph that is closer to the neighboring cluster instead of the actual distance.

Corollary 1. *The routing scheme can be implemented by storing $O(\log n)$ bits per node in the network.*

We remark that the above addressing scheme as well as the neighboring information can be computed efficiently by restricted flooding during the initialization stage. It only increases the construction cost slightly over that of DCH's.

2.3 Efficient Information Brokerage Using Hierarchical Decompositions

Given some fixed HD H, we now show how to achieve efficient distance-sensitive information brokerage based on the routing scheme described above. Let $\Sigma = \{\sigma_1, \sigma_2, \ldots, \sigma_m\}$ denote the discrete set of all data items possibly produced or queried in the sensor net. Some properties of a desirable information brokerage system include:

- **load balance:** no node should have the burden of providing lookup-information for many different types of data items;
- **efficiency:** references to a certain type of data should be stored at a small number of nodes, and the time required for node w to access data produced by node u should be proportional to the distance between u and w.
- **robustness:** failure of nodes or links should only increase the time to store or retrieve a certain data item, but not make storage/retrieval impossible.

Our information brokerage system exploits the routing structure described above and to meet these desiderata.

First assume that we have a hash function $\mu : \Sigma \times HD \rightarrow S$, such that given any data item $\sigma \in \Sigma$ and a cluster $L \in HD$, we can compute a unique sensor node $\mu(\sigma, L) \in S$ within this cluster. Furthermore, $\mu(\sigma, L)$ can be accessed from any node in cluster L (of diameter D) within $2D$ hops using our routing structure (we will describe one such hash function at the end of this section). We call $\mu(\sigma, L)$ the *information server* of data item σ in cluster L. Our information brokerage system relies on collecting and distributing some synopses of data items to a small set of information servers.

Information diffusion. Suppose a node (producer) u has data item $\sigma \in \Sigma$. Recall that u is contained in h clusters of the tree H (i.e., its ancestors), one in each level. Let $L(u, d)$ be the cluster containing u at level d, and $L_d^1, \ldots, L_d^l$ the neighboring clusters of $L(u, d)$. The producer u sends a message (containing its own address and some synopses of σ) to the information server $\mu(\sigma, L_d^j)$ associated with each of these L_d^j's for all $0 \le d < h$, see Figure 1. If the maximum number of neighbors at each level and the maximum degree of H are constants, then a producer will store a synopses of σ at $O(h) = O(\log n)$ nodes. Since the routing structure already specifies how to access all these neighboring clusters, no extra per-node storage is required to implement the diffusion process.

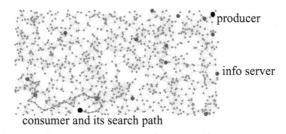

Fig. 1. There are exponentially fewer information servers away from the producer (upper right). The consumer (bottom center) looks for information at information servers in exponentially bigger clusters containing itself.

The length of the paths to the information servers decreases geometrically with decreasing level in the hierarchy. Hence the total number of hops to send synopses to all information servers, i.e., the *communication cost* for the producer, is dominated by the length of the paths to the information servers in the highest level of the hierarchy. For graphs of constant doubling dimension, this cost is in the order of the diameter of the network. We summarize this in the following lemma:

Lemma 2. *In the above diffusion scheme a producer of some data item $\sigma \in \Sigma$ distributes σ to $O(\log n)$ nodes in the network at a total cost of $O(D)$ hops, where D denotes the diameter of the network.*

Information retrieval. When a consumer w wants to access some data item σ, it will look for it in growing neighborhoods, namely, in clusters $L(w, i)$'s in increasing order of i, where $L(w, i)$ denotes the ancestor of w at level i, see Figure 1. More precisely, it starts from w, and visits information servers $\mu(\sigma, L(w, 1))$, $\mu(\sigma, L(w, 2))$, ..., in sequential order to check whether the data item σ is there. Note that unlike the producer which sent out messages in different branches to various information servers, the consumer will only follow one path, and return as soon as it finds the information sought. The following lemma guarantees the distance sensitivity of our method (proof omitted from this extended abstract).

Lemma 3. *If node w wants to retrieve the data item $\sigma \in \Sigma$ associated with node u, this request can be completed in $O(d_{uw})$ time steps, where d_{uw} denotes the distance between u and w.*

We would like to emphasize that the constant in the O-notation experienced in practice is very close to 1. Furthermore it frequently happens that when a node observes an event, nearby nodes also observe the same event. To prevent multiple hashing for the same data item, each node can attempt to first retrieve the same item from its local neighborhood using the information retrieval process. If the item is not found, it can start the information diffusion process.

Approximate range counting and reporting. Let $RC(w, r; \sigma)$ denote the number of occurrences of data item σ detected by some sensor at most r hops away from w. Our information brokerage system can also be used to perform approximate range counting or range reporting for a consumer. In particular, we have the following lemma (proof omitted).

Lemma 4. *Let s be the number of distinct messages about σ received at node $\mu(\sigma, L(w, d))$, where $d = \lceil \log(r/\alpha) \rceil$, then $RC(w, r; \sigma) \leq s \leq RC(w, 4r; \sigma)$.*

In other words, by visiting the information server $\mu(\sigma, L(w, d))$ directly, a consumer w is guaranteed to collect all sources that have information about data item σ within roughly distance 2^d to itself. If the consumer only wants to know the number of such sources (range counting), it can simply return. Otherwise, if it wishes to report all such data (range reporting), it can then route to each of these sources and fetch the data.

Hash function $\mu(\sigma, L)$. We still have to define the hash function $\mu(\sigma, L)$ that maps any given data item σ to an information server in a particular cluster L. Ideally, in order to have a good load balance, this function should distribute the information servers uniformly to all nodes contained in L for various $\sigma \in \Sigma$.

One possible choice of this function is deployed in the GLS approach, where each sensor node in the sensor network has a unique id (an integer). Given an data item $\sigma \in \Sigma$, assume there is a function that maps σ to one of this id, say ID_σ, randomly. The information server $\mu(\sigma, L)$ is then defined as the node $s \in L$ with smallest id that is greater than ID_σ. However, in order to identify and reach node s, one has to build extra structure (which needs about $O(h \log n)$ bits per-node memory) on top of whatever routing structure exploited.

A simpler way to define $\mu(\sigma, L)$ is to use σ as the seed of some pseudo-random function and traverse the HD tree downward randomly according to that function. $\mu(\sigma, L)$ is then defined as the leaf node reached in the process. Note that this definition of $\mu(\sigma, L)$ also gives a routing path to access it. We can first route to any node in L, using σ to determine the pseudo-random subcluster L' of L that contains $\mu(\sigma, L)$, then recursively route toward $\mu(\sigma, L') = \mu(\sigma, L)$. The number of hops to route from any node in L to $\mu(\sigma, L)$ is obviously bounded by $\sum_{i=1}^{d} \alpha \cdot 2^i \le \alpha \cdot 2^{d+1}$.

Robustness. The routing components of our information brokerage scheme inherit the robustness properties of the routing scheme; in particular, messages that are unable to make progress due to node or link failures can recover by simple *local* rules and be eventually forwarded to their destinations. Furthermore, recovery after failure of an information server is possible by querying the information server one level higher, incurring only a constant factor overhead.

3   Experimental Results

We implemented the discrete center hierarchy in java to experimentally evaluate the performance of our proposed schemes. Currently our implementation simulates the network at the graph level only. While it does not mimic network attributes like packet loss or delay, we are quite confident that the results reported here give a good indication about the usefulness of our approach in practical scenarios.

3.1   Data Generation and Implemented Algorithms

All our measurements were carried out on a unit disk graph of nodes that were spread *uniformly at random* in a unit square, and subsequently nodes may have been removed to simulate the presence of large holes in the network topology. Note that this "unit disk graph" is not required in our approach— our system can take an arbitrary graph as input. We also want to emphasize that in this model, as long as the average node degree is below about 10, a large fraction of the unit disk graph is not connected, and the dilation factor is rather large, i.e. the shortest path in the graph between two nodes is much larger than their Euclidean distance. This is quite different in the 'skewed grid' model where the node positions are determined by randomly perturbing points on a grid by some rather small amount. There the unit-disk graph is almost always connected for any grid-width slightly smaller than 1 (which corresponds to an average degree of 4 or more) and the geometric dilation is very small. See Figure 2 for an example. While our scheme performs much better in the skewed grid model as compared to the random model, we feel the latter provide more insight on realistic scenarios, and present all our results in this random model.

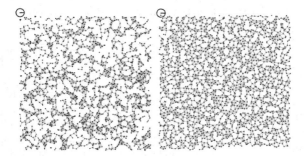

Fig. 2. Random sensor field (left) has degree 6.26 and has many holes, some of which are quite large. Skewed grid field (right) has lower degree, 5.18, yet is much more regular.

In the following subsections, we compare our routing scheme with GPSR, and our information brokerage scheme with GHT by assuming that the geographic locations of sensors are known (although not used in our approach). The underlying planar graph for GPSR is the Gabriel graph.

3.2 Evaluation of Routing Strategies

Routing quality. In the first experiment we evaluate the quality of the paths produced by our routing scheme. We fix a sensor field of 2000 nodes and vary the average degree from roughly 6 to 12. We then select at random 1000 pairs of nodes from the largest connected component of the sensor field. We compute the paths between the nodes in these pairs using the HD as well as using GPSR. For each of such path, its *quality* is defined as the ratio between its length and the shortest distance between its two endpoints. We show in Table 1 (a) the average and standard deviation of the quality of these 1000 paths. Note that our HD routing scheme always produces near-optimal paths regardless of the node density. The practical constant is much better than the worst case bound of 4 we could prove.

Network initialization and routing scalability. Here, we fix the average degree at roughly 6, vary the number of nodes from 200 to 20000 and compute the per-node storage for the HD routing structure. As expected, the number of entries in the routing table needed at each node grows very slowly, see Figure 3 (a). In particular, the maximal storage at a node in the network is quite reasonable, merely about twice the average per-node storage.

We note that the cost of initializing the network, i.e., the number of messages sent to establish our hierarchical decomposition, is directly proportional to the storage at each node. As the storage cost is low, the cost of network initialization is also low.

Hot spots. Even though our implementation of the HD uses cluster heads, they are not special nodes in the network In a typical route, the moment a package heading towards a cluster reaches any of its nodes, the package starts heading towards a different cluster. So these cluster heads do not form a backbone

Table 1. (a) Quality of paths from HD and from GPSR. (b) Performance of brokerage.

Avg.	Qual. of HD		Qual. of GPSR	
deg.	avg	std	avg	std
6.21	1.08	0.18	4.91	6.79
6.80	1.06	0.11	4.04	7.41
7.39	1.05	0.09	3.25	7.02
7.93	1.09	0.16	2.04	3.10
8.53	1.07	0.12	1.59	2.22
8.94	1.07	0.10	1.51	2.42
9.82	1.07	0.10	1.35	1.62
10.2	1.06	0.09	1.31	1.80
10.8	1.05	0.09	1.44	2.90
12.0	1.06	0.11	1.15	1.30

(a)

Size	# Info. servers		Query time		Total time	
	avg	std	avg	std	avg	std
200	10.8	0.42	1.22	0.82	2.28	1.31
400	17.0	0.21	0.93	0.63	1.91	1.06
600	28.7	1.36	0.74	0.59	1.73	0.92
1000	34.8	4.36	0.74	0.49	1.71	0.76
2000	50.9	8.22	0.73	0.45	1.76	0.70
4000	65.2	11.8	0.75	0.39	1.69	0.56
6000	75.7	13.5	0.78	0.43	1.79	0.64
8000	83.5	14.8	0.79	0.43	1.83	0.65
10000	85.9	15.5	0.77	0.38	1.77	0.60

(b)

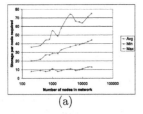
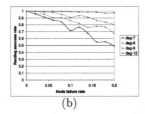
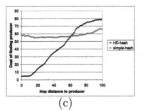

(a)   (b)   (c)

Fig. 3. (a) The storage required growths slowly when the network becomes large. (b) The success rate of routing vs. node depletion for sensor fields with various average degree. (c) Number of hops to information server using HD and number of hops in a shortest path to an ideally random information server.

structure, nor do they create bottlenecks in network traffic. Figure 4 (c) gives an example where two routes with nearby sources and destination nodes stay separate during their course. On the other hand, when large holes are present in the network, nodes close to holes will naturally have a heavier burden, as our HD paths approximate the shortest paths well. Still, our paths do not hug the holes in the sensor net as tightly as GPSR paths do, as shown in Figure 4 (a) and (b) for a sensor field with 2000 nodes and average degree 9.5. Our HD scheme produces many fewer higher load nodes (larger dots in the picture).

Robustness. To measure the performance of our routing scheme under node depletions, we start with a graph with average degree of $7, 8, 9$, or 12, build the HD routing structure on top of it, and then randomly remove a small percentage of sensors (from 2% to 20%) from the sensor field. We then pick 1000 pairs of live nodes at random, and show the success rate of routing between these pairs in Figure 3 (b). During the routing process, if a node finds that the next sensor on its shortest path to some cluster L is dead, it locally floods a neighborhood of nodes at most 5 hops away from itself to find a node with smaller distance

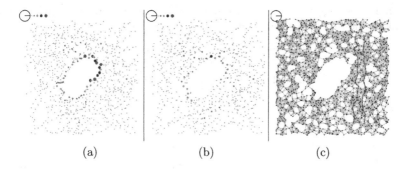

(a)   (b)   (c)

Fig. 4. Hotspots comparisons for (a) GPSR and (b) HD scheme. Larger dots are nodes with higher traffic loads. In (c), two paths generated by HD scheme with nearby sources and destinations.

to L. The result shows that the performance is gracefully degraded when the node failure rate increases.

3.3   Evaluation of the Information Brokerage Scheme

Brokerage quality. The efficiency of a brokerage system includes both the number of messages (i.e., # information servers) that a producer needs to replicate, and the number of hops that a consumer needs to access before locating the data it needs (i.e., the query time). In Table 1 (b), we vary the size of the sensor field from 200 to 10, 000 nodes. For each sensor field, we randomly choose 1, 000 producer/consumer pairs, with each pair producing/requesting a random data item. Columns 2,3 show the average number and standard deviation of information servers for each producer. Columns 4 and 5 show the quality of the path from the consumer to the information server, defined as the ratio between query time using our scheme (i.e., the path length to the respective information server) and the shortest hop distance between the producer and the consumer. We see from the table that this ratio is always close to 1.0 (in fact, in most cases it is smaller than 1.0, since the information server can be even closer than the producer). If the data is large, the producers may not replicate their data and only leave their addresses at the information servers. In that case, a consumer after locating the desired information server must further route to the producers to get the data itself. The quality of the brokerage path is then defined as the ratio between the path length from the consumer to the producer obtained using our scheme over the shortest path length between the consumer and producer. Column 6 and 7 in Table 1 (b) show this ratio in our experiment. In all cases, it is always small, around 2.0.

   While the number of replications used by a particular producer is higher in our system than in the GHT approach, the query time can be much smaller, especially when the consumer is closer to the producer. This phenomenon is illustrated in Figure 3 (c), where we compare the query time in our scheme (lower curve) with the length of the shortest path to an ideally random information server (upper curve). Note that the latter is in fact a lower bound of the query

Table 2. The average/standard deviation of the number of times that a sensor serves as an information server for some data item for various network sizes

Size	200	400	600	1000	4000	6000	8000
Avg.	10.0	16.0	25.5	31.5	47.2	61.2	69.7
Std.	14.4	28.1	36.4	52.7	75.4	90.8	99.7

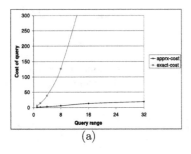
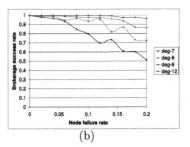

(a)   (b)

Fig. 5. (a) Approximate query cost is very low compared to the naive flooding. (b) The success rate of information brokerage under nodes depletions.

time for any scheme using GHT for information brokerage. The query time for GHT using GPSR as the underlying routing scheme may be much longer than this shortest path, due to the path quality returned by GPSR. In short, our system is attractive for scenarios where there are multiple queries for the same data, as the overhead for the producer is then amortized.

It is also important to keep the distribution of information servers for different data items as uniform as possible. To test this, we let each sensor in the network produce a different data item, and record for each node the number of times that it serves as an information server for some data item. The results are in Table 2. The distribution of information servers observed is reasonably good compared to a distribution obtained by a centralized uniformly random hash function.

Approximate range counting. One important application of our information brokerage system is for approximate range counting, such as reporting all horses detected within some distance r from a particular sensor. When r is quite small, flooding is simple and effective. However, the number of nodes accessed in the flooding approach increases quadratically as r increases. This is illustrated in Figure 5 (a) where the query cost in our approach (lower curve) increase in a linear manner, while that for flooding (upper curve) increases quadratically. The size of the sensor field in this example is 2,000 with an average degree of 6.1.

Robustness. Again, we fix a sensor field of 2000 nodes with various average degree, compute the HD routing structure, and remove a portion of sensors randomly (from 2% to 20%) from the field. We then randomly choose a set of producer/consumer pair, each generating/seeking a random data item. The resulting success rates are shown in Figure 5 (b). The brokerage system is slightly more robust than the routing scheme, which is not surprising: as the robustness of our information brokerage system comes partly from the robustness of the routing scheme, and partly from the fact that even if a query fails to route to a particular information server, it can simply go to the information server one level up.

4 Conclusion and Discussion

In this paper we have presented a unified framework for efficient routing and distance-sensitive information brokerage based on augmented hierarchical decompositions of communication graphs. In particular, for communication graphs of constant doubling dimensions, the guarantee for almost optimal routing paths comes at an additional cost of only $O(\log n)$ bits of storage per network node. Our routing scheme does not rely on a dedicated backbone or hub structure, and hence performs rather well when some network nodes fail while providing a natural load balance between routing paths.

Our novel *distance-sensitive information brokerage* scheme is built based on the above routing scheme. We showed how information producers can diffuse pointers to their information to $O(\log n)$ other locations and in turn how information consumer can exploit this stored data to retrieve the information they want in a distant sensitive manner. As an application, our brokerage infrastructure allows for range queries with specified radius that take time proportional to the radius instead of time proportional to the area of the relevant range region. All our procedures come with rigorous proofs of their worst-case performance guarantees, and the experimental results for both routing and information brokerage show considerably better performance than the worst case considerations in the theoretical analysis, (but the latter in some way explain this good behavior in practice).

In future work, it might be interesting to view the problem also from a producer's perspective. In particular, we can try to trade off the producer's effort to make its information available against the consumer's effort to obtain that information. The exact tradeoff can depend on the relative frequencies of data collection operations vs. queries, in the style of [13]. Furthermore, even though the main focus of this paper has been the static case where sensor nodes do not move over the lifetime of the network, it might be interesting to extend our approach to allow for efficient routing and information brokerage in the presence of mobile sensor nodes. Also we believe that the use of our distance-sensitive range queries can lead to interesting new in-network data-aggregation and processing algorithms.

Acknowledgements

The authors would like to thank the anonymous referees for their constructive comments. The authors gratefully acknowledge the support of DoD Multidisciplinary University Research Initiative (MURI) program administered by the Office of Naval Research under Grant N00014-00-1-0637, NSF grants FRG-0354543 and CNS-0435111, and the Max Planck Center for Visual Computing. The 2nd author also wishes to thank Scott Shenker for many useful discussions.

References

1. E. M. Belding-Royer. Multi-level hierarchies for scalable ad hoc routing. *Wireless Networks*, 9(5):461–478, 2003.
2. P. Bose, P. Morin, I. Stojmenovic, and J. Urrutia. Routing with guaranteed delivery in ad hoc wireless networks. *Wireless Networks*, 7(6):609–616, 2001.
3. H. T.-H. Chan, A. Gupta, B. M. Maggs, and S. Zhou. On hierarchical routing in doubling metrics. In *SODA '05: Proceedings of the sixteenth annual ACM-SIAM symposium on Discrete algorithms*, pages 762–771, Philadelphia, PA, USA, 2005. Society for Industrial and Applied Mathematics.
4. B. Chen and R. Morris. L+: Scalable landmark routing and address lookup for multi-hop wireless networks, March 2002.
5. Q. Fang, J. Gao, and L. J. Guibas. Landmark-based information storage and retrieval in sensor networks. In *Proc. of the 25th Conference of the IEEE Communication Society (INFOCOM'06)*, April 2006.
6. Q. Fang, J. Gao, L. J. Guibas, V. de Silva, and L. Zhang. GLIDER: Gradient landmark-based distributed routing for sensor networks. In *24th Conference of the IEEE Communication Society (INFOCOM)*, 2005.
7. W. F. Fung, D. Sun, and J. Gehrke. COUGAR: The network is the database. In *SIGMOD Conference*, 2002.
8. J. Gao, L. J. Guibas, J. Hershberger, and L. Zhang. Fractionally cascaded information in a sensor network. In *3rd Int'l. Sympos. Information Processing in Sensor Networks (IPSN)*, pages 311–319, 2004.
9. J. Gao, L. J. Guibas, and A. Nguyen. Deformable spanners and applications. In *Proc. of the 20th ACM Symposium on Computational Geometry (SoCG'04)*, pages 179–199, June 2004.
10. B. Greenstein, D. Estrin, R. Govindan, S. Ratnasamy, and S. Shenker. DIFS: A distributed index for features in sensor networks. In *1st IEEE International Workshop on Sensor Network Protocols and APplications Anchorage*, 2003.
11. A. Gupta, R. Krauthgamer, and J. R. Lee. Bounded geometries, fractals, and low-distortion embeddings. In *Proc. IEEE Symposium on Foundations of Computer Science*, 2003.
12. M. Herlihy and Y. Sun. Distributed transactional memory for metric-space networks. In *Proc. International Symposium on Distributed Computing (DISC 2005)*, pages 324–338, 2005.
13. Y. Huang and H. Garcia-Molina. Publish/subscribe in a mobile enviroment. In *MobiDe '01: Proceedings of the 2nd ACM international workshop on Data engineering for wireless and mobile access*, pages 27–34, New York, NY, USA, 2001. ACM Press.
14. C. Intanagonwiwat, R. Govindan, D. Estrin, J. Heidemann, and F. Silva. Directed diffusion for wireless sensor networking. *IEEE/ACM Trans. Netw.*, 11(1):2–16, 2003.
15. B. Karp and H. T. Kung. GPSR: Greedy perimeter stateless routing for wireless networks. In *6th ACM Int'l. Conf. Mobile Computing and Networking (MobiCom)*, pages 243–254, 2000.

16. F. Kuhn, R. Wattenhofer, Y. Zhang, and A. Zollinger. Geometric ad-hoc routing: of theory and practice. In *PODC '03: Proceedings of the twenty-second annual symposium on Principles of distributed computing*, pages 63–72, New York, NY, USA, 2003. ACM Press.

17. F. Kuhn, R. Wattenhofer, and A. Zollinger. Asymptotically optimal geometric mobile ad-hoc routing. In *Proc. Int. Workshop on Discrete Algorithms and Methods for Mobile Computing and Communications (Dial-M). ACM Press*, pages 24–33, 2002.

18. J. Kulik, W. Heinzelman, and H. Balakrishnan. Adaptive Protocols for Information Dissemination in Wireless Sensor Networks. In *5th ACM MOBICOM*, Seattle, WA, August 1999.

19. J. Li, J. Jannotti, D. S. J. D. Couto, D. R. Karger, and R. Morris. A scalable location service for geographic ad hoc routing. In *6th ACM Int'l. Conf. Mobile Computing and Networking (MobiCom)*, pages 120–130, 2000.

20. X. Li, Y. J. Kim, R. Govindan, and W. Hong. Multi-dimensional range queries in sensor networks. In *SenSys '03: Proceedings of the 1st international conference on Embedded networked sensor systems*, pages 63–75, New York, NY, USA, 2003. ACM Press.

21. J. Newsome and D. Song. GEM: Graph embedding for routing and data-centric storage in sensor networks without geographic information. In *1st Int'l Conf. Embedded networked sensor systems*, pages 76–88, 2003.

22. R. Ramanathan and M. Steenstrup. Hierarhically-organized, multihop mibile wireless networks for quality-of-service support. *Mobile Networks and Applications*, 3(1):101–119, 1998.

23. A. Rao, C. Papadimitriou, S. Shenker, and I. Stoica. Geographic routing without location information. In *9th ACM Int'l. Conf. Mobile Computing and Networking (MobiCom)*, pages 96–108, 2003.

24. S. Ratnasamy, B. Karp, L. Yin, F. Yu, D. Estrin, R. Govindan, and S. Shenker. GHT: A geographic hash table for data-centric storage in sensornets. In *1st ACM Workshop on Wireless Sensor Networks and Applications*, pages 78–87, 2002.

25. N. Sadagopan, B. Krishnamachari, and A. Helmy. Active query forwarding in sensor networks. *Ad Hoc Networks*, 3(1):91–113, 2005.

26. A. Savvides, C. C. Han, and M. B. Strivastava. Dynamic fine-grained localization in ad-hoc networks of sensors. In *7th ACM Int'l. Conf. Mobile Computing and Networking (MobiCom)*, pages 166–179, 2001.

27. S. Shenker, S. Ratnasamy, B. Karp, R. Govindan, and D. Estrin. Data-centric storage in sensornets. *ACM SIGCOMM Computer Communication Review*, 33(1):137–142, 2003.

28. P. F. Tsuchiya. The landmark hierarchy: a new hierarchy for routing in very large networks. In *Proceedings on Communications architectures and protocols*, pages 35–42, 1988.

29. F. Zhao and L. Guibas. *Wireless Sensor Networks: An Information processing approach*. Elsevier/Morgan-Kaufmann, 2004.

Efficient In-Network Processing Through Local Ad-Hoc Information Coalescence*

Onur Savas, Murat Alanyali, and Venkatesh Saligrama

Department of Electrical and Computer Engineering,
Boston University, Boston, MA, 02215
{savas, alanyali, srv}@bu.edu

Abstract. We consider in-network processing via local message passing. The considered setting involves a set of sensors each of which can communicate with a subset of other sensors. There is no designated fusion center; instead sensors exchange messages on the associated communication graph to obtain a global estimate. We propose an asynchronous distributed algorithm based on local fusion between neighboring sensors. The algorithm differs from other related schemes such as gossip algorithms in that after each local fusion one of the associated sensors ceases its activity until it is re-activated by reception of messages from a neighboring sensor. This leads to substantial gains in energy expenditure over existing local ad-hoc messaging algorithms such as gossip and belief propagation algorithms. Our results are general and we focus on some explicit graphs, namely geometric random graphs, which have been successfully used to model wireless networks, and d-dimensional lattice torus with n nodes, which behave exactly like mesh networks as n gets large. We quantify the time, message and energy scaling of the algorithm, where the analysis is built upon the coalescing random walks. In particular, for the planar torus the completion time of the algorithm is $\Theta(n \log n)$ and energy requirement per sensor node is $O((\log n)^2)$ and for 3-d torus these quantities are $\Theta(n)$ and $O(\log n)$ respectively. The energy requirement of the algorithm is thus scalable, and interestingly there appears little practical incentive to consider higher dimensions. Furthermore, for the planar torus the algorithm exhibits a very favorable tradeoff relative to gossip algorithms whose time and energy requirements are shown here to be $\Omega(n)$. Also, the proposed algorithm can be generalized to robustify against packet losses and permanent node failures without entailing significant energy overhead. The paper concludes with numerical results.

1 Introduction

Wireless sensors bear a vast potential as they can be networked to form amorphous systems that are far more capable than their parts. This potential is

---
* This research was supported by Presidential Early Career Award N00014-02-100362, NSF CAREER Programs ANI-0238397, ECS-0449194 and NSF programs CCF-0430983, and CNS-0435353.

P. Gibbons et al. (Eds.): DCOSS 2006, LNCS 4026, pp. 252–265, 2006.
© Springer-Verlag Berlin Heidelberg 2006

accompanied by substantial technical challenges, namely such systems call for distributed models of operation that comply with requirements that arise due to energy limitations of sensors, packet losses along wireless links, sensor failures, and possibly uncontrollable network topologies.

In this paper we study ad-hoc distributed computation of a wide class of functions of network-wide measurements. The topic of distributed computation has received significant interest recently in the context of sensor networks (see [17, 8, 5 ,13, 14, 15, 11, 9, 12, 2, 3, 4, 1] and included references). Distributed computation and optimization arises in a number of different applications ranging from signal processing such as distributed localization/detection/estimation/tracking to load balancing and self-organization in communication networks [8, 7]. Previously proposed techniques can be broadly categorized into two groups: The fusion-centric approach (see [17, 11, 9] and included references) assumes that each sensor has a communication link to a data fusion center and each sensor node computes a local decision, which is then communicated to the fusion center. The fusion center then forms a global estimate of the desired function based on local decisions. The ad-hoc approach, on the other hand, involves no designated fusion center but focuses on establishing consensus within the network via local message exchanges. This approach is appears to be more suitable to address energy issues in large-scale networks and also appears to have robustness advantages. This is because: (a) it requires much less energy to communicate to neighboring sensors; (b) in-network processing through locally fusing information results in compression; (c) no fusion center implies no single point of failure; (d) they are well suited for asynchronous operation and hence robust to packet losses and node failures. Specifically, consider the so called gossip algorithm [5] for computing the average of all the sensor observations. Gossip algorithms accomplish this task by randomly choosing two neighboring sensor nodes at each time and replacing their current values by their average. It turns out that this process over time converges to the average of all sensor values at all the sensors, i.e., all the sensors achieve a consensus. Such consensus algorithms have been recently explored in other contexts such as detection [2, 3] and control [13].

Nevertheless, these algorithms have fundamental disadvantages from an energy efficiency perspective. In particular, if n nodes are uniformly distributed in a given area and E_b is the average communication energy-per-message required to communicate information to a neighboring node, then the energy-per-node required to achieve consensus scales as $\Omega(nE_b)$, which can be significant for a large sensor network. The fundamental reason is that energy efficiency resulting from in-network processing is offset by ad-hoc message passing that results in redundant computations, i.e., the same set (or largely similar set) of nodes repeatedly fuse their information at different points in time. In a related problem involving distributed detection, the significant energy scaling can be attributed to the loopy nature of the network where messages sent from one node repeatedly arrive at the node at different points in time. In order to ensure that no information from any node is forgotten, each node must re-inject their messages in to the network to reinforce their information [3]. At a fundamental level the significant scaling of energy arises due to the slow mixing rate of large networks, which can be attributed to rather large second eigenvalues of certain connectivity matrices associated with the underlying communication graph. An immediate solution to reduce such redundant computations is to design a communication tree so that message from the leaf nodes arrive at accumulation points (or clusterheads), which fuse the information and forward this information to a parent node. However, such a construction requires centralized planning and is inconsistent with the requirements of ad-hoc sensor network operation, where packet losses and node failures are common.

To address these issues we present a novel distributed computing approach in this paper, where not only is the ad-hoc network operation preserved but where the energy-related disadvantages of the existing local message passing resulting from repeated redundant computations is also minimized. The emphasis of the paper is on two important but conflicting figures of merit for wireless sensor networks, namely the time and the energy required to complete the computation. We introduce an asynchronous distributed algorithm that is based on autonomous pairwise communications between pairs of neighboring sensors in the underlying communication topology. The algorithm forces one of the communicating nodes to cease transmitting new messages until it is re-activated due to reception of a message from a neighboring node. This results in exponential energy gains compared other similar schemes. We adopt messaging complexity as a proxy for energy requirement (as described earlier) and quantify the trade-off between this quantity and the completion time. For the d-dimensional lattice torus with n sensors, we show that the completion time is $\Theta(n(\log n)^\alpha)$ and energy requirement per sensor node is $O((\log n)^{\alpha+1})$ where $\alpha = 1$ for $d = 2$ and $\alpha = 0$ for $d \geq 3$. The algorithm thus has scalable energy requirement, furthermore its performance is almost insensitive to changes in the network connectivity represented by different values of $d \geq 2$, hinting at the possibility of predictable performance over mesh topologies. Section 5 focuses on averaging of sensor measurements as a case study for comparison with gossip algorithms. We show that both the time and the per-node energy requirement of the gossip algorithm of [5] are $\Omega(n)$ on the 2-dimensional torus with n nodes, irrespective of the choice of the parameters of this algorithm. The algorithm introduced in the present paper thus exhibits a very favorable trade-off between the two performance measures for $d = 2$, but the gains in energy requirements come at the expense of more substantial compromise in the completion time for larger d.

With regard to other important practical considerations, the algorithm is robust against packet losses provided that reliable link-layer protocols are employed in wireless exchanges. The algorithm is also resilient against permanent node failures, furthermore the tolerable number of node failures can be provisioned and the energy requirement of the algorithm increases linearly with this number.

The rest of the paper is organized as follows. The distributed computation problem is formulated in Section 2 and the proposed algorithm is specified in Section 3. Section 4 is devoted to the analysis of this algorithm on d-dimensional tori for $d \geq 1$. A comparative study with gossip algorithms, specifically on geometric random graphs, is given in Section 5, and the paper concludes with final remarks in Section 6.

2 Problem Definition

Let Λ be a set that is closed under operation f. We shall assume that f is commutative and associative, and that Λ has an identity element with respect to f, that is there exists $e \in \Lambda$ such that $f(\lambda, e) = \lambda$ for $\lambda \in \Lambda$. Let $F_2 = f$. Given $n \geq 3$ and a permutation π of $\{1, 2, \cdots, n\}$ define the mapping $F_n : \Lambda^n \mapsto \Lambda$ recursively by

$$F_n(\lambda_1, \lambda_2, \cdots, \lambda_n) = f(\lambda_{\pi(1)}, F_{n-1}(\lambda_{\pi(2)}, \lambda_{\pi(3)}, \cdots, \lambda_{\pi(n)})), \tag{1}$$

for $\lambda_1, \lambda_2, \cdots, \lambda_n \in \Lambda$. The notational dependence on the permutation π is dropped here because the mapping F_n does not depend on π due to the commutativity and associativity of f.

We will be concerned with distributed computation of $F_n(\lambda_1, \lambda_2, \cdots, \lambda_n)$ in cases when each λ_i is known to a distinct agent. In typical applications that motivate this work such an agent represents one of n sensors involved in a statistical inferencing procedure, λ_i represents a measurement taken by sensor i or a function thereof that reflects a local estimate, and $F_n(\lambda_1, \lambda_2, \cdots, \lambda_n)$ represents a global estimate or a sufficient statistic of the measurements. For example, in the simplistic case of computing the maximum value of the sensor measurements one may take $\Lambda = \mathbb{R} \cup \{-\infty\}$, $e = -\infty$, and $f(\lambda_1, \lambda_2) = \max(\lambda_1, \lambda_2)$. Representation of weighted vector averages, which are of interest in finding least-squares estimates, is illustrated in the following example:

Example 1. Let $k \geq 1$ be an integer and let $M^{k \times k}$ be the set of k-dimensional positive definite matrices. For $i = 1, 2, \cdots, n$ let $x_i \in \mathbb{R}^k$ and let $w_i \in M^{k \times k}$. The weighted average $(\sum_{i=1}^{n} w_i)^{-1} \sum_{i=1}^{n} w_i x_i$ can be expressed in the form (1) by choosing $\Lambda = \{\mathbb{R}^k \times M^{k \times k}\} \cup \{e\}$ with $\lambda_i = (x_i, w_i)$, and by setting

$$f(\lambda, \lambda') = ((w + w')^{-1}(wx + w'x')\,,\; w + w'),$$

for $\lambda = (x, w), \lambda' = (x', w') \in \Lambda - \{e\}$.

We consider computation of $F_n(\lambda_1, \lambda_2, \cdots, \lambda_n)$ under communication constraints that are specified by an undirected graph $G = (V, E)$ where each node in V denotes a sensor (hence $|V| = n$) and an edge $(i, j) \in E$ indicates a bidirectional communication link between sensors i and j. To avoid trivialities we shall assume that G is connected. Note that since F_n admits flexible decomposition in terms of the atomic operation f, there exist a sequence $(e_1, e_2, \cdots, e_{n-1})$ of distinct edges in G such that $F_n(\lambda_1, \lambda_2, \cdots, \lambda_n)$ can be computed by sequentially executing f on the values at the two ends of each link in the given order. Furthermore the edge set $\{e_1, e_2, \cdots, e_{n-1}\}$ forms a spanning tree for G, and some of the aforementioned operations can be executed in parallel so that the overall computation can be completed in time proportional to the diameter of G. Rather than such centralized algorithms, our focus here is on decentralized algorithms that require neither global information about G nor network-wide synchronization.

3   Information Coalescence

Consider the following distributed, asynchronous algorithm which requires each sensor to be aware of only the sensors that it can communicate directly:

Algorithm COALESCENT(f, λ, G): Each sensor maintains a variable *value* which is an element of the set Λ, and a variable *status* which is either '*carrier*' or '*idle*'. We denote the value of sensor $i \in V$ at time $t \geq 0$ by $v_i(t)$, and indicate the status of the sensor via $\xi_i(t)$ defined by

$$\xi_i(t) = \begin{cases} 1 & \text{if status of sensor } i \text{ is 'carrier' at time } t, \\ 0 & \text{else.} \end{cases}$$

Initially $(v_i(0), \xi_i(0)) = (\lambda_i, 1)$ and these variables evolve as follows: Each sensor has an independent Poisson clock that ticks at unit rate. When the local clock of sensor i ticks, say at time t_o, the sensor does not take any action if its current status is 'idle' (i.e. $\xi_i(t_o^-) = 0$). Otherwise, if $\xi_i(t_o^-) = 1$, the sensor chooses a neighbor at random, sends its current value $v_i(t_o^-)$ to that neighbor, and sets $(v_i(t_o), \xi_i(t_o)) = (e, 0)$ (in particular sets is status to 'idle'). The selected neighbor, say sensor j, sets $v_j(t_o) = f(v_j(t_o^-), v_i(t_o^-))$ and $\xi_j(t_o) = 1$. A pseudo-code for the algorithm is given in Figure 1.

Procedure Initialize()	Procedure Send()	Procedure Receive(message)
$\{v_i \leftarrow \lambda_i;$	if(status=='carrier'){	$\{v_i \leftarrow f(v_i, \text{message});$
status←'carrier';}	choose neighbor;	status←'carrier';}
	send(v_i);	
	$v_i \leftarrow e;$	
	status←'idle';}	

Fig. 1. Three subroutines that specify Algorithm COALESCENT(f, λ, G) at node i. Send() is activated by the local Poisson clock, and Receive() is activated by message reception from another sensor.

Algorithm COALESCENT(f, λ, G) is based on sequential execution of f on edges of G, but these edges are selected in a distributed and randomized manner, without particular regard to any optimality notion. Note that when an idle node receives a message the value at the originating node simply passes onto the idle node, whereas if the receiving node is also a carrier then its value becomes the image of the two values under f, thereby executing a step towards computation of $F_n(\lambda_1, \lambda_2, \cdots, \lambda_n)$. The following proposition points out a sample-path property that is useful in proving correctness of the algorithm.

Proposition 2. *Under Algorithm* COALESCENT(f, λ, G),

$$F_n(v_1(t), v_2(t), \cdots, v_n(t)) = F_n(\lambda_1, \lambda_2, \cdots, \lambda_n),$$

for all $t > 0$.

We say that a *coalescence* occurs whenever a message is transmitted from a carrier sensor to another carrier sensor. Note that the number of carrier sensors $|\xi(t)| \triangleq \sum_{i=1}^{n} \xi_i(t)$ is non-increasing in time t and it decreases by 1 at the times of coalescence in the network. For each integer $k \leq n$ define the random time σ_k as

$$\sigma_k = \inf\{t \geq 0 : |\xi(t)| = k\}.$$

That is, σ_k is the first time that k carrier nodes remain in the network. The random variable σ_1 is of particular interest since $v_i(t) = e$ for each sensor i such that $\xi_i(t) = 0$, and thus by Proposition 2 for $t \geq \sigma_1$ there exists a unique sensor $i(t)$ such that $v_{i(t)}(t) = F_n(\lambda_1, \lambda_2, \cdots, \lambda_n)$. We therefore regard σ_1 as the stopping time of Algorithm COALESCENT(f, λ, G).

Issues in implementation and robustness. The basic form of the algorithm can be modified to recognize the termination time σ_1 by maintaining local counters that keep track of how many coalescence operations were involved in obtaining the present value of each sensor. Note that this can be implemented in a distributed manner by including the local counter as part of each transmitted message. The algorithm is robust against packet losses provided that reliable protocols such as those based on handshaking are employed at the link layer. A more serious mode of failure is permanent failure of sensor nodes, since if a sensor dies when its status is 'carrier', then in addition to its initial value λ_i a set of other such sensor values are also lost. The impact of such failures is more pronounced in later stages of the algorithm when each carrier node is the unique bearer of typically many sensor values. This issue can be mitigated by running multiple independent instances of the algorithm simultaneously in the network. The resulting cost in the energy expenditure is a constant factor, which is the number of such instances.

For more insight on σ_1 define $\xi(t) = (\xi_i(t) : i \in V)$ and note that $(\xi(t) : t \geq 0)$ is a time-homogeneous Markov process with state space $\{0, 1\}^V$. It can be verified that $(\xi(t) : t \geq 0)$ can be constructed as follows: At time 0 simultaneously start n simple symmetric random walks, one at each node of the graph G. Namely each random walk jumps at the ticks of an independent Poisson clock of unit rate, to a neighboring node chosen at random. Let distinct random walks evolve independently until two of them occupy the same node, and coalesce these two random walks into one (that is, bind these two walks together so that they make the same moves) from that time on. Finally, set $\xi_i(t) = 1$ if there is a random walk occupying node i at time t, and set $\xi_i(t) = 0$ otherwise. The process $(\xi(t) : t \geq 0)$ is known as the coalescing random walk, and has been extensively studied as dual process for voter models of interacting particle systems.

The observation of the previous paragraph will be useful in the analysis of the algorithm in the following section. Here we note that the algorithm terminates almost surely on any finite graph:

Lemma 3. $P(\sigma_1 < \infty) = 1$.

Despite its close relationship to random walks, drawing more detailed conclusions about the complexity of Algorithm COALESCENT(f, λ, G) on arbitrary graphs appears difficult. In the next section we pursue this goal for the special case of d-dimensional torus for which substantial understanding of the coalescing random walk has been developed in the applied probability literature.

4   Time and Energy Requirements on the d-Dimensional Torus

Given integers $d, N \geq 1$ let $\mathbb{T}_N^d$ denote the d-dimensional lattice torus with N^d nodes. The graph $\mathbb{T}_N^d$ can be formally defined by identifying its nodes by members of $\{1, 2, \cdots, N\}^d$ and its edges by pairs in $\{1, 2, \cdots, N\}^d$ that are at Hamming distance 1 under modulo arithmetic with respect to N. In this section we analyze the complexity of Algorithm COALESCENT$(f, \lambda, \mathbb{T}_N^d)$ by examining the coalescing random walk process $(\xi(t) : t \geq 0)$.

We start with considering the termination time σ_1. It appears reasonable to expect σ_1 to be stochastically increasing in the network size; in fact for fixed d the following result of Cox [10] provides the precise growth rate of each σ_k with N. Let

$$s_N = \begin{cases} N^2 & \text{if } d = 1 \\ N^2 \log N & \text{if } d = 2 \\ N^d & \text{if } d \geq 3, \end{cases} \quad \text{and} \quad Q = \begin{cases} 1/6 & \text{if } d = 1 \\ 2/\pi & \text{if } d = 2 \\ \gamma(d) & \text{if } d \geq 3, \end{cases}$$

where each $\gamma(d)$ is a finite and strictly positive constant as identified in [10, Equation (1.2)].

Theorem 4. ([10, Theorem 6]) Let $G = \mathbb{T}_N^d$. For each integer $k \geq 1$ there exists a random variable σ_k^* such that σ_k / s_N converges in distribution to σ_k^*. Furthermore $\lim_{N \to \infty} E[\sigma_1 / s_N] = E[\sigma_1^*] = Q$.

Distributions of the limiting random variables σ_k^* are also obtained in [10]. The characterization of σ_k^* therein reveals an interesting and somewhat surprising relationship between $(|\xi(t)| : t \geq 0)$ and a far simpler random process that provides substantial insight for the present analysis. Namely, let $(D_t : t \geq 0)$ be the Markovian pure death process where for each integer state $m \geq 1$ a transition from m to $m - 1$ occurs at rate $\binom{m}{2}$. For $t \geq 0$ and integers n, k let

$$q_{n,k}(t) = P(D_t = k | D_0 = n).$$

The exact form of $q_{n,k}(t)$ is not immediately relevant to the present discussion, however we note that $\lim_{n \to \infty} q_{n,k}(t)$ exists for each $k \geq 1$ and $t \geq 0$, and denote the limit value by $q_{\infty,k}(t)$. Explicit expressions for $q_{n,k}(t)$ and $q_{\infty,k}(t)$ can be found in [10, 16].

Theorem 5. ([10, Theorem 6]) For each integer $k \geq 1$ and $t \geq 0$, $P(\sigma_k^* \leq t) = \sum_{l=1}^{k} q_{\infty,l}(2t/Q)$.

In informal terms, Theorems 4 and 5 suggest that for large values of N the time scaled process $(|\xi(ts_N)| : t \geq 0)$ behaves almost like the death process $(D_{tQ/2} : t \geq 0)$. It can be verified via straightforward comparison of generators that $(D_t : t \geq 0)$ represents the number of distinct random walks in a coalescing random walk process when G is completely connected. Hence in the limit of large N the spatial dependence of the process $(|\xi(ts_N)| : t \geq 0)$ completely washes out; in the new time-scale distinct walks coalesce with randomly chosen counterparts, at rates that do not depend on locations. This intuition, however, should be treated with caution, since like Theorems 4–5 it is valid only after the number of remaining distinct walks reduce to bounded values (for example, Theorem 5 does not provide any insight on $\sigma_{N/3}$, i.e. the time required to have $N/3$ carriers left).

Remark 6. In search of some insight about the conclusions drawn in previous paragraphs, suppose for the moment that at each time t the $|\xi(t)|$ carrier nodes are uniformly distributed over $\mathbb{T}_N^d$, independently of the history prior to t. Note that there is no particular reason for this to hold for $t > 0$. The instantaneous rate of decrease of $|\xi(t)|$ is proportional to the expected number of edges that connect two carrier nodes; thus if the above assumption were correct, then this rate would be equal to

$$\sum_{\text{edges } l \in E} P(l \text{ connects carrier nodes})$$

$$= \left(\frac{dN^d}{2}\right) \frac{|\xi(t)|(|\xi(t)| - 1)}{N^d(N^d - 1)}$$

$$\approx \left(\frac{|\xi(t)|}{2}\right) \frac{d}{N^d};$$

and in turn one would expect $(|\xi(tN^d)| : t \geq 0)$ to behave like the death process $(D_{td} : t \geq 0)$. Theorems 4–5 indicate that this conclusion is not too far off for dimensions $d \geq 3$.

We next turn to the energy requirement of the algorithm. Let

$$\eta(t) = \text{Total number messages sent in the network by time } t.$$

Note that each carrier node transmits messages at rate 1 and idle nodes do not engage in message transmission; hence $|\xi(t)|$ is the rate of message generation in the network at time t. The mean number of messages transmitted in the network before the termination of the algorithm is thus given by

$$E[\eta(\sigma_1)] = \sum_{k=2}^{N^d} kE[\sigma_{k-1} - \sigma_k] = E\left[\int_0^{\sigma_1} |\xi(t)|dt\right]. \qquad (2)$$

It is appealing to apply Theorem 5 and compute $E[\eta(\sigma_1)]$ for large values of N by computing the limiting expectation of each σ_k, however only tail of the

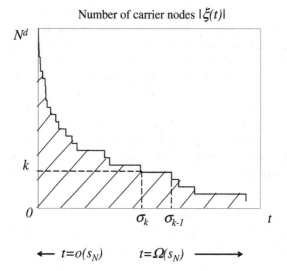

Fig. 2. The figure illustrates a sample path of the number of carrier nodes in the network. This trajectory has qualitatively different statistics for small and large values of t/s_N. The mean of the shaded area is the mean aggregate number of transmitted messages in the network.

integral in equality (2) can be computed in this manner. More explicitly, the algorithm has two qualitatively different phases as illustrated by Figure 2. For $t = o(s_N)$ there are many carrier nodes in the network, but their number diminish very quickly due to their high density. For $t = \Omega(s_N)$ only a bounded number of carrier nodes exist and σ_k can be approximated by Theorem 5 in this time interval. The former phase is short but it involves a high rate of message transmissions, whereas the latter phase lasts long but fewer messages are transmitted per unit time.

An estimate of the message complexity over the interval $t = \Omega(s_N)$ may be obtained via the following heuristic argument. Since in this interval $(|\xi(t s_N)| : t \geq 0)$ is informally approximated by a death process that dies at a rate that is roughly proportional to the square of its current value, use the solution of

$$\dot{y}_t = -y_t^2, \qquad y_0 = N^d,$$

as a proxy to $(|\xi(t s_N)| : t \geq 0)$. In particular $y_t = (t + N^{-d})^{-1}$. One may then expect

$$\int_0^{\sigma_1} |\xi(t)| dt = s_N \int_0^{\sigma_1/s_N} |\xi(t s_N)| dt$$

$$\approx s_N \int_0^{\sigma_1/s_N} y_t dt$$

$$= s_N \ln(t + N^{-d})|_0^{\sigma_1/s_N}.$$

Since $\sigma_1/s_N = O(1)$ by Theorem 4 this argument suggests that $O(s_N \log N)$ messages are transmitted in the considered interval. This intuition turns out to be correct and in fact the bound applies to both intervals. A formal statement is provided by the next theorem:

Let

$$m_N = \begin{cases} N^2 & \text{if } d = 1 \\ N^2 (\log N)^2 & \text{if } d = 2 \\ N^d \log N & \text{if } d \geq 3. \end{cases}$$

Theorem 7. *Under Algorithm* COALESCENT$(f, \lambda, \mathbb{T}_N^d)$

$$\limsup_{N \to \infty} E[\eta(\sigma_1)/m_N] < \infty, \quad \text{for } d \geq 1,$$

$$\liminf_{N \to \infty} E[\eta(\sigma_1)/m_N] > 0, \quad \text{for } d = 1,$$

$$\liminf_{N \to \infty} E[\eta(\sigma_1)/s_N] > 0, \quad \text{for } d \geq 2.$$

5   A Comparative Case Study

This section compares the time and energy requirements of the proposed algorithm and a gossip algorithm that has been previously studied for distributed computation of averages. Namely, the task here is to obtain the algebraic average of N^d numbers $x_1, x_2, \cdots, x_{N^d}$ each of which is known to a distinct node on the torus $\mathbb{T}_N^d$. As noted by Example 1 this task can be accomplished by Algorithm COALESCENT$(f, \lambda, \mathbb{T}_N^d)$ by proper choice of Λ, f and λ_i (namely by choosing $\lambda_i = (x_i, 1)$ for node i).

In broad terms, gossip algorithms refer to distributed randomized algorithms that are based on pairwise relaxations between randomly chosen node pairs. In the context of the present section a pairwise relaxation refers to averaging the two values available at the associated nodes. For completeness we next give a full description of this algorithm as studied in [5]. The algorithm is specified by a stochastic matrix $P = [P_{ij}]_{N^d \times N^d}$ such that $P_{ij} > 0$ only if nodes i and j are neighbors in $\mathbb{T}_N^d$:

Algorithm GOSSIP-AVE(P)**:** Each sensor i maintains a real valued variable with initial value $z_i(0) = x_i$. Each sensor has a local Poisson clock that tick independently of other such clocks. At the tick of this clock, say at time t_o, sensor i chooses a neighbor j with respect to the distribution $(P_{ij} : j = 1, 2, \cdots, N^d)$ and both nodes update their internal variables as $z_i(t_o) = z_j(t_o) = (z_i(t_o^-) + z_j(t_o^-))/2$.

Let the vector $z(t) = (z_1(t), z_2(t), \cdots, z_{N^d}(t))$ denote the sensor values at time t, $\bar{x}$ denote the average of $x_1, x_2, \cdots, x_{N^d}$, and $\mathbf{1}$ denote the vector of all 1s. Define τ_k as the kth time instant such that some local clock ticks and thereby triggers messaging in the network. For $\varepsilon > 0$ let the deterministic quantity $K(\varepsilon, P)$ be defined by

$$K(\varepsilon, P) = \sup_{z(0)} \inf \left\{ k \; : \; Pr\left(\frac{\|z(\tau_k) - \bar{x}\mathbf{1}\|_2}{\|z(0)\|_2} \geqslant \varepsilon \right) \leqslant \varepsilon \right\}.$$

Table 1. Comparison of the two algorithms on the d-dimensional torus with N^d nodes, $\mathbb{T}_N^d$

	Energy requirement per node		Termination time	
	COALESCENT$(f, \lambda, \mathbb{T}_N^d)$	GOSSIP-AVE(P)	COALESCENT$(f, \lambda, \mathbb{T}_N^d)$	GOSSIP-AVE(P)
$d = 1$	$\Theta(N)$	$\Omega(N^2)$	$\Theta(N^2)$	$\Omega(N^2)$
$d = 2$	$O((\log N)^2)$	$\Omega(N^2)$	$\Theta(N^2 \log N)$	$\Omega(N^2)$
$d \geq 3$	$O(\log N)$	$\Omega(N^2)$	$\Theta(N^d)$	$\Omega(N^2)$

In [5] $K(\varepsilon, P)$ is considered as a termination time for Algorithm GOSSIP-AVE(P) and minimization of $K(\varepsilon, P)$ is sought by proper choice of P. While the choice of $K(\varepsilon, P)$ as a stopping criterion might be questioned ($\|z(\tau_{K(\varepsilon,P)}) - \bar{x}\mathbf{1})\|_\infty / |\bar{x}|$ may be much larger than ε), here we adopt the same interpretation for this quantity for comparison purposes. The following theorem determines the order of $K(\varepsilon, P)$ uniformly for all P on the torus $\mathbb{T}_N^d$:

Theorem 8. *For fixed $\varepsilon > 0$, $K(\varepsilon, P) = \Omega(N^{d+2})$ uniformly for P such that $P_{ij} > 0$ only if (i, j) is an edge in $\mathbb{T}_N^d$.*

Note that $K(\varepsilon, P)$ is a termination criterion in terms of the transmitted messages in the network. Specifically, $K(\varepsilon, P)$ represents half of the aggregate number of messages transmitted in the network before Algorithm GOSSIP-AVE(P) terminates (to be precise, every time a clock ticks two messages are transmitted in opposite directions on some edge). This observation translates to a termination-time estimate since the point process $(\tau_1, \tau_2, \tau_3, \cdots)$ is Poisson with rate N^d and it takes roughly $K(\varepsilon, P)/N^d$ time units to produce a total of $2K(\varepsilon, P)$ messages. Theorem 8 thus has the following corollary:

Corollary 9. *Algorithm GOSSIP-AVE(P) terminates within $\Omega(N^2)$ time on $\mathbb{T}_N^d$.*

The obtained complexity results are summarized in Table 1.

Interestingly, there appears little practical incentive to consider higher dimensions for COALESCENT$(f, \lambda, \cdot)$. Namely Table 1 indicates that in arranging n nodes on $\mathbb{T}_{\sqrt[d]{n}}^d$, the marginal gain in either performance measure by going from d to $d + 1$ is a factor of $O(\log n)$ for $d = 2$, and a factor of $O(1)$ for larger values of d. This observation hints at favorable properties of the 3-dimensional torus and suggests that it is unlikely to experience substantial performance losses due to edge failures. A formal statement of this intuition does not seem straightforward. However the suggested insensitivity to nodal degrees is likely to have important practical implications with regard to robustness, especially in situations where the network topology cannot be planned or controlled.

We next provide numerical results on geometric random graphs, which have received substantial interest as suitable models for wireless ad-hoc networks. See for example [18] and [5]. A geometric random graph $\Gamma = (V_\Gamma, E_\Gamma)$ is obtained by uniformly distributing a set V_Γ of nodes, hence $|V_\Gamma| = n$, on the unit square and drawing an edge between any pair of nodes that fall within distance $r(n) = \sqrt{2 \log n / n}$ of each other. We have first created geometric random graphs and

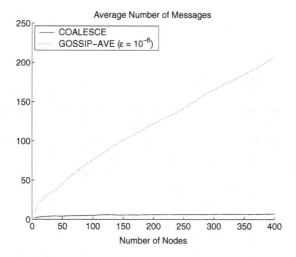

Fig. 3. The average number of transmitted messages per-node on geometric random graphs with n nodes. $\epsilon = 10^{-6}$ for GOSSIP-AVE.

discarded the disconnected ones in order to avoid trivialities. Then we have run both COALESCE and GOSSIP-AVE on the created geometric random graphs 50 times in order to quantify the number of transmitted messages. We have taken $\epsilon = 10^{-6}$ for GOSSIP-AVE. The results are plotted in Figure 3. It is observed from Figure 3 that COALESCE favors significantly against GOSSIP-AVE in terms of number of messages transmitted. The asymptotic properties of COALESCE on geometric random graphs as the number of nodes gets large is still ongoing research.

Remark 10. Time and message complexities of Algorithm GOSSIP-AVE(P) are inherently related to mixing times of random walks on the torus, and thereby to the spectral gap of the stochastic matrix P. The conclusions of this section should be expected to hold for other distributed algorithms, such as those in [13, 3, 15], that are based on powers of the incidence matrix of an underlying connectivity graph and have similar stopping criteria.

6   Conclusions

This paper concerns distributed algorithms for in-network computation of decomposable functions in sensor networks. Such algorithms typically need to comply with two conflicting requirements: On the one hand expedited convergence of the algorithm is desirable from an application viewpoint. On the other hand limitations due to energy-limited sensors impose frugal usage of energy as an indispensable feature. The emphasis of this paper is the trade-off between time and energy requirements.

We introduced and analyzed a distributed algorithm that is based on coalescing random walk on the communication graph of a sensor network. The algorithm is asynchronous, requires local information for each sensor node, and it is resilient against packet losses and node failures. In informal terms, the main theme of this algorithm is parsimonious message transmissions. Namely, upon transmitting a message a node enters a quiescent state which lasts until the node receives a message from a neighbor. This operational mode leads to substantial energy savings as the long-term rate of message transmissions at any node decreases in time and tends to the reciprocal of the total number sensors in the network. We show that the algorithm terminates with the exact value of interest, regardless of the network topology. We pursue more detailed analysis on the d-dimensional torus with n nodes and show that the completion time of the algorithm is $\Theta(n(\log n)^\alpha)$ and energy requirement per sensor node is $O((\log n)^{\alpha+1})$ where $\alpha = 1$ for $d = 2$ and $\alpha = 0$ for $d \geq 3$. The algorithm thus has a scalable energy requirement, furthermore its performance fairly insensitive to the dimension d of the torus so long as $d \geq 2$. This latter observation may prove useful in estimating the performance of the algorithm on less regular network topologies.

We also studied the relative time and energy requirements of the algorithm with respect to other in-network processing algorithms based on powers of the network connectivity graph. In particular we focused on a gossip algorithm for computing averages, and showed that both time and per-node energy requirements scale as $\Omega(n)$ on the 2-dimensional torus irrespectively of the choice of distribution for neighbor selection for pairwise relaxation. Hence the proposed algorithm achieves a factor of $\Omega(n/(\log n)^2)$ gain in energy at the expense of a factor of $O(\log n)$ loss in completion time.

References

1. O. Savas, M. Alanyali and V. Saligrama, Randomized Sequential Algorithms for Data Aggregation in Sensor Networks, CISS 2006, Princeton, NJ.
2. M. Alanyali, V. Saligrama, O. Savas, and S. Aeron. Distributed bayesian hypothesis testing in sensor networks. In *American Control Conference*, Boston, MA, July 2004.
3. V. Saligrama, M. Alanyali and O. Savas. Distributed detection in sensor networks with packet losses and finite capacity links. *IEEE Transactions on Signal Processing*, to appear, 2005.
4. M. Alanyali and V. Saligrama, "Distributed target tracking on multi-hop networks," *IEEE Statistical Signal Processing Workshop*, Bordeaux, France, July 2005.
5. S. Boyd, A. Ghosh, B. Prabhakar and D. Shah. Gossip algorithms: Design, analysis and applications. In *Proceedings of IEEE INFOCOM 2005*, 2005.
6. Fan R. K. Chung. *Spectral Graph Theory*. American Mathematical Society, 1997.
7. D. Bertsekas, R. Gallager *Data Networks*. Pearson Education, 1991.
8. D. P. Bertsekas and J. N. Tsitsiklis, *Parallel and Distributed Computation: Numerical Methods*. Prentice Hall, 1989, republished by Athena Scientific, 1997.
9. J. F. Chamberland and V. V. Veeravalli. Decentralized detection in sensor networks. *IEEE Transactions on Signal Processing*, 2003.
10. J. T. Cox. Coalescing random walks and voter model consensus times on the torus in $\mathbb{Z}^d$. *The Annals of Probability*, 17(4):1333-1366, 1989.

11. A. Giridhar and P. R. Kumar. Computing and communicating functions over sensor networks. *IEEE JSAC Special Issue on Self-Organizing Distributed Collaborative Sensor Networks*, 23, 2005.

12. A. O. Hero and D. Blatt. Sensor network source localization via projection onto convex sets(pocs). In *IEEE International Conference on Acoustics, Speech, and Signal Processing*, Philadelphia, PA, March 2005.

13. R. Olfati-Saber and R. M. Murray. Consensus problems in networks of agents with switching topology and time-delays. *IEEE Transactions On Automatic Control*, vol. 49, pp. 1520-1533, Sep., 2004.

14. M. G. Rabbat and R. D. Nowak. Distributed optimization in sensor networks. In *Third International Symposium on Information Processing in Sensor Networks*, Berkeley, CA, April 2004.

15. D.S. Scherber and H.C. Papadopoulos. Distributed computation of averages over ad hoc networks. *IEEE Journal on Selected Areas in Communications*, vol. 23(4), 2005.

16. S. Tavaré. Line-of-descent and genealogical processes, and their applications in population genetics models. *Theoret. Population Biol.*, vol. 26, pp. 119-164, 1984.

17. J. N. Tsitsiklis. Decentralized detection. *in Advances in Statistical Signal Processing, H. V. Poor and J. B. Thomas Eds*, 2, 1993.

18. P. Gupta and P. Kumar, The capacity of wireless networks, IEEE Trans. on Information Theory, 46(2):388-404, March 2000.

Distributed Optimal Estimation from Relative Measurements for Localization and Time Synchronization

Prabir Barooah[1], Neimar Machado da Silva[2], and João P. Hespanha[1]

[1] University of California, Santa Barbara, CA 93106, USA
{pbarooah, hespanha}@ece.ucsb.edu
[2] Federal University of Rio de Janeiro, Rio de Janeiro, Brazil
neimarms@gmail.com

Abstract. We consider the problem of estimating vector-valued variables from noisy "relative" measurements. The measurement model can be expressed in terms of a graph, whose nodes correspond to the variables being estimated and the edges to noisy measurements of the difference between the two variables. This type of measurement model appears in several sensor network problems, such as sensor localization and time synchronization. We consider the optimal estimate for the unknown variables obtained by applying the classical Best Linear Unbiased Estimator, which achieves the minimum variance among all linear unbiased estimators.

We propose a new algorithm to compute the optimal estimate in an iterative manner, the *Overlapping Subgraph Estimator* algorithm. The algorithm is distributed, asynchronous, robust to temporary communication failures, and is guaranteed to converges to the optimal estimate even with temporary communication failures. Simulations for a realistic example show that the algorithm can reduce energy consumption by a factor of two compared to previous algorithms, while achieving the same accuracy.

1 Introduction

We consider an estimation problem that is relevant to a large number of sensor networks applications, such as localization and time synchronization. Consider n vector-valued variables $x_1, x_2, \ldots, x_n \in \mathbb{R}^k$, called node variables, one or more of which are known, and the rest are unknown. A number of noisy measurements of the difference between certain pairs of these variables are available. We can associate the variables with the nodes $\mathbf{V} = \{1, 2, \ldots, n\}$ of a directed graph $\mathbf{G} = (\mathbf{V}, \mathbf{E})$ and the measurements with the edges $\mathbf{E}$ of it, consisting of ordered pairs (u, v) such that a noisy "relative" measurement between x_u and x_v is available:

$$\zeta_{uv} = x_u - x_v + \epsilon_{uv}, \tag{1}$$

P. Gibbons et al. (Eds.): DCOSS 2006, LNCS 4026, pp. 266–281, 2006.
© Springer-Verlag Berlin Heidelberg 2006

where the ϵ_{uv}'s are uncorrelated zero-mean noise vectors with known covariance matrices. That is, for every edge $e \in \mathbf{E}$, $P_e = \mathrm{E}[\epsilon_e \epsilon_e^T]$ is known, and $\mathrm{E}[\epsilon_e \epsilon_{\bar{e}}^T] = 0$ if $e \neq \bar{e}$. The problem is to estimate all the unknown node variables from the measurements. We call $\mathbf{G}$ a *measurement graph* and x_u the *u-th node variable*. The node variables that are known are called the *reference variables* and the corresponding nodes are called the *reference nodes*. The relationship of this estimation problem with sensor network applications is discussed in section 1.1.

Our objective is to construct an optimal estimate $\hat{x}_u^*$ of x_u for every node $u \in \mathbf{V}$ for which x_u is unknown. The *optimal estimate* refers to the estimate produced by the classical Best Linear Unbiased Estimator (BLUE), which achieves the minimum variance among all linear unbiased estimators [1]. To compute the optimal estimate directly one would need all the measurements and the topology of the graph (cf. section 2). Thus, if a central processor has to compute the $\hat{x}_u^*$s, all this information has to be transmitted to it. In a large ad-hoc network, this burdens nodes close to the central processor more than others. Moreover, centralized processing is less robust to dynamic changes in network topology resulting from link and node failures. Therefore a distributed algorithm that can compute the optimal estimate while using only local communication will be advantageous in terms of scalability, robustness and network life.

In this paper we propose a new distributed algorithm, which we call the *Overlapping Subgraph Estimator* (OSE) algorithm, to compute the optimal estimates of the node variables in an iterative manner. The algorithm is distributed in the sense that each node computes its own estimate and the information required to perform this computation is obtained from communication with its one-hop neighbors. We show that the proposed algorithm is correct (i.e., the estimates converge to the optimal estimates) even in the presence of faulty communication links, as long as certain mild conditions are satisfied. The OSE algorithm asymptotically obtains the optimal estimate while simultaneously being scalable, asynchronous, distributed and robust to communication failures.

1.1   Motivation and Related Work

Optimal Estimation. The estimation problem considered in this paper is motivated by sensornet applications such as time synchronization and location estimation. We now briefly discuss these applications.

In a network of sensors with local clocks that progress at the same rate but have unknown offsets between them, it is desirable to estimate these offsets. Two nodes u and v can obtain a measurement of the difference between their local times by exchanging time stamped messages. The resulting measurement of clock offsets can be modeled by (1)(see [2] for details). The problem of estimating the offset of every clock with respect to a single reference clock is then a special case of the problem considered in this paper. Karp *et. al.* [3] have also investigated this particular problem. The measurement model used in [3] can be seen as an alternative form of (1). In this application, the node variable x_u is the offset of u's local time with respect to a "reference" clock, and is a scalar variable.

Optimal Estimation from relative measurements with *vector-valued* variables was investigated in [4, 5]. Localization from range and bearing measurements is an important sensor network application that can be formulated as a special case of the estimation problem considered in this paper. Imagine a sensor network where the nodes are equipped with range and bearing measurement capability. When the sensors are equipped with compasses, relative range and bearing measurement between two nodes can be converted to a measurement of their relative position vector in a global Cartesian reference frame. The measurements are now in the form (1), and the optimal location estimates for the nodes can now be computed from these measurements (described in section 2). In this application the node variables are vectors.

Several localization algorithms have been designed assuming only relative range information, and a few, assuming only relative angle measurement. In recent times combining both range and bearing information has received some attention [6]. However, to the best of our knowledge, no one has looked at the localization problem in terms of the noisy measurement model (1). The advantage of this formulation is that the effect of measurement noise can be explicitly accounted for and filtered out to the maximum extent possible by employing the classical Best Linear Unbiased Estimator(BLUE). This estimator produces the minimum variance estimate, and hence is the most accurate on average. Location estimation techniques using only range measurement can be highly sensitive to measurement noises, which may introduce significant errors into the location estimate due to flip ambiguities [7]. The advantage of posing the localization problem as an estimation problem in Cartesian coordinates using the measurement model (1) is that the *optimal* (minimum variance) estimates all node positions in a connected network can be unambiguously determined when only one node that knows its position. A large number of well placed beacon nodes that know their position and broadcast that to the network – a usual requirement for many localization schemes – are not required.

Distributed Computation. Karp *et. al.* [3] considered the optimal estimation problem for time synchronization with measurements of pairwise clock offsets, and alluded to a possible distributed computation of the estimate, but stopped short of investigating it. In [5], we have proposed a distributed algorithm for computing the optimal estimates of the node variables that was based on the Jacobi iterative method of solving a system of linear equations. This Jacobi algorithm is briefly discussed in section 2. Although simple, robust and scalable, the Jacobi algorithm proposed in [5] suffered from a slow convergence rate. The OSE algorithm presented in this paper has a much faster convergence rate than the Jacobi algorithm. Delouille *et. al.* [8] considered the minimum mean squared error estimate of a different problem, in which absolute measurements of random node variables (such as temperature) were available, but the node variables were correlated. They proposed an Embedded Polygon Algorithm (EPA) for computing the minimum mean squared error estimate of node variables in a distributed manner, which was essentially a block-Jacobi iterative method for solving a set of linear equations. Although the problem in [8] was quite different from the problem investigated in this paper, their EPA algorithm could be adapted to apply to our problem. We will call it the *modified EPA*. Simulations show that the OSE algorithm converges faster than the modified EPA.

Energy Savings. Since OSE converges faster, it requires fewer iterations for the same estimation error, which leads to less communication and hence saves energy in ad-hoc wireless networks. Here estimation error refers to the difference between the optimal estimate and the estimate produced by the algorithm. It is critical to keep energy consumption at every node at a minimum, since battery life of nodes usually determines useful life of the network. The improved performance of OSE comes from the nodes sending and processing larger amounts of data compared to Jacobi and modified EPA. However, the energy cost of sending additional data can be negligible due to the complex dependence of energy consumption in wireless communication on radio hardware, underlying PHY and MAC layer protocols, network topology and a host of other factors.

Investigation into energy consumption of wireless sensor nodes has been rather limited. Still, we can get an idea of which parameters are important for energy consumption from the studies reported in [9, 10, 11]. It is reported in [9] that for very short packets (in the order of 100 bits), transceiver startup dominates the power consumption; so sending a very short message offers no advantage in terms of energy consumption over sending a somewhat longer message. In fact, in a recent study of dense network of IEEE 802.15.4 wireless sensor nodes, it is reported in transmitted energy per bit in a packet *decreases monotonically* upto the maximum payload [10]. One of the main findings in [11] was that in highly contentious networks, "transmitting large payloads is more energy efficient". On the other hand, receive and idle mode operation of the radio is seen to consume as much energy as the transmit mode, if not more [12]. Thus, the number of packets (sent and received) appear to be a better measure to predict energy consumption than the number of bits.

In light of the above discussion, we used the number of packets transmitted and received as a coarse measure of the energy consumed by a node during communication. With number of packets as the energy consumption metric, simulations indicate that the OSE algorithm can cut down the average energy consumption for a given estimation accuracy as much by a factor of two or more (compared to Jacobi and modified EPA).

1.2   Organization

The paper is organized as follows. In section 2, the optimal estimator for the problem at hand is described. In section 3, we describe three algorithms to compute the optimal estimate iteratively - Jacobi, modified EPA and the Overlapping Subgraph Estimator (OSE) and discuss correctness and performance. Simulation studies are presented in section 4. The paper concludes with a summary in section 5.

2   The Optimal Estimate

Consider a measurement graph $\mathbf{G}$ with n nodes and m edges. Recall that k is the dimension of the node variables. Let $\mathbf{X}$ be a vector in $\mathbb{R}^{nk}$ obtained by stacking together all the node variables, known and unknown, i.e., $\mathbf{X} := [x_1^T, x_2^T, \ldots, x_n^T]^T$. Define $\mathbf{z} := [\zeta_1^T, \zeta_2^T, \ldots, \zeta_m^T]^T \in \mathbb{R}^{km}$ and $\boldsymbol{\epsilon} := [\epsilon_1^T, \epsilon_2^T, \ldots, \epsilon_m^T]^T \in \mathbb{R}^{km}$. This stacking together of variables allows us to rewrite (1) in the following form:

$$\mathbf{z} = \mathcal{A}^T \mathbf{X} + \boldsymbol{\epsilon}, \tag{2}$$

where $\mathcal{A}$ is a matrix uniquely determined by the graph. To construct $\mathcal{A}$, we start by defining the *incidence matrix* A *of the graph* $\mathbf{G}$, which is an $n \times m$ matrix with one row per node and one column per edge defined by $A := [a_{ue}]$, where a_{ue} is nonzero if and only if the edge $e \in \mathbf{E}$ is incident on the node $u \in \mathbf{V}$. When nonzero, $a_{ue} = -1$ if the edge e is directed towards u and $a_{ue} = 1$ otherwise. The matrix $\mathcal{A}$ that appears in (2) is an "expanded" version of the incidence matrix A, defined by $\mathcal{A} := A \otimes I_k$, where I_k is the $k \times k$ identity matrix and $\otimes$ denotes the Kronecker product.

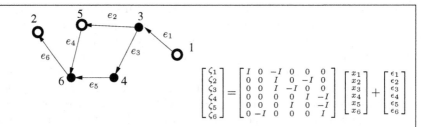

Taking out the rows $1 : k, k+1 : 2k$ and $4k+1 : 5k$ (corresponding to the reference nodes $1, 2, 5$), we construct $\mathcal{A}_r$; the remaining rows constitute $\mathcal{A}_b$. For this example, $\mathcal{A}_r^T \mathbf{x}_r = [x_1^T, -x_5^T, 0, x_5^T, 0, -x_2^T]^T$ and so eq. (3) becomes

$$\underbrace{\begin{bmatrix} \zeta_1 - x_1 \\ \zeta_2 + x_5 \\ \zeta_3 \\ \zeta_4 - x_5 \\ \zeta_5 \\ \zeta_6 + x_2 \end{bmatrix}}_{\tilde{\mathbf{z}}} = \underbrace{\begin{bmatrix} -I & 0 & 0 \\ I & 0 & 0 \\ I & -I & 0 \\ 0 & 0 & -I \\ 0 & I & -I \\ 0 & 0 & I \end{bmatrix}}_{\mathcal{A}_b^T} \underbrace{\begin{bmatrix} x_3 \\ x_4 \\ x_6 \end{bmatrix}}_{\mathbf{x}} + \underbrace{\begin{bmatrix} \epsilon_1 \\ \epsilon_2 \\ \epsilon_3 \\ \epsilon_4 \\ \epsilon_5 \\ \epsilon_6 \end{bmatrix}}_{\boldsymbol{\epsilon}}.$$

In the case when every measurement covariance is equal to the identity matrix, eq. (4) becomes

$$\begin{bmatrix} 3I & -I & 0 \\ -I & 2I & -I \\ 0 & -I & 3I \end{bmatrix} \begin{bmatrix} \hat{x}_3^* \\ \hat{x}_4^* \\ \hat{x}_6^* \end{bmatrix} = \begin{bmatrix} -\tilde{\zeta}_1 + \tilde{\zeta}_2 + \tilde{\zeta}_3 \\ -\tilde{\zeta}_3 + \tilde{\zeta}_5 \\ -\tilde{\zeta}_4 - \tilde{\zeta}_5 + \tilde{\zeta}_6 \end{bmatrix},$$

whose solution gives the optimal estimates of the unknown node variables x_3, x_4 and x_6.

Fig. 1. A measurement graph $\mathbf{G}$ with 6 nodes and 6 edges. Nodes $1, 2$ and 5 are reference nodes, which means that they know their own node variables.

By partitioning $\mathbf{X}$ into a vector $\mathbf{x}$ containing all the unknown node variables and another vector $\mathbf{x}_r$ containing all the known reference node variables: $\mathbf{X}^T = [\mathbf{x}_r^T, \mathbf{x}^T]^T$, we can re-write (2) as $\mathbf{z} = \mathcal{A}_r^T \mathbf{x}_r + \mathcal{A}_b^T \mathbf{x} + \boldsymbol{\epsilon}$, where $\mathcal{A}_r$ contains the rows of $\mathcal{A}$ corresponding to the reference nodes and $\mathcal{A}_b$ contains the rows of $\mathcal{A}$ corresponding to the unknown node variables. The equation above can be further rewritten as:

$$\bar{\mathbf{z}} = \mathcal{A}_b^T \mathbf{x} + \boldsymbol{\epsilon}, \tag{3}$$

where $\bar{\mathbf{z}} := \mathbf{z} - \mathcal{A}_r^T \mathbf{x}_r$ is a known vector. The optimal estimate (BLUE) $\hat{\mathbf{x}}^*$ of the vector of unknown node variables $\mathbf{x}$ for the measurement model 3 is the solution to the following system of linear equations:

$$\mathcal{L}\hat{\mathbf{x}}^* = \mathbf{b}, \tag{4}$$

where $\mathcal{L} := \mathcal{A}_b \mathscr{P}^{-1} \mathcal{A}_b^T$, $\mathbf{b} := \mathcal{A}_b \mathscr{P}^{-1} \bar{\mathbf{z}}$, and $\mathscr{P} := \mathrm{E}[\boldsymbol{\epsilon}\boldsymbol{\epsilon}^T]$ is the covariance matrix of the measurement error vector [1]. Since the measurement errors on two different edges are uncorrelated, $\mathscr{P}$ is a symmetric positive definite block diagonal matrix with the measurement error covariances along the diagonal: $\mathscr{P} = \mathrm{diag}(P_1, P_2, \ldots, P_m) \in \mathbb{R}^{km \times km}$, where $P_e = \mathrm{E}[\epsilon_e \epsilon_e^T]$ is the covariance of the measurement error ϵ_e.

The matrix $\mathcal{L}$ is invertible if and only if every weakly connected component of the graph $\mathbf{G}$ has at least one reference node [4]. A directed graph $\mathbf{G}$ is said to be weakly connected if there is a path from every node to every other node, not necessarily respecting the direction of the edges. In a weakly connected graph, the optimal estimate $\hat{\mathbf{x}}^*$ for every node u is unique for a given set of measurements $\mathbf{z}$. The error covariance of the optimal estimate $\boldsymbol{\Sigma} := \mathrm{E}[(\mathbf{x} - \hat{\mathbf{x}}^*)(\mathbf{x} - \hat{\mathbf{x}}^*)^T]$ is equal to $\mathcal{L}^{-1}$ and the $k \times k$ blocks on the diagonal of this matrix gives the estimation error covariances of the node variables.

Figure 1 shows an example of a measurement graph and the relevant equations.

3   Distributed Computation of the Optimal Estimate

In order to compute the optimal estimate $\hat{\mathbf{x}}^*$ by solving the equations (4) directly, one needs all the measurements and their covariances $(\mathbf{z}, \mathscr{P})$, and the topology of the graph $(\mathcal{A}_b, \mathcal{A}_r)$. In this section we consider iterative distributed algorithms to compute the optimal estimates for the measurement model (2). These algorithms compute the optimal estimate through multiple iterations, with the constraint that a node is allowed to communicate only with its neighbors. The concept of "neighbor" is determined by the graph $\mathbf{G}$, in the sense that two nodes are *neighbors* if there is an edge in $\mathbf{G}$ between them (in either direction). This implicitly assumes bidirectional communication. We describe three algorithms - Jacobi, modified EPA, and OSE, the last being the novel contribution of this paper. We will see that OSE algorithm converges even when communication faults destroy the bidirectionality of communication.

3.1   The Jacobi Algorithm

Consider a node u with unknown node variable x_u and imagine for a moment that the node variables for all neighbors of u are exactly known and available to u. In this case, u could compute its optimal estimate by simply using the measurements between itself and its 1-hop neighbors. This estimation problem is fundamentally no different than the original problem, except that it is defined over the much smaller graph $\mathbf{G}_u(1) = (\mathbf{V}_u(1), \mathbf{E}_u(1))$, whose node set $\mathbf{V}_u(1)$ include u and its 1-hops neighbors and the edge set $\mathbf{E}_u(1)$ consists of only the edges between u and its 1-hops neighbors. We call $\mathbf{G}_u(1)$ the *1-hop subgraph of* $\mathbf{G}$ *centered at* u. Since we are assuming that the node variables of all neighbors of u are exactly known, all these nodes should be understood as references.

In the Jacobi algorithm, at every iteration, a node gathers the estimates of its neighbors from them by exchanging messages and updates it own estimate by solving the optimal estimation problem in the 1-hop subgraph $\mathbf{G}_u(1)$ by taking the estimates of its neighbors as the true values (reference variables). It turns out that this algorithm corresponds exactly to the Jacobi algorithm for the iterative solution to the linear equation (4) and is guaranteed to converge to the true solution of (4) when the iteration is done in a synchronous manner [5]. When done asynchronously, or in the presence of communication failures, it is guaranteed to converge under additional mild assumptions [5]. The algorithm can be terminated at a node when the change in its recent estimate is seen to be lower than a certain pre-specified threshold value, or when a certain maximum number of iterations are completed. The details of the Jacobi algorithm can be found in [2, 5].

Note that to compute the update $\hat{x}_u^{(i+1)}$, node u also needs the measurements and associated covariances ζ_e, P_e on the edges $e \in \mathbf{E}_u(1)$ of its 1-hop subgraph. We assume that after deployment of the network, nodes detect their neighbors and exchange their relative measurements as well as the associated covariances. Each node uses this information obtained initially for all future computation.

3.2   Modified EPA

The Embedded Polygon Algorithm (EPA) proposed in [8] can be used for iteratively solving (4); since it is essentially a block – Jacobi method of solving a system of linear equations, where the blocks correspond to non-overlapping polygons. The special case when the polygons are triangles has been extensively studied in [8]. We will not include here the details of the algorithm, including triangle formation in the initial phase, the intermediate computation, communication and update. The interested reader is referred to [8]. It is not difficult to adapt the algorithm in [8] to the problem considered in this paper. We have implemented the modified EPA algorithm (with triangles as the embedded polygons) and compared it with both Jacobi and OSE. Results are presented in section 4.

3.3   The Overlapping Subgraph Estimator Algorithm

The Overlapping Subgraph Estimator (OSE) algorithm achieves faster convergence than Jacobi and modified EPA, while retaining their scalability and robustness properties. The OSE algorithm is inspired by the multisplitting and Weighted Additive Schwarz method of solving linear equations [13].

The OSE algorithm can be thought of as an extension of the Jacobi algorithm, in which individual nodes utilize larger subgraphs to improve their estimates. To understand how this can be done, suppose that each node broadcasts to its neighbors not only is current estimate, but also all the latest estimates that it received from his neighbors. In practice, we have a simple two-hop communication scheme and, in the absence of drops, at the ith iteration step, each node will have the estimates $\hat{x}_v^{(i)}$ for its 1-hop neighbors and the (older) estimates $\hat{x}_v^{(i-1)}$ for its 2-hop neighbors (i.e., the nodes at a graphical distance of two).

Under this information exchange scheme, at the ith iteration, each node u has estimates of all node variables in the set $\mathbf{V}_u(2)$ consisting of itself and all its 1-hop and 2-hop neighbors. In the OSE algorithm, each node updates its estimate using the *2-hop subgraph centered at* u $\mathbf{G}_u(2) = (\mathbf{V}_u(2), \mathbf{E}_u(2))$, with edge set $\mathbf{E}_u(2)$ consisting all the edges of the original graph $\mathbf{G}$ that connect element of $\mathbf{V}_u(2)$. For this estimation problem, node u takes as references the variables of the nodes at the "boundary" of its 2-hop subgraph: $\mathbf{V}_u(2) \setminus \mathbf{V}_u(1)$. These nodes are at a graphical distance of 2 from u. We assume that the nodes use the first few rounds of communication to determine and communicate to one another the measurements and associated covariances of their 2-hop subgraphs. The *OSE algorithm* can be summarized as follows:

1. Each node $u \in \mathbf{V}$ picks arbitrary initial estimates $\hat{x}_v^{(-1)}$, $v \in \mathbf{V}_u(2) \setminus \mathbf{V}_u(1)$ for the node variables of all its 2-hop neighbors. These estimates do not necessarily have to be consistent across the different nodes.

2. At the ith iteration, each node $u \in \mathbf{V}$ assumes that the estimates $\hat{x}_v^{(i-2)}$, $v \in \mathbf{V}_u(2) \setminus \mathbf{V}_u(1)$ (that it received through its 1-hop neighbors) are correct and solves the corresponding optimal estimation problem associated with the 2-hop subgraph $\mathbf{G}_u(2)$. In particular, it solves the following linear equations: $\mathcal{L}_{u,2}\mathbf{y}_u = \mathbf{b}_u$, where $\mathbf{y}_u$ is a vector of node variables that correspond to the nodes in its 1-hop subgraph $\mathbf{G}_u(1)$, and $\mathcal{L}_{u,2}, \mathbf{b}_u$ are defined for the subgraph $G_u(2)$ as $\mathcal{L}, \mathbf{b}$ were for $\mathbf{G}$ in eq. (4). After this computation, node u updates its estimate as $\hat{x}_u^{(i+1)} \leftarrow \lambda y_u + (1 - \lambda)\hat{x}_u^{(i)}$, where $0 < \lambda \le 1$ is a pre-specified design parameter and y_u is the variable in $\mathbf{y}_u$ that corresponds to x_u. The new estimate $\hat{x}_u^{(i+1)}$ as well as the estimates $\hat{x}_v^{(i)}$, $v \in \mathbf{V}_u(1)$ previously received from its 1-hop neighbors are then broadcasted by u to all its 1-hop neighbors.

3. Each node then listens for the broadcasts from its neighbors, and uses them to update its estimates for the node variables of all its 1-hop and 2-hop neighbors $\mathbf{V}_u(2)$. Once all updates are received a new iteration can start.

The termination criteria will vary depending on the application, as discussed for the Jacobi algorithm. As in the case of Jacobi, we assume that nodes exchange measurement and covariance information with their neighbors in the beginning, and once obtained, uses those measurements for all future time.

As an illustrative example of how the OSE algorithm proceeds in practice, consider the measurement graph shown in figure 2(a) with node 1 as the single reference. Figure 2(b) shows the 2-hop subgraph centered at node 4, $\mathbf{G}_4(2)$, which consists of the following nodes and edges: $\mathbf{V}_4(2) = \{1, 3, 5, 4, 6, 2\}$ and $\mathbf{E}_4(2) = \{1, 2, 3, 4, 5, 6\}$. Its 2-hop neighbors are $\mathbf{V}_4(2) \setminus \mathbf{V}_4(1) = \{1, 2, 5\}$. After the first round of inter node communication, node 4 has the estimates of its neighbors 3 and 6: $x_3^{(0)}, x_6^{(0)}$ (as well as the measurements ζ_3, ζ_5 and covariances P_3, P_5). After the second round of communication, node 4 has the node estimates $x_1, \hat{x}_3^{(1)}, \hat{x}_5^{(0)}, \hat{x}_6^{(1)}, \hat{x}_2^{(0)}$ (and the measurements $\zeta_1, \ldots, \zeta_6$ and covariances $P_1, \ldots, P_6$). Assuming no communication failures, at every iteration i, node 4 uses $x_1, \hat{x}_3^{(i-2)}$ and $\hat{x}_5^{(i-2)}$ as the reference variables and computes "temporary" estimates y_3, y_4, y_6 (of x_3, x_4 and x_6) by solving the optimal estimation problem in its 2-hop subgraph. It updates its estimate as : $\hat{x}_4^{(i+1)} \leftarrow \lambda y_4 + (1 - \lambda)\hat{x}_4^{(i)}$, and discards the other variables computed.

Note that all the data required for the computation at a node is obtained by communicating with its 1-hop neighbors. Convergence to the optimal estimate will be discussed in section 3.5.

Remark 1 (h-hop OSE algorithm). One could also design a *h-hop OSE* algorithm by letting every node utilize a h-hop subgraph centered at itself, where h is some (not very large) integer. This would be a straightforward extension of the 2-hop OSE just described, except that at every iteration, individual nodes would have to transmit larger amounts of data than in 2-hop OSE, potentially requiring multiple packet transmissions at each iteration. In practice, this added communication cost will limit the allowable value of h. ☐

The Jacobi, EPA and OSE algorithms are all iterative methods to compute the solution of a system of linear equations. The Jacobi and EPA are similar in nature, EPA essentially being a block-Jacobi method. The OSE is built upon the Filtered Schwarz method [2], which is a refinement of the Schwarz method [13]. The OSE algorithm's gain in convergence speed with respect to the Jacobi and modified EPA algorithms comes from the fact that the 2-hop subgraphs $\mathbf{G}_u(2)$ contain more edges than the 1-hop subgraphs $\mathbf{G}_u(1)$, and the subgraphs of different nodes are overlapping. It has been observed that a certain degree of overlap may lead to a speeding up of the Schwarz method [13].

Improving Performance Through Flagged Initialization. One can further improve the performance of OSE (and also of Jacobi and modified EPA) by providing a better initial condition to it, which does not require more communication or computation. After deployment of the network, the reference nodes initialize their variables to their known values and every other node initializes its estimate to ∞, which serves as a flag to declare that it has no estimate. In the subsequent updates of a node's variables, the node only includes in its 1- or 2-hop subgraph those nodes that have finite estimates. If none of the neighbors have a finite estimate, then a node keeps its estimate at ∞. In the beginning, only the references have finite estimates. In the first iteration, only the neighbors of the references compute finite estimates by looking at their 1-hop subgraphs. In the next iteration, their neighbors do the same by looking at their 2 hop subgraphs and so on until all nodes in the graph have finite estimates. In general, the time it takes for all nodes to have finite estimates will depend on the radius of the network (the minimum graphical distance between any node and the nearest reference node). Since the flagged initialization only affects the initial stage of the algorithms, it does not affect their convergence properties.

3.4   Asynchronous Updates and Link Failures

In the previous description of the OSE algorithm, we assumed that communication was essentially synchronous and that all nodes always received broadcasts from all their neighbors. However, the OSE algorithm also works with *link failures* and *lack of synchronization* among nodes. To achieve this a node broadcasts its most recent estimate and waits for a certain time-out period to receive data from its neighbors. It proceeds with the estimate update after that period, even if it does not receive data from all of its neighbors, by using the most recent estimates that it received from its neighbors. A node may also receive multiple estimates of another node's variable. In that case, it uses the most recent estimate, which can be deduced by the time stamps on the messages. The Jacobi and the modified EPA algorithms can similarly be made robust to link failures [5, 8].

In practice nodes and communication links may fail temporarily or permanently. A node may simply remove from its subgraph those nodes and edges that have failed permanently (assuming it can be detected) and carry on the estimation updates on the new subgraph. However, if a node or link fails permanently, the measurement graph changes permanently, requiring redefinition of the optimal estimator. To avoid this technical difficulty, in this paper we only consider temporary link failures, which encompasses temporary node failures.

3.5   Correctness

An iterative algorithm is said to be *correct* if the estimate produced by the algorithm $\mathbf{x}^{(i)}$ converges to the true solution $\hat{\mathbf{x}}^*$ as the number of iterations i increase, i.e., $\|\mathbf{x}^{(i)} - \hat{\mathbf{x}}^*\| \to 0$ as $i \to \infty$. The assumption below is needed to prove correctness of the OSE algorithm, which is stated in Theorem 1.

Assumption 1. *At least one of the statements below holds:*

1. *All covariance matrices P_e, $e \in \mathbf{E}$ are diagonal.*
2. *All covariance matrices P_e, $e \in \mathbf{E}$ are equal.* $\square$

Theorem 1 (Correctness of OSE). *When assumption 1 holds, the OSE algorithm converges to the optimal estimate $\hat{\mathbf{x}}^*$ as long as there is a number ℓ_d such that the number of consecutive failures of any link is less than ℓ_d.* $\square$

We refer the interested reader to [2] for a proof of this result.

Remark 2. Assumption 1 is too restrictive in certain cases. In particular, the simulations reported in Section 4 were done with covariance matrices that did not satisfy this assumption, yet the algorithm was seen to converge in all the simulations. The reason is that this assumption is needed so that sufficient conditions for convergence are satisfied [2], and is not necessary in general.

3.6   Performance

Since minimizing energy consumption is critically important in sensor networks, we choose as performance metric of the algorithms the total energy consumed by a node before a given normalized error is achieved. The *normalized error* $\varepsilon^{(i)}$ is defined as

$$\varepsilon^{(i)} := \|\hat{\mathbf{x}}^{(i)} - \hat{\mathbf{x}}^*\|/\|\hat{\mathbf{x}}^*\|$$

and is a measure of how close the iterate $\hat{\mathbf{x}}^{(i)}$ is to the correct solution $\hat{\mathbf{x}}^*$ at the end of iteration i. We assume that nodes use broadcast communication to send data.

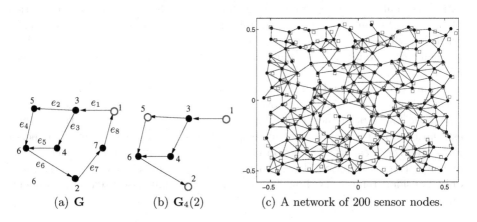

(a) **G**   (b) $\mathbf{G}_4(2)$ (c) A network of 200 sensor nodes.

Fig. 2. (a) A measurement graph **G** with node 1 as reference, and (b) a 2-hop subgraph $\mathbf{G}_4(2)$ centered at node 4. While running the OSE algorithm, node 4 treats 1, 5 and 2 as reference nodes in the subgraph $\mathbf{G}_4(2)$ and solves for the unknowns x_3, x_4 and x_6. (c) A sensor network with 200 nodes in a unit square area. The edges of the measurement graph are shown as line segments connecting the true nodes positions, which are shown as black circles. Two nodes with an edge between them have a noisy measurement of their relative positions in the plane. The little squares are the positions estimated by the (centralized) optimal estimator. A single reference node is placed at $(0, 0)$.

As discussed in section 1.1, we take the number of packets transmitted and received by a node as a measure of energy consumption. Let $N_{tx}^{(i)}(u)$ be the number of packets a node u transmits to its neighbors during the ith iteration.

The energy $E^{(i)}(u)$ expended by u in sending and receiving data during the ith iteration is computed by the following formula:

$$E^{(i)}(u) = N_{tx}^{(i)}(u) + \frac{3}{4} \sum_{v \in \mathcal{N}_u} N_{tx}^{(i)}(v), \tag{5}$$

where N_u is the set of neighbors of u. The factor $3/4$ is chosen to account for the ratio between the power consumptions in the receive mode and the transmit mode. Our choice is based on values reported in [10] and [14]. The average energy consumption $\bar{E}(\epsilon)$ is the average (over nodes) of the total of energy consumed among all the nodes till the normalized error reduces to ϵ. For simplicity, eq. (5) assumes synchronous updates and perfect communication (no retransmissions). When packet transmission is unsuccessful, multiple retransmissions maybe result, making the resulting energy consumption a complex function of the parameters involved [11, 10].

In one iteration of the Jacobi algorithm, a node needs to broadcast its own estimate, which consists of k real numbers. Recall that k is the dimension of the node variables. Assuming a 32 bit encoding, that amounts to $4k$ bytes of data. In the OSE algorithm, a node with d neighbors has to broadcast data consisting of $4d$ bytes for its neighbors' IP addresses, $4k(d+1)$ bytes for the previous estimates of itself and its neighbors, and $3d$ bytes for time stamps of those estimates. This leads to a total of $(7 + 4k)d + 4k$ bytes of data, and consequently the number of packets in a message becomes

$$N_{tx}(u) = \left\lceil \frac{(7 + 4k)d + 4k}{max\_databytes\_pkt} \right\rceil, \tag{6}$$

where $max\_databytes\_pkt$ is the maximum number of bytes of data allowed in the payload per packet. In this paper we assume that the maximum data per packet is 118 bytes, as per IEEE 802.15.4 specifications [15]. For comparison, we note that the number of bytes in a packet transmitted by MICA motes can vary from 29 bytes to 250 bytes depending on whether B-MAC or S-MAC is used [16]. If the number of data bytes allowed is quite small, OSE may require multiple packet transmission in every iterations, making it more expensive.

4   Simulations

For simulations reported in this section, we consider location estimation as an application of the problem described in this paper. The node variable x_u is node u's position in 2-d Euclidean space. We present a case study with a network with 200 nodes that were randomly placed in an area approximately 1×1 area (Figure 2(c)). Some pairs of nodes u, v that were within a range of less than $r_{max} = 0.11$ were allowed to have measurements of each others' relative distance r_{uv} and bearing θ_{uv}. Node 1, placed at $(0, 0)$ was the only reference node. Details of the noise corrupting the measurements and the resulting covariances can be found in [2]. The locations estimated by the (centralized) optimal estimator are shown in Figure 2(c) together with the true locations.

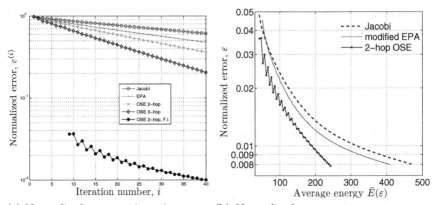

(a) Normalized error vs. iteration num-   (b) Normalized error vs. average energy
ber.   consumed.

Fig. 3. Performance comparison of the three algorithms. (a) shows the reduction in
normalized error with iteration number for the three algorithms, and also the drastic
reduction in the error with flagged initialization for the 2-hop OSE (the legend "F.I."
refers to flagged initialization). The simulations without flagged initialization were
done with all initial node position estimates set to $(0,0)$. (b) shows the Normalized
error vs. average energy consumption of 2-hop OSE, modified EPA and Jacobi with
broadcast communication. Flagged initialization was used in all the three algorithms.
All simulations shown in (a) and (b) were done in Matlab.

Simulations were done both in Matlab and in the network simulator package GTNetS [17]. The Matlab simulations were done in a synchronous manner. The purpose of the synchronous simulations was to compare the performance of the three algorithms – Jacobi, modified EPA and OSE – under exactly the same conditions. Synchronous Matlab simulations with link failure
were conducted to study the effect of communication faults (in isolation from
the effect of asynchronism). The GTNetS simulations were done to study OSE's
performance in a more realistic setting, with asynchronous updates and faulty
communication. For all OSE simulations, λ was chosen (somewhat arbitrarily)
as 0.9.

Figure 3(a) compares the normalized error as a function of iteration number
for the three algorithms discussed in this paper - Jacobi, EPA and the OSE.
Two versions of OSE were tested, 2-hop and 3-hop. It is clear from this figure
that the OSE outperforms both Jacobi and modified EPA. As the figure shows,
drastic improvement was achieved with the flagged initialization scheme. With
it, the 2-hop OSE was able to estimate the node positions within 3% of the
optimal estimate after 9 iterations. For the flagged OSE, the normalized error
is not defined till iteration number 8, since some nodes had no estimate of their
positions till that time.

The Performance of the three algorithms - Jacobi, modified EPA and 2-hop OSE
are compared in terms of the average energy consumption $\bar{E}$ in Figure 3(b). Flagged
initialization was used in all three algorithms. To compute the energy consumption

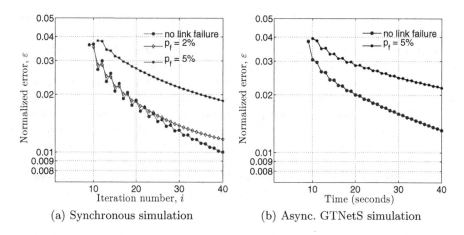

(a) Synchronous simulation   (b) Async. GTNetS simulation

Fig. 4. (a)Normalized error as a function of iteration number in the presence of link failures. Two different failure probabilities are compared with the case of no failure. (b) Normalized error vs. Time (seconds) for asynchronous 2-hop OSE simulations conducted in GTNetS, with and without link failure. As expected, performance in the asynchronous case is slightly poorer than in the corresponding synchronous case.

for the 2-hop OSE, we apply (6) with $k = 2$ and $max\_databytes\_pkt = 118$ to get $N_{tx}(u) = \lceil (15d_u + 8)/118 \rceil$. The average node degree being 5, the number of packets broadcasted per iteration in case of the OSE algorithm was 1 for almost all the nodes. For Jacobi, the number of packets broadcasted at every iteration was 1 for every node. For the modified EPA algorithm, the number of packets in every transmission was 1 but the total number of transmissions in every iteration were larger (than Jacobi and OSE) due to the data exchange required in both the EPA update and EPA solve steps (see [8] for details). The normalized error against the average (among all the nodes) total energy consumed $\bar{E}$ is computed and plotted in Figure 3(b). Comparing the plots one sees that for a normalized error of 1%, the OSE consumes about 70% of the energy consumed by modified EPA and 60% of that by Jacobi. For slightly lower error, the difference is more drastic: to achieve a normalized error of 0.8%, OSE needs only 60% of the energy consumed by EPA and about half of that by Jacobi.

Note that the energy consumption benefits of OSE become more pronounced as one asks for higher accuracy, but less so for low accuracy. This is due to flagged initialization, which accounts for almost all the error reduction in the first few iterations.

To simulate faulty communication, we let every link fail independently with a probability p_f that is constant for all links during every iteration. Figure 4(a) shows the normalized error as a function of iteration number (from three representative runs) for two different failure-probabilities: $p_f = 0.025$ and 0.05. In all the cases, flagged initialization was used. The error trends show the algorithm converging with link failures. As expected, though, higher failure rates resulted in deteriorating performance.

The OSE algorithm was also implemented in the GTNetS simulator [17], and the results for the 200 node network are shown in Figure 4(b). Each node sleeps until it receives the first packet from a neighbor, after which it updates its estimate and sends data to its neighbors every second. Estimates are updated in an asynchronous manner, without waiting to receive data from all neighbors. Time history of the normalized error is shown in Figure 4(b). Both failure-free and faulty communication (with $p_f = 0.05$) cases were simulated. Even with realistic asynchronous updates and link failures, the OSE algorithm converges to the optimal estimate. Since the nodes updated their estimates every second, the number of seconds (x-axis in Figure 4(b)) can be taken approximately as the number of iterations. Comparing Figure 4(a) and (b), we see that the convergence in the asynchronous case is slightly slower than in the synchronous case.

5   Conclusions

We have developed a distributed algorithm that iteratively computes the optimal estimate of vector valued node variables, when noisy difference of variables between certain pairs of nodes are available as measurements. This situation covers a range of problems relevant to sensor network applications, such as localization and time synchronization. The optimal estimate produces the minimum variance estimate of the node variables from the noisy measurements among all linear unbiased estimates. The proposed Overlapping Subgraph Estimator (OSE) algorithm computes the optimal estimate iteratively. The OSE algorithm is distributed, asynchronous, robust to link failures and scalable. The performance of the algorithm was compared to two other iterative algorithms – Jacobi and modified EPA. The OSE outperformed both of these algorithms, consuming much less energy for the same normalized error. Simulations with a simple energy model indicate that OSE can potentially cut down energy consumption by a factor of two or more compared to Jacobi and modified EPA.

There are many avenues of future research. Extending the algorithm to handle correlated measurements and developing a distributed algorithm for computing the covariance of the estimates are two challenging tasks that we leave for future work.

Bibliography

[1] Mendel, J.M.: Lessons in Estimation Theory for Signal Processing, Communications and Control. Prentice Hall P T R (1995)

[2] Barooah, P., da Silva, N.M., Hespanha, J.P.: Distributed optimal estimation from relative measurements: Applications to localizationa and time synchronization. Technical report, Univ. of California, Santa Barbara (2006)

[3] Karp, R., Elson, J., Estrin, D., Shenker, S.: Optimal and global time synchronization in sensornets. Technical report, Center for Embedded Networked Sensing,Univ. of California, Los Angeles (2003)

[4] Barooah, P., Hespanha, J.P.: Optimal estimation from relative measurements: Electrical analogy and error bounds. Technical report, University of California, Santa Barbara (2003)

[5] Barooah, P., Hespanha, J.P.: Distributed optimal estimation from relative measurements. In: 3rd ICISIP, Bangalore, India (2005)

[6] Chintalapudi, K., Dhariwal, A., Govindan, R., Sukhatme, G.: Ad-hoc localization using ranging and sectoring. In: IEEE Infocom. (2004)

[7] Moore, D., Leonard, J., Rus, D., Teller, S.: Robust distributed network localization with noisy range measurements. In: Proceedings of the Second ACM Conference on Embedded Networked Sensor Systems. (2004)

[8] Delouille, V., Neelamani, R., Baraniuk, R.: Robust distributed estimation in sensor networks using the embedded polygon algorithms. In: IPSN. (2004)

[9] Min, R., Bhardwaj, M., Cho, S., Sinha, A., Shih, E., Sinha, A., Wang, A., Chandrakasan, A.: Low-power wireless sensor networks. In: Keynote Paper ESSCIRC, Florence, Italy (2002)

[10] Bougard, B., Catthoor, F., Daly, D.C., Chandrakasan, A., Dehaene, W.: Energy efficiency of the IEEE 802.15.4 standard in dense wireless microsensor networks: Modeling and improvement perspectives. In: Design, Automation and Test in Europe (DATE). (2005)

[11] Carvalho, M.M., Margi, C.B., Obraczka, K., Garcia-Luna-Aceves, J.: Modeling energy consumption in single-hop IEEE 802.11 ad hoc networks. In: IEEE ICCCN. (2004)

[12] Shih, E., Cho, S., Fred S. Lee, B.H.C., Chandrakasan, A.: Design considerations for energy-efficient radios in wireless microsensor networks. Journal of VLSI Signal Processing **37** (2004) 77–94

[13] Frommer, A., Schwandt, H., Szyld, D.B.: Asynchronous weighted additive Schwarz methods. Electronic Transactions on Numerical Analysis **5** (1997) 48–61

[14] Ye, W., Heidemann, J., Estrin, D.: An energy-efficient mac protocol for wireless sensor networks. In: Proceedings of the IEEE Infocom. (2002)

[15] IEEE 802.15 TG4: IEEE 802.15.4 specifications (2003) http://www.ieee802.org/15/pub/TG4.html.

[16] Ault, A., Zhong, X., Coyle, E.J.: K-nearest-neighbor analysis of received signal strength distance estimation across environments. In: 1st workshop on Wireless Network Measurements, Riva Del Garda, Italy (2005)

[17] Riley, G.F.: The Georgia Tech Network Simulator. In: Workshop on Models, Methods and Tools for Reproducible Network Research (MoMeTools). (2003)

GIST: Group-Independent Spanning Tree for Data Aggregation in Dense Sensor Networks

Lujun Jia, Guevara Noubir, Rajmohan Rajaraman, and Ravi Sundaram

College of Computer and Information Science
Northeastern University, Boston, MA 02115, USA
{lujunjia, noubir, rraj, koods}@ccs.neu.edu

Abstract. Today, there exist many algorithms and protocols for constructing agregation or dissemination trees for wireless sensor networks that are optimal (for different notions of optimal, i.e. under different cost metrics). However, all these schemes differ from one common failing - they construct an optimal tree for a given *fixed* subset of the sensors. In most practical scenarios, the sensor group is continuously and dynamically *varying* - consider for example the set of sensors scattered in a forest that are sensing temperatures above some specified threshold, during a wildfire. Given the limited computational and energy resources of sensor nodes it is impossible to either prestore the optimal tree for every conceivable group or to dynamically generate them on the fly.

In this paper we propose the novel approach of constructing a *single* group-independent spanning tree (GIST) T for the network and then letting any sensor group S use the subtree induced by S on T, T_S as its group aggregation tree. The important question is, how does the quality of the subtree T_S compare to the optimal tree, OPT_S, across different groups S. We consider two well-accepted measures - aggregation cost (sum over all links) and delay (diameter). We show that in polynomial time we can construct a GIST that simultaneously achieves $O(\log n)$-approximate aggregation cost and $O(1)$-approximate delay, for all groups S.

To the best of our knowledge GIST is the first construction with a nontrivial and provable performance guarantee that works for all groups. We provide a practical and distributed protocol for realizing GIST that requires only local knowledge. We show an $\Omega(n)$ lower bound for commonly accepted solutions such as MST and SPT (i.e. there exists a group for which the induced subtree performs poorly) and demonstrate by simulation that GIST is good not just in the worst case - it outperforms SPT and MST by between 30 and 60 per cent in realistic random scenarios. GIST is an overlay construction and for the special case of grids we present GRID-GIST, a physical tree that uses only grid edges and achieves the same performance bounds.

1 Introduction

Wireless sensor networks have emerged at the forefront of applications involving the measurement of physical phenomena, environmental monitoring, medical

P. Gibbons et al. (Eds.): DCOSS 2006, LNCS 4026, pp. 282–304, 2006.
© Springer-Verlag Berlin Heidelberg 2006

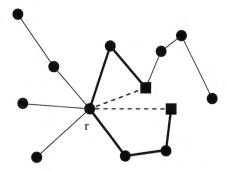

Fig. 1. In the above network, a spanning tree T is depicted by solid edges. A subset S consists of the two square nodes. A subtree of T induced by S is depicted by the thick edges. Clearly, the subtree is far from optimal, which includes the two dotted edges.

instrumentation, and building monitoring in warehouses and homes. A wireless sensor network is comprised of hundreds or thousands of sensor nodes networked through wireless links for collecting and processing environmental data. Sensor networks have stringent energy restrictions, and are typically deployed in high density to ensure coverage and fault tolerance.

With the advent of large-scale sensor network applications, there is considerable interest in a database abstraction for sensor networks, in which users program the sensors using a high-level declarative language [31,17,38]. A user issues a query to the network, and every sensor node that meets the criteria defined in the query replies with the desired data reading. In the reply phase, data can be aggregated in-network to reduce communication complexity, and hence energy consumption [31,28]. Data aggregation is usually performed using a reverse multicast tree, in which each intermediate node receives packets from its children, aggregates the information, and sends one packet to its parent. As shown in multiple studies, this can lead to considerable savings in energy consumption over an approach that does not use in-network aggregation [31,17,28].

Consider a sensor network deployed in a large forest for monitoring forest fires. A user may issue the following query (written in a variant of SQL) to the whole network:

```
SELECT MAX(temperature), location FROM sensors
WHERE temperature > threshold
DURATION (now,now+3600s)
EVERY 30s
```

In the above declarative query, the user asks for the maximum temperature and its location, every 30 seconds for a duration of an hour, if this maximum temperature is above a certain threshold. Due to temperature changes in different areas of the forest, the group of sensors satisfying the threshold criteria may change continually. Therefore, the aggregation group (the group of sensors that

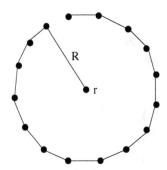
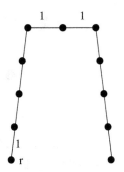

Fig. 2. An SPT for this network consists of all the "spoke" edges. Let root node r be the center node, and subset S consist of all the nodes on the circle. Clearly, the worst case stretch is $\Omega(n)$ if the SPT is used as GIST.

Fig. 3. An MST for this network consists of all the length 1 edges. Let root node r be the bottom left node, and subset S consist of only the bottom right node. Clearly, the worst case stretch is $\Omega(n)$ if the MST is used as GIST.

need to send readings to the user) changes in an unpredictable manner over time, and the total number of such groups is large. In such a scenario, the following dilemma seems unavoidable:

- Since the number of different groups is large – exponential in the number of sensors, in the worst case – it is prohibitively expensive for the energy-constrained, distributed and multi-hop sensor nodes to compute efficient aggregation trees for each group.
- On the other hand, if a communication-optimized data aggregation structure is not used, the aggregation itself can be expensive.

Most existing algorithms and protocols for constructing data aggregation trees have drawbacks in that one aggregation tree has to be constructed and maintained for one group of sensors [4, 20, 27, 31]. Such aggregation protocols are not suitable for large-scale sensor networks serving aggregation queries at high frequencies. We refer to these protocols as group-dependent data aggregation protocols. Group-dependence, or group-awareness, makes the optimization of communication cost possible, however only for a single group. Therefore, in sensor applications where the sensor groups of interest evolve constantly or different queries are issued on a frequent basis, new solutions are desired to support effective in-network aggregation.

We propose GIST (Group-Independent Spanning Tree) for data aggregation in sensor networks. A GIST, T is a single spanning tree that is "oblivious" to any aggregation group in the sense that the aggregation structure adopted for any group is simply the subtree of T induced by the group. The performance of GIST for a given group is measured by comparing the cost of this induced subtree with the cost of an optimal aggregation tree for that group. Figure 1 illustrates a group-independent spanning tree and its subtree induced by a subset of nodes. It is not clear a priori that it is possible to find such a group-independent tree with good guarantee. Two natural candidates for GIST are the minimum spanning tree (MST) and the shortest-paths tree (SPT). However as shown in Figure 2 and 3, they both have $\Omega(n)$ worst case performance.

1.1 Main Results

- We propose the group-independent tree paradigm for providing effective underlying structures to support data aggregation in wireless sensor networks that performs well in terms of aggregation cost and delay. The worst-case performance of common tree structures such as MST and SPT is shown to be $\Omega(n)$ times the optimal in terms of aggregation cost.
- We propose an algorithm for constructing GIST in sensor networks, such that the cost of the tree induced by any group is within $O(\log n)$ factor of the cost of the optimal cost for that group, and delay is within $O(1)$ factor of the optimal delay. We also present a distributed protocol for constructing our GIST, performing data aggregation, and for maintaining GIST as nodes join or leave the network.
- We prove that our upper bound of $O(\log n)$ is nearly tight by presenting a lower bound of $O(\log n/\log\log n)$ for aggregation cost, for any polynomial time algorithm that approximates the problem.
- Through extensive simulations, we show that our GIST algorithm outperforms both MST and SPT by $30-60\%$ in terms of both the communication cost as well as average delay.

Our GIST protocol reduces the tree construction overhead, and each node in GIST only needs to memorize a single parent. Also, our GIST protocol has a provable $O(\log n)$ performance guarantee for aggregation cost and the delay is within a constant factor of all induced subtrees.

The remainder of this paper is organized as follows. In Section 2, we survey related work. Section 3 presents our models and basic assumptions. In Section 4, we give the formal definition of GIST and present the GIST algorithm. In Section 5, we give a distributed implementation of our GIST algorithm. In Section 6, we evaluate the performance of our GIST algorithm by simulations. In Section 7, we discuss future work and some limitations of our scheme, and propose an algorithm for constructing physical GIST on grid networks.

2 Related Work

A number of data aggregation algorithms and protocols have been proposed for wireless sensor networks over the past several years. Directed diffusion [21] is proposed as a data centric communication paradigm for sensor networks. In directed diffusion, subscriptions use flooding to spread interest. Initially, the data is sent to the sink along multiple paths; however, better aggregation paths are gradually enforced. SAFE [25] uses geographically limited flooding to forward query to nodes. Due to expensive flooding operations, it is not suitable for large sensor networks. TTDD [39] exploits local flooding within a local cell of a grid to facilitate large scale data dissemination. However, when the sink moves out of the cell, the dissemination path has to be reconstructed. In [19], a regional-flooding based multicast scheme for the problem of mobicast is proposed, where high sensor network density is exploited for delivery guarantee as well as satisfying certain temporal requirements.

In [28], data-centric routing protocols were compared with traditional address-centric protocols, and the authors showed that data-centric routing offers significant performance gains in a wide range of operational scenarios. In [24, 3], the authors consider the problem of data dissemination from a source node to multiple mobile sinks. In their algorithm, a mobile sink is attached to a static sensor access point close to it, and a multicast group consists of a set of such access points that request the same information from a sensor region. Their algorithm is based on the construction of minimum Steiner trees. In [27, 31], data aggregation algorithms are designed to reduce the rounds of transmitted data from sensors to sink. In-network data aggregation has also been considered in [27, 31, 4]. We however note that none of the proposed schemes is able to achieve a provable performance guarantee.

In [22], the authors developed a polynomial time algorithm for computing a spanning tree with $O(\log^4 n/\log\log n)$ stretch on arbitrary metrics. For Euclidean metrics, an $O(\log n)$-stretch spanning tree is presented. The $O(\log n)$-stretch spanning tree construction of [22] requires the whole network to be known *a priori*, and the algorithm is centralized and not suitable for a distributed network. In addition, the algorithm only specifies the initial construction of the tree, and does not consider the maintenance of the tree in presence of node failures. Note that the scheme of dividing the Euclidean plane into recursively small regions is a commonly used technique, and has been considered in [30]. Schemes that take into account both edge cost and network diameter are also considered in [32].

Multicast communication algorithms are also considered in the context of ad hoc networks. The authors in [20] construct a publish/subscribe spanning tree across the network. A location-aware Steiner tree based algorithm is proposed in [11] for multicast in ad hoc networks. Geocasting [26], where multicast algorithms are tailored for sending information to a geographical area, is studied in [26]. In [15], hierarchical multicast routing protocols are proposed. The authors adopted an overlay-driven approach for supporting hierarchical routing.

A number of overlay multicast protocols have been proposed in the context of IP multicast. In [37], a proactive approach is considered for reconstructing overlay multicast trees. In [5], the authors consider the trade-off between path length and the load on nodes in overlay multicasting, and proposed an application-layer mininum delay multicast algorithm. Related studies identify network structures that optimize other metrics. In [9, 10], the authors proposed algorithms for constructing routing structures that minimize network congestion at nodes. In [8], the problem of routing for minimizing transmission energy under the relay model is studied.

3   Models and Assumptions

3.1   Network Model

Unit disk graphs. We model a wireless sensor network as an undirected unit disk graph $G = (V, E)$ on the Euclidean plane, where V is the set of nodes, and E is a set of edges on V. (We remark that our GIST algorithm can be easily extended to 3-dimensional space.) A edge (u, v) exists between $u, v \in V$ if both nodes are capable of exchanging messages over the distance $|uv|$. We assume that each node is aware of its own geographical location. This can be achieved by either a GPS device, or a location service such as the position estimator in [1, 6].

Node density. The focus of this paper is on large-scale dense sensor networks, where scalability is the main challenge for data aggregation. We assume a deployment of n sensors randomly placed in the 2-D plane, with a density of $\Omega(\log n)$ nodes per unit area, assuming unit transmission range for all nodes. It is well-known that an n-node network is disconnected (assuming unit transmission range) with high probability if the density is $O(\log n)$ [16, 33].

As discussed in Section 1, our main performance measures are the total communication cost and the average delay. The total communication cost for an aggregation is measured by total number of hops used in the aggregation, while the delay is the number of hops between the root and the farthest node being aggregated. Through standard probabilistic analysis, it has been shown that if the density is $\Omega(\log n)$ (for a suitably chosen constant hidden in the Ω notation), each cell of a sufficiently small constant area is occupied [16]. In such a dense network, the network hop distance is well-approximated by the Euclidean distance (within a small constant factor). This correlation has also been established by means of an in-depth experimental study in [19]. In our performance analysis, we adopt the Euclidean distance as our measure of the hop metric since these are equivalent up to small constant factors and the Euclidean metric is more conducive to our analysis for establishing provable bounds.

We emphasize that the correctness of our protocol does not rely on any assumptions about the sensor node density; our protocol works for *any* connected network.

Routing. Our general GIST algorithm constructs an overlay tree on the underlying physical graph $G = (V, E)$. An edge in the overlay tree corresponds to a path in the underlying graph. Therefore, an underlying routing service from a source to a destination is required. Note that the idea of using overlay trees for data dissemination and aggregation is not new and has been considered in [24, 3, 7] for wireless sensor networks and in [36, 14] in the context of ad hoc networks. Such underlying paths can be shortest paths, or paths computed by geographic routing protocols [23, 29], etc.

In Section 7, we propose an algorithm, GRID_GIST, which is not dependent on any underlying routing service. In a GRID_GIST, all edges are physical edges in the underlying graph G. In location-based services, the deployment regions can be divided into a number of small geographic areas such that two nodes in neighboring areas can communicate with each other. In each area, a leader can be selected, and a GRID_GIST can be constructed for all the leaders. Whenever a node want to participate in the data aggregation process, it first sends messages to its leader in the area, then the GRID_GIST is employed to transport and aggregate data to the sink.

3.2   Aggregation Functions

In this study, we assume distributive aggregation functions [31, 28], where intermediate nodes compute and transmit one single output packet as the result of aggregating over multiple input packets. In such aggregates, the size of the partial state record is the same as the final aggregate. Thus, each transmitted packet is of the same size. Computing *maximum, minimum, average, sum, count* are examples of this class of aggregation functions. Assuming distributive aggregates, it can be easily seen that given a group S of sensor nodes and a root node (information sink), the minimum cost (total number of transmissions) data aggregation tree is a minimum Steiner tree, where the Steiner set consists of the information sink and all the data sources involved [24, 3, 11]. Therefore, the cost of using an aggregation tree can be measured by simply counting the number of edges on the tree, since each node transmits only once.

3.3   Simulation Model

In our simulation, we first evaluate our algorithm on dense sensor networks where all nodes in the network are capable of sensing, aggregating and transmitting data. We refer to this model as SNM (Sensor Network Model). We compare the performance of GIST, MST and SPT under two metrics: aggregation cost and aggregation delay. As discussed in the preceeding section, aggregation cost can be measured by the number of edges on the tree, assuming distributive aggregation functions. We measure aggregation delay by the maximum tree depth rooted at the sink node, since this represents the maximum number of hops for a packet to reach the sink. Note that delays caused by packet collision are not considered.

In addition to the above model, we consider sensor networks consisting of regular sensor nodes, as well as a dense collection of cooperative "relay" nodes [18,34]. Relay nodes are only capable of routing and forwarding; they do not participate in data aggregation and other higher-layer sensor network applications. In [18], the authors show that given a certain energy budget, it is more efficient to deploy additional relay nodes than increase energy in existing nodes in order to extend the life time of the network. Our bound on the performance of GIST requires a dense network of the combined sensor and relay deployment. We refer to this model as SRNM (Sensor and Relay Network Model).

4   A Provably Near-Optimal Group-Independent Spanning Tree

The optimal data aggregation tree problem can be reduced to the classic minimum Steiner tree problem, assuming distributive aggregate functions. The minimum Steiner tree problem is NP-complete [13], but can be easily approximated to within a constant factor in polynomial time, given the fixed group of vertices that need to be connected [35]. As discussed in Section 1, however, in many sensor network applications, the group of relevant sensors that needs to be aggregated may evolve constantly. Thus, it is infeasible for the resource-constrained sensor nodes to compute efficient multicast trees for the many groups (potentially, exponential in the number of sensors) on-line. Motivated by this [22], one can consider the following natural variation of the Steiner tree problem, *universal Steiner tree* problem: Given a root node $r \in V$, is there a spanning tree T connecting r to all nodes in $V - \{r\}$, such that for any subset S of V the cost of tree induced by $S + \{r\}$ on T is "close" to that of an optimal Steiner tree for the set $S + \{r\}$? If a "good" universal Steiner tree exists and can be computed efficiently, it is an excellent candidate for group-independent data aggregation, with the information sink as the root. We now present a formal definition of the universal Steiner tree problem [22].

Definition 3.1. An instance of the universal Steiner tree problem is a triple $\langle V, d, r \rangle$ where (V, d) forms a metric space, and r is a distinguished vertex in V that we refer to as the *root*. Let $||T||$ denote the cost of tree T. For any spanning tree T of V, define the *stretch* of T as $\max_{S \subseteq V} ||T_{S+\{r\}}|| / ||Opt_{S+\{r\}}||$, where $T_{S+\{r\}}$ denote the tree induced by $S + \{r\}$ on T and $Opt_{S+\{r\}}$ is a minimum Steiner tree for subset $S + \{r\}$. The goal is to determine a spanning tree with minimum stretch.

Two natural candidates for a universal Steiner tree are the minimum spanning tree (MST) and the shortest-paths tree (SPT). In Section 4.1, we show that both MST and SPT have $\Omega(n)$ stretch, thus making them poor choices for group-independent aggregation in the worst-case. The main focus of this section is a new algorithm for constructing GIST that achieves $O(\log n)$ stretch in Euclidean metrics, and can be efficiently implemented in wireless sensor networks.

4.1   Lower Bounds for MST and SPT

MST and SPT [12] are polynomial time computable structures that are often used for optimizing the overall tree cost or individual cost of each node communicating with the root. However, when used as a GIST, MST and SPT can perform arbitrarily bad. We have the following,

Theorem 1. *The worst-case stretch of both SPT and MST in Euclidean plane is $\Omega(n)$, where n is the number of vertices in the metric.*

Due to space constraints, the proof of the above theorem is included in Appendix A.

4.2   GIST for Wireless Sensor Networks

Our GIST algorithm will adopt the following approaches to address the issues discussed in the preceding section. First, we assume that each sensor node is aware of its geographical location. This can be achieved by adopting a location service such as the position estimators in [1, 6]. We also assume that each node also knows the location of the root node (information sink). Given such location information, our GIST algorithm divides the whole deployment region into recursively small regions (*levels*), and tree construction computation is limited to the subset of nodes in each such region. Second, our GIST algorithm constructs an *overlay* tree, i.e., an edge in this overlay tree can be a path in the original unit disk graph. Each node in the overlay tree represents a geographical region it resides in, and is referred to as a *leader* of this region. This enables GIST to be constructed without full knowledge of the network. Finally, leader election, as well as node joining/leaving the tree will be implemented on the overlay tree structure, which will also be the basis for handling communication failures in unreliable sensor networks.

Algorithm 1. $GIST(v, (x_1, y_1), (x_2, y_2))$

1: **if** $|x_2 - x_1| > R/\sqrt{5}$ **then**
2:    Divide the square region marked by (x_1, y_1) and (x_2, y_2) into 9 equally-sized smaller square regions (Figure 4);
3: **else**
4:    return;
5: **end if**
6: Within each square region, select any node not selected before to be a leader, except for the square region where v resides (for which v shall still be the leader);
7: Output an edge between each selected leader ℓ and v;
8: For each square region, invoke $GIST(\ell, (x_1', y_1'), (x_2', y_2'))$, where ℓ is the leader of the square region, (x_1', y_1') and (x_2', y_2') are the coordinates marking the square region;

The value $\sqrt{5}$ is used to ensure communication between two nodes at the opposite end of a 2×1 rectangle (consisting of two neighboring squares). Let R be the transmission range of sensor nodes, and let D be the maximum distance between any node in $V - \{r\}$ and r. (Recall that r is the root node.) Without loss of generality, let the coordinate of r be $(0, 0)$. Algorithm 1 presents the formal definition of our GIST algorithm. It takes as input a root node and a square region specified by two diagonally-opposite locations, and computes a spanning tree connecting the root node to nodes within the square. To compute a final GIST tree, $GIST(r, (-D, -D), (D, D))$ is invoked. Our algorithm adopts a top-down approach: In each recursion, a set of new leaders representing certain geographical regions are selected and connected to their parent; each new invocation of the algorithm divides a square region into smaller pieces and repeats the same leader selection process. Note that a selected leader for a square region will also be the leader for one of the 9 smaller square regions. This applies to root node r. Let T be the GIST computed by our algorithm. For any subset $S \subseteq V$, we use *aggregation cost* to refer to the total edge distance of the tree induced by $S + \{r\}$ on T, and use *aggregation delay* to denote the length of the path from a node to the root. We have the following theorem for the aggregation delay,

Theorem 2. *(Aggregation Delay) Let T be a GIST generated by Algorithm 1. The distance from any node $v \in V$ to r on T is within a constant factor of the Euclidean distance between v and r.*

For the aggregation cost, we have the following.

Theorem 3. *(Aggregation Cost) Under the Euclidean metric, Algorithm 1 computes an GIST with stretch $O(\log n)$ in polynomial time.*

The proof of the above two theorems is included in Appendix B.

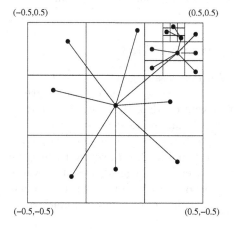 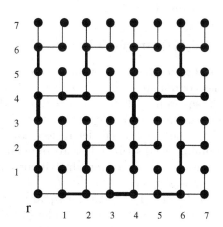

Fig. 4. Algorithm 1 divides the network into recursively small regions, and constructs an overlay tree with stretch $O(\log n)$

Fig. 5. Algorithm 3 construct a GRID_GIST on grid networks. The stretch is $O(\log n)$.

One interesting property of Algorithm 1 is that it eliminates dependence on knowledge of the network topology, since each node in the overlay tree represents a geographical region. Nodes can join (e.g., powered on) or leave (e.g., due to failure) the network in a dynamic fashion, while geographical regions are relatively stable. This forms the basis for our tree maintenance and fault tolerance mechanisms (Section 5). Another property of Algorithm 1 is that the distance between any node $v \in V$ and root r on the computed GIST is within a constant factor of the minimum distance between v and r. This implies that the aggregation delay using the induced tree of $GIST$ is within a constant factor of the minimum aggregation delay.

4.3   Lower Bound for GIST

Our bound on the aggregation cost of GIST is almost tight. We can establish the following lower bound on GIST(please refer to Appendix C for the proof).

Theorem 4. *No algorithm can build a GIST with stretch better than* $\Omega(\frac{\lg n}{\lg \lg n})$ *for any sensor network on Euclidean plane.*

5   A GIST Based Data Aggregation Protocol

In this section, we present a distributed implementation of Algorithm 1 for constructing GIST. Our protocol proceeds in rounds by selecting leader among nodes in small regions in the first round, and electing leaders among leaders selected in the previous round in larger regions, and so on. We also define tree maintenance and fault tolerance mechanisms, as well as a brief description on data aggregation using GIST. Our protocol is a bottom-up implementation of Algorithm 1.

5.1   A Distributed Protocol for Constructing GIST

Let (x_v, y_v) be the coordinate of $v \in V$. Without loss of generality, root node r has coordinate $(0,0)$. Let $c = R/\sqrt{5}$ be the smallest square size (side length of the square). As discussed in Section 4, the output of Algorithm 1 is a hierarchical overlay tree, and a selected leader ℓ of a larger square region (higher level) is also the leader for the smaller region (lower level) in which ℓ resides in. Each node running the distributed protocol maintains a variable Cur_Level indicating the largest square region for which it currently is a leader. The size of a square where v is a leader is equal to $3^{Cur\_Level-1}c$, i.e., a leader for the square with size c has $Cur\_Level = 1$, a leader for squares with size $3c$ has $Cur\_Level = 2$, and so on. By default, each node is a leader of level 0.

The protocol for constructing GIST proceeds in rounds. In each round, a leader is elected by exchanging LEADER_ELECTION packets. The LEADER_ELECTION packet contains the following information: $[L, SQ\_X, SQ\_Y, id]$, where id is an integer uniquely identifying a node who is participating in the leader election process. The first field, L, represents the level (also the size of the square region) for which a leader is *to be* elected. The level number L, together with SQ_X, SQ_Y, defines the geographical region where a leader is elected. SQ_X and SQ_Y are integers that are defined as follows,

$$SQ\_X = \lfloor x_v/(3^{L-1} \cdot c) \rfloor,$$
$$SQ\_Y = \lfloor y_v/(3^{L-1} \cdot c) \rfloor,$$

where (x_v, y_v) is the coordinate of node v. Thus, a node is able to compose a packet by filling in the leader selection level, SQ_X, SQ_Y, and its id number.

Upon receiving a LEADER_ELECTION packet, a node u can decide whether it is contained in the square region defined by the packet's L, SQ_X and SQ_Y: if

$$SQ\_X \cdot 3^{L-1}c \leq x_u < (SQ\_X + 1) \cdot 3^{L-1}c,$$
$$SQ\_Y \cdot 3^{L-1}c \leq y_u < (SQ\_Y + 1) \cdot 3^{L-1}c,$$

then u is in the region; otherwise, u is not.

In each round, each node v broadcasts its own LEADER_ELECTION packet with $[Cur\_Level + 1, SQ\_X, SQ\_Y, id]$, if v has not heard of any $Cur\_Level + 1$ LEADER_ELECTION packet with lower node id, or any $Cur\_Level + 2$ or higher level LEADER_ELECTION packet. When v broadcasts a packet, it also records the level number and its own id in a local database. The root node r can fill in a negative id number in r's broadcast packet to ensure that it is elected a leader in each round. Since by default a node is a level 0 leader, the leader election starts with level 1. A node u upon receiving a LEADER_ELECTION packet pkt will invoke Algorithm 2.

Algorithm 2. A distributed protocol for constructing GIST

1: **if** u is not contained in pkt's region **then**
2:     u discards pkt;
3: **end if**
4: **if** u is contained in pkt's region, and u's Cur_Level is equal to $L - 1$ **then**
5:     **if** u's database has a record indexed by pkt's L and pkt's id field is lower than the record's id field, or u does not have such a record yet **then**
6:         Record pkt's L and id, and re-broadcast pkt;
7:     **else**
8:         u discards pkt;
9:     **end if**
10: **else if** u is contained in pkt's region, and u's Cur_Level is smaller than $L - 1$ **then**
11:     **if** u's database does not have a record indexed by pkt's L, or u has such a record and pkt's id field is lower than the record's id field **then**
12:         u re-broadcast pkt and record L and id;
13:     **else**
14:         u discards pkt;
15:     **end if**
16: **else if** u is contained in pkt's region, and u's Cur_Level is larger than $L - 1$ **then**
17:     u broadcasts a LEADER_ELECTION packet with $[Cur\_Level + 1, SQ\_X, SQ\_Y, id]$;
18: **end if**

The if-block of Line 4 handles the case where v actively participates in the leader selection process, since $Cur\_Level = L - 1$. The if-block of Line 10 is only for forwarding leader selection packets for higher level election, since $Cur\_Level < L - 1$ and v is aware of the fact that it cannot participate in this leader election. The if-block of Line 17 handles the case where v is already a leader for a certain region while another node is attempting leader election. Combined with Line 10 in Algorithm 2, the purpose of Line 17 is to suppress such leader election requests. However, different policies can be considered here:

e.g., an old leader can allow a new leader to be elected if it decides that its residual energy is not sufficient for reliable communication any more. The local database used for recording level number and node id helps in reducing the broadcast traffic in the network.

Each node u keeps a timer with time out value τ. When u sends out a LEADER_ELECTION packet attempting to be a level $Cur\_Level + 1$ leader, u starts the timer. When the timer fires after τ, if no better leader is detected, u increment its Cur_Level variable by 1, indicating that it is now a leader; otherwise, u picks the node id indexed by $Cur\_Level + 1$ in its database to be its parent. After a node u has picked its parent, u sends a register packet to the parent.

Algorithm 2 ensures that at least one leader is elected within the contended region in each level. Note that there are several cases in which multiple leaders may be elected for the same region and same level:

- The timeout value τ is too small for the broadcast packets to reach all destined nodes; or some broadcast packets are destroyed due to collision while they are being forwarded. The value τ is thus a adjustable protocol parameter.
- If there are multiple connected network components within the region for which a leader is to be elected, multiple leaders will be elected.

Note that the presence of multiple leaders within the same region on the same level, though it may have impact on the performance of the GIST, does not affect the correctness of Algorithm 2. This is because 1) each level i leader is able to promote itself to a level $i + 1$ leader (possibly after timeout) and compete in the next round for a level $i + 2$ leader, which eventually leads to root r; 2) each node will have one and only one parent, due to Line 5 of Algorithm 2.

The tie-breaking scheme in the above process is based on node ID. However, other tie-breaking schemes can easily be adopted by putting extension fields in the LEADER_ELECTION packet. For example, to elect leaders with higher residual energy, an extension field containing the residual energy reading can be used.

5.2   Data Aggregation with GIST

After a GIST T is constructed, the information sink r can broadcast queries and collect data readings using T. This process is similar to most other data aggregation schemes, e.g., [31]. Due to space constraints, we only give a brief description of the query distribution and data aggregation phase, and omit detailed discussions of protocol parameters.

In the query distribution phase, the root can send a query to its direct children. Each direct child sets a timer of τ_w (a function of the *Epoch* time) for awaiting replies from its children, and includes this information in the query passing down the tree. The next level children will set their own timer (smaller than their parent) and pass down the query, and so on. The choice of timer τ_w is related to the hop distance between a child and its parent. In our scheme, the region is divided into 9 smaller regions in each recursion. Consequently, the parent-child distance decreases by a factor of 3 each time. Thus, the timeout values for the leaders along a query distribution path can be approximated by a geometric series by $\tau_w^{child} = c \cdot \tau_w^{parent}$, where $0 < c < 1$. We omit the details for selecting c here, and consider it in our future study. In the data aggregation phase, each intermediate node waits for the timeout, aggregates all the received data readings, and sends the packet to its parent.

5.3   Tree Maintenance and Fault Tolerance

We now consider the maintenance of a GIST in case of node failures. We assume an independent protocol for failure detection. For example, each child can check the status of its parent using periodic ping messages. If a node u concludes that its parent has failed, it sends out a LEADER_ELECTION packet with $[Cur\_Level + 1, SQ\_X, SQ\_Y, id]$ to be a level $Cur\_Level + 1$ leader. At the same time, other children of the failed parent may also broadcast such LEADER_ELECTION packets. Such operations can easily be handled by the same procedure as in Algorithm 2. This is because Algorithm 2, though presented in a synchronous fashion, is in fact an asynchronous protocol. Multiple leaders can appear in this process. However as we have discussed in preceeding sections, the correctness of the protocol is not affected. Note that if a level i leader u failed, only the several level $i - 1$ leaders in the region where u resides will initiate LEADER_ELECTION packets. Other nodes in that region only forward such packets (Line 10 in Algorithm 2). In a similar manner, the operations for new nodes joining the tree can also be handled by Algorithm 2. This way, our scheme can potentially be combined with activity scheduling protocols [7] to achieve more energy savings.

In the data aggregation phase, fault tolerance can be achieved by constructing two independent trees. Each time a node sends its data reading, it sends two copies. For fault tolerance in the data aggregation phase, we can adopt the techniques of [31].

6   Experimental Evaluation

In this section, we compare the performance of our GIST algorithm with MST and SPT under the aggregation cost and aggregation delay metrics as discussed in Section 3. We simulate under two different settings: SNM (sensor network model) and SRNM (sensor relay network model). Recall that SNM models the situation where each sensor node in the network is capable of sensing, aggregating and transmitting; while in the SRNM model, the relay node is only capable of forwarding traffic.

6.1   SNM Performance Evaluation

In our simulations, the network topology is generated by randomly distributing nodes in the deployment region. The density (average number of neighbors within a node's transmission range), is around 10 in each experiment. In our first experiments, we study the aggregation cost when the event sources are picked within a strip close to the border of the sensor deployment region. This is useful

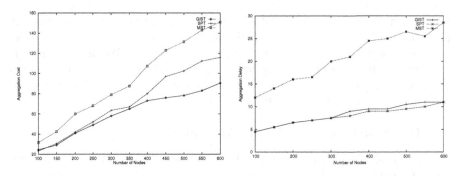

Fig. 6. Aggregation cost with increasing network size under SNM model. Event sources are picked within a strip.

Fig. 7. Aggregation delay with increasing network size under SNM model. Event sources are picked within a strip.

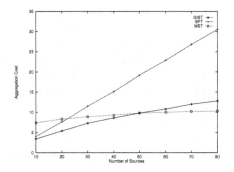

Fig. 8. Aggregation cost with increasing number of random sources under the SRNM model. 200 sensor nodes.

Fig. 9. Aggregation delay with increasing number of random sources under the SRNM model. 200 sensor nodes.

in applications where the sensor nodes observing common phenomena are along a line, for example when a group of intruders approaching the border in a battle field trigger the sensors. Figure 6 demonstrates the result. In this simulation, GIST performs better than SPT, which is better than MST. The reason that GIST is better than SPT is because when the event sources are picked along a line strip, a large number of event sources may traverse non-overlapping paths to the sink, thus reducing the chance for data aggregation in intermediate nodes. Figure 7 illustrates the aggregation delay in this experiment, in which GIST demonstrates close to optimal (SPT) delay.

6.2   SRNM Performance Evaluation

In this section, we evaluate the performance of GIST, MST and SPT under the SRNM model, where the network consists of sensor nodes as well as relay nodes.

We simulate a network that consists of 200 sensor nodes, and a dense network of relay nodes. The density is such that the probability that a underlying routing path between two sensor nodes containing another sensor node is small. This case can be thought of as the opposite of the SNM model, where the relay density is 0. In this model, the SPT overlay tree from the sensor nodes to the sink exhibits a star-like structure. Therefore, the aggregation in intermediate sensor nodes is minimal.

Figure 8 and Figure 9 illustrate the aggregation cost and delay in the experiments with $10, 20, \ldots$, or 80 random event sources. Note that the values in the figures are calculated by normalizing the network deployment region to a 1×1 region. The aggregation cost (induced tree cost) of SPT is the worst among the 3 algorithms as expected, due to the lack of aggregation. MST and GIST both have much better performance than SPT. When the number of sources is small, an induced aggregation tree of MST costs more than that of GIST, because some event sources may traverse long routes to the root, as can be observed in the aggregation delay in Figure 9. When the number of random sources increases, such long paths may contain more and more other sources so that aggregation can be performed.

7   Conclusion and Future Work

In this paper, we proposed a novel approach to data aggregation based on the concept of group-independent spanning tree GIST, and show that such a tree can be found in polynomial time with $O(\log n)$ performance guarantee. Specifically, we have designed an algorithm for constructing an GIST for dense sensor networks such that for any group, the cost of the induced subtree of our GIST is within a logarithmic factor of the optimal solution for the group. We have also shown that traditional spanning tree algorithms MST and SPT are extremely poor in the worst case. An important aspect of our GIST algorithm is its simplicity and amenability to distributed implementation. We have presented a protocol for constructing and maintaining GIST and for performing data aggregation over the tree.

Our algorithm for constructing GIST yields an overlay tree and assumes an underlying routing mechanism. This assumption can limit the application of our protocol. We propose an algorithm (Appendix D) for constructing GRID_GIST for grid sensor networks that does not require an underlying routing service. In this algorithm, each edge selected into the physical GRID_GIST tree has to be an existing edge in grid networks. Routing from any node in G to the root is specified during tree contruction phase, thus eliminating the dependence on an underlying routing service as required by GIST. Such physical trees on grids can be useful for densely deployed sensor networks. Due to space constraints, we present our algorithm GRID_GIST in Appendix D. One limitation to the GRID_GIST scheme is that it assumes regularity of the network. An important direction for future research is to determine the best "physical" GIST for arbitrary network topologies.

In our simulations, we have studied two variants of the sensor network model, one consisting purely of sensor nodes, and the other consisting of sensors placed in a dense relay field. Our simulations validate our theoretical work by demonstrating that our GIST outperforms both MST and SPT. As part of our future work, we will compare our scheme with other data aggregation schems in this literature in addition to MST and SPT.

References

1. J. Albowicz, A. Chen, and L. Zhang. Recursive position estimation in sensor networks. In *IEEE ICNP*, November 2001.
2. N. Alon and Y. Azar. On-line steiner trees in euclidean plane. In *Proceedings of the 8th Annual ACM Symposium on Computational Geometry*, pages 337–343, 1992.
3. S. Bhattacharya, H. Kim, S. Prabh, and T. Abdelzaher. Energy-conserving data placement and asynchronous multicast in wireless sensor networks. In *International Conference on Mobile Systems, Applications, and Services (MobiSys)*, May 2003.
4. B. Bonfils and P. Bonnet. Adaptive and decentralized operator placement for in-network query processing. In *Proceedings of Information Processing in Sensor Networks*, April 2003.
5. E. Brosh and Yuval Shavitt. Approximation and heuristic algorithms for minimum delay application-layer multicast trees. In *INFOCOM*, March 2004.
6. N. Bulusu, J. Heidemann, and D. Estrin. Gps-less low cost outdoor localization for very small devices. In *IEEE Personal Communications, Special Issue on Smart Space and Environments*, October 2000.
7. U. Cetintemel, A. Flinders, and Y. Sun. Power-efficient data dissemination in wireless sensor networks. In *ACM MobiDE*, September 2003.
8. J. Chen, L. Jia, X. Liu, G. Noubir, and R. Sundaram. Minimum energy accumulative routing in wireless networks. In *In Proceedings of IEEE INFOCOM 2005, The 24th Annual Joint Conference of the IEEE Computer and Communications Societies*, 2005.
9. J. Chen, R. Kleinberg, L. Lovász, R. Rajaraman, R. Sundaram, and A. Vetta. (Almost) tight bounds and existence theorems for confluent flows. In *Proceedings of the 36th Annual ACM Symposium on Theory of Computing (STOC)*, pages 529–538, June 2004.
10. J. Chen, R. Rajaraman, and R. Sundaram. Meet and merge: Approximation algorithms for confluent flows. In *Proceedings of the 35th Annual ACM Symposium on Theory of Computing (STOC)*, pages 373–382, June 2003. This paper has been accepted for publication in the special issue of the Journal of Computer and System Sciences (JCSS).
11. K. Chen and K. Nahrstedt. Effective location-guided tree construction algorithm for small group multicast in manet. In *IEEE INFOCOM*, June 2002.
12. T. H. Cormen, C. E. Leiserson, and R. L. Rivest. *Introduction to Algorithms*. MIT Press, Cambridge, MA, 1990.
13. M. R. Garey and D. S. Johnson. *Computers and Intractability: A Guide to the Theory of NP-Completeness*. Freeman, New York, 1979.
14. C. Gui and P. Mohapatra. Efficient overlay multicast for mobile ad hoc networks. In *IEEE WCNC*, April 2003.
15. C. Gui and P. Mohapatra. Scalable multicasting in mobile ad hoc networks. In *INFOCOM*, March 2004.

16. P. Gupta and P. Kumar. Capacity of wireless networks. *IEEE Transactions on Information Theory*, IT-46:388–404, 2000.
17. J. Hellerstein, W. Hong, S. Madden, and K. Stanek. Beyond average: Towards sophisticated sensing with queries. In *Intl Workshop on Information Processing in Sensor Networks (IPSN)*, 2003.
18. Y.T. Hou, Y. Shi, H. Sherali, and S. Midkiff. On energy provisioning and relay node placement for wireless sensor networks. *IEEE Transactions on Wireless Communications*, 4, 2005.
19. Q. Huang, C. Lu, and R. Gruia-Catalin. Spatiotemporal multicast in sensor networks. In *SenSys*, November 2003.
20. Y. Huang and H. Garcia-Molina. Publish/subscribe tree construction in wireless ad-hoc networks. In *Proceedings of the 4th International Conference on Mobile Data Management*, January 2003.
21. C. Intanagonwiwat, R. Govindan, and D. Estrin. Directed diffusion: A scalable and robust communication paradigm for sensor networks. In *ACM MobiCom*, August 2000.
22. L. Jia, G. Lin, G. Noubir, R. Rajaraman, and R. Sundaram. Universal approximations for TSP, Steiner tree, and set cover. In *Proceedings of the Thirty-Seventh ACM Symposium on Theory of Computing (STOC)*, May 2005.
23. B. Karp and H. T. Kung. GPSR: greedy perimeter stateless routing for wireless networks. In *Proceedings of ACM Symposium on Mobile Computing and Networking*, pages 243–254, August 2000.
24. H. Kim, T. Abdelzaher, and W. Kwon. Minimum-energy asynchronous dissemination to mobile sinks in wireless sensor networks. In *ACM SenSys*, November 2003.
25. S. Kim, S.H. Son, J.A. Stankovic, S. Li, and Y. Choi. Safe: A data dissemination protocol for periodic updates in sensor networks. In *Workshop on Data Distribution for Real-Time Systems (DDRTS)*, May 2003.
26. Y.B. Ko and N.H. Vaidya. Geocasting in mobile ad hoc networks: Location-based multicast algorithms. In *WMCSA*, Feburary 1999.
27. B. Krishnamachari, d. Estrin, and S. Wicker. The impact of data aggregation in wireless sensor networks. In *Proceedings of the 22nd International Conference on Distributed Computing Systems Worshops*, July 2002.
28. B. Krishnamachari, D. Estrin, and S. Wicker. Modelling data-centric routing in wireless sensor networks. In *IEEE INFOCOM*, June 2002.
29. F. Kuhn, R. Wattenhofer, and A. Zollinger. Worst-case optimal and average-case efficient geometric ad-hoc routing. In *ACM Mobihoc*, June 2003.
30. J. Li, J. Jannotti, D. De Couto, D. Karger, and R. Morris. A scalable location service for geographic ad-hoc routing. In *Proceedings of the 6th ACM International Conference on Mobile Computing and Networking (MobiCom '00)*, pages 120–130, August 2000.
31. S. Madden, M.J. Franklin, J.M. Hellerstein, and W. Hong. Tag: A tiny aggregation service for ad hoc sensor networks. In *OSDI*, December 2002.
32. Madhav V. Marathe, R. Ravi, R. Sundaram, S. S. Ravi, Daniel J. Rosenkrantz, and Harry B. Hunt III. Bicriteria network design problems. In *Automata, Languages and Programming*, pages 487–498, 1995.
33. M. Penrose. On fk-connectivity for a geometric random graph. *Random Structures and Algorithms*, 15:145–164, 1999.
34. B. Sirkeci-Mergen and A. Scaglione. A continuum approach to dense wireless networks with cooperation. In *INFOCOM*, March 2005.

35. V. Vazirani. *Approximation Algorithms.* Springer-Verlag, 2003.
36. J. Xie and R. Talpade. AMRoute: Ad hoc multicast routing protocol. *ACM Mobile Networks and Applications*, 7, December 2002.
37. M. Yang and Z. Fei. A proactive approach to reconstructing overlay multicast trees. In *INFOCOM*, March 2004.
38. Y. Yao and J. Gehrke. The Cougar approach to in-network query processing in sensor networks. *Sigmod Record*, 31(3), September 2002.
39. F. Ye, H. Luo, J. Cheng, S. Lu, and L. Zhang. A two-tier data dissemination model for large-scale wireless sensor networks. In *MOBICOM*, September 2002.

A   Proof of Lower Bounds for MST and SPT

Proof of Theorem 1 (SPT): Consider Figure 2, which depicts a Euclidean metric consisting of a center node (the root), and $n-1$ nodes evenly distributed on a circle. Let R be the radius. The cost of an optimal Steiner for a subset S of the $n-1$ peripheral nodes (plus the root) is at most $R + 2\pi R = (1 + 2\pi)R$, as shown in Figure 2. The SPT for this metric consists of all the "spoke edges", i.e., the $n-1$ edges from the peripheral nodes to the center. If we use SPT as GIST for the corresponding network, the induced tree cost for subset S is $(n-1)R$. Therefore, SPT has a worst case stretch of $\Omega(n)$.   ■

Proof of Theorem 1 (MST): Consider the Euclidean metric in Figure 3. We have two lines of nodes, each node in a line is separated from its neighbors in the line by a distance of 1, and the two lines are separated by a distance of at least 2. At the top end of the line, we have an additional node between the two end-points of the line, at distance of 1 from each of two end-points. We fix the root to be an end-point at the bottom-end of a line. It is easy to see that the unique MST consists of all the distance 1 edges. If the group consists of just the opposite bottom end-point, then the induced subtree of the MST has a cost of $n-1$, while the optimal Steiner tree cost is just 2.   ■

B   Proof of Aggregation Delay and Aggregation Cost

Proof of Theorem 2: Consider Figure 4, and assume that the side length of the square region is equal to 1. We first consider the case where v is not in the $\frac{1}{6} \times \frac{1}{6}$ square where r resides. Let $|rv|$ denote the Euclidean distance between r and node v. Let $\{e_1, e_2, \ldots, e_k\}$ be the set of edges on the path from r to v. By our algorithm, it can be shown that $|e_1| \leq 6\sqrt{2}|rv|$. This is because r is in the center of the topology, therefore $|rv|$ is at least $\frac{1}{6}$. Since $6\sqrt{2}|rv| \geq \sqrt{2}$, the length of the diagonal of the 1×1 square, e_1 is at most $6\sqrt{2}|rv|$. The algorithm divide a square into 9 equally sized smaller squares in each recursion. Therefore, $|e_2| \leq \frac{1}{3} \cdot 6\sqrt{2}|rv|$, $|e_3| \leq \frac{1}{3}^2 6\sqrt{2}|rv|$, and so on. The length of the path of r to v on T is thus $\sum_{j=1}^{k} |e_j| \leq \sum_{j=1}^{k} \frac{1}{3}^{k-1} \cdot 6\sqrt{2}|rv| \leq \frac{3}{2} \cdot 6\sqrt{2}|rv| = 9\sqrt{2}|rv|$.

The argument is similar for the case where v is in the $\frac{1}{6} \times \frac{1}{6}$ square where r resides. Since Algorithm 1 is a recursive construction, the smaller $|rv|$ is, the smaller $|e_1|$ will be. This completes the proof.   ■

Proof of Theorem 3: It can be easily seen that the number of iterations in dividing the square region into smaller ones are bounded by $O(\log_3(D/R))$, where an iteration denotes all the invocations of Algorithm 1 in which the square size are the same. Since $D \leq R \cdot n$ (assuming the network is connected), we have that the number of steps is bounded $O(\log n)$. We now show that for any subset of nodes S, the total length of the edges selected in each iteration of the algorithm on the induced subtree is within a constant factor of the optimal Steiner tree cost of for S.

Consider any iteration, and let H denote the square edge length in this iteration. Let t denote the number of leaders selected in this iteration that are also on the induced subtree of S. It can be easily seen that the cost of the edges on the induced subtree of S that are selected in this iteration is bounded by $O(t \cdot H)$, since each edge is within a constant factor of H. We now place a lower bound on the optimal Steiner tree cost of the subset S. Clearly, each one of the t leaders selected in the iteration corresponds to a distinct leaf node on the induced subtree, all of which are in subset S. It can also be seen that each leaf node is contained in a distinct square with edge length H. Therefore, there exists at least $t/9$ leaf nodes (assuming $t > 9$), where the distance between any two of them is at least H. This is because on the Euclidean plane, each square is surrounded by 8 other squares of the same size. Therefore, the optimal Steiner tree cost to connect the $t/9$ leaf nodes and the root r is at least $(\frac{t}{9} \cdot H)$. Combined with the upper bound of $O(t \cdot H)$, we showed that the cost of the selected edges in any iteration that are on the induced subtree of S is within a constant factor of the optimal Steiner tree cost of subset S.

For the case where $t \leq 9$, the total cost of the selected edges in this iteration is bounded by $O(9H)$. An lower bound for the Steiner tree cost of the corresponding t leaf nodes (and root r) can be easily established since each leaf node has at least a distance of $H/2$ to r. This is because r is always selected as a leader in each iteration, and all the other t leaf nodes are in different squares than the one r resides in. Therefore, the cost of the selected edges in this iteration is again bounded by a constant factor of the optimal Steiner tree cost of subset S. Since we have show that the number of iterations is at most $O(\log n)$, the proof of the theorem is completed. ∎

C   Proof of Lower Bound of GIST

Proof of Theorem 4: The $O(\log n)$ bound for our GIST algorithm is almost tight since no spanning tree can achieve a stretch better than the lower bound of $\Omega(\log n/\log\log n)$. This is due to the lower bound for on-line Steiner tree problem in [2], where nodes are exposed one at a time to be connected instead of being given all at once. The lower bound of $\Omega(\log n/\log\log n)$ is established for the on-line Steiner tree problem on a $n \times n$ grid. If there is an algorithm A that can construct a UST with stretch s, we then can obtain an online algorithm for the online Steiner tree problem that has approximation ratio s. This is because an UST can be built without even knowing the sequence of nodes that are exposed except for the first one, and such UST can be used as a solution for any subset (as an on-line Steiner tree node sequence) and the approximation ratio is s. According to [2], the stretch s is at least $\Omega(\log n / \log \log n)$.   ∎

D GRID_GIST

Let G be an $n \times n$ grid, and r be a root node on G, as illustrated in Figure 5. We show that the stretch on the computed GRID_GIST on G is $O(\log n)$. The problem for constructing GRID_GIST on general unit disk graphs is still open. Each node on G has coordinate (i, j) ($i, j \in I$ can be negative). Without loss of generality, the root node r has coordinate (0, 0). We use (i, j) to denote both the node and its coordinate, and use $((i_1, j_1), (i_2, j_2))$ to denote an edge between node (i_1, j_1) and (i_2, j_2). We assume that each edge $e \in G$ has a unit cost of 1.

Our algorithm computes a GRID_GIST T that is a union of four distinct trees $T_{\mathrm{TR}}, T_{\mathrm{TL}}, T_{\mathrm{BL}}$ and T_{BR}, where T_{TR} connects all the nodes in $\{(i, j) | i \geq 0, j \geq 0\}$ to r, and similarly, T_{TL} for $\{(i, j) | i < 0, j \geq 0\}$, T_{BL} for $\{(i, j) | i < 0, j < 0\}$, and T_{BR} for $\{(i, j) | i \geq 0, j < 0\}$. The algorithm for constructing T_{TR} can be used to obtain $T_{\mathrm{TL}}, T_{\mathrm{BL}}$ and T_{BR} by simply rotating and shifting T_{TR}, i.e., $T_{\mathrm{TL}}, T_{\mathrm{BL}}$, or T_{BR} can be obtained by rotating T_{TR} by 90°, 180° or 270°, then shifting to (-1, 0), (-1, -1), or (0, -1) respectively. Note that root r can be any node on the grid, which implies that size of the top right region can be different from other regions.

Algorithm 3. GRID_GIST for constructing T_{TR}

1: $T_{\mathrm{TR}} \leftarrow \Phi; k \leftarrow 0;$
2: **repeat**
3:      **for all** (i, j) such that $i \mod 2^k == 0$ and $j \mod 2^k == 0$ **do**
4:          **if** ($i \mod 2^{k+1} == 0$ and $j \mod 2^{k+1} == 2^k$) **then**
5:              $T_{\mathrm{TR}} \leftarrow T_{\mathrm{TR}} + ((i, j), (i, j\text{-}1))$
6:          **else if** ($i \mod 2^{k+1} == 2^k$ and $j \mod 2^{k+1} == 2^k$) **then**
7:              $T_{\mathrm{TR}} \leftarrow T_{\mathrm{TR}} + ((i, j), (i, j\text{-}1))$
8:          **else if** ($i \mod 2^{k+1} == 2^k$ and $j \mod 2^{k+1} == 0$) **then**
9:              $T_{\mathrm{TR}} \leftarrow T_{\mathrm{TR}} + ((i, j), (i\text{-}1, j))$
10:         **end if**
11:     **end for**
12:     $k \leftarrow k + 1;$
13: **until** (no edge is picked in this round)

We now describe our algorithm for T_{TR}. Let $BSQ_k = \{(i, j) | i'2^k \leq i < (i' + 1)2^k, j'2^k \leq j < (j' + 1)2^k\}$ denote a level k *bounding square*, where $0 \leq k \leq \lceil \log n \rceil$. The number k is refered to as the *level number* of BSQ_k. Clearly, larger level number k indicates larger square region, which contains smaller regions with level number smaller than k. The bottom left node of each bounding square BSQ is selected as the leader for BSQ. In the special case of $k = 0$, each BSQ_0 is a single node, and its leader is the node itself. If ℓ_k is the leader of BSQ_k, then the leader node ℓ_{k+1} of BSQ_{k+1} containing BSQ_k is the *parent node* of ℓ_k, and ℓ_k is connected to ℓ_{k+1} using a shortest path on the grid. Note that in the GRID_GIST, ℓ_k does not connect to ℓ_{k+1} using a dedicated link, as in the GIST tree. Instead, ℓ_k connects to a single node in ℓ_{k+1}'s BSQ_k region using a single edge (ℓ_{k+1} is a leader bounding squares in level $k + 1$ as well as $k, k-1, \ldots$).

Among the bounding squares that a node v acts as a leader at different levels, we use $MaxLevel_v$ to denote the maximum level number of these bounding squares. Our Algorithm D proceeds by linking leader ℓ_k with its parent leader ℓ_{k+1} , first walking down then walking left till ℓ_{k+1} is met, excluding edges on this path that is already selected. Formal definition of the algorithm is presented in Algorithm D. Finally, our algorithm GRID_GIST for computing a GRID_GIST on G is as follows: Output each of the four subtree; output three additional edges $((-1, 0), (0, 0)), ((-1, -1), (-1, 0)), and ((0, -1), (0, 0))$, if such edges exist (subtree is not empty). Figure 5 is illustration of Algorithm D for computing T_{TR}, where the thickness of an edge e indicates the level of the bounding squares e connectes. For example, $((1, 1), (1, 0))$ and $((1, 0), (0, 0))$ connect BSQ_0 squares; $((2, 2), (2, 1))$ and $((2, 0), (1, 0))$ connects BSQ_1 squares; $((4, 4), (4, 3))$ and $((4, 0), (3, 0))$ connectes BSQ_2 squares. For Algorithm D, we have the following theorem,

Theorem 5. *Algorithm GRID_GIST computes a universal Steiner tree with $O(\log n)$ stretch in polynomial time.*

To prove Theorem 5, we first establish the following two lemmas.

Lemma D1. *Algorithm GRID_GIST finishes in polynomial time, and it computes a spanning tree on G.*

Proof: It can be easily seen that the algorithm terminates in polynomial time. Since in each for-loop, at least one edge is picked; otherwise, the outer repeat-loop would stop.

We now show that T_{TR} is a tree connected to root r, and T_{TR} spans all the nodes in $\{(i, j)|i \geq 0, j \geq 0\}$. The proofs for other sub-trees are similar, except that each sub-tree is connected to r through one of the three additional edges. In Algorithm D, when k increases from 0, any combination of integers i and j will satisfy the condition in Line 3 at least once. If one of the contitions in Lines 4, 6 or 8 is further satisfied by (i, j), an edge is added for the node (i, j); furthermore, (i, j) will not satisfy Line 3 any more after k increases. If none of the conditions in Line 4, 6 or 8 is satisfied, (i, j) will satisfy Line 3 again when k increases. For example, $(4, 4)$ satisfies Line 3 when $k = 1$, but does not satisfy either Line 4, 6 or 8; however, $(4, 4)$ will satisfy Line 3 again when $k = 2$, and will further satisfy Line 6, which leads to the addition of edge $((4, 4), (4, 3))$. Thus, each node in $\{(i, j)|i \geq 0, j \geq 0\}$ is connected to exactly one other node, which eventually leads to the root, and there no loop is formed (k is increasing). ∎

Lemma D2. *Let T be a GRID_GIST tree computed by Algorithm D. Given $S \subseteq V$, the inducted tree $T_{S+\{r\}}$ of $S + \{r\}$ on T has a cost of at most $O(\log n) \cdot Opt_{S+\{r\}}$.*

Proof: We first place a bound on the length of the path between two leaders on T. Let ℓ be the leader of bounding square BSQ_k, and ℓ' be the leader of bounding square $BSQ_{k'}$, where BSQ_k contains $BSQ_{k'}$. Therefore, ℓ is an ancestor of ℓ' on T. We have the following

$$d_T(\ell, \ell') = \sum_{e \in P} |e| \leq \sum_{i=1}^{k} (2 \cdot 2^{i-1}) \leq 2 \cdot 2^k. \tag{1}$$

Let $\{v_1, v_2, \ldots, v_t\}$ be the set of leaders on $T_{S+\{r\}}$, such that $\forall 1 \leq i \leq t, MaxLevel_{v_i} = k$. As described in our algorithm, the path from any node v to r on T consists of at most $\lceil \log n \rceil$ sub-paths, each of which is a path from the leader of a smaller bounding square to its parent leader of a larger bounding square. Therefore in order to show $T_{S+\{r\}} \leq O(\log n) \cdot Opt_{S+\{r\}}$, we only need to prove that the cost of connecting the set of nodes $\{v_1, v_2, \ldots, v_t\}$ to their parent leaders is bounded by $O(Opt_{S+\{r\}})$. Since $T_{S+\{r\}}$ is the induced tree of S on T, each node v_i in $\{v_1, v_2, \ldots, v_t\}$ can identify a node $u_i \in S$ such that the set of nodes $\{u_1, u_2, \ldots, u_t\}$ are distinct and are all leaf nodes of $T_{S+\{r\}}$. Since the path connecting v_i to its parent leader on T is at most $2 \cdot 2^k$, we have an upper bound of $2t \cdot 2^k$ on the total cost of connecting the set of nodes $\{v_1, v_2, \ldots, v_t\}$ to their parent leaders.

We now place a lower bound on $Opt_{S+\{r\}}$. Let $q > 4$ be any constant, and let $H \subseteq \{v_1, v_2, \ldots, v_t\}$ be any maximal subset such that: 1) $\forall v, v' \in H$, $d_G(v, v') \geq q \cdot 2^k$, and 2) $\forall u \in \{v_1, v_2, \ldots, v_t\} - H$, u can find a node $v \in H$ with $d_G(u, v) < q \cdot 2^k$. Since $\forall 1 \leq i, j \leq t, d_G(v_i, v_j)$ is at least 2^k, we have that for any $v \in \{v_1, v_2, \ldots, v_t\}$, the number of other nodes in $\{v_1, v_2, \ldots, v_t\}$ that is within distance $q \cdot 2^k$ of v is at most q^2. Therefore, the cardinality of H is at least t/q^2. From Equation 1, we have that $d_G(u_i, v_i) \leq 2 \cdot 2^k$, since $d_G(u_i, v_i) \leq d_T(u_i, v_i)$ and $MaxLevel_{v_i} = k$. This directly leads to the fact that $d_G(u_i, u_j)$ is at least $(q - 2 \cdot 2) \cdot 2^k$, for any $v_i, v_j \in H$. Combined with the cardinality lower bound of H, we established a lower bound of $(t/q^2) \cdot (q - 4) \cdot 2^k$ for $Opt_{S+\{r\}}$.

Given the upper bound $2t \cdot 2^k$ on the total cost of connecting the set of nodes $\{v_1, v_2, \ldots, v_t\}$ to their parent leaders, the lower bound of $(t/q^2) \cdot (q - 4) \cdot 2^k$ on $Opt_{S+\{r\}}$, and the fact that there are at most $\lceil \log n \rceil$ level numbers, we completed the proof of this lemma. However, we note that the approximation factor is not optimized in this proof. ∎

Theorem 5 follows as a direct corollary of Lemma D1 and Lemma D2.

Distributed User Access Control in Sensor Networks*

Haodong Wang and Qun Li

Department of Computer Science
College of William and Mary
Williamsburg, VA 23187-8795, USA
{wanghd, liqun}@cs.wm.edu

Abstract. User access control in sensor networks defines a process of granting user the access right to the information and resources. It is essential for the future real sensor network deployment in which sensors may provide users with different services in terms of data and resource access. A centralized access control mechanism requires base station to be involved whenever a user requests to get authenticated and access the information stored in the sensor node, which is inefficient, not scalable, and is exposed to many potential attacks along the long communication path. In this paper, we propose a distributed user access control under a realistic adversary model in which sensors can be compromised and user may collude. We split the access control into local authentication conducted by the sensors physically close to the user, and a light remote authentication based on the endorsement of the local sensors. Elliptic Curve Cryptography (ECC), a public key cryptography scheme, is used for local authentication. We implement the access control protocols on a testbed of TelosB motes. Our analysis and experimental results show that our scheme is feasible for real access control requirement.

1   Introduction

Access control defines a process of identifying user and granting user the access right to information or resources. Sensor network is a computing platform for users to collect data, transmit data, and process data. The access control pertaining to sensor network predominantly aims to protect the network usage and collected data. Unauthorized user should not be allowed to use the network since network bandwidth is very limited and, more importantly, the battery power of each node may be depleted after malicious users aggressively effuse messages to the network. The data collected or processed, many times, is classified so that data of different classifications requires security clearance for authorized access. For example, a high rank officer may need to know more information about the

---

* This work was partially supported by the U.S. National Science Foundation under grant CCF-0514985.

P. Gibbons et al. (Eds.): DCOSS 2006, LNCS 4026, pp. 305–320, 2006.
© Springer-Verlag Berlin Heidelberg 2006

field deployment than a soldier. In another scenario, information may be sensitive compartmented so that users have to be denied of access to the data that is beyond his access right. An example would be a user is authorized to access the data from the sensors in his office, but not other people's offices.

To achieve access control, it is essential for sensor nodes to authenticate the identities of the requesters. This paper aims to explore an efficient and secure authentication scheme for the sensor nodes. A natural way for the authentication check is to use a centralized mechanism. After receiving a request, the sensor node sends the user information to the base station. Then the base station decides whether the access is granted or not and replies the result to the sensor node. This solution may yield a good security result because of the fact that the base station is considered secure, and the communication channels between sensors and the base station are assumed secure. However, this scheme suffers two major problems. First, the centralized authentication requires at least one round-trip communication between the sensor and the base station. If a number of users are accessing the network at the same time, the authentication traffic may easily cause network congestion. Second, this authentication pattern is vulnerable to adversary's DoS attacks. The sensor nodes have no knowledge about user access right until they get replies from the base station. The adversary can easily launch DoS attacks by forging a large number of user access requests, which will in-turn trigger the same amount of authentication traffic. The consequence will severely saturate the network and quickly deplete the sensor node power.

This paper gives a thorough exploration for sensor network data access control problem in a general setting. We consider a data access scenario that a user can access in-network stored data at any location from anywhere in the network, which includes local data access from user's nearby sensors and remote data access. Moreover, we consider access control problem in a much harsher environment in which the users may collude and sensors may be compromised. Compromised sensors can get the information from the user authentication process and may disclose this information to an adversary, which may potentially help the adversary to gain more access privileges. Colluding users may analyze their information and design a scheme to counteract the access control system. Besides, we also addresses node duplication attack and DoS attack by inundating authentication messages to the network.

It is our belief that our more general data access model and realistic adversary threat model define a very realistic problem for future sensor deployment. Our work has following four contributions. First, we propose a practical and scalable certificate-based local authentication based on ECC. Public key cryptography eliminates the complicated key management and pre-distribution required by symmetric key schemes, and provides a very clean interface between the user and sensors. The advantage of certificate-based authentication is that sensors do not need the storage for user's public keys or a third party for public key verification. User public keys can be constructed from user certificates and published system information. Second, we propose a novel group endorsement scheme to authenticate a user locally by a group of sensors and transfer the endorsement to the remote sensor. This scheme is resilient to limited number of compromised sensors and the DoS attack launched in the form of remote authentication. Third, our scheme eliminates the possibility of user collusion attack. The polynomial based secret sharing scheme proposed in [18] suffers user collusion attack. The collusion by a number of users can easily reconstruct the secret polynomial and reveal the system secrecy. Our certificate-based authentication is resilient to any user collusion attack. Fourth, we show our scheme is feasible in real sensor network deployment. We have implemented both local authentication and remote authentication on TelosB motes, which are based on our implementation of 160-bit ECC security primitives. Since the TelosB hardware multiplier is disabled in TinyOS, the computation is longer that it should be. It takes 3.1s to generate a public key and 10.8s to conduct local authentication.

2 Related Work

We believe that, with fast expanding sensor network technologies, more services will be available to allow direct interactions between users and sensor nodes. Obviously, the new communication paradigm poses more security challenges for small and power constrained sensor nodes. Different from the security problem in user access control we address in this paper, most related researches focus on secure and resilient communication links and resource management inside the networks.

Perrig *et al.* [12] construct μTesla and introduce the asymmetric mechanism through a delayed symmetric keys disclosure: the base station broadcasts an encrypted message first, and then releases the secret key in scheduled time frame. Although KDC-based schemes suffer the scalability problem, broadcasting is still the basic, efficient to distribute or revoke secret keys in sensor networks.

Eschenauer and Gligor propose a random graph based key pre-distribution scheme [7]. The scheme assigns each sensor a random subset of keys from a large key pool, and allows any two nodes to find one common key and use that key as their shared symmetric key. Based on their contribution, a number of researches [3, 5] have delivered to strengthen the security and improve the efficiency. Since each sensor node only needs to store a small number of keys, the random graph based schemes have the advantage of scalability. However, in a sparse network or non-uniform distributed network, the key establishment could be difficult because a number of sensor pairs may not successfully finish pairwise key establishment.

Besides the above two types of security schemes, a number of research teams focus on the group key and authentication problems [17, 15, 1, 14, 6, 2]. Ye *et al.* [17] design a Statistical En-Route Filtering (SEF) mechanism to detect and drop false reports. The idea is to use probabilistic key sharing to authenticate the legitimate messages on the routing path. However, SEF cannot be used to authenticate the message sender because the remote sensor does not have enough knowledge (as the sink) to verify the message source.

Zhang *et al.* [18] propose several schemes to restrict and revoke the access privilege of a mobile sink. Their approaches are based on Blundo's scheme to establish secret key between the mobile sink and sensor nodes, and then use Merkle tree technique to reduce the overhead. The limitation of the scheme is that the mobile sink's moving track has to be predetermined by the base station. Compared with our scheme, we address a more general user/sensor communication problem. The mobile sink can be regarded as one type of special users in our scheme.

3   System Model

We consider a large scale wireless sensor network deployed in a variety of environments, e.g., at a hostile battlefield, in an office building, or in a national park. Data access to the stored data on each node is protected according to the attributes of the data, e.g., data type (temperature, light, noise, etc.), data location, data collection time, and so on. For a certain data, only authorized user can access the data from the storing node. Since the data is distributed in the entire network instead of in a central position, data protection by relying on a powerful sink node with all data access authorization information and computational power is not possible. Instead, data access authorization should be done in a distributed fashion accordingly. After the data access has been authorized, data access is granted to the user and data is transfered to the user.

A user equipped with a powerful computing device, such as a PDA, interacts with the sensor network for data query and retrieval and maybe network control such as network reconfiguration or sensing mode change. The PDA is the interface for the user to talk to the sensor network. The computing device is more powerful than the sensor nodes, so it is capable of more computationally intensive tasks. User can query data at any location of the network through sensor node relay. The data access capability, however, must be granted by a central authorization center before data access. A data access list is associated with the user about the types, locations, and the durations of the authorized data access. This information is encrypted in a way that the user is unable to forge and can be authenticated by the sensor holding the requested data.

The sensor network is managed by a Key Distribution Center (KDC), which is responsible for generating all security primitives (i.e., random numbers, one-way hash function, message authentication code (MAC), access list) and revoking users' access privilege if necessary. KDC distributes secret keys through the base stations. To access the sensor network, users need to apply for the access permission from KDC. KDC maintains a user access list pool and associated user identifications. The access list defines the user's access privilege. A typical access list is composed of *uid* and *user access privilege mask*. *uid* is a unique number to identify the user. *user access privilege mask* is a number of binary bits; each bit represents a specific information or service. An access list example is shown in Fig 1. The information stored at the sensor nodes is divided into multiple access privilege levels. The user with a lower access privilege is not allowed to get the

64 : 23 : 00 : 07 : E9 : 26 : F1 : A5

uid   privilege mask   timestamp

Fig. 1. An example of user access list. The access list is composed of three parts: *uid*, *access privilege mask*, and *timestamp*. *uid* is a unique number assigned to each user. *access privilege mask* is to define the user's access privilege to the system information. *timestamp* specifies the access list is only valid in a certain time frame.

information that requires the higher privilege. We assume the users can securely acquire their access lists from KDC through out-of-band secure communication channels. Once a user passes the authentication check, the sensor nodes provide their local information to the user. If the required information is not available locally, for the reason we will discuss later, a group of sensor nodes have to collaborate and request the information from the remote sensor which holds the information.

An adversary is assumed to use all possible means to access the data that is not authorized to him. He can eavesdrop message transmission to extract transmitted information or carry out message replay. Message eavesdrop and replay are easy to handle, as discussed by many papers, by using regular message encryption and including message sequence or time information. More hazard is created when nodes are compromised by the adversary who is able to garner all the information stored in the sensors. It is even worse that the adversary may inject his own program to the compromised sensors, which, under the control of the adversary, pretend to be trustworthy gaining as much information as possible. A user may also collude with the adversary for mutual benefit by attacking the access control system. The base station and the central authorization center cannot be compromised, however.

We mainly consider the following two potential attacks. First, Compromised sensors may capture much information and give to an unauthorized user so that that user may access data by impersonating another user. Second, user collusion may help users to subvert the system and gain more access right than that of anyone among the colluding users. We assume that at most t sensors can be compromised. The assumption is reasonable because compromising sensors takes time and effort. On the other hand, we assume unbounded number of users can collude since it is not hard for mischievous users to share information and orchestrate an aggregated analysis to the collected information. The fact that a compromised sensor is hard to identify prevents a user from trusting any of the sensors. A user may have to disclose information for authentication, but the revealed information has to be specific to the sensor in contact and should not be used for authentication at another sensor.

We do not explicitly address the introduction of duplicated compromised sensors. However, since the duplicated compromised sensors do not introduce more information to the adversary, our carefully designed protocols do not enable the adversary to access the data from an uncompromised sensor.

4   Proposed Access Control Schemes

The user may request data stored locally or in a distant sensor. We first define following two types of sensor nodes. The sensor nodes which are directly within the contact range of the user are called *local* sensor nodes. The sensor nodes which cannot establish direct communication link with the user but hold the requested information are called *remote* sensor nodes.

In this section, we first propose a public-key cryptography based local access control scheme. Then we develop a remote access control approach (we assume that the ID of the remote sensor for data access is known by some scheme that is beyond the scope of this paper, e.g., resource discovery or geographic or location-based routing). Finally, we provide the security analysis for both schemes.

4.1   PKC Based Local Authentication

Public-key cryptography has been used extensively in data encryption, digital signature, user authentication, etc. Compared with the popular symmetric key cryptography widely used in sensor network, public-key cryptography provides a more flexible and simple interface requiring no complicated key pre-distribution and management as in symmetric-key schemes. It is a popular belief, however, in sensor network research community that public-key cryptography is not practical because the required computational intensity is not suitable for resource constrained sensor nodes. The nascent exploration seems to disabuse of the misconception. The recent progress in 160-bit Elliptic Curve Cryptography (ECC) implementation [9] on Atmel ATmega128, a CPU of 8Hz and 8 bits, shows that an ECC point multiplication takes less than one second, which proves public-key cryptography is feasible for sensor network security related applications.

We present our ECC based local authentication scheme as follows. KDC selects a particular elliptic curve over a finite field $GF(p)$ (where p is a prime), and publishes base point P with order q (where q is also a large prime). KDC picks a random number $x \in GF(q)$ as the system private key, and publishes its corresponding public key $Q = x \times P$. Given point P and Q, it is computationally infeasible to get system secret x.

A straightforward user authentication scheme can be described as follows. The user uses her private key to sign her access list and sends to the sensors. The sensors just verify the signature by using user's public key. However, it is difficult for the sensors to find an authorized third party to certify that the user is who she claims to be. To solve this problem, we adopt the certificate-based authentication in our local authentication scheme. To access the sensor network, the user has to present her certificate first. Based on the certificate, the sensors generate user's public key, and then use the derived public key to encrypt a random number as the challenge. If the user can successfully decrypt the message, then the *local* sensors are convinced that the user's certificate is legitimate.

Initially, the user comes to KDC to apply for an access list to visit the sensor network. KDC picks a random number $c_A \in GF(p)$, and then calculates the user's public key constructor $C_A = c_A \times P$. Based on the user's request, KDC issues a proper access control list ac_A, and attaches it to public constructor C_A as the certificate, denoted as T_A. Meanwhile, a digest e_A is generated for T_A, where $e_A = H(T_A)$ (H is a $\{0,1\}^* \rightarrow \{0,1\}^q$ hash function). Then, KDC constructs Alice's private key $q_A = e_A c_A + x$ and public key $Q_A = e_A \times C_A + Q$. Note q_A and Q_A satisfy $Q_A = q_A \times P$. Finally, Alice holds q_A, Q_A and T_A. We assume above procedure is conducted at an out-of-band secure channel.

The user authentication protocol is illustrated in Fig. 2. We denote s_l as a local sensor. When the user approaches a sensor node s_l, she sends her access request with certificate T_A. Given certificate T_A, s_l constructs user's public key $Q_A = e_A \times C_A + Q$. To verify the user indeed holds private key q_A, node s_l uses the challenge as follows. s_l selects a random number $r \in GF(p)$ (to be used as the session key with the user), and calculate its hash $H(r)$ over GF(p). Node s_l then generates temporary public key $Y_r = H(r) \times P$, and computes $Z_r = H(r) \times Q_A$. Next, s_l encrypts the session key by doing $r \oplus X(Z_r)$, where $X(Z_r)$ is the X coordinate of point Z_r. Finally, s_l sends ciphertext $\langle z_r, Y_r \rangle$ to the user, attached with the MAC of a nonce (N_A), $MAC(r, N_A)$.

With private key q_A, the user can regenerate Z_r because $q_A \times Y_r = q_A \times H(r) \times P = H(r) \times Q_A = Z_r$. She then decrypts session key $r = z_r \oplus X(Z_r)$, and verifies if $Y_r = H(r) \times P$. If yes, She uses r as the session key to generate MAC for nonce N_A concatenated with her access privilege ac_A, and sends to s_l.

Local sensor s_l decrypts the MAC message and verifies N_A and ac_A. A successful verification proves that the user is the owner of certificate T_A. Finally, s_l replies the information requested by the user, which again is encrypted by session key r.

$$
\begin{array}{rl}
user \rightarrow s_l : & T_A = (C_A | ac_A) \\
s_l \text{ computes } : & Q_A = e_A \times C_A + Q \\
: & \text{picks a random } r \in GF(p) \\
: & Z_r = H(r) \times Q_A, \\
: & Y_r = H(r) \times P, \\
: & z_r = r \oplus X(Z_r), \\
: & MAC(r, N_A). \\
s_l \rightarrow usesr : & z_r, Y_r, MAC(r, N_A) \\
user \text{ computes } : & q_A \times Y_r = q_A \times H(r) \times P = Z_r \\
: & X(Z_r) \oplus z_r = r \\
: & \text{decrypts } MAC(r, N_A) \\
user \rightarrow s_l : & MAC(r, N_A | ac_A) \\
s_l \rightarrow user : & MAC(r, reply)
\end{array}
$$

Fig. 2. User access list authentication protocol. We let s_l be the local sensor, T_A be the user certificate, which includes a public-key constructor C_A and an access list ac_A.

4.2   Remote Access Control

In remote access control, the *remote* sensor node cannot directly contact the user due to the limitation of radio transmission range. Therefore, the user queries have to travel multiple hops to reach the *remote* sensor. With this communication pattern, the authentication schemes used in local access control cannot be applied on remote access control. In other words, it is improper for the user to directly contact the *remote* sensor. Otherwise the adversary can easily take the advantage and launch the bogus data injection attack to deplete the sensor network. With the above security concern in mind, we develop a remote access scheme that uses *local* sensors to endorse the user query to the *remote* sensor. Since it is widely accepted [11, 12] that a single sensor node cannot be trusted, the user's remote access request has to be endorsed by k local sensor nodes, where k is a system parameter. We assume the adversary cannot compromise k sensors at a time. Any user remote access query without k local endorsements will be dropped immediately by either forwarding sensor nodes or the *remote* sensor. A caveat is that some sensors may be compromised if a valid user cannot be authenticated by a group of sensors. In that case, the user can move to find another group of sensors for authentication or report the failure to the base station for analysis.

The requirement of *local* sensor endorsement raises a new security challenge: how does the remote sensor verify that the user is indeed endorsed by k *local* sensors? If each local endorsing sensor can share a secret with the *remote* sensor, then the endorsement can be easily verified by the *remote* sensor. We use polynomial-based scheme for secret sharing between the local and remote sensors. More specifically, the KDC randomly generates a bivariate t-degree polynomial $f(x, y) = \sum_{i,j=0}^{t} a_{ij}x^i y^j$ over a finite field $GF(q)$, where q is a prime number and $a_{ij} = a_{ji}$. The polynomial has the symmetric property such that $f(x, y) = f(y, x)$. In practice, we select $t = k - 1$ so that the polynomial can not be reconstructed by the adversary with the assumption that the adversary cannot compromise up to k sensors. To endorse a user access list, each local sensor can encrypt the access list with the key shared with the remote sensor, computed by substituting x and y with the sensor IDs. This scheme, however, has to provide the remote sensor with the IDs of the local sensors for verification, which leads to a long message. In order to reduce the message size, before the deployment, sensor nodes are divided into k groups $\{g_1, g_2, \cdots, g_k\}$, where g_j $(1 \leq j \leq k)$ is a group ID. Besides the group ID, each sensor i has its unique sensor ID s_i. From now on, we also denote a sensor node as s_i^j, where s_i is the sensor ID, and j means it is belong to group g_j. During configuration procedure, each sensor s_i^j is pre-loaded with two shares of polynomial, $f(x, s_i)$ and $f(x, g_j)$. Given the *remote* sensor ID s_r, a local sensor $s_{i_1}^{j_1}$ can establish a pairwise key with the *remote* sensor by plugging $s_r^{j_r}$ in $f(x, g_{j_1})$. And, the remote sensor can also generate the pairwise key by plugging group ID g_{j_1} in its $f(x, s_r)$. To use group ID instead of sensor ID, we can achieve a shorter message due to a small number of groups. For the remote sensor to check the authentication list, we attach a bitmap for the groups in the message showing which group IDs are

user finds k local sensors s_i^j with different j

   $user \to s_1, \cdots, s_{k'} (k' \geq k)$: bcast. request

   $s_1, \cdots, s_{k'} \to user$: group id

   $user \to s_{p_1}, \cdots, s_{p_k}$: confirm request

for (each sensor $s_{p_i}^{g_i}, i = 1, 2, \cdots, k$)

   $s_{p_i}^{g_i}$ authenticate user access list T_A

   $s_{p_i}^{g_i} \to user : mac_i = MAC(f(s_r, g_i), ac_A)$

user computes: $mac = H(mac_1 || \cdots || mac_k)$

$user \to s_r : MAC(mac, ac_A || N_A) || ac_A ||$ group list

s_r : compute $f(g_1, s_r), \cdots, f(g_k, s_r)$

s_r : reconstruct $mac = Hash(mac_1 || \cdots || mac_k)$

s_r : decrypt and verify ac_A

$s_r \to user : MAC(mac, reply || N_A || N_B)$

Fig. 3. The polynomial based remote access control protocol

used for authentication. We incorporate the remote sensor ID in the polynomial computation rather than the group ID of the remote sensor to avoid the attack due to the scenario that a compromised sensor has the same group ID with the remote sensor and then can decode the shared keys between the local sensors and the remote sensor.

The remote access control protocol is described in Fig. 3. To start a remote access procedure, the user has to find k endorsing sensors s_i^j such that no two sensors have the same group ID. The user first broadcasts the remote access request, and the local sensors receiving the request reply with their group ID. The user then select k local sensors with different group ID to form an endorsing sensor group. Note the user may have to broadcast the request several times due to the possible transmission collisions. Then, each endorsing sensor conducts the local authentication as described previously. After the user has been authenticated, sensor s_i^j computes the pairwise key $f(s_r, g_i)$ with the remote sensor, and uses the key to encrypt user's access list ac_A. Note only the access list part of certificate T_A is encrypted because the *remote* sensor does not need user's public key constructor C_A. The user collects k MACs from the endorsing sensors and generates a hash digest, $mac = H(mac_1 || \cdots || mac_k)$, where $g_1 < g_2 < \cdots < g_k$.

After computing the hash digest, the user encrypts her access list ac_A and N_A with mac. Again, N_A is a nonce to guarantee the message freshness. Then, the user sends it along with her access list ac_A and the local endorsing sensor group list, to the *remote* sensor.

When a *remote* sensor (denoted as s_r) receives the access request from the user, s_r retrieves the information in the group list and user access list to reconstruct the MAC digest as shown in the protocol, and then decrypts the user's access list ac_A. If the decrypted access list matches the one provided by the user, it proves that the user has already been authenticated by k local sensors. Sensor s_r replies the user with the requested information, along with nonce N_B randomly picked by s_r. Again, all data is encrypted by *mac*.

4.3   Security Analysis

In both access control schemes, the authentication messages are encrypted by MAC algorithm in the access control protocol, except the user certificate. As long as the MAC algorithm is secure (such as RC5[13]), and the secret key is large enough (at least 64 bits), any number of compromised sensors cannot break the ciphertext in the messages.

In the local authentication, the sensor nodes can not capture any secret from the user, nor can the user gain more access privilege than granted due to the nice security features of public-key cryptography. The 160-bit elliptic-curve crypto-system is considered to have the same security level as 1024-bit RSA. Given an elliptic curve E over finite field F, to find system secret x from the relation $Q = xP$ (where P, Q are published system parameters) is equivalent to solve the discrete logarithm problem, which is considered computationally infeasible. During the local authentication procedure, user's certificate T_A including access list ca_A is transmitted in plaintext. The malicious sensors may duplicate the user certificate, or the adversary may capture the certificate by eavesdropping. The certificate information, however, can not help the adversary to impersonate the user and get the data service. The reason is that the local sensors use user's public key to encrypt the challenge (random number r). It is easy for the adversary to calculate the public key given the stolen certificate, but it is computationally infeasible to acquire the associated private key. As the result, the adversary is not able to correctly respond the challenge, so her access request will be rejected by any local sensor. Due to the same reason, the user cannot forge or alter her access list to acquire higher level access privileges or to extend the allowed access time period. Otherwise, the user will not be able to decrypt the challenge message from the local sensor because she does not have the private key associated to the certificate she claims. More importantly, the certificate-based local authentication effectively defends against user collusion attacks. The collusion among any number of users does not jeopardize the system secret for the reason explained above.

The security features of our remote access scheme lie on the local sensor group endorsement. The combination of our local endorsement scheme with existing false report filtering schemes, such as SEF [17] and IHA [14], can effectively prevent the potential DoS attacks. In our scheme, users are not allowed to send requests directly to the remote sensor. Any remote access request has to be endorsed by k local sensors. Since the adversary can not compromise up to k sensors (the system assumption), there is no way for an illegitimate user to get k genuine MACs to access the remote sensor. If the adversary attempts to forge k MACs, the bogus request will be immediately dropped by forwarding sensors in false report filtering. Again, the user still can not alter or forge her access list in the remote access request. The local endorsing sensors generate the MACs using authenticated user access list. If the user forges her access list in the remote access request, the MAC verification at the remote sensor will fail, and the remote request will be rejected.

5   Experimental Results

To evaluate the proposed access control schemes, we have implemented both local access control and remote access control scheme on TelosB (TPR2420) motes, the latest research oriented mote developed by UC Berkeley. TelosB is powered by MSP430 microcontroller. MSP430 incorporates an 8MHz, 16-bit RISC CPU, 48K bytes flash memory (ROM) and 10K RAM. The RF transceiver on TelosB is IEEE 802.15.4/ZigBee compliant, and can have 250kbps data rate. To simplify the experiments, we have implemented the user module on TelosB motes instead of PDAs.

5.1   Metrics and Methodology

We use four metrics: authentication time, computation cost, communication cost, and power consumption, to evaluate the performance of access control protocols. The authentication time measures user perceived waiting time from sending out the access request to receiving the authentication confirmation. Computation cost is the amount of energy consumed in data processing. Similarly, communication cost is the energy used by RF transceiver. The power consumption is the total amount of energy used by all participating sensor nodes to assist one user access request.

Table 1. The amount of current draw on different operations for TelosB motes

Operation	Normal	Max
MCU On, Radio Off	$1.8mA$	$2.4mA$
MCU On, Radio Rx	$21.8mA$	$23mA$
MCU On, Radio Tx	$19.5mA$	$21mA$

The energy consumption E can be calculated by $E = U \cdot I \cdot t$, where U is the voltage, I is the current and t is the time duration. TelosB motes are powered by two AA batteries, so U is approximately equal to 3 volts. The current value varies in different operations as shown in Table 1 (abstracted from [4]). We use authentication time as the time duration for MCU data processing. And communication time can be estimated by following way. Given 250kbps radio transmission rate, and 38 bytes in each packet, it takes one sensor node 38 × $8bits/250kbps = 1.2ms$ to transmit or receive a data packet. Without considering message loss and retransmission, the total transmission time is the product of $1.2ms$ with the number of packets.

5.2   Experiment of Local Access Control

We have implemented 160-bit ECC cryptosystem on TelosB motes. We choose SECG recommended 160-bit elliptic curve, secp160r1, in our ECC implementation because large integer multiplication and reduction over prime number finite field can be more effectively optimized than those over binary finite field. The most expensive operation in ECC exponentiation is point multiplication. To achieve the better performance as possible, we have adopted a number of techniques including hybrid multiplication, modular reduction over pseudo-Mersenne prime field, Great Division and mixed Jacobian Coordinate. Due to the space limit, we omit the detail implementation and corresponding optimization of our ECC implementation on TelosB motes. Interested readers may refer to [16] for detail explanation. On average, it takes 3.1 seconds for a TelosB sensor mote to do a fixed point multiplication, and 3.5 seconds to do a random point multiplication. Note this performance is achieved under the circumstance that TelosB micro-controller's hardware multiplier is disabled in TinyOS.

Our local access control implementation strictly follows the protocol presented in section IV except that the data encryption/decryption part is not implemented due to the reason that TinySec (which provides block-cipher module) does not work with CC2420 radio module on TelosB, but it does not affect our performance evaluation because encryption/decryption overhead is negligible (e.g., in RC5, the most expensive step (key setup) only costs 4ms on ATmega128 [8]) compared with ECC exponentiation.

The user certificate T_A has 48 bytes, including 40-byte public key constructor and 8-byte access list. The challenge from sensor nodes has 80 bytes, including a 40 byte ECC point, 20 byte z_r and a 20 byte ciphertext. Since one TelosB packet only has 28 byte payload, the user has to use multiple packets to deliver the certificate. In total, user needs to send four messages (three messages to deliver user certificate, the forth one to response sensor's challenge). Similarly, the local sensor also needs to send four messages to deliver the challenge. We use challenge generation time as our authentication delay. The challenge generation time is user perceived delay from sending out the access request to receive the challenge from the sensor. We exclude the user response time from the authenticate delay because the user usually carry much more powerful devices in the real world, so the response time is negligible compared with sensor processing time.

Our experiment results show that a challenge generation costs 10.8 seconds on average. Obviously, computation delay dominates communication delay in this procedure. Recall that a sensor node needs to perform two ECC random point multiplications and one fixed point multiplication to generate a challenge. The three point multiplications combined already contribute 10.1 second delay. The communication delay to send/receive 8 packets only has $8 \times 1.2ms = 9.6$ milliseconds. The power consumption for the computation is 58.3mJ, while the energy cost for the communication is 0.59mJ.

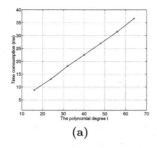 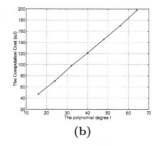

(a)   (b)

Fig. 4. (a). The time consumption to generate a pairwise key from the polynomial.
(b). The power consumption to generate a pairwise key.

5.3   Experiment of Remote Access Control

The essential part of the experiment of remote access control is the polynomial
based local endorsement scheme and MAC recovery at the remote sensor. We
are particularly interested in the performance of the t-degree polynomial computation in sensors. Given a share of the polynomial $f(x) = a_0 + a_1 x + \cdots + a_t x^t$
over $GF(q)$, the computation of $f(x)$ requires t modular multiplications and t
modular additions, plus the computation of values $x^2, \cdots, x^t$. A typical cryptosystem (e.g., RC5) suggests q should be at least 64 bits. Therefore, t 64-
bit $\times$ 64-bit modular multiplications are required to compute the polynomial.
On TelosB's 16-bit CPU platform, each 64-bit $\times$ 64-bit multiplication costs 16
word multiplications. To reduce the computational cost, we adopt the simplification proposed in [10]. The simplification is based on the fact that variable
x is either sensor ID or group ID, which is normally a 16-bit integer. We can
use another finite field $GF(q')$ for $x, x^2, \cdots, x^t$. Therefore, the modular multiplication in polynomial $f(x)$ is always performed between a 64-bit integer
and 16-bit integer. As the result, the cost of multiplication is reduced by four
times.

The modular reduction operation is as important as multiplication. Each multiplication must be followed by a reduction operation. To further reduce the
computational cost, we pick a pseudo-Mersenne prime as q because modular
reduction cost on field of a pseudo-Mersenne prime can be optimized to a negligible amount. A pseudo-Mersenne prime can be represented as $q = 2^m - \omega$,
where $\omega \ll 2^m$. Given a $2m$-bit multiplication result $B = (b_1, b_0)$, (b_1, b_0 are two
m-bit halves), the reduction can be computed based on the congruence $2^m \equiv \omega$:
$(b_1, b_0) = b_1 * \omega + b_0 \rightarrow (b_1', b_0')$. Repeat this process until $b_1' = 0$, the result is
$B = b_0' \bmod q$.

In our experiment, we choose $q = 2^{64} - 2^8 - 1$, $q' = 2^{16} - 2^4 - 1$. We test the
average time delay and power consumption for computing the polynomial with
different t values. In each test, we randomly generate $t + 1$ 64-bit coefficients and
a 16-bit variable x, we repeat 20 times to get the average time delay. The test
results are shown in Fig. 4.

The test results show the polynomial computation is efficient in low-power
sensor nodes. The figure shows that the time consumption for generating a pair-

wise key is only 8.8ms, 17.1ms, and 36.8ms, given the polynomial degree of 16, 32, and 64, respectively.

To evaluate the remote access control procedure, we divide the experiment into two parts. The first part includes local sensor discovery, local sensor authentication and MAC collecting. In the second part, we perform the MAC reconstruction and verification at the remote sensor. The message routing between the user and the remote sensor is a typical communication process that has been investigated extensively and the time delay is very small, so in our experiment we omit the message routing between the user and the remote sensor.

During the experiment, we assume the sensor field is dense enough so that the user can reach *local* sensors from different groups without moving. To acquire the endorsements from local sensors, the user first broadcasts a remote access request. Each local sensor replies the user with its group ID. The user picks those sensors from different groups to fill in her endorse list. Due to the message collision, some replying messages are corrupted, so the user may not find enough endorsing sensors with one broadcast. As the result, the user may have to broadcast several times to find all k endorsing nodes. Our experiments show the user has to broadcast at least twice if $k \geq 6$. After successfully finding k endorsing sensors, the user unicasts an endorse acknowledge to each of the k sensors. The endorsing sensors processes the user authentication in parallel. The user first broadcasts her certificate, and then sequentially receives and responses each local sensor's challenge. A simple scheduling algorithm can be used for the endorsing sensors to send challenges without packet collision. In our implementation, we arrange the endorsing sensors to send the challenge in ascending order of their group IDs. If the user is successfully authenticated, then each endorsing sensor generates the MAC and returns it to the user. After collecting all k MACs from endorsing sensors, the user finally generates a MAC digest and sends the access request to the remote sensor. We perform the experiment with k changing from 2 to 16. The result of endorsing time consumption is shown in Fig. 5(a). Note that the time duration includes the time for user's broadcasts for request, receiving the group ID reply from sensors, unicasts to sensors for acknowledging receiving their group IDs, and sensor nodes' data processing time to generate the MACs.

We first perform a separate experiment just to test the time delay to find k sensors only (without local authentication and MAC generation). The result is shown as the dotted line in the same Fig. 5(a). It is interesting to find that it takes $105ms$ to find just 2 endorsing sensors and considerable time for discovering 4, 8, and 16 sensors, which is surprisingly slow, considering $1ms$ transmitting/receiving delay. Two factors contribute to the long delay. First, as discussed in previous section, the user may not get all information from local endorsing sensors after the first broadcast. The user may have to broadcast the request more than twice. Second, more importantly, a timer is set between any two broadcasts in our implementation to regulate the packet transmission and reception. Every time the timer fires, the user checks whether the endorsing list is complete. If not complete, the user will do broadcast again. The time delay

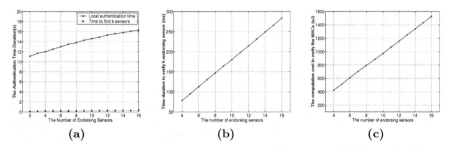

Fig. 5. (a). The solid line shows the time duration for the user to get authenticated by k local sensor, k is changing from 1 to 16. The dotted line reveals the time delay for the user to find k endorsing sensors; (b). The time duration for remote sensor to verify k endorsing local sensors; (c). The energy cost for the remote sensor to verify k endorsing local sensors.

between the fires of the timer predominantly accounts for the sensor discovery delay. We can reduce this time duration by setting a higher timer frequency.

The total endorsing time is presented in Fig. 5(a) (solid line). Apparently, the expensive local authentication dominates other delays. However, because k local sensor authenticate the user in parallel, the total endorsing time is practical and not much longer than the local authentication delay. When $k = 16$, it only takes 16.7 seconds for the user to get all endorsements.

Once receiving user's remote access request, the remote sensor has to verify whether the user is endorsed by k local sensors. To do so, the remote sensor reconstructs k MACs by plugging the group ID into its own share of polynomial. After k MACs are reconstructed, the remote sensor then generates and verifies the digest. In the experiment, we measure the time duration for the remote sensor to do the verification with $k = 4, 5, \cdots, 16$ endorsing sensors. The experiment results are shown in Fig. 5(b)(c).

Finally, we estimate the total time for a user to be authenticated for remote data access. Suppose the network requires the user to get 16 endorsing sensors to access a remote sensor. First, the user has to get local authentication by all 16 local sensors and receive corresponding MACs. This procedure costs 16.7 seconds according to Fig. 5(a). Then, the remote sensor needs $283ms$ to reconstruct and verify 16 MACs. In total, a remote access with 16 local sensor endorsement will cost around 17 seconds. Note that our estimation does not include the message traveling time from the user to the remote sensor and then back to the user.

6   Conclusion

In this paper, we show our effort in designing access control scheme for sensor networks. We describe our local access control and remote access control under a very realistic adversary model. We implement the protocols on a TelosB mote testbed. The security and performance analysis and the experimental results show that our access control is feasible for real application. We are currently in the process of doing more experiments and designing more schemes for access control for comparison.

References

1. D. Balfanz, G. Durfee, N. Shankar, D. Smetters, J. Staddon, and H. Wong. Secret handshakes from pairing-based key agreements. In *2003 IEEE Symposium on Security and Privacy*, Berkeley, CA, May 2003.
2. H. Chan and A. Perrig. Pike: Peer intermediaries for key establishment in sensor networks. In *INFOCOM 2005*, Miami, FL, March 2005.
3. H. Chan, A. Perrig, and D. Song. Random key predistribution schemes for sensor networks. In *In IEEE Symposium on Security and Privacy*, pages 197–213, Berkeley, California, May 2003.
4. Moteiv Co. Telos datasheet. *http://www.moteiv.com /products/docs /tmote-sky-datasheet.pdf.*
5. W. Du and J. Deng. A pairwise key pre-distribution scheme for wireless sensor networks. In *ACM CCS 2003*, 2003.
6. Wenliang Du, Jing Deng, Yunghsiang S. Han, Shigang Chen, and Pramod Varshney. A key management scheme for wireless sensor networks using deployment knowledge. In *IEEE INFOCOM'04*, Hong Kong, March 2004.
7. L. Eschenauer and V.D. Gligor. A key-management scheme for distributed sensor networks. In *In Proceedings of the 9th ACM conference on Computer and Communication Security*, November 2002.
8. Prasanth Ganesan, Ramnath Venugopalan, Pushkin Peddabachagari, Alexander Dean, Frank Mueller, and Mihail Sichitiu. Analyzing and modeling encryption overhead for sensor network nodes. In *WSNA03*, San Diego, CA, Sept 2003.
9. Nils Gura, Arun Patel, Arvinderpal Wander, Hans Eberle, and Sheueling Chang Shantz. Comparing elliptic curve cryptography and rsa on 8-bit cpus. In *CHES*, Boston, Aug. 2004.
10. D. Liu and P. Ning. Establishing pairwise keys in distributed sensor networks. In *CCS'03*, Washington, DC, October 2003.
11. A. Perrig, J. Stankovic, and D. Wagner. Security in wireless sensor networks. *Communications of The ACM*, 47(6):53–57, June 2004.
12. A. Perrig, R. Szewczyk, V. Wen, D. Culler, and D. Tygar. Spins: Security protocols for sensor networks. *ACM/Kluwer Wireless Networks Journal (WINET)*, September 2002.
13. Ronald L. Rivest. The rc5 encryption algorithm. In *Proceedings of the 1994 Leuven Workshop on Fast Software Encryption (Springer 1995)*, pages 86–96, Springer, 1995.
14. S. Jajodia S. Zhu, S. Setia and P. Ning. An interleaved hop-by-hop authentication scheme for filtering of injected false data in sensor networks. In *In Proc. IEEE Symposium on Security and Privacy*, Oakland, CA, May 2004.
15. Harald Vogt. Exploring message authentication in sensor networks. In *1st European Workshop on Security in Ad-Hoc and Sensor Networks (ESAS 2004)*, Heidelberg, Germany, August 2004.
16. H. Wang, B. Sheng, and Q. Li. Telosb implementation of elliptic curve cryptography over primary field. In *Technical Report*, Dec 2005.
17. F. Ye, H. Luo, S. Lu, and L. Zhang. Statistical en-route filtering of injected false data in sensor networks. In *INFOCOM 2004*, 2004.
18. W. Zhang, H. Song, S. Zhu, and G. Cao. Least privilege and privilege deprivation: Towards tolerating mobile sink compromises in wireless sensor networks. In *MobiHoc'05*, Chicago, IL, May 2005.

Locating Compromised Sensor Nodes Through Incremental Hashing Authentication

Youtao Zhang[1], Jun Yang[2], Lingling Jin[2], and Weijia Li[1]

[1] Computer Science Department, University of Pittsburgh, Pittsburgh, PA 15260
[2] Computer Science and Engineering Department,
University of California at Riverside, Riverside, CA 92507

Abstract. While sensor networks have recently emerged as a promising computing model, they are vulnerable to various node compromising attacks. In this paper, we propose COOL, a COmpromised nOde Locating protocol for detecting and *locating* compromised nodes once they misbehave in the sensor network. We exploit a proven collision-resilient *incremental hashing* algorithm and design secure steps to confidently locate compromised nodes. The scheme can also be combined with existing en-route false report filtering schemes to achieve both early false report dropping and accurate compromised nodes isolation.

1 Introduction

The sensor network has recently emerged as a promising computing model for many applications e.g. patient status monitoring in a hospital, and target tracking in a battlefield. However, its unattended nature makes the network vulnerable to varying forms of security attacks such as a compromised node dropping true data reports [9] or injecting false reports [18, 22]. Without being detected, compromised nodes may prevent the sink from reaching a correct or optimal decision. In addition, routing false reports wastes the energy of relay nodes, which reduces the lifetime of the network.

The previous work proposed either to locate compromised nodes through en-network detection [12, 15] or to filter false reports early in routing [18, 22]. While they are effective in many cases, both approaches have limitations — the former suffers from low accuracy due to possible collusion attacks and the latter cannot exclude the compromised nodes. In this paper we propose COOL, a COmpromised nOde Locator for locating malicious nodes if they send out false data reports or drop real reports. Our design is based on an intuitive observation — for any well-behaved node in the sensor network, the set of outgoing messages should be equal to the set of incoming and locally generated or dropped messages [1]. We exploit a proven collision-resilient incremental hashing scheme — AdHASH [1] and show how to securely collect the AdHASH values and confidently locate compromised nodes. We incrementally extend the testing so as to capture an inconsistency when a bad link is included. A bad link is a hop between

---

[1] Some messages may be lost due to weak connection in the sensor network. It is also considered as one type of fault. We detect such links and let the sink decide if the involved nodes should be excluded.

P. Gibbons et al. (Eds.): DCOSS 2006, LNCS 4026, pp. 321–337, 2006.
© Springer-Verlag Berlin Heidelberg 2006

two nodes in which at least one is compromised. For such links, we drop both nodes achieving an upper bound of $2m$ excluded nodes if there are m malicious ones.

The remainder of the paper is organized as follows. We describe the problem, and the network and attack models in Section 2. The COOL protocol is then presented in Section 3 with optimizations presented in Section 4. We evaluate proposed schemes and show the results in Section 5. Section 6 discusses the related work. Section 7 concludes the paper.

2   Problem Statement

2.1   The Network Model

We assume that the sink assign a unique ID and a unique secret key to each sensor before deployment. Sensors are left unattended after deployment and monitor events of interests. While some may be compromised, we assume that the majority of sensing nodes for any single event are trustworthy.

We adopt a cluster-based multi-hop routing scheme due to its energy efficiency [6, 19]. Sensing readings (including the timestamps [18]) are first sent to the cluster head (CH) at which the are aggregated to a data report. By taking the majority of the readings, the CH includes the selected sensing node IDs and their MACs (message authentication code, discussed next) in the content of the aggregated report. The CH also appends its own ID and MAC to the report. After generating the report, the CH forwards it along the routing path to the sink. Messages from the sink are first sent to CH and then broadcasted within the cluster.

At the cluster head level, the routing graph is built using directed diffusion protocol [2]. Paths are set up to monitor different *interests*. We assume reports are forwarded according to the routing path in one epoch. Each node checks the received report and drops it if not for a cached interest.

2.2   The Attack Model

After compromising a sensor node, the adversary can retrieve all security information including the secret key. (S)He can then inject false reports ([22, 18]), or drop some of its received reports ([9]). The adversary knows the COOL or other security enhancement algorithms, and may strive to send back data targeting at defeating the protection.

The injection attack or the dropping attack may occur at a sensing node; at a source CH; or at a relay CH. In this paper we address all these types except the dropping attack at a source CH node. Dropping at a source CH is more difficult to defend since a compromised CH may refuse to form a report even after receiving several sensing readings. On the other hand, a CH is usually granted the power to *legally* drop some readings when constructing the report (to shield random erroneous readings). If it is a concern, then each sensing reading could be sent to more than one source CH nodes resulting in increased routing overhead as we will discuss later.

A compromised node is located if and only if its node id is known to the sink who can then securely notify other sensors (using broadcast authentication [13]). Without being located, the compromised node can be elected as a CH node and continuously inject or drop reports. After being located, the network is free from its injection and dropping attacks since others know it is excluded. Of course additional mechanism might be needed to prevent it from malicious signal collision or changing its id.

Reports may be lost due to weak connections. This is one type of faults that should also be identified. Identifying a weak connection is beneficial since it gives the accurate location where a problem occurs. Based on the frequency of a faulty link, the sink can always make the decision whether or not to exclude the involved nodes. Since it is straightforward to detect/eliminate such links, we will focus on the report loss due to security attacks in the rest of the paper.

2.3 The Design Objectives

For a sensor network with above settings and models, our design goal is to effectively identify those compromised nodes and then exclude them from the network. The proposed algorithm meets the following requirements.

- The sink has the ability to discriminate the false reports;
- The scheme can defend both true report dropping at the relay nodes and all types of injection attacks;
- The algorithm can locate compromised nodes;
- The algorithm is effective with small overhead introduced to existing clustering and routing algorithms.

3   The COOL Protocol

In this section we present the basic design of the COOL protocol. We first discuss the incremental hash function, and then describe the high-level idea of malicious node detection using a simple example. The details of the systematic protocol operations are then discussed, followed by security analyses of the protocol.

3.1 The Incremental Hash Function

Fig. 1 illustrates the concept of the incremental hash [1]. It computes a cryptographic hash value for a finite set of elements. Each element is first concatenated with a unique *id* and then hashed by a standard cryptographic hash function e.g. MD5 or SHA [14]. Those intermediate hash values are then combined by a combining operator to get the incremental hash value.

In this paper, we use the AdHASH introduced in [1] (abbreviated as AH(...)):

$$AH_M^h(x_1, x_2, .., x_n) = \sum_{i=1}^{n} h(\langle i \rangle . x_i) \bmod M$$

where h is a standard cryptographic hash function and M is a very large integer value with k bits. The $\langle i \rangle$ is an *id* assigned to each message such that the concatenation of them is unique in the entire set. When we apply the AdHASH in our sensor network, each report is assigned with the sensor's ID and a local report sequence number. Therefore, each report received by the forwarding cluster head is unique. As we can see, the AH computed by the cluster head is independent of the order at which reports are received. This incremental hash function has the following properties that are useful in our design.

– Compression. It compresses inputs of larger size into k bits such that each incremental hash value can be stored using small number of bits in each node.
– Incrementality. The AH of a larger set can be computed incrementally from the AH of its subset. In particular, when a new item is inserted to the set, the new AH can be computed from the old value and the $h()$ value of the new item. i.e. $AH_M^h(x_1,...,x_{n+1}) = (AH_M^h(x_1,..,x_n) + h(\langle n+1\rangle.x_{n+1})) \bmod M$
– Efficiency. The computation of an AH hash value just needs several additions and one modulation in addition to the standard hashing. Particularly, for the insertion of a new item, the computation overhead is one addition and one modulation only (the width of the h and AH is of the same order). This is important as most hash values are to be maintained by resource constrained relay nodes in our design.
– Proven collision-resilience. It is computationally infeasible to forge another set of items that can result in a same hash value [1]. This gives us a solid security ground for designing security enhancement schemes for sensor networks.

We selected h to be MD5 [14] and k to be 128 in the design. The selection of MD5 is independent and can be substituted if for example security is a concern [17, 16]. The security of AdHASH requires that the number of reports should be greater than k [1]. This is generally not a restriction — the protocol can start after the network has been warmed up.

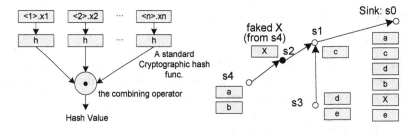

Fig. 1. An incremental hash function   **Fig. 2.** Locate compromised nodes using incremental hashing

3.2 The Basic Design — A Simple Example

We next show how an incremental hash function can be applied to authenticate messages and in particular how to locate malicious nodes in a sensor network.

The design is based on an intuitive observation, i.e. the set of outgoing (forwarded) messages of a well-behaved node equals the set of the incoming (received) and locally generated/dropped messages. Unfortunately, these message sets are maintained on different sensor nodes across the network making it impractical to pass them around and compare. Luckily with the incremental hash function, we only need to compare the hash values of different sets while keeping sufficient confidence to claim that *their hash values match indicates that the message sets also match*, i.e. Fig. 2(a).

No node being compromised

$\Leftrightarrow$

$\{\text{msg}_{out}\} = \{\text{msg}_{in}\} \cup \{\text{msg}_{local}\}$
$\Leftrightarrow$ (with sufficient confidence)
$\text{AH}(\{\text{msg}_{out}\}) = (\text{AH}(\{\text{msg}_{in}\}) + \text{AH}(\{\text{msg}_{local}\})) \bmod M$

To see how this principle is applied in our sensor network, let us look at a simple example (Fig. 2). Here we show four cluster head sensors (s1-s4) and one sink (s0). The messages are labeled in letter a, b etc. Suppose the compromised node s2 injects a false message X and pretends that X is sent by s4 (in this section, we will also discuss what if X is forged as if it is sent by s2). All messages are forwarded to the sink, but the AH values are calculated and kept locally. Specifically, s4/s3 calculates outgoing AHs for $(a,b)/(d,e)$; s1 calculates two incoming AHs for (d,e) and (a,b,X) respectively, and one outgoing AH for (a,b,c,d,e,X); s2, as a compromised node, can fake the incoming or outgoing AHs for either (a,b) or (a,b,X). Note that s2 will not produce another incoming AH for X as s2 tries to hide itself from being detected immediately (X is "originated" from s4). As we will elaborate later that the AHs for locally generated legitimate messages are computed at the sink.

The sink receives all messages including X. It can immediately identify that X is false since s2 does not have the secret key of s4 and cannot generate the consistent MAC for X. Next, the sink tries to locate the sender of X, i.e., the node that has been compromised. At this time, the sink collects all the AHs. We can assume that they all arrive correctly as simple endorsement using secret keys can assure this, and there are fixed number of them for a given routing so that no AHs are dropped. The sink then starts to check the "node consistency", i.e., if

$$AH(incoming \cup generated) = AH(outgoing \cup dropped) \qquad (1)$$

holds for every node. Note that due to the additivity of the AH function, $AH(incoming \cup generated) = (AH(incoming) + AH(generated)) \bmod M$ (and the same for the right hand side). It is easy to see those conditions for s1, s3 and s4 satisfy. For s2, as we mentioned, there are two options —$AH(a,b)$ or $AH(a,b,X)$— for both the incoming and outgoing AH values, forming four possibilities. If s2 choses the different values for incoming and outgoing AHs, it would immediately be identified as compromised as equality (1) would not satisfy. Let us assume s2 is intelligent enough not to expose itself too easily, and thus chose a consistent incoming and outgoing AH pair. Thus, it will pass at least the "node consistency" check.

Next the sink starts a "link consistency" check in which the AH of the outgoing message on a link should equal the AH of the upstream incoming message. This can be easily checked for links without the node s2. Let us assume that the s2's incoming and outgoing hashes are both $AH(a,b,X)$. Then, as an outgoing AH, $AH(a,b,X)$ is consistent with one of the incoming AHs for s1. However, as an incoming AH, $AH(a,b,X)$ is inconsistent with the outgoing AH for s4 which is $AH(a,b)$. In other words, s2 chose that value to lie that s4 had given it a false incoming message. The sink now cannot distinguish who is the real compromised node, but can at least conclude that one of them is flawed. A new routing graph will be generated excluding both s4 and s2 after detecting the link inconsistency.

Notice that node s2 can destroy s1 in the same way by producing $AH(a,b)$ for link consistency checking, or even destroy all the adjacent nodes by forging an arbitrary AH making none of the links consistent. It would be unnecessarily conservative if all involved nodes in such a scenario are eliminated since the faults are due to only one evil axial node. Instead, our protocol removes one link at a time, removing the axial node in the first place and saving the other nodes being circumvented.

Let us discuss what if X is forged as if it is sent by s2. The sink can still identify X as a false report since each legal one should have multiple sensing node MACs — s2 cannot construct these sensing node MACs as it does not have their secret keys. Old readings cannot be replayed since a timestamp is included in the reading. Notice that the sink computes local AH values only from legitimate reports, that is, it excludes X from generating the local AH value for s2. Hence, no matter which incoming or outgoing AH values s2 chooses (either $AH(a,b)$ or $AH(a,b,X)$), there bounds to be a node or link inconsistency.

Before presenting the COOL protocol, we summarize the benefits from excluding compromised nodes from the network.

- Communication energy savings. Once the compromised nodes have been excluded, no false reports can be injected into the network. As a result, the energy drained by forwarding false reports can be saved. This is different from en-route filtering schemes in which false reports are still forwarded several hops ([18, 22]) before being detected and dropped.
- Computation energy savings. In the basic COOL scheme, we do not perform en-route packet authentication but rather only update incremental hash values. We can afford less frequent authentication by excluding compromised nodes and form a network with trustworthy nodes.

3.3   The COOL Protocol

The goal of the COOL protocol is straightforward: we securely collect AH hash values from the network and send them to the sink; we drop the identified node if a node inconsistency is found, and drop both nodes if an inconsistent link is found. The detailed protocol contains the following phases.

(1) In the initialization phase, we assign a unique ID and a secret symmetric key to each sensor node. The sensor nodes are deployed thereafter.
(2) In the routing graph discovery phase, we broadcast *hello* messages along the downstream routing path and collect the current routing graph from replies.
(3) In the report forwarding phase, we endorse each report by the secret key of the sender and have it forward along the routing path. Each node maintains the AH hash values for each of its incoming and outgoing links.
(4) In the hash value collection phase, the sink sends out a request to collect hash values from the path where the false message belongs to.
(5) In the compromised node detection phase, the sink finds inconsistent nodes and links using the incremental hash function AH.
(6) In the routing graph fix phase, the sink replaces the excluded cluster heads with newly selected ones.

Next, we elaborate each of the phases with more details.

Phase 1: Initialization. Each sensor is assigned a unique integer ID and a secret symmetric key before being deployed into the field. Both the ID and the key are known to the sink. This is similar to [22] but we do not request any key sharing among sensor nodes. The node ID occupies two bytes (16 bits) which can distinguish 64K sensors in a network. Larger networks can adaptively adjust this bit width to accommodate their needs. The key is used to generate a MAC by a sensing node for the sensing reading sent to the source CH node, and by a source CH node for the report sent to the sink. By checking the MACs using the secret keys, the sink can detect tampered reports or injected false reports from compromised source CH or relay CH nodes.

Phase 2: Routing graph discovery. The discovery starts at the beginning of an epoch, e.g. cluster heads are reselected and a new routing graph is constructed according to [2]. The sink forms an entire routing graph through collecting information from distributed cluster heads. It sends out a timestamped "hello" message to all its adjacent cluster heads who then forward this message downstream until it reaches the leaf nodes. All the nodes then respond with their node IDs as well as their adjacent node IDs. For those messages, a MAC using the local secret key is also attached to ensure their integrity.

Each node only collect replies from its downstream nodes. To prevent malicious report dropping, the reply is collected in order — each node first collects the replies from all its children nodes and then sends out its own reply. The sink finally assembles the complete routing graph from all replies.

To ensure "hello" messages are not abused, a broadcast authentication [13] is applied. In addition, a selected cluster head may try to find a different routing path if it does not receive a "hello" message within a certain time interval after its election to avoid being isolated from the network.

Phase 3: Report endorsement and forwarding. Fig. 3 illustrates the report generation, endorsement and forwarding in the network. We also list the related authentication actions and discuss why injected false reports and dropped legitimate reports can be detected.

To check equation 1, each node maintains its incoming AHs and outgoing AHs. The local AH (generated from locally generated reports) is computed at the sink rather than the individual node. This is because false reports should not be used to compute the local AH (otherwise there is no inconsistency) and the sink knows what those false reports are. The sink generates a local AH for each node using only the legitimate reports whose source IDs are that node, discarding all false reports discriminated. **For example**, a node m may forge a false report using its own ID as the source ID. This false report can be identified by the sink since it does not have enough legal sensing node MACs. After identifying X as a false report, the sink will exclude it from updating the local AH of m, which creates a node inconsistency if node m updated the false report into its outgoing AH, or a link inconsistency otherwise.

The drop AH (generated from locally dropped reports) is not used in the basic scheme i.e. we assume a non-compromised node does not drop reports intentionally.

	Sensing node	Cluster head (CH) node	
		Data aggregation	Report relay
Report generation and endorsement	An event is detected by at least M surrounding sensing nodes. Each of them e.g. m sends the sensing reading and the MAC (generated using m's secret key) to C CH nodes.	By taking the majority of received readings, the source CH constructs a report containing the sensing value and received MACs. It also appends its ID and a unique sequence number. A unique MAC is generated for the report using its own secret key. Both the report and the MAC are forwarded along a multi-hop route to the sink.	Since the keys to attached MACs are only known to the sink, relay nodes do not perform any enroute checks in the basic scheme. A relay node receives the report from one downstream link and forwards it along one upstream link.
Authentication	An AH value is maintained for the outgoing link of m. The value is updated with the sensing reading and the MAC each time when m sends them out.	A source CH maintains one outgoing AH and in the case it is also a relay node, it maintains several AH values with one for each of its incoming/outgoing links respectively. After sending the report generated by itself, it updates the outgoing AH value with the report and the MAC.	A relay CH node maintains a different AH value for each of its incoming and outgoing links respectively. For the forwarded report, it updates two AH values — the incoming AH value for the link from which the report is received and the outgoing AH value for the outgoing link.
Detecting injection attack	A source CH should receive readings from at least M sensing nodes. If some $(< M/2)$ are compromised and send back false readings, these readings will not affect the report generation at the CH node.	A compromised source CH may forge false reports. It cannot accumulate $M/2$ legal MACs for the false reading. A report with such a value can be detected at the sink. Old readings and MACs cannot be replayed since the sensing reading has a timestamp indicating when the event happens [18]. Different sensing node MACs are expected even for the same sensing reading but at a different time.	If a relay node forges a report with the source node as itself, the false report can be detected as the data aggregation case. If a relay node forges a report with the source node id as one of its downstream nodes, it does not have the secret key of the faked sender to generate a matching MAC. The report will be detected by the sink. The compromised node is then located using our algorithm.
Detecting dropping attack	Dropping readings at some $(< M/2)$ compromised sensing nodes does not affect the generation of the legitimate report at a source CH node.	[Discussion only: not the focus in the paper] We can elect more than one CH to perform data aggregation. If some but not majority of them refuse to generate reports from received readings, the sink can still receive the legitimate report and thus detect packet dropping in the network. In the rest of the paper we assume that for each event one CH node is elected for data aggregation.	If a relay node drops some reports, the sink can check the sequence number from a source CH and reveal the dropping from non-contiguous numbers. If the compromised relay node chooses to drop all following reports from a CH, the sink can periodically collects all AH values to detect the dropping attack.

Fig. 3. Actions taken by different nodes

A report is dropped only by compromised nodes who always try to conceal themselves as much as possible. Consequently, the AHs for dropped reports are never created.

Either false report injection or legitimate report dropping creates AH inconsistency for some nodes or links. The difficulty is how to expose the inconsistency. The technique presented in phase 4 handles this problem.

Phase 4: Hash value collection. Hash value collection is triggered by any of the following two conditions: (i) the sink has detected one or multiple false reports; (ii) a preset timer has elapsed. The former is to detect injected false report attack while the latter is to detect report dropping attack. In the first case, AH values are collected only from the path where the erroneous report belongs to. Such a path can be identified correctly as we explained earlier that a spoofed report can reach the sink only if it is generated from a downstream node. Collecting the hash values along a path greatly reduces the number of messages introduced to the network. In the second case, however, all the hash values in the network are queried as the sink has no clue of where reports could be dropped. Next we show how to collect AHs from the erroneous path. It is trivial to extend the scheme to all nodes.

To collect the hash values, the sink sends out an inquiry message onto the erroneous path. For example, in Fig. 2, to detect the compromised node on path s4-s2-s1-s0, we only collect hash values on this path but not from the link s1-s3 (but in phase 5 the sink still needs to compute the AH values for these reports injecting to the path from s3). We must be very careful in this collection process as the forwarded AHs may be altered by compromised nodes as well. Thus, we treat all the AH values as *normal data reports* and send them upstream starting from the leaf cluster nodes. The only constraint is: the outgoing AH on a link does not update the incoming AH on the same link since this would result in link inconsistency and make the hash collection process and later checking too convoluted. As a result, the node consistency checking would be adjusted to accommodate this exception. We will give formal derivation later.

Phase 5: Identify compromised nodes. The sink performs two types of tests: the node consistency and the link consistency test. The first is to test the matching of incoming and outgoing AH values for each node on the erroneous paths. The AHs for the incoming links not on erroneous paths, e.g, the s1-s3 path in our example, and for locally generated reports are calculated by the sink directly; other AHs are from the returned AH reports. If there is a mismatch, the node is tagged as a compromised node. For example, in Fig. 2, the sink tests s1 using

$$\underbrace{(AH_{2\to 1}}_{collected} + \underbrace{AH(c) + AH(d,e))}_{calculated\ by\ the\ sink}\ mod\ M\ =?\underbrace{AH_{1\to 0}}_{collected}$$

The second type is to test if the outgoing and incoming AHs are consistent on all links. Each hash value should match with the one reported by the other end of the link. If any inconsistency is found, the sink tags both nodes as problematic as it is now hard to flag one node with 100% confidence. e.g. we test

$$(\underbrace{AH_{1\to 0}}_{collected} =? \underbrace{AH_{0\leftarrow 1}}_{computed\ at\ the\ sink}\ )\ and\ (\underbrace{AH_{1\to 2}}_{collected} =?\underbrace{AH_{2\leftarrow 1}}_{collected})$$

Phase 6: Excluding compromised nodes and routing graph fix. Once any nodes are tagged as suspicious, they should be excluded from the sensor network immediately. To do so, the sink broadcasts the IDs of the tagged nodes across the network, particularly to those nodes around the compromised ones. This can be done using broadcast authentication algorithms e.g. μTESLA [13]. This packet also initiates the selection of new cluster heads to replace the excluded ones, and then incorporates new heads into the routing graph. The newly joined nodes send back their IDs to the sink to check if they are allowed to join the network. The sink acknowledges back with the most up-to-date AH values for the new cluster heads.

3.4   The Security Analyses

In this section we give the major security analysis results. Their proof details can be found in [20].

Theorem 1. *Any injection attacks can be detected by the COOL protocol.*

Corollary 1. *The report dropping attack at the relay nodes can also be detected by the COOL protocol.*

Corollary 2. *If there are m compromised nodes, our scheme removes at most 2m nodes including those m compromised nodes.*

Theorem 2. *The AH value collection process is secure: correct AH values can be retrieved from the received AH report; no AH value may be compromised or dropped without being detected.*

4   Optimizations

In this section, we discussion two optimizations to the basic design.

4.1   Drop-COOL: Combining En-route Filtering Schemes

In the basic COOL protocol false reports are forwarded all the way to the sink. While the sink can detect these false reports, it is just too late since the energy has already been consumed along the routing paths. It may become even worse if a lot of false reports are injected before the COOL protocol is activated to collect AH values. We therefore propose Drop-COOL, a hybrid scheme that integrates an en-routing filtering scheme [18, 22].

The Drop-COOL scheme works as follows. The system initializes according to both the basic COOL and the SEF [18] protocols. In addition, each node is assigned an integer threshold which is the maximal number of false reports that the node can forward in one round. In the packet forwarding phase, detected false reports up to this threshold are forwarded while following ones are discarded. An additional AH hash value — *drop* hash value, is maintained on each node that drops false reports. It is updated incrementally each time when a false report is detected and dropped. To improve the effectiveness of Drop-COOL, a node may try to forward these false reports with different source CH IDs, which can trigger collecting and detecting more paths and thus expose more compromised nodes in one round.

The Drop-COOL protocol combines the advantages of both COOL and SEF protocols. It detects and excludes compromised nodes while saves the energy from routing less false reports. The energy spent to route a small number of false reports is small compared to the savings after excluding compromised nodes. It removes the worst case overhead that the basic COOL protocol has on routing false reports.

We next illustrate that the Drop-COOL does not affect the ability to locate compromised nodes although random false reports are dropped in the middle. Due to the introduction of the drop hash value, we have *for a well-behaved node in the routing graph, the set of forwarded and dropped messages should be the same as the set of received and locally generated messages.* By collecting and comparing the AH hash values of these message sets, we can adjust the node test to

$$(AH(MESSAGE_{forwarded} + AH(MESSAGE_{dropped}) \bmod M = (AH(MESSAGE_{received} + AH(MESSAGE_{local}) \bmod M$$

The link test is unaffected and an inconsistent node or link test result exposes at least one compromised node.

4.2 Hi-COOL: A Hierarchical Authentication Scheme

To further reduce the overhead, we propose Hi-COOL, a hierarchical approach which groups multiple adjacent nodes in the routing graph as a *super node*. We only collect hash values with respect to this super node, and refine the collection if a super node is found problematic.

The Hi-COOL scheme works as follows. First the sink picks up an integer number l and forwards this integer with the "hello" message (for collecting the routing graph). The integer value is decremented for each hop downstream along the routing path and reset after reaching zero. A node sets itself to be the head of a super node if it receives l and be the leaf of a super node if it receives zero. If a node receives two values from two upstream nodes, it picks up the smallest one. In Fig. 5 nodes s1 to s5 form a super node in which s2, s4, and s5 are leaf nodes while s1 is the head node. In the phase to collect the AdHASH values, only the incoming hash values to this super node and the outgoing hash values from this super node are collected. For example hash values AH_{23}, AH_{32} are omitted. The link test at the super node level is processed the same as the basic COOL — a failed link test removes two involved nodes. However an inconsistent (super) node test results in one additional round of hash value collection such that we can determine the exact location of the compromised node within the super node. In the second round, we only collect hash values from these inconsistent super nodes.

To ensure high level security, in the routing graph collection phase, each node should reply with its received integer number. In addition, the head of each super node may need to collect all internal hash values. More details can be found in [20].

5   Limitations of the COOL Protocols

The limitations of the COOL protocol are: (1) It is possible that a subregion is isolated from the sink. Without having the information about a cluster head in the routing graph

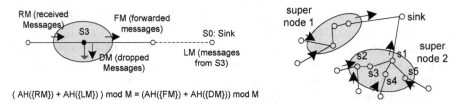

Fig. 4. Reducing routing energy through combined en-route filtering

Fig. 5. Hierarchical authentication with super nodes

collection phase, the sink cannot identify its status and decide if it is compromised or not. (2) Signal blocking or collision is another source of attack, if normal communication cannot be ensured between two nodes. Both of two involved nodes are excluded while the nodes themselves may not be hacked.

6   Performance Evaluation

6.1   Settings

To evaluate the effectiveness of the proposed COOL protocols, we simulate a sensor network with 450 cluster head nodes uniformly distributed in a field of 400x400m^2 area. Each sensor node is Mica2 running TinyOS [7] operating at 19.2Kbps data rate, with battery voltage 3V. It takes 16.25/12.5 μJ to transmit/receive a byte [18]. This will be referred as the baseline setting in the rest of the paper. The sink is located at (20,20) and the communication range of each node is 40m. These sensor nodes form a multihop routing network using the directed diffusion routing algorithm [2]. A normal packet is of 24 bytes long, a MAC is of 8 bytes (64 bits), and an incremental hash value is of 16 bytes (128 bits). The evaluation is based on false report injection attacks. All results are averaged from 100 different runs.

6.2   The Overhead

The protocol overhead comes from four sources: (i) AH hash value computation overhead; (ii) hash value collection overhead; (iii) routing graph discovery overhead; and (iv) routing false reports overhead. Next we study them in more detail.

(i) Computation overhead. The computation overhead is for updating incremental hash values at each sensor node. As the incremental hash is maintained per link based, a received report updates two AH values on the relay node. The updates are done incrementally with the overhead mainly from computing the standard hashing MD5() on the input report. MD5() intermediate result is used to update both AH values. Our simulation results show that the incremental hash computation overhead is about twice of the overhead of one RC5 [14] computation (used in [18, 7]), that is, 30μJ per node. It is small and thus omitted in the rest of the discussion.

(ii) Hash value collection overhead. The main overhead of the COOL protocol comes from collecting hash values across the network. We first induce a theoretical formula

about this overhead. Assume the routing graph is a b-nary balanced routing tree with height $O(log_b(N))$ where N is the total number of nodes in the graph. Since the hash value collection is per problematic routing path based, the number of node-to-node transmission T for one path is,

$$T = (1+2+3+...+H) \cdot (2 \cdot k + M) = \frac{H \cdot (H+1)}{2} \cdot (2 \cdot k + M) \qquad (2)$$

$$H = log_b(N)$$

where H is the height of the b-nary tree, N is total number of sensor in the field, k is the length of the incremental hash value, and M is the length of the MAC value. From the equation, T is in the range of $O((logN)^2)$. Since we need to exclude all m compromised nodes, we may need to collect from m disjoint paths or detect in m rounds. Therefore the worst case hash value collection cost is $O(m \cdot (logN)^2)$.

We next present the average number of rounds to exclude all compromised nodes in Fig. 6. In the experiment, false reports are randomly injected from the compromised nodes and we start a new detection round if 30 false reports are received. As expected it requires more rounds when there are a larger number of compromised nodes. On average we can detect and exclude more than 10 nodes in one round when more than 30 nodes are compromised.

Fig. 7 illustrates the total energy overhead for collecting hash values in multiple rounds. From equation 3, the AH collection cost is proportional to the number of compromised nodes and to the square of the tree height, these two factors are used as the x and y axes respectively. To change the tree height, we deploy in a different square field with the same node density, e.g. 1000 nodes are distributed in a field of 600x600 m^2. The tree height varies from 9.2 to 28.5.

The trend confirms what we observed from equation 3. For example, the energy overhead is about 0.87J with 20 compromised nodes and the tree height 13.6. It increases about 2 times to 1.56J if the number of compromised nodes increases 2 times to 40; It increases about 4 times to 3.89J if the tree height increases about 2 times to 28.4.

We also present the results using the Hi-COOL scheme. It effectively reduces the hash value collection cost. For example, when 20 nodes are compromised, the Hi-COOL overhead is 0.99J or 61% of the baseline setting (1.61J).

In collecting the results, we perform a simple optimization. For the two nodes of each link, they may report the same AH() value e.g. both nodes are healthy nodes. We therefore only need to transmit one AH hash value with two MACs to the sink.

(iii) Routing graph overhead. Fig. 7 also illustrates the energy overhead to collect the routing graph. Compared to the hash value collection overhead, it is usually small ranging from 0.5J when the tree height is 9.2 to 2.0J when the tree height is 28.5.

(iv) The overhead to route false reports. Fig. 8 illustrates the wasted routing energy in forwarding false reports. As we discussed, the sink has the option to start the hash value collection phase after accumulating a number of false reports. Clearly if we increase the threshold, the routing overhead increases as well. The benefit of accumulating a reasonable larger number of false reports is that we increase the chance that these false reports are from more problematic paths. They can then detect multiple paths and reduce the total number of rounds. For example, if we detect after receiving one false report, we may need m rounds to exclude all m compromised nodes; on the other hand, we

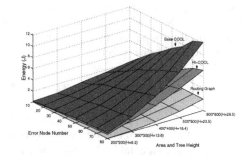

	number of compromised nodes							
	10	20	30	40	50	60	70	80
Rounds	2.1	2.4	3.5	4.0	4.5	5.3	6.1	6.5

Fig. 6. The number of rounds to exclude all compromised nodes

Fig. 7. Hash value collection overhead

may need only m/2 round if we set the threshold to be 2 and these two false reports are always from 2 different compromised nodes on different routing paths. However, our experiments show that the difference is not significant (with one or two rounds difference). In addition, if the threshold is set larger than 30, the number of rounds does not change much but the wasted energy increases drastically. Therefore we set the threshold to 30 in the paper.

6.3   The Savings

We next study the benefits from applying COOL protocols and compare it with an en-route false report filtering scheme [18]. As discussed, the savings comes from two sources: the communication savings and the computation savings. The latter is omitted as it is usual very small.

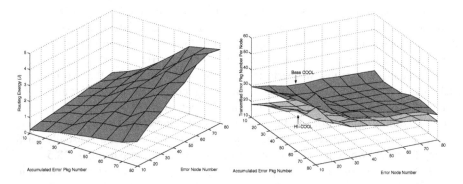

Fig. 8. The overhead to route false reports

Fig. 9. Comparing the overhead to en-route filtering schemes

Let us compare the overhead from different protocols. The overhead in the COOL protocol contains the energy to collect hash values, to discover routing graph, and to route false reports to the sink. The overhead in the SEF protocol is from routing false reports before detecting and dropping them. The exact number of hops varies with the key sharing scheme and the number of compromised nodes. For comparison purpose, we assume on average each false report is dropped after 5 hops in the paper.

In Figure 9, x and y axes are the detection trigger threshold and the number of compromised nodes in the network respectively. Each point in the figure is a number of false reports averaged to each compromised node. It is a break even point at which the overhead to route and drop this amount of false reports in SEF equals the overhead in the COOL protocol. For example, with 20 compromised nodes and trigger threshold set at 30, the number is 25 reports meanings that, if SEF routes and drops 500 reports (=25 reports/node×20node) in 5 hops, the energy it wastes is the same as all of the overhead in the COOL protocol.

In addition, consider that the scheme needs to spend 3.3 rounds and a round is triggered at 30 false reports, there are another 100 injected false reports (=30 reports/round × 3.3 rounds). Therefore the COOL protocol outperforms the SEF if each compromised node injects more than 30 reports (=25 reports/node + 100 reports/20 nodes). This is very small. For example, as suggested in [21], a compromised node may inject a faked report every 10 seconds. At this rate, we outperform SEF in 300 seconds or 5 minutes. With the Hi-COOL optimization, it is further reduced to 4 minutes, a 20% reduction. Of course, depending on the pattern that the false reports are injected, this number may vary but the results show that the COOL overhead is very modest.

7   Related Work

Extensive research has been done on sensor network security. Karlof *et al.* [9] identified several attacks for a multihop routing based sensor network.

Marti *et al.* proposed to monitor each node by a neighboring *watchdog* node. Wang *et al.* [15] improves the scheme through the collaborative decision of neighbors around a suspicious node. Both schemes have limitations [12] as the watchdog node may be compromised as well and multiple compromised nodes can collude to attack. The schemes to locate compromised nodes share some similarity with the approaches to detect faults in sensor networks [8, 11, 4]. However the significant difference lies in that a faulty node always returns a wrong report while a compromised node is smarter e.g. it may inject false reports but communicate normally with the sink.

En-route false data filtering schemes [18, 22] are proposed to actively detect and drop false reports early in the routing minimizing the impact of false report. In these schemes, each data report is attached with several MACs generated from different keys that are distributed probabilistically [18], set up before routing [22], or refreshed periodically [21]. However, those schemes also have limitations as compromised nodes are left undetected, which causes severe consequences in the long run.

Algorithms have been proposed to securely manage keys in sensor network. Eschenauer and Gligor [5] proposed a key pre-distribution scheme in which each sensor randomly selects a subset of all keys before deployment; Protocols were then developed for shared key discovery and path-key establishment. Improvements were later proposed for enhancing security [3] and achieving higher probability of key establishment [10]. Zhang and Cao [21] proposed to represent keys as group key polynomials whose shares are distributed around neighbors. New keys can be re-generated collaboratively by neighbors from these shares achieving better security and resilience.

8 Conclusion

In this paper we introduced the COOL protocol and its optimizations based on the provably secure incremental hash function AdHASH to detect and locate compromised nodes effectively. We first discussed how to securely maintain and collect AdHASH values on sensor nodes, and then use these values to perform node and link tests to expose the compromised nodes. Our experimental results showed that the COOL protocols are very effective and introduce very small overhead to the network.

Acknowledgment

This work is partially supported by the U.S. National Science Foundation under grants CCF CAREER 0447934 and CCF 0430021.

References

1. M. Bellare and D. Micciancio, "A New Paradigm for Collision-free Hashing: Incrementality at reduced cost," In *Eurocrypt'97*, LNCS 1233, 1997.
2. C. Intanagonwiwat, R. Govindan, and D. Estrin, "Directed Diffusion: a Scalable and Robust Communication in Wireless Sensor Networks," In *5th IEEE/ACM Mobicom*, pages 174-185, 1999.
3. H. Chan, A. Perrig, and D. Song, "Random Key Predistribution Schemes for Sensor Networks," In *IEEE Symposium on Security and Privacy"*, 2003.
4. P. Chew and K. Marzullo, "Masking Failures of Multidimensional Sensors," In *Proc. of the 10th Symposium on Reliable Distributed Systems*, pages 32-41, 1991.
5. L. Eschenauer and V. D. Gligor, "A Key-Management Scheme for Distributed Sensor Networks," In *Proc. of the 9th ACM Conference on Computer and Communication Security*, pages 41-47, November 2002.
6. W.R. Heinzelman, A. Chandrakasan, and H. Balakrishnan, "An Application-Specific Protocol Architecture for Wireless Microsensor Networks," *IEEE Transactions on Wireless Communications*, vol 1:4, pages 660-670, 2002.
7. J. Hill, R. Szewczyk, A. Woo, S. Hollar, D. Culler, K. Pister, "System Architecture Directions for Networked Sensors," In ASPLOS IX, 2000.
8. C. Jaikaeo, C. Srisathapornphat, and C. Shen, "Diagnosis of Sensor Networks," In *IEEE international Conference on Communications*, June 2001.
9. C. Karlof and D. Wagner, "Secure Routing in Wireless Sensor Networks: Attacks and Countermeasures," In *IEEE international workshop on Sensor Network Protocols and Applications*, pages 113-127, 2003.
10. D. Liu and P. Ning, "Establishing Pairwise Keys in Distributed Sensor Networks," In *Proc. ACM CCS*, 2003.
11. K. Marzullo, "Tolerating Failures of Continuous-valued Sensors," In *ACM Transactions on Computer Systems*, November 1990.
12. S. Marti, T.J. Giuli, K. Lai, and M. Baker, "Mitigating Routing Misbehavior in Mobile Ad Hoc Networks,", In *MOBICOM*, 2000.

13. A. Perrig, R. Szewczyk, V. Wen, D.E. Culler, and J.D. Tygar, "SPINS: security protocols for sensor networks," In *Proc. of Seventh Annual International Conference on Mobile Computing and Networks*, 2001.

14. B. Schneier, "Applied Cryptography," 2nd Edition, John Wiley & Sons, 1996.

15. G. Wang, W. Zhang, G. Cao, and T.L. Porta, "On Supporting Distributed Collaboration in Sensor Networks," In *IEEE MILCOM*, 2003.

16. X. Wang, Y. Yin, H. Yu, "Finding Collisions in the Full SHA-1 Collision Search Attacks on SHA1," In *Crypto'05*, 2005.

17. X. Wang, D. Feng, X. Lai, H. Yu, "Collisions for Some Hash Functions MD4, MD5, HAVAL-128, RIPEMD," In *Crypto'04*, 2004.

18. F. Ye, H. Luo, S. Lu and L. Zhang, "Statistical En-route Detection and Filtering of Injected False Data in Sensor Networks," In *IEEE INFOCOM 2004*, 2004.

19. O. Younis, and S. Fahmy, "Distributed Clustering in Ad-hoc Sensor Networks: A Hybrid, Energy-Efficient Approach," In *INFOCOM*, 2004.

20. Y. Zhang, J. Yang, L. Jin, and W. Li, "Locating Compromised Sensor Nodes through Incremental Hashing Authentication," Technical Report, University of Pittsburgh, 2006.

21. W. Zhang and G. Cao, "Group Rekeying for Filtering False Data in Sensor Networks: A Predistribution and Local Collaboration-Based Approach," In *INFOCOM*, 2005.

22. S. Zhu, S. Setia, S. Jajodia, P. Ning, "An Interleaved Hop-by-Hop Authentication Scheme for Filtering of Injected False Data in Sensor Networks," In *Proceedings of IEEE Symposium on Security and Privacy*, Oakland, California, May 2004.

COTA: A Robust Multi-hop Localization Scheme in Wireless Sensor Networks

Yawen Wei, Zhen Yu, and Yong Guan

Department of Electrical and Computer Engineering
Iowa State University, Ames, Iowa 50011, USA
{weiyawen, yuzhen, yguan}@iastate.edu

Abstract. In wireless sensor networks, multi-hop localization schemes are very vulnerable to various attacks such as wormholes and range modification attacks. In this paper, we propose a robust multi-hop localization scheme, namely COTA, to mitigate various attacks in wireless sensor networks. In this scheme, each localized sensor generates a COnfidence TAg to quantify the quality of its estimated location and broadcasts both the confidence tag and the estimated location in reference messages. When receiving sufficient number of references, an unlocalized sensor filters out bad references and weighs the remaining ones according to their tags. Once location determination is done, the sensor generates its own confidence tag from the indicator of its localization error. By properly setting the filtering metrics and computing the confidence tags, COTA can prevent the proliferation of location errors and achieves accurate location estimations and high localized percentage for sensor networks. To our knowledge, ours is the first work to address the security-aware multi-hop localization problem. We finally present security analysis and simulations to evaluate the effectiveness of COTA.

1 Introduction

Localization in wireless sensor network is very important for many applications such as environment monitoring, target tracking, and geographic routing. The traditional localization approaches require to equip sensor nodes with expensive GPS devices, which are not affordable in most cases. Hence, many localization schemes [1], [3], [7], [10], [11], [15], [18], [19], [20], [21], [27] assume that there exists some special nodes called anchors who know their positions through GPS devices or manual configuration. These schemes can be classified into one-hop and multi-hop ones. In one-hop schemes, anchor density is high, thus each sensor can use anchors' positions as location references to localize. In Multi-hop schemes, anchor density is low and many sensors have to determine their locations depending on other localized sensors.

Multi-hop localization schemes have the following advantages: (1) they are very economical, especially for the large-scale outdoor networks, because they require very small number of anchors; (2) they can be used in special environment such as indoor fire-fighting system [8], because neither GPS devices nor manual

P. Gibbons et al. (Eds.): DCOSS 2006, LNCS 4026, pp. 338–355, 2006.
© Springer-Verlag Berlin Heidelberg 2006

configuration is feasible in such situations. However, the multi-hop schemes are more vulnerable than the one-hop ones because they rely on location propagation of sensor nodes, thus a small number of wrongly localized sensors will affect the localization of a large number of sensors. Current secure localization schemes mainly [4], [15], [16], [17] focus on securing one-hop localization, and no research has been done about security-aware multi-hop localization schemes.

In this paper we propose a robust multi-hop localization scheme COTA. In COTA, each localized sensor bears a confidence tag that indicates the accuracy of its estimated position. The confidence tag and the estimated location will be combined and sent as location references to other unlocalized sensors, who will filter out bad references and weigh the remaining ones to compute optimal solutions for their locations. Finally, they calculate some indicators to estimate their localization errors and translate the errors to proper confidence tags using the *tag-generation function*. COTA is a complete tag-based localization system and consists of two main phases: the localization phase and the tag-generation phase. We did comprehensive experiments and prove that COTA can achieve low localization errors and high localized percentage in sensor networks. In non-adversary scenarios, COTA can also effectively mitigate the error accumulation problem caused by noisy range measurements.

The rest of this paper is organized as follows. The next section describes the related work. Section 3 provides the system model and threat model. Section 4 describes the COTA scheme in details. Section 5 is the security analysis. Section 6 presents our simulation results. Finally, we conclude and lay out some future work in Section 7.

2   Related Work

A large number of localization schemes have been proposed for sensor network. Most of them assume that some special sensor nodes called *anchors* know their positions via GPS or manual configuration, while other sensors use anchors' positions and the connectivity/distance information between sensors to localize. We can classify them into *one-hop scheme* and *multi-hope scheme* (Table 1):

- **Definition 1 (*one-hop scheme*).** In one-hop schemes, anchors are densely deployed and each sensor can receive enough number of beacon messages from anchors to localize.
- **Definition 2 (*multi-hop scheme*).** In multi-hop schemes, some sensors cannot receive enough number of or even no beacon messages from anchors, such that they refer to the locations of other localized sensors to localize.

Because localization in sensor network is vulnerable to many malicious attacks, some secure localization schemes have been proposed. To defend against wormhole attacks [13], L. Hu et al. proposed a scheme [12] in which each sensor is equipped with directional antennas, thus the messages from innocent neighboring nodes should be sent and received in opposite antenna sectors but the

Table 1. Classification of Localization Schemes

	One-hop Scheme	Multi-hop Scheme
Range-based	RADAR[1] Active Bat [10]	APS-Euclidean [20], AHLos [22] Trilateration Graph[7], Quad [18]
Range-free	Active Badge [27], Centroid [3] APIT [11], Serloc[15]	DV-hop [21] Amorphous [19]

wormholed messages do not have this property in most cases. Lazos et al. proposed *SeRLoc* [15], in which anchors send beacon messages through directional antennas, and sensors can detect wormholes when messages from far-apart anchors or from different sectors of one anchor are received simultaneously. Capkun and Hubaux proposed *verifiable multilateration* to verify sensors' locations that are inside the verifier triangles. Their work is based on the distance-bounding protocol and requires RF signal to obtain the upper bounds of the distances between sensors and verifiers. Liu [17] suggested a greedy algorithm to find out the maximum subset of consistent references by checking if the mean square error drops below a reasonable threshold. Zhang [16] introduced Least Median Square (LMS) estimation to improve the filtering capacity, especially when the outliers take up a large percentage (e.g. 50%) among all the references.

3   System Model and Threat Model

3.1   System Model

In COTA system, there are a small number of anchors who broadcast their locations in beacon messages. The sensors who can directly receive sufficient beacon messages localize themselves and serve as references points to other sensors. We assume that anchors are densely deployed within a local area, thus some sensors can firstly localize themselves and the location propagation process can get started. Current propagation techniques include DV-based method [19], [21], APS-Euclidean method [20], and Distributed Trilateration method [7], [22]. We use distributed trilateration as the location propagation method of COTA in this paper, leaving the study of other methods to our future work. We assume the communications between sensors are bidirectional and are protected by pairwise secret keys, thus the compromised nodes can not impersonate others and send multiple wrong position messages. The key distribution can be achieved through probabilistic or online algorithms.

3.2   Threat Model

Localization in sensor network is subject to various attacks [4], [13], [16], including cheating report, wormhole, range reduction, and range enlargement attacks. The attackers can be classified into (1) internal attackers who compromise sensors' key materials and can authenticate themselves to other nodes, and (2) external attackers who do not compromise nodes and cannot authenticate themselves. Traditional security mechanisms such as encryption and authentication are not sufficient to defend against external attackers.

- Cheating reports: Attackers send wrong position messages to other sensors.
- Wormholes: Attackers record the position messages received at one location, transmit and replay them at another location.
- Range enlargement: Attackers jam the signals between sensors and reply them at a later time if using Time of Arrival (TOA) for ranging, or they attenuate the signal strength if using the signal strength for ranging.
- Range reduction: Attackers speedup the signals if using Time of Arrival (TOA) for ranging, or they jam and replay the signal with higher power if using signal strength for ranging.

The cheating reports can only be sent by internal attackers; wormholes are generally launched by external attackers; the range modifications can be performed by either internal or external attackers. According to their impacts on localization, we consider them as *false position* and *false range* attacks (as shown in Table. 2). Note that wormhole attack belong to false position attack, since the end-point transmitter of the wormhole replays the location of the source-point node. All these attacks can produce false position references and mislead sensors to estimate wrong locations. COTA is designed to defend against all above attacks, and can achieve satisfiable localization performance for sensor networks.

Table 2. Classification of Threat Models

	False Position		False Range	
	Cheating Reports	Wormholes	Range Enlargement	Range Reduction
Internal Attacker	•		•	•
External Attacker		•	•	•

4   Proposed Robust Multi-hop Localization Scheme

4.1   Overview

In this paper, we propose a robust multi-hop localization scheme COTA. As shown in Fig. 1, during the localization process, each sensor stays in one of the states including waiting state, localizing state, and transmitting state. A sensor initially stays in waiting state and receives reference messages from its neighbors. As soon as the adequate number (which is three if using trilateration technique) of references are received, it enters the localizing state and performs COTA scheme. COTA consists of a localization phase and a tag-generation phase. In the first phase, a sensor filters out bad references according to the absolute and relative metrics, then checks if the remaining references are more than the minimum number. If no, it goes back to waiting state for more references; otherwise, it calculates its position through weighed optimization mechanism. In the second phase, the sensor firstly computes the statistical or geographical indictors, then derives its localization error and translates it into a confidence tag. Finally the sensor enters transmitting state, combines its estimated position and the confidence tag into a reference message and sends it to others.

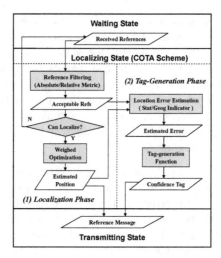

Fig. 1. Sensor States & COTA Scheme

4.2 Confidence Tag

Confidence Tag indicates the reliability of a sensor's estimated position, thus we provide the following definitions.

Definition 3 *(Tag-Generation Function).* Let p_e and p_t be a sensor's estimated position and true position, thus $e = |p_e - p_t|$ is its localization error. We call function $t = f_t(e)$ the *tag-generation function*, and call t the *confidence tag* of the sensor. $t \leq T$ is a nonnegative integer, where T is the highest confidence tag, e.g., $T = 8$.

Definition 4 *(Inverse-tag Function).* Let e and t be the localization error and the confidence tag of a sensor, and $\hat{e}$ is an upper bound of e, we call function $\hat{e} = f_e(t)$ the *inverse-tag function*.

We note that function f_e is not the exact inverse function of f_t, because instead of returning a sensor's localization error e from its tag t, f_e returns an upper bound $\hat{e}$. The coefficients of these functions can be computed (section 4.4) and stored in sensors memories, and they are used through COTA scheme.

4.3 Localization Phase

The localization phase of COTA consists of two function blocks: reference filtering and weighed optimization. We use the tuple (p_i, t_i, d_{ij}) to denote the location reference sent by node i and received by node j, where p_i and t_i are the claimed position and the confidence tag of node i, and d_{ij} is the mutual distance measured by sensor j.

Filtering Metrics. We provide two filtering metrics: the absolute metric and the relative metric. When using the absolute metric, sensor simply filters out

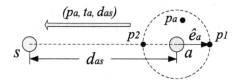

Fig. 2. COTA Filtering Metric: Relative Metric

bad reference whose confidence tag $t < t_0$. The threshold t_0 can be obtained through training. Let e_{max} be the maximum localization error of sensors in non-adversary scenarios, then $t_0 = f_t(e_{max})$ is the minimum reasonable confidence tag. If t_0 is set very high, then most references will be filtered out and a sensor may not acquire adequate number of references to localize. Therefore, we need to consider the tradeoff between sensors' localization error and the localized percentage. Actually, how to set a reasonable threshold is application-specific.

The relative metric is computed by $u = f_e(t)/d$. A sensor filters out bad reference if $u > u_0$, where u_0 is a preset threshold. We use a figure to illustrate the underlying idea. In Fig. 2, sensor s has a reference message (p_a, t_a, d_{as}). We can compute an upper bound of node a's localization error by function $\hat{e}_a = f_e(t_a)$. Namely, a's claimed position p_a should be within distance $\hat{e}_a$ of its true position. Hence, the distance from position p_a to sensor s should be no larger than $d_{as} + \hat{e}_a$ and no smaller than $d_{as} - \hat{e}_a$. The relative metric $u = f_e(t)/d = \hat{e}/d = \Delta d/d$ essentially indicates the relative accuracy of the mutual distance, and the references with higher uncertainties than the threshold will be dropped. We can obtain u_0 through training. If sensors' maximum localization error in non-adversary scenarios is e_{max} and the communication range between sensors is R, then we set the threshold by $u_0 = e_{max}/R$.

The two filtering metrics can be used separately or together. In our simulation, it shows that the absolute metric outperforms the relative one in providing stronger filtering capacity, but realizes less localized percentage. However, when both metrics are applied, sensors can be localized with better performance than that when using a single metric.

Weighed Optimization. After filtering out the bad references, each sensor uses the remaining references and performs the trilateration technique to compute its position. Because of the noisy rang measurements and various attacks to the location references, the unique solution may not exist to satisfy all the constraints. In COTA, we use Weighed Least Square Estimation (WLSE) mechanism to compute the optimal solution for each sensor.

Assume sensor s has n references (p_i, t_i, d_{is}), where $p_i = (x_i, y_i)$, $0 \le i \le n$. Sensor s weighs the references by their confidence tags and computes the optimal solution (x_0, y_0) by minimizing the following equation:

$$Min \sum_{i=1}^{n} t_i^2 \cdot (\sqrt{(x_i - x_0)^2 + (y_i - y_0)^2} - d_{is})^2 \qquad (1)$$

This nonlinear LSE optimization problem can be solved by many standard methods, e.g., the MMSE matrix solution [9] or Kalman filter method [2], [23]. In COTA, we adopt the MMSE technique and transmit the nonlinear problem into linear equations:

$$t_1 \cdot \sqrt{((x_1 - x_0)^2 + (y_1 - y_0)^2)} = t_1 \cdot d_{1s}$$
$$t_2 \cdot \sqrt{((x_2 - x_0)^2 + (y_2 - y_0)^2)} = t_2 \cdot d_{2s}$$
$$\cdots \quad \cdots \tag{2}$$
$$t_n \cdot \sqrt{((x_n - x_0)^2 + (y_n - y_0)^2)} = t_n \cdot d_{ns}$$

After squaring and rearranging terms on each side of equations 2, we obtain:

$$2t_1^2 x_1 x_0 + 2t_1^2 y_1 y_0 = t_1^2(x_1^2 + y_1^2 - d_{1s}^2 - x_0^2 - y_0^2)$$
$$2t_2^2 x_2 x_0 + 2t_2^2 y_2 y_0 = t_2^2(x_2^2 + y_2^2 - d_{2s}^2 - x_0^2 - y_0^2)$$
$$\cdots \quad \cdots \tag{3}$$
$$2t_n^2 x_n x_0 + 2t_n^2 y_n y_0 = t_n^2(x_n^2 + y_n^2 - d_{ns}^2 - x_0^2 - y_0^2)$$

Then we compute the average of above equations:

$$C_x \cdot x_0 + C_y \cdot y_0 = C_d - x_0^2 - y_0^2, \tag{4}$$

$$C_x = 2 \sum_{i=1}^n t_i^2 \cdot x_i^2 / \sum_{i=1}^n t_i^2,$$
$$C_y = 2 \sum_{i=1}^n t_i^2 \cdot y_i^2 / \sum_{i=1}^n t_i^2,$$
$$C_d = \sum_{i=1}^n t_i^2 \cdot (x_i^2 + y_i^2 - d_{is}^2) / \sum_{i=1}^n t_i^2$$

We multiply equation (4) by t_i^2 and subtract it from equations in (3). Finally, we get following standard linear least square equation:

$$A \cdot [x_0 \ y_0]^T = B, \tag{5}$$

$$A = \begin{bmatrix} 2t_1^2 \cdot (x_1 - C_x) & 2t_1^2 \cdot (y_1 - C_y) \\ 2t_2^2 \cdot (x_2 - C_x) & 2t_2^2 \cdot (y_2 - C_y) \\ \vdots & \vdots \\ 2t_n^2 \cdot (x_n - C_x) & 2t_n^2 \cdot (y_n - C_y) \end{bmatrix}, \quad B = \begin{bmatrix} t_1^2(x_1^2 + y_1^2 - d_{1s}^2 - C_d) \\ t_2^2(x_2^2 + y_2^2 - d_{1s}^2 - C_d) \\ \vdots \\ t_n^2(x_n^2 + y_n^2 - d_{ns}^2 - C_d) \end{bmatrix}$$

Then the optimal matrix solution can be given by:

$$[x_0 \ y_0]^T = (A^T A)^{-1} A^T B \tag{6}$$

4.4   Tag-Generation Phase

In the tag-generation phase, a sensor estimates its localization error then generates a confidence tag. Since sensor has no knowledge of its true position, it needs some indicators to derive its localization error. In this subsection, we firstly propose two indicators: the statistical indicator and geographical indicator, then discuss the construction of the tag-generation function and inverse-tag function.

Statistical Indicator. We use the weighed sum of error squares, called residual r, as the statistical indicator of sensor's localization error:

$$r = \sum_{i=1}^{n} t_i^2 \cdot (\sqrt{(x_i - x_0)^2 + (y_i - y_0)^2} - d_{is})^2 / \sum_{i=1}^{n} t_i^2 \qquad (7)$$

Consider that all the references are correct and accurate, namely, both the references' positions and mutual distances are correct, then the residual r will be minimized to zero and the sensor will be localized at its true position. It suggests that small residuals may imply consistent references and accurate location estimations for sensors. Thus, we intend to obtain an increasing function f_s which can map the residual r to sensor's localization error $\tilde{e}$ by $\tilde{e} = f_s(r)$.

We do experiments in non-adversary scenarios to explore the relationship between residual r and sensor's localization error e. However, we notice that small residuals sometimes lead to large localization errors. Such results are caused by the noisy distance measurements. In presence of measuring errors, incorrect solution may sometimes better minimize the residual than the true one. This *flex ambiguity* problem in localization is studied by David Moore et al. in [18], in which the authors also proposed a geographical constraint called *Robust quadrilateral* to the location propagation process. Localized sensors and one unlocalized sensor form several triangles. If all triangles satisfy $b \cdot sin^2\theta > d_{min}$ (where b is the length of the shortest side, θ is the smallest angle, d_{min} is a preset threshold), this quadrilateral is considered to be robust and can be used for location propagation. We add this constraint to the simple trilateration technique and find that the relationship turns more regular. In our simulation setup, we calculate the function coefficients as $\tilde{e} = f_s(r) = 2.5 * r + 1.5$.

Geographical Indicator. The statistical indicator requires the coefficients of function f_s computed offline and stored in sensors' memories. In this subsection, we propose a geographical indicator d_{max} computed online.

As shown in Fig. 3, sensor s has three references: (p_a, t_a, d_{as}), (p_b, t_b, d_{bs}), (p_c, t_c, d_{cs}). The location error of each reference node can be estimated by $\hat{e}_i = f_e(t_i)$, and the location uncertainty $\hat{e}_a$, $\hat{e}_b$, $\hat{e}_c$ can be translated into the uncertainty of the mutual distance, given that the claimed positions p_a, p_b, p_c are correct. Also consider the distance measurement error δd_{is}, we can obtain the overall distance error by $\Delta d_{is} = \hat{e}_i + \delta d_{is}$. Therefore, sensor s should reside inside several rings, each of which centers p_i and whose inside/ouside radiuses are $d_{is} - \Delta d_{is} / d_{is} + \Delta d_{is}$. We call the overlapping region of all these rings as *residing area* (the shadow area). Since s's true position is within this area, thus the maximum distance d_{max} from its estimated position p_s to this area is an upper bound of its localization error, which we take as the geographical indicator.

In what follows, we discuss how to obtain the geographical indicator d_{max} efficiently. Since the computation is expensive to determine *residing area* based on the intersection of rings, we replace each ring with a "frame" and simplify the computation process in following three steps (Fig. 4).

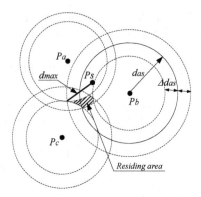

Fig. 3. COTA Localization Error Indicator: Geographical Indicator

Step 1: Replace each ring with a frame, whose outer/inner squares are tangent with the outer/inner circles of the ring.

Step 2: Obtain the overlapping region of all the outer squares, called *pre-area*. Its lower-left and upper-right coordinates are (X_l, Y_l) and (X_r, Y_u):

$$(X_l, Y_l) = (\max_{i=1\cdots n} \{x_i - \Delta d_{is}\}, \max_{i=1\cdots n} \{y_i - \Delta d_{is}\})$$
$$(X_r, Y_u) = (\min_{i=1\cdots n} \{x_i + \Delta d_{is}\}, \min_{i=1\cdots n} \{y_i + \Delta d_{is}\})$$

However, because of the noisy distance measurements and attacks, we may have $X_l > X_r$ or $Y_l > Y_u$ in above equations. Thus, we make following relaxations to the coordinates:

$$If\ X_l > X_r,\ X_l' = \min_{i=1\cdots n} \{x_i - \Delta d_{is}\}, X_r' = \max_{i=1\cdots n} \{x_i + \Delta d_{is}\}$$
$$If\ Y_l > Y_u,\ Y_l' = \min_{i=1\cdots n} \{y_i - \Delta d_{is}\}, Y_u' = \max_{i=1\cdots n} \{y_i + \Delta d_{is}\}$$

Step 3: Exclude from the *pre-area* any parts inside the inner squares, resulting in the *residing area*. Since the maximum distance from a point to a convex

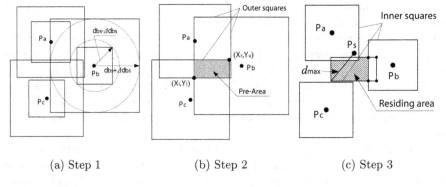

(a) Step 1   (b) Step 2   (c) Step 3

Fig. 4. COTA Geographical Localization Error Indicator: Computation Steps

polygon is between the point and one of the polygon's points, we only record the points' coordinates. When the *pre-area* intersects with an inner square, their overlapping region should be excluded (Fig. 4(c)), thus some points will be deleted and some new points will be introduced. On comparing the distances from p_s to each of the points, we can obtain the maximum distance d_{max}.

Tag-generation function. Each sensor uses the tag-generation function $f_t(e)$ to translate its estimated localization error into a proper confidence tag. $f_t(e)$ should have the following properties:

- Decreasing function: if $x_1 > x_2$, then $f_t(x_1) < f_t(x_2)$. This guarantees high localization errors will generate low confidence tags.
- The domain of $f_t(e)$ are sensors' localization errors, namely an infinite rational range $(0, +\infty)$.
- The range of $f_t(e)$ are sensors' confidence tags, namely a set of discrete nonnegative integers $\{0, 1, \cdots, T\}$.

Since decreasing function $f_t(e)$ maps an infinite interval to a finite set of nonnegative integers, thus when $e > e_0$, the function should output the lowest confidence tag zero. We can construct a linear function as following:

$$t = f_t(e) = \max(0, \lfloor -\frac{T}{e_0} \cdot e \rfloor + T), \tag{8}$$

where e_0 is the boundary value that $f_t(e_0) = 0$, T is the highest confidence tag.

We perform simulations in non-adversary scenarios to obtain a reasonable boundary value e_0. Since more than 90% sensors can be localized with $e < 10m$, we set $e_0 = 10m$ and believe that a sensor with localization error larger than that has a high probability to be attacked. We set $T = 8$ in our experiments. The value of T is application-specific, and higher T generally leads to more delicate differentiation between localization accuracies. The inverse-tag function $f_e(t)$ is closely related to $f_t(e)$. We provide the following construction:

$$\hat{e} = f_e(t) = -\frac{e_0}{T} \cdot (t - T), \tag{9}$$

where $\hat{e}$ is an upper bound of localization error e.

5   Security Analysis

Besides launching the known attacks to COTA as to other localization schemes, the adversaries may crab COTA scheme especially by manipulating sensors' confidence tags. We must consider the situation that compromised nodes or wormholes provide some references whose tags can not correctly represent the reliability of the location information. There are four types of attacks that an adversary can perform, and we list them from simple ones to sophisticated ones:

- Decrease-tag Attack: A compromised sensor can broadcast its location reference with a smaller tag than the true one. Since it's worth nothing to decrease the confidence tag of a false reference, which may cause it to be filtered out or bear small weight during localization, the adversary probably launch this attack to the correct location references.
- Remain-tag Attack: A compromised sensor or wormhole transmitter can produce incorrect location references whose confidence tags are kept unchanged.
- Increase-tag Attack: This attack is the opposite to the decrease-tag attack, in which the adversary broadcasts location references with higher confidence tags. Besides compromised nodes, wormholes can also perform this attack. They replay the location reference of the source-point node, namely, the false reference will bear the source-point node's tag.
- Invert-tag Attack: Sophisticated adversaries can invert tags: set low tags to correctly localized sensors and high tags to wrongly localized sensors.

For the decrease-tag attack, the resulting small-tag references will probably be filtered out before participating in localization, thus a sensor may fail to localize itself for lack of enough valid references. The adversary performs this attack to launch the denial of service (DOS). However, since they can simply jam the communication between sensors or destroy sensor nodes to launch DOS, delicately tampering the confidence tags cannot achieve any extra benefit.

In the remain-tag and increase-tag attacks, a sensor may estimate a false position, because incorrect references are not filtered out and may be high-tag attached. However, if the sensor being attacked can generate a low confidence tag to its location reference, it will be dropped by its neighbors and will not affect many other nodes. Therefore, COTA is robust in the sense that it prevents local damage from proliferating to other areas. The reason why wrongly localized sensors can generate low confidence tags relies in the effective indicators: as long as there are some benign references inconsistent with the incorrect ones, the statistical indicator (residual) will produce big values, and the geographical indicator will result in large residing area and a long distance d_{max}. Then the tag-generation function (decreasing function) can compute low tags from the large indicators. The increase-tag attack is more severe than the remain-tag attack, thus we simulate it in our simulations. We further assume that the adversaries always generate the highest confidence tag to the false references. Experiment results show that COTA can survive through various attacks and provide accurate location estimations.

The invert-tag attack is the most sophisticated, in which the adversary not only contaminate correct references by attaching low tags, but also disguise false references with high tags. The impact is that the victim sensor not only wrongly localizes itself but also computes a high confidence tag to its reference. However, to mount this attack, an attacker should be very resourceful to jam/tamper all the benign references and launch multiple false references. Furthermore, the attacker needs to launch invert-tag attack to each victim sensor, otherwise the victim sensor will performs as increase-tag attacker who can affect only the direct neighboring nodes.

Table 3. Simulation Parameters and Default Values

	Meaning	Default
FM	the filtering metric	Combination
EI	the localization error indicator	Statistical
D	the damage degree	20m
n	the number of false references	1
P	the percentage of sensors being attacked	10%
n_f	the noise factor of distance measurement	1%

6   Simulation Results

6.1   Simulation Setup

In our simulation, 600 sensors are deployed uniformly and randomly in a square field of $300m \times 300m$. The sensor-to-sensor and anchor-to-sensor communication range are both $R = 25m$, thus each sensor can hear 12 neighbor nodes on average. Within the center square area, whose lower-left/upper-right coordinates are $(100, 100)/(200, 200)$, we randomly deploy 10 anchors, thus about 18 (3%) sensors can receive adequate number of beacon messages to be firstly localized. Anchors' positions bear the highest confidence tag $T = 8$. We perform the non-protected multi-hop scheme and COTA in every deployment, and average the results over 100 independent deployments. The distance measurement error model is $\widetilde{d} = d + d * x * n_f$, where $\widetilde{d}$ and d are measured and real distance, x is uniformly distributed within $[-1, 1]$, n_f is the noise factor. E.g., if $n_f = 1\%$, then $|\widetilde{d} - d| \leq 1\% \cdot d$.

Various attacks (cheating report/wormholes/rang enlargement/range reduction) will result in either *false position* or *false range* attack. To launch the former attack, we change one of sensor s's location references from (p_i, t_i, d_{is}) to (p'_i, T, d_{is}). To launch *false distance*, we replace the reference with (p_i, T, d'_{is}), where d'_{is} is smaller than the communication range R, otherwise it will be easily detected. We perform the tag-increase attack to COTA by appending the highest tag T to each false reference. As we have discussed, this attack is both powerful

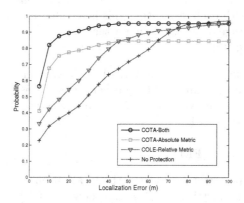

Fig. 5. CDF of Localization Error using different FM

and easy to perform. We also perform non-colluding and colluding attacks by launching different numbers of false references to target sensors. In the former scenario, only one of the references that sensor s uses is contaminated; in the latter scenario, multiple contaminated references will collude to mislead s to localize at another position.

6.2   Criteria and Parameters

The first criteria we use to evaluate the performance of COTA is sensor's local-ization error. The average localization error $\overline{L_e}$ is defined as:

$$\overline{L_e} = \frac{1}{N} \sum_{i=1}^{N} \sqrt{(\tilde{x}_i - x_i)^2 + (\tilde{y}_i - y_i)^2}, \tag{10}$$

where N is the total number of localized sensors, $(\tilde{x}_i, \tilde{y}_i)$ and (x_i, y_i) are the esti-mated and true positions of sensor i. We also study the Cumulative Distribution Function (CDF) of sensors' localization errors. CDF shows the percentage of localized sensors whose localization error is smaller than some value, thus tells us whether local damages have proliferated to other areas. The localized per-centage L_p is computed by $L_p = N_l/N$, where N_l is the number of sensors localized.

To evaluate the effectiveness of COTA, we test on the following parameters.

1. FM is the filtering metric, including absolute metric and relative metric.
2. EI is the localization error indicator, including the statistical and geograph-ical indicator.
3. D is the damage degree:
   Definition 5 (Damage Degree). Sensor s has a false reference (p_i, t_i, d_{is}). One or more of the items in the triple are contaminated. If the distance between the claimed position p_i and s's true position is d_t, then we call $|d_{is} - d_t|$ the damage degree of this false reference.
4. n is the number of false references received by the target sensor. When $n = 1$, this attack is non-colluding; when $n > 1$, multiple contaminated references are consistent with each other, thus this is a colluding attack.
5. P is the percentage of sensors being attacked.
6. n_f is the noise factor of the measured distances.

We summarize the parameters in TABLE 3. In the simulation, we vary one or two parameters at one time and keep others constant as their default values: FM is the combination of two metrics, EI is the statistical indicator, $D = 20m$, $n = 1$, $P = 10\%$, and $n_f = 1\%$.

6.3   Filtering Metric (FM)

The goal of this experiment is to study the localization performance of COTA with different filtering metrics (FM). Fig. 5 shows the CDF of sensors' localiza-tion errors when using the unprotected multi-hop scheme and COTA. We can

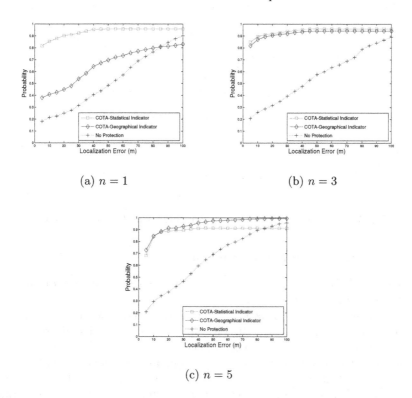

(a) $n = 1$ (b) $n = 3$

(c) $n = 5$

Fig. 6. CDF of Localization Error using Different Localization Error Indicators (EI)

see that without protection, more than 50% sensors are localized with $L_e > 30m$, which indicates that local damage has proliferated and impacted many sensors. When using COTA, the localization performance is effectively improved, e.g., more than 80% sensors are localized with $L_e < 10m$ when the two filtering metrics are applied together.

The two filtering metrics have different performances. Firstly, absolute metric can localize more sensors with small errors, e.g., about 70% sensors can be localized with $L_e < 10m$ when using the absolute metric, but only 40% when using the other. The reason is that absolute metric makes a sensor to discard all inaccurate references whose location errors are higher than e_{max}, but the relative metric fails to do so if the false reference contains a large mutual distance, which leading to a small $u = f_e(t)/d < u_0$. Secondly, the absolute metric has a low localized percentage: when the CDF curve turns flat, only 84% sensors are localized. By looking at sensor field snapshot, we found it is because many peripheral sensors are unlocalized for lack of enough valid references.

When the two metrics are used together, COTA has the best performance and achieves both small L_e and high L_p. Therefore, we use the two metrics together to filter out bad references in the rest of our studies.

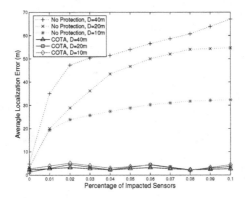

Fig. 7. $\overline{L_e}$ under Different Attack Percentages and Damage Degrees

6.4 Localization Error Indicator (EI)

The goal of this experiment is to study how the localization error indicator (EI) affects the performance of COTA.

Fig. 6 shows the CDF of sensors' localization errors when the number of false references are $n = 1, 3, 5$. In Fig. 6(a), the statistical indicator outperforms the geographical one: more sensors can be localized with the same localization error; when $n = 3$, they have very similar performances; when $n = 5$, the geographical indicator performs a little better. For example, in Fig. 6(c) about 92% sensors can be localized with $L_e < 20m$ when using the geographical indicator, higher than the 88% when using the statistical one. The reason is that when there are overwhelming consistent false references, the sensor will localize itself at a wrong position with a small residual, thus the statistical indicator can not properly indicate sensor's localization error. But the geographical indicator always produces large values of the residing area and d_{max}, as long as benign references can be received. Therefore, we conclude that the statistical indicator is more effective in defending against non-colluding attacks, but the geographical indicator performs slightly better against colluding attacks. We adopt the statistical indicator as default in our simulations.

6.5 Attack Percentage (P) and Damage Degree (D)

In this subsection, we test the robustness of COTA under different attack percentages (P) and damage degrees (D). We observe in Fig. 7 that when damage degree is set as $D = 20m$ and the attack percentage P increases from 1% to 10%, the average localization error $\overline{L_e}$ of non-protected localization scheme increases quickly from 20m to 50m. Secondly, $\overline{L_e}$ also increases with the damage degrees. E.g., when the attack percentage is set as $P = 2\%$, and $D = 10m$, 20m, 40m, the localization error $\overline{L_e}$ is 22m, 28m, 48m respectively. COTA effectively improves the localization performance: $\overline{L_e}$ is less than 5m in all the cases.

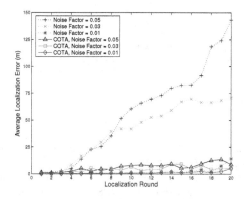

Fig. 8. $\overline{L_e}$ under Different Noise Factors

6.6   Noise Factor n_f

The goal of this experiment is to test COTA's property of mitigating error accumulation problem caused by noisy distance measurements in non-adversary scenarios. We vary the noise factor n_f and study the average localization error.

The simulation results are shown in Fig. 8. We compute $\overline{L_e}$ of sensors that are localized in each round. "Round" roughly indicates the localizing sequence of sensors. In the first round, the sensors who can receive adequate number of beacon messages from anchors are localized. Then the second-round sensors will be localized using location references from the first-round sensors and the anchors, and so forth. We can see from the figure that when noise factor is 1%, error accumulation problem is not severe that sensors in all rounds can be localized with small errors. However, when $n_f = 3\%$, L_e grows quickly as the round increases; when $n_f = 5\%$, the problem becomes more severe. COTA can greatly mitigate the error accumulation problem, that even in the last round, the average localization error is less than $10m$, which means the sensors who reside around the peripheral areas in the sensor field can still be accurately localized.

7   Conclusion and Future Work

In this paper, we proposed a robust multi-hop localization scheme COTA based on the novel notion of confidence tag. We evaluated the localization performance of COTA through simulations. It shows that COTA can effectively prevent local location errors from proliferating to other sensors and can provide accurate position estimations. COTA takes the first step toward robust multi-hop localization. For future work, we are planning to perform implementations and experimental evaluations in a real wireless sensor network testbed.

Acknowledgment

This research was supported in part by NSF under contract number DUE-0313837, DTO/ARDA under contract number NBCHC030107, and Carver Trust Foundation.

References

1. Bahl, P., Padmanabhan, V.: RADAR: An In-Building RF-based User Location and Tracking System , IEEE INFOCOM (2000)
2. Brown, R., Hwang, P.: Introduction to Signals and Applied Kalman Filtering (97)
3. Bulusu, N., Heidemann, J., Estrin, D.: GPS-less Low Cost Outdoor Localization for Very Small Devices, IEEE Personal Communications Vol.7 No.5, pp. 28 C34 (2000)
4. Capkun, S., Hubaux, J.: Secure Positioning in Sensor Networks, Technical report EPFL/IC/200444 (2004)
5. Doherty,L., Pister, K., Ghaoui, L.: Convex position estimation in wireless sensor networks, IEEE INFOCOM (2001)
6. Du,W., Fang,L., Ning,P.: LAD: Localization anomaly detection for wireless sensor networks, IEEE IPDPS (2005)
7. Eren,T., Goldenberg,D., Whiteley,J., Yang,Y., Morse,A., Anderson,B., Belhumeur, P.: Rigidity, Computation, and Randomization in Network Localization, IEEE INFOCOM (2004)
8. Fok,C., Roman,G., Lu,C.: Mobile Agent Middleware for Sensor Networks: An Application Case Study, IPSN (2005)
9. Greene,W., Econometric Analsis, third edition, Prince Hall (1997)
10. Harter,A., Hopper,A.,Steggles, P.,Ward,A.,Webster,P.: The anatomy of a context-aware application, ACM Mobicom (1999)
11. He,T., Huang,C., Blum,C., Stankovic,J., Abdelzaher,T.: Range-Free Localization Schemes in Large Scale Sensor Network, ACM MobiCom (2003)
12. Hu,L., Evans,D.: Using directional antennas to prevent wormhole attacks, Proceedings of the 11th Network and Distributed System Security Symposium, pages 131-141 (2003)
13. Hu,Y., Perrig,A., Johnson,D.: Packet Leashes: A Defense against Wormhole Attacks in Wireless Ad Hoc Networks, IEEE INFOCOM (2003)
14. Ji,X., Zha,H.: Sensor positioning in wireless ad-hoc sensor networks using multidimensional scaling, IEEE INFOCOM (2004)
15. Lazos,L.,Poovendran,R.: SeRLoc: Secure Range-Independent Localization for Wireless Sensor Networks, ACM Workshop on Wireless Security (WiSe)(2004)
16. Li,Z., Trappe,W., Zhang,Y.,Nath,B.: Robust Statistical Methods for Securing Wireless Localization in Sensor Networks, IPSN (2005)
17. Liu,D., Peng,N., Du,W.: Attack-Resistant Location Estimation in Sensor Networks, IPSN (2005)
18. Moore,D., Leonard,J., Rus,D., Teller,S.: Robust distributed network localization with noisy range measurements, ACM SenSys (2004)
19. Nagpal,R., Shrobe,H., Bachrach,J.: Organizing a Global Coordinate System from Local Information on an Ad Hoc Sensor Network, IPSN (2003)
20. Nicolescu,D.,Nath,B.: Ad-Hoc Positioning Systems (APS), IEEE GLOBECOM (2001)

21. Niculescu,D.,Nath,B.: Dv based positioning in ad hoc networks, Journal of Telecommunication Systems (2003)
22. Savvides,A., Han,C.,Srivastava,M.: Dynamic fine-grained localization in ad-hoc networks of sensors, ACM MobiCom (2001)
23. Savvides,A., Park,H., Srivastava,M.: The Bits and Flops of the N-hop Multilateration Primitive for Node Localization Problems, ACM WSNA (2002)
24. Shang,Y., Ruml,W., Zhang,Y., Fromherz,M.: Localization from mere connectivity, ACM MobiHoc (2003)
25. Shang,Y., Ruml, Y. Zhang, Fromherz, M.: Improved MDS-Based Localization, IEEE INFOCOM (2004)
26. So,A., Yu,Y.: Theory of Semidefite Programming for Sensor Network Localization, ACM-SIAM Symposium on Discrete Algorithms (SODA)(2005)
27. Want,R., Hopper,A., Falcao,V., Gibbons,J.: The Active Badge Location System, ACM Transactions on Information Systems, vol. 10, No. 1, pp. 91-102 (1992)

Contour Approximation in Sensor Networks

Chiranjeeb Buragohain[1], Sorabh Gandhi[1],
John Hershberger[2], and Subhash Suri[1]

[1] Dept. of Computer Science, University of California, Santa Barbara, CA 93106*
{chiran, sorabh, suri}@cs.ucsb.edu
[2] Mentor Graphics Corp., 8005 SW Boeckman Road, Wilsonville, OR 97070
john_hershberger@mentor.com

Abstract. We propose a distributed scheme called ADAPTIVE-GROUP-MERGE for sensor networks that, given a parameter k, approximates a geometric shape by a k-vertex polygon. The algorithm is well suited to the distributed computing architecture of sensor networks, and we prove that its approximation quality is within a constant factor of the optimal. We also show through simulation that our scheme outperforms several other alternatives in preserving important shape features, and achieves approximation quality almost as good as the optimal, centralized scheme. Because many applications of sensor networks involve observations and monitoring of physical phenomena, the ability to represent complex geometric shapes faithfully but using small memory is vital in many settings.

1 Introduction

We consider the problem of approximating polygonal paths and cycles in the context of a sensor network. The problem of representing complex geometric shapes using small memory is fundamental in many sensor net applications: sensor networks observe, measure, and track physical phenomena, which often involves representing and communicating a geometric shape. The problem arises, for example, in the application of computing contour lines on a field of sensor measurements [8]. Suppose that a geographically distributed set of sensors measures some physical parameter, say temperature, that varies smoothly over the sensor region. An analyst is interested in the rough shape of the temperature distribution, but does not care about the exact values measured by all the sensors. A collection of *isocontours*—cycles along which the measured and interpolated sensor values are constant—can be a useful summary of the distribution.

Contour lines reduce the data to be reported from two dimensions (the full set of sensors) to one dimension (dependent on only those sensors near the contour). However, even this reduction may not be enough. Communication is arguably the most important resource in a sensor net, and a one-dimensional contour whose feature size depends on the spacing of the sensors may contain too much

* The research of C. Buragohain, S. Gandhi and S. Suri was supported by grants from the National Science Foundation (CCF-0514738) and Army Research Organization (DAAD19-03D0004).

© Springer-Verlag Berlin Heidelberg 2006

data to send through the network back to the analyst. Therefore, it is important to consider methods for simplifying a one-dimensional contour that approximate the original data well and can be computed by a distributed network.

We use "contour approximation" as a guiding application, but our treatment of the problem is at an abstract level: distributed algorithms to compute a bounded-memory approximation of a polygonal curve embedded in a sensor field. Because sensor networks are envisioned as distributed "spatial instruments" that take measurements in a physical space but have limited resources (bandwidth, power, etc.), the ability to represent complex geometric shapes faithfully but using small memory is vital to sensor networks. In particular, significant improvement in system lifetime is possible if the network performs local computation to build compact approximations instead of sending the entire raw data to a centralized location. Indeed, a number of techniques have been proposed recently for "in-network aggregation" of sensor data [8, 12, 16]. The focus of these papers is on numerical summaries, such as min, max, average, or median, while the main focus of our paper is a fundamental form of *spatial summary*. Imagine, for instance, a physical phenomenon, such as a structural fault, that is evolving with time, and an analyst who wants to receive a periodic snapshot of the general shape of the phenomenon. Another possible application is building a compact representation of the boundary of the entire sensor field, which can be broadcast efficiently throughout the network so that each node knows the overall geographical coverage of the system. Awareness of the sensor field's shape can be useful in data storage schemes like Geographical Hash Tables (GHT) that associate data with geometric locations.

The problem of contour approximation was considered by Hellerstein et al. [8] in a sensor net setting. They proposed an algorithm in which a contour is initially approximated by its axis-aligned bounding box, and then the approximation is successively refined. At each step the approximate polygon encloses the original contour. Each refinement step deletes from the current approximation the maximum-area rectangular notch that lies outside the original contour. The refinement stops when the approximating polygon reaches some target complexity (number of vertices). This approach, while a useful heuristic, has several liabilities: (1) the restriction to rectangular approximation imposes an axis-dependence where none is required by the data; (2) the greedy maximization of area removed at each step does not ensure that the approximating polygon is near the original; and (3) the algorithm is a heuristic, with no proof of approximation quality. In [17], Singh, Bakshi, and Prasanna consider the problem of producing topographic maps over a sensor field using a quadtree-based approach, but they do not focus on constructing a compact representation of the map.

Approximating polygons is a fundamental problem that has been considered in many fields, including geographic information systems (GIS), computer vision, and computational geometry. In these settings the computational model favors centralized computation, in which all the input data are available to a single computational engine. Performance is measured in terms of approximation quality (in any of a variety of metrics) and running time/memory usage as a function of the input size n and the output size k. Typical algorithms include dynamic programming (which can optimize most metrics in roughly $O(n^2k)$ time [10, Chapter 3]) and the Douglas–Peucker algorithm (which provides good practical approximation quality in $O(n \log n)$ time [4, 9]). Because of the centralized computation requirement, however, these algorithms are ill-suited for use in a sensor net setting without significant adaptation.

Our Contribution

We make the following contributions in this paper: (1) We propose a new distributed algorithm, called ADAPTIVE-GROUP-MERGE (AGM), for polygon approximation with a worst-case constant factor *approximation guarantee*. (2) We develop a *distributed* wavelet-based scheme as a natural, simple alternative to AGM. (3) We show through simulation that AGM significantly outperforms the wavelet scheme in approximation quality. (4) Our experiments show that, in fact, AGM performs almost as well as the centralized, dynamic-programming-based optimal scheme. Thus, our new scheme is able to combine the virtues of the two extreme alternatives: it delivers the approximation quality of the centralized optimal scheme, but it incurs a computational and communication cost comparable to the wavelet scheme.

One of the most attractive features of our algorithm is its *locality*, which makes it highly suitable for heterogeneous multi-tiered sensor architectures, such as Tenet [7, 19]. These networks include a small number of high-powered (tier 1) nodes that act as clusterheads for many low-powered, mote-caliber (tier 2) devices. The motes simply collect and send their data to a neighboring clusterhead—the application software runs only on the tier 1 nodes. Using AGM, each tier 1 node can approximate its own portion of the contour *without jeopardizing* the global approximation quality. These partial contour approximations then can be exchanged among the tier 1 nodes to compute the final approximation. By contrast, centralized schemes such as dynamic programming or Douglas-Peucker require global knowledge of the data to decide which portions of the contour to keep, and thus are not amenable to distributed computation.

2   Preliminaries

We make the following assumptions about the sensors in the network: each sensor has a fixed radio range r, it knows its geographical location by using some localization technique [1, 14] and every sensor knows its neighbors' positions (other sensors within a circle of radius r). These assumptions, though somewhat idealized—radio ranges are not disks in practice [11, 20], and localization is nontrivial—are fairly standard in sensornet research, and allow us to focus on the approximation problem of interest. At the same time, we make no assumptions about the distribution of sensors in the field, or the shape of the field, so our results apply to an *arbitrary* collection of sensors.

Many different metrics have been used to measure the quality of a polygonal approximation. Two common choices are the L_p metrics and the Hausdorff metric. Given a polygonal curve S (a *polyline*) whose vertex sequence is $(p_1, p_2, \ldots, p_n)$, let $A = (a_1, a_2, \ldots, a_k)$ be a k-point approximation of S. To measure the approximation using the L_p metric, let $S' = (p'_1, p'_2, \ldots, p'_n)$ be the points on the polyline A closest to the corresponding vertices of S. Define the point coordinates to be $p_i = (x_i, y_i)$ and $p'_i = (x'_i, y'_i)$. Then the L_p approximation error of A is

$$\varepsilon_p \equiv ||S - S'||_p \equiv \left(\sum_i (|x_i - x'_i|^p + |y_i - y'_i|^p) \right)^{1/p}.$$

In particular ε_2 is the Euclidean mean squared error and ε_∞ is the maximum error. To define the Hausdorff approximation error, let $d(p, Q)$ be the minimum Euclidean distance from a point p to a polyline Q. Then the Hausdorff distance between S and A is

$$H(S, A) \equiv \max(\max_{0 \leq i < n} d(p_i, A), \max_{0 \leq j < k} d(a_j, S)).$$

Given the above definition of the distance $d(p, Q)$ between a point p and a polyline Q, we can think of the Haussdorff error as follows. The Hausdorff error between two polylines is the maximum distance of a point on either of the two polylines from the other polyline.

We will evaluate the efficiency of our algorithms primarily in terms of total communication complexity (also known as *message complexity*). If an algorithm requires N message transmissions, with each message of size m, then the communication complexity of the algorithm is defined to be $O(Nm)$. We will also consider total work (the sum over all processors of the running time they use) and overall running time (the elapsed time between the start and end of an algorithm). Overall running time helps us measure how much of the *computational parallelism* present in a sensor network we are able to exploit.

We assume that the isocontour (or the shape) to be approximated is already available to the network. The problem of determining an isocontour from raw sensor data is a well-studied problem, and many (distributed) algorithms are available. An interested reader may consult [3, 15, 17] for various approaches to constructing the contour boundary. Thus, we assume that a subset of the sensors, namely, $s_1, s_2, \ldots, s_n$, collectively stores the detailed representation of the isocontour, and the goal of our algorithm is to build a provable-quality approximation that fits in a given memory size. There has not been significant previous work on this *data reduction* aspect of isocontour construction.

3   Algorithms for Shape Approximation

We assume that the isocontour to be approximated is a polygonal curve embedded in the two-dimensional plane, and a sequence of sensor nodes $s_1, s_2, \ldots, s_n$

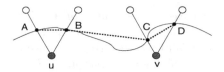

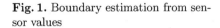

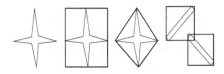

Fig. 1. Boundary estimation from sensor values

Fig. 2. Low approximation quality using bounding box or convex hull

collectively stores the contour. Specifically, each node s_i stores a consecutive subsequence of the contour polygon so that the concatenation of the chains stored at nodes $s_1, s_2, \ldots, s_n$ results in the full contour. We allow each sensor node to contribute arbitrarily complex portions of the isocontour because, in general, sensors can use complex and collaborative algorithms to compute the contour boundary. As an example (see Fig. 1), the contour detection algorithm may use interpolation to decide that the points A, B, C and D lie on the contour. Points A and B may be stored at node u, while C and D may be stored at node v. In order to keep the presentation of our algorithm simple, however, we will assume that each sensor s_i has only one vertex p_i of the contour. (The location of the contour vertex p_i does not necessarily coincide with the sensor s_i.) However, it will be clear from the description that our algorithm extends easily to the general case where each sensor may store a contiguous portion of the contour boundary.

We assume that adjacent sensor nodes storing the contour boundary are within the communication range of each other; that is, each node s_i is within one hop of s_{i-1} and s_{i+1}. Given a user-specified parameter k, where typically $k \ll n$, we wish to compute a k-vertex approximation of S. Of course, a trivial approach is to communicate all the vertices of S to a central node, and build the approximation there. This scheme, however, has message complexity $\Theta(n^2)$, and we seek more efficient alternatives.

In the following three subsections, we describe contour approximation schemes with which we will compare our new scheme ADAPTIVE-GROUP-MERGE. In Section 3.1, we briefly mention two naïve schemes, which are simple to compute but are too crude to be useful. In Section 3.2, we design a distributed two-dimensional wavelet-based scheme that takes advantage of the signal compression abilities of wavelets. This scheme is easy to implement in the distributed environment of the sensor network, though it lacks good theoretical bounds on the approximation quality. In Section 3.3 we describe a dynamic programming based algorithm that can compute an optimal contour approximation. The dynamic programming algorithm gives optimal approximation, but requires centralized computation, and so is ill-suited for an efficient implementation in sensor networks. It serves, however, as the ultimate benchmark for approximation quality.

3.1 Bounding Boxes and Convex Hulls

One of the simplest representations of any polygon is its *bounding box*, the smallest axis-aligned rectangle containing the polygon. The bounding box of S can be computed with $O(n)$ message complexity and time. Another simple representation is the convex hull of the polygon vertices, which can be computed exactly with $O(n^2)$ message complexity, or approximated to within any fixed relative error with $O(n)$ message complexity (using an approximation technique due to Dudley [2,5]). These approximations can be very poor, as shown in Fig. 2: they are too coarse, fail to highlight significant boundary features, and may lose important topological properties—the approximations of widely separated contours may intersect (Fig. 2, right side).

3.2 Wavelets

Wavelet transforms [13] have been used extensively in signal processing, image analysis and database operations. They represent a signal as a linear combination of normalized wavelet basis functions. A wavelet transform takes a one-dimensional signal sampled at n points $\{f_1, f_2, \ldots, f_n\}$ and outputs n coefficients $\{c_1, c_2, \ldots, c_n\}$ for a given set of basis functions. Given a parameter $k < n$, we construct a size-k approximation of the signal by retaining just the k coefficients with largest absolute magnitudes, and truncate the rest to zero. Let $\tilde{c}$ and $\tilde{f}$, respectively, denote the approximate wavelet coefficient vector and the reconstructed signal. Then the L_2 error of the approximation is given by

$$\sum_i (f_i - \tilde{f}_i)^2 = \sum_i (c_i - \tilde{c}_i)^2 = \sum_{i \in \text{truncated}} c_i^2.$$

We now describe a natural way to use wavelets for approximating a polygon embedded in the two-dimensional plane, and a distributed scheme to implement it. Suppose the coordinates of a point p_i are given by (x_i, y_i). We decompose S into two vectors S_x and S_y such that $S_x = (x_1, x_2, \ldots, x_n)$, $S_y = (y_1, y_2, \ldots, y_n)$. We carry out independent wavelet transforms on S_x and S_y, and achieve a compact representation of the curve by keeping only the k most important wavelet coefficients. We can implement this computation in a distributed fashion, with every sensor forwarding a single message to its neighbor in the sequence. The message from s_i to s_{i+1}, which has size $O(k + \log i)$ for the specific case of Haar wavelets [6], contains the top k wavelet coefficients of the sequence $p_1, p_2, \ldots, p_i$. Sensor s_{i+1} integrates its own coordinates into the wavelet transform and forwards the new message to s_{i+2} and so on. Sensor s_1 initiates the computation and when the message reaches s_n the algorithm terminates. Summing up the message sizes $\sum_i (k + \log i)$, we see that the total communication complexity of the DISTRIBUTED-WAVELET algorithm is $O(n(k + \log n))$.

Unfortunately, this algorithm does not exploit the parallelism available in the sensor network. In the full version of this paper, we describe a pipelined version of the distributed wavelets algorithm that completes the computation in optimal $O(n)$ time. We state this as a theorem.

Theorem 1. *There is a distributed implementation of the two-dimensional Haar wavelet approximation that takes $O(n)$ time, with total communication complexity $O(nk + n \log n)$.*

Two key disadvantages of the wavelet representation of a polygon are that it tries to minimize L_2 error, rather than the more important Hausdorff error, and it uses a fixed, nonadaptive set of basis functions. In Section 5, we will show some examples where these disadvantages lead to very poor approximations. This motivates us to consider approximation schemes that attempt to minimize the Hausdorff error.

3.3   Optimal Approximation Using Dynamic Programming

Our goal is to partition the polygonal curve $S = \{p_1, p_2, \ldots, p_n\}$ into k fragments $S_1, S_2, \ldots, S_k$, with associated approximating line segments $A_1, A_2, \ldots, A_k$. Each fragment consists of a subsequence $\{p_i, p_{i+1}, \ldots, p_j\}$ of S, with consecutive fragments sharing a common vertex. Each fragment and its approximating segment have an associated error value, and the error of a partition is the maximum error over all fragments in the partition. An optimal partition $OPT_k(S)$ is defined as a partition $Q(S)$ such that the error is minimum over all possible partitions of S. If the optimum approximating segment for a fragment depends only on the points of the fragment, then an optimal partition $OPT_k(S)$ can be computed using dynamic programming as follows. Let T be a $k \times n$ table, where $T(\alpha, j)$ contains the optimal (minimum) error for approximating the polygonal curve $\{p_1, p_2, \ldots, p_j\}$ using α segments, where $\alpha \leq k$. We wish to compute $T(k, n)$. The key insight is that the optimal α-segment approximation of $\{p_1, \ldots, p_j\}$ consists of two pieces: the optimal $(\alpha - 1)$-segment approximation of a prefix curve $\{p_1, \ldots, p_i\}$ for some $i < j$, and a single approximating segment for the fragment $\{p_i, p_{i+1}, \ldots, p_j\}$. This leads to the following recurrence:

$$T(\alpha, j) = \min_{1 \leq i < j} \max \left(T(\alpha - 1, i), \ e(i, j) \right), \tag{1}$$

where $e(i, j)$ is the error of the optimum single-segment approximation for $\{p_i, p_{i+1}, \ldots, p_j\}$.

We fill in the entries of T in increasing order of α, and for each α in order of increasing j. Since the table has nk entries and computing each entry using Eqn. 1 takes $O(n)$ time, the dynamic program runs in $O(n^2 k)$ time once the $e(i, j)$ values are known. The general recurrence of Eqn. 1 can be used to compute optimal approximations under several different error metrics. We use the following two in this paper:

1. *Fixed-Segment Error*: A fragment $S_\alpha = \{p_i, \ldots, p_j\}$ is approximated by the line segment $\overline{p_i, p_j}$. The error $e(i, j)$ is defined to be the maximum distance of any point in the fragment from $\overline{p_i, p_j}$, which is nothing but the Hausdorff error.

2. *Floating-Segment Error*: A fragment $S_\alpha = \{p_i, \ldots, p_j\}$ is approximated by the bisector of the *minimum bounding rectangle* (MBR) of the points in the fragment. The error $e(i, j)$, the maximum distance between any point of S and the approximating segment, is half the width of the MBR, which is also the width of S_α.

The floating segment model allows the approximating polygon to use arbitrary points in the plane as vertices, which can potentially improve the approximation quality. However, the approximating segments for neighboring fragments do not necessarily meet at a common point, and so additional segments may be needed to patch them into a connected polyline.

A third approximation model, which we may call the *Min-Link* model, allows the approximating polygon to use arbitrary vertices (not just vertices of S), but requires the approximating segments for neighboring fragments to share a common vertex. The optimum approximation for the Min-Link model *cannot* be computed by dynamic programming, because the optimum approximating segment for a fragment depends on points outside the fragment. Nevertheless, the optimum k-segment approximation under the floating-segment model has error no larger than the optimum k-segment Min-Link approximation (which has half as many vertices).

4   Adaptive-Group-Merge (AGM): Provable-Quality Contour Approximation

We now describe the main result of this paper: a new, efficient, distributed contour approximation algorithm that delivers a worst-case guarantee on the approximation quality. In particular, we show that whatever approximation quality the optimal (centralized) scheme achieves with k segments, our algorithm is able to achieve that with at most $2k$ segments.

We prove this guarantee using the Floating-Segment model of error, described in the previous section. That is, given an input polyline S, we consider an approximation A consisting of k *possibly-disconnected* segments. The polyline S is partitioned into k polyline *fragments* $S_1, \ldots, S_k$, each associated with an approximating segment A_i. The Hausdorff distance between S_i and A_i is ε_i, and the maximum ε_i over all i is the error ε of the approximation A. Because the segments of A are independent of each other, the error ε_i depends only on S_i. By choosing A_i to be the bisector of the MBR of S_i, we achieve error ε_i equal to half the width of the fragment S_i. (The width of a set is the minimum separation of two parallel lines that sandwich the set between them. The approximating segment A_i lies parallel to and halfway between these lines.)

Let us define the *width* of a partition of S into fragments to be the maximum width of a fragment S_i. Let us denote a partition of S by $Q(S)$, and its width by $width(Q(S))$. We call a partition optimal if it has the minimum width among all partitions of size k and denote it by $OPT_k(S)$.

In order to reason about the approximation quality of a partition, we define the *min-merge property*. A partition $Q(S)$ has the min-merge property if merging any two adjacent fragments results in a fragment with width at least as large as $width(Q(S))$.

One algorithm that produces a partition with the min-merge property is GREEDY-MERGE: starting with the trivial partition of S into n segments (all fragments with zero width), repeatedly merge the adjacent pair of fragments whose merge product has minimum width, until the partition consists of k fragments. It is easy to prove by induction that this algorithm produces a partition with the min-merge property. Likewise, applying GREEDY-MERGE to a partition with the min-merge property preserves the property. However, GREEDY-MERGE is not the only way to produce a partition with the min-merge property, as we will see.

We now argue that *any* partition into $2k$ fragments with the min-merge property has width no greater than that of $OPT_k(S)$.

Lemma 1. *Let $Q(S)$ be a partition of the path S into $2k$ fragments that has the min-merge property. Then $width(OPT_k(S)) \geq width(Q(S))$.*

Proof. Any partition of S into k fragments splits at most $k-1$ fragments of a $2k$ fragment partition. Therefore $Q(S)$ will have at least $k+1$ of its fragments unsplit. By the pigeonhole principle, there exists some fragment S_i of $OPT_k(S)$ that contains at least two unsplit fragments of $Q(S)$. By definition, $width(OPT_k(S)) \geq width(S_i)$, which is in turn at least as large as the the width of the union of the two unsplit fragments. By the min-merge property, this is at least $width(Q(S))$.

The preceding lemma assumes that S is a path, with distinct endpoints p_1 and p_n. If S is in fact a cycle, as in an isocontour application, then the proof can be modified to show that $width(OPT_{k-1}(S)) \geq width(Q(S))$. This difference in approximation quality between paths and cycles is minor, and we will ignore it in the remainder of this paper.

The GREEDY-MERGE algorithm maintains the min-merge property, as noted above. However, implementing GREEDY-MERGE in a distributed setting would require global minimization at each step, and thus would suffer from a serialization bottleneck. We propose an alternative hierarchical merging algorithm, and prove that it also preserves the min-merge property.

In the ADAPTIVE-GROUP-MERGE algorithm we divide the original curve S into n/k *groups*, each with k fragments of size 1 each. The total number of fragments is n. The algorithm proceeds in rounds that reduce the number of groups, maintaining the invariant that each group contains k fragments. In each round we split the current sequence of g groups into $\lfloor g/2 \rfloor$ disjoint pairs of adjacent groups (possibly with one group left over unpaired). For each pair we combine the two groups into one group of $2k$ fragments, then run GREEDY-MERGE on the combined group until the total number of fragments is k. We repeat this for $\log(n/k)$ rounds until the total number of fragments is k.

For this algorithm to work we need to argue that each of the groups it produces has the min-merge property. This is true for the initial groups of segments; the following lemma establishes the fact inductively.

Lemma 2. *Let Q and Q' be two adjacent groups of fragments of S, each containing k fragments and each with the min-merge property. If we apply GREEDY-MERGE to the union of Q and Q' until k fragments remain, the resulting group has the min-merge property.*

Proof. Without loss of generality assume $width(Q) \geq width(Q')$. It follows that $width(Q \cup Q') = width(Q)$. As long as the min-merge property does *not* hold, GREEDY-MERGE produces fragments with width less than $width(Q)$.

Thus the min-merge property starts to hold *just before* GREEDY-MERGE first produces a fragment with width at least $width(Q)$. In particular, if GREEDY-MERGE produces a fragment that includes two original fragments of Q, the min-merge property must have held prior to that round of GREEDY-MERGE. After $k + 1$ rounds of GREEDY-MERGE, at least $k + 2$ fragments of $Q \cup Q'$ are contained inside GREEDY-MERGE products, including at least two fragments from Q. Thus the min-merge property holds after k rounds of GREEDY-MERGE, if not before.

If n is not a multiple of k, at least one of the original fragment groups does not have k members, violating the precondition of Lemma 2. However, this is easy to overcome: to take up the slack we create one group of segments with size s in the range $k \leq s < 2k$, and greedily merge it to size k before the main AGM algorithm begins.

To implement this algorithm in a distributed fashion, we need to keep track of the widths of the new fragments after every merge operation. The simplest way to achieve this is to maintain the convex hull of the points in each fragment [18]. When two neighboring fragments are merged, the convex hull of the resulting fragment is the convex hull of the union of the convex hulls of the old fragments. Thus when a merge operation occurs, the merging fragments need to exchange information about their individual convex hulls.

In the worst case the convex hull of n points can have $\Theta(n)$ vertices. This would give a message complexity of $\Theta(n^2 \log(n/k))$ for AGM, which is more than we would like. Fortunately, we can approximate each convex hull H using only a constant number of points [2,5], such that the width of the approximation satisfies

$$(1 - \delta)width(H) \leq width(approx(H)) \leq width(H)$$

for any desired $0 < \delta < 1$. This degrades the approximation quality of the result by a small relative error, but allows the algorithm to run much faster. (The proof of correctness appears in the full paper.)

Using this convex hull approximation, we can implement the GREEDY-MERGE algorithm on each group of $2k$ fragments by sending all the fragment data (of total size $O(k)$) to a coordinator node within the group, and let it run GREEDY-MERGE locally. If the group encompasses N segments of S, this takes $O(kN)$ message complexity and $O(N)$ time. Summing over all groups in all rounds of the algorithm, we get total message complexity $O(kn \log(n/k))$ and total time $O(n)$. Putting it all together, we have the following theorem:

Theorem 2. *Given an n-vertex polyline stored in a neighbor-connected sequence of sensors, the algorithm* ADAPTIVE-GROUP-MERGE *computes an approximation by k segments whose approximation error is at most $(1 + \delta)$ times the error of the optimum approximation by k/2 segments, for any $0 < \delta < 1$. The algorithm has total message complexity $O(kn \log(n/k))$ and total running time $O(n)$, with the constant factor dependent on δ.*

The ADAPTIVE-GROUP-MERGE algorithm's approximation of S consists of k disconnected segments: adjacent segments do not necessarily meet. Thus the output is neither a polygon, nor directly comparable with other schemes like wavelets or Douglas–Peucker, because k disjoint segments require $2k$ vertices to describe the output. Of course, we could simply join each pair of adjacent segments, but that naïve scheme always produces a $(2k-1)$-segment approximation. In practice this may be improved: whenever joining two consecutive segments at the intersection of their supporting lines would not degrade the local approximation quality beyond the worst-case bound for the whole partition, we can omit a connecting segment. Our simulation results (cf. Section 5) show that indeed the size of the resulting polyline approximation remains close to k.

Note that the best polyline approximation by k segments has error at least as large as the error of the best approximation by k disconnected segments. This allows us to convert the Floating-Segment approximation guarantees of this section to bounds in the Fixed-Segment or Min-Link models, with the loss of another worst-case factor of two.

5   Experiments and Results

In this section we experimentally demonstrate the quality of approximations obtained by DISTRIBUTED-WAVELET and ADAPTIVE-GROUP-MERGE. We use dynamic programming as the optimal reference approximation.

The implementation of ADAPTIVE-GROUP-MERGE computes an approximation by k disconnected segments, then heuristically reduces the number of vertices in the final approximation by linking consecutive segments at the intersection of their supporting lines whenever that does not increase the error of the overall approximation. In our experiments we found that this heuristic reduced the size of the final approximation from $2k$ vertices to around $1.2k$.

Our implementation does not use the Dudley approximation [2, 5] as described in Section 4, because we did not want to introduce another parameter into the experimental setup. We computed the width based on the full convex hulls of the fragments. Using the Dudley approximation might degrade our approximation quality slightly. Interestingly enough, the full hulls were very small in practice—none contained more than eight vertices. This suggests that a practical implementation should be coded with a threshold, so that it uses the Dudley approximation only when the true convex hull has too many points.

We believe that the key measure of approximation quality in practice is the error associated with a given output size. Thus when we compare AGM against other algorithms whose output size is fixed, we choose an input parameter k for AGM that produces the same output size. In a practical setting, the user would specify an input parameter k, with the knowledge that the output will contain slightly more than k vertices. Because the algorithms produce polylines as output, we use dynamic programming with the Fixed-Segment error as our benchmark of quality.

In the following subsections we first compare the approximation performance of the three algorithms, then give a few brief vignettes focused on the approximation behavior of individual algorithms.

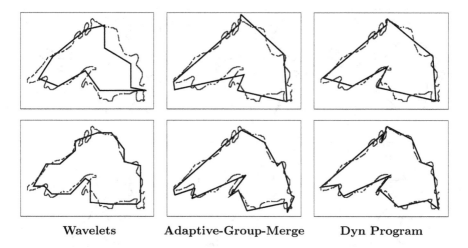

| Wavelets | Adaptive-Group-Merge | Dyn Program |

Fig. 3. Approximations for the Lake Superior dataset. The top row shows outputs for $k = 8$, the bottom row for $k = 16$.

5.1   Overall Approximation Quality

We compare the approximation performance of the algorithms on a GIS data set that digitizes the boundary of Lake Superior into 1024 points. Fig. 3 shows the approximations obtained with $k = 8$ and 16. Because wavelets aim to minimize L_2 error, the wavelet approximations cut off the extreme points and round them out. ADAPTIVE-GROUP-MERGE does better, and the dynamic programming reference algorithm gives the best results, as expected. The trend was similar for other data sets (Lake Huron boundary and Death Valley), and values of k ranging from 8 to 64.

Next we show the approximations obtained by ADAPTIVE-GROUP-MERGE on GIS datasets digitizing the boundaries of India and England into 1383 and 1213 points respectively. Fig. 4 shows the approximations obtained with $k = 32$, 48 and 64 points for both these boundaries. ADAPTIVE-GROUP-MERGE captures these complex boundaries faithfully using a relatively small amount of memory; approximation quality improves as k increases.

Next we evaluate quantitatively the approximations that are obtained by the three schemes. We measure the algorithms using the Hausdorff error and the *relative area error* ε_A we define as follows: if A_{diff} is the area of the symmetric difference between the regions enclosed by the original and approximate curves, and A_S is the area enclosed by the original curve, then $\varepsilon_A = A_{diff}/A_S$. We again work with the Lake Superior dataset to analyze the approximation performance of the three algorithms. In Figs. 5 and 6 we show the Hausdorff and ε_A errors respectively. ADAPTIVE-GROUP-MERGE consistently and significantly outperforms wavelets, and is typically very close to the optimal dynamic programming solution. Results for the L_1 and L_2 metrics were intermediate between those for the Hausdorff and ε_A errors.

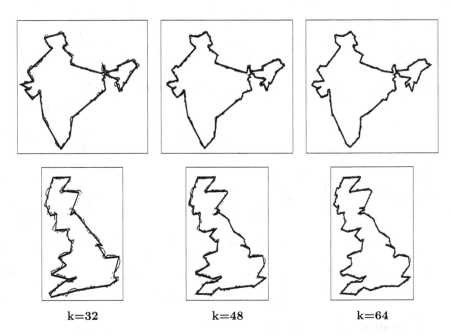

k=32   k=48   k=64

Fig. 4. AGM approximates complex shapes faithfully. The first row shows approximations for the boundary of India, and the second row shows approximations for the boundary of England.

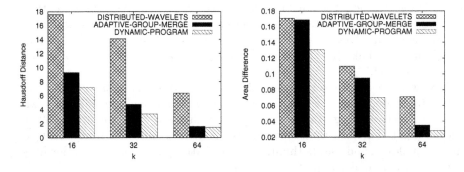

Fig. 5. Hausdorff distances from S to A **Fig. 6.** Fraction of area missed (ε_A)

5.2   Wavelets and the Effects of Sparse Sampling

Because wavelet approximations try to minimize mean squared error instead of maximum error, they can miss some important features, as illustrated in Fig. 7. The figure shows a hand-crafted point set of size 64 and the 8 point approximations obtained by DISTRIBUTED-WAVELET (Fig. 7(a)) and ADAPTIVE-GROUP-MERGE (Fig. 7(b)). The wavelet approximation tends to weigh all 64 points in the original curve equally, and so large errors for a few extreme points are offset

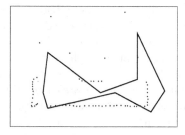 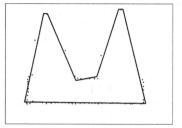

Fig. 7. Bad approximation by wavelets in low density regions

by small errors for the rest of the points. On the other hand the ADAPTIVE-GROUP-MERGE algorithm seeks to minimize maximum error and thus produces a much more acceptable approximation. This shortcoming of wavelet approximations is seen in more realistic data sets as well.

5.3 Approximation Quality in Theory and Practice

Although our primary concern is approximation quality as a function of the number of vertices in a polyline approximation, our provable quality bounds in Section 4 use the Floating-Segment error. We compared the Hausdorff errors of ADAPTIVE-GROUP-MERGE and dynamic programming in the Floating-Segment model to see how tight our bounds are. See the table below for the numerical results on the Lake Superior data set. As predicted, ADAPTIVE-GROUP-MERGE results for k segments are somewhat worse than the optimum k-segment approximation, but better than the optimum $(k/2)$-segment approximation.

# segments	ADAPTIVE-GROUP-MERGE	Optimum
8	15.8	13.3
16	7.43	5.14
32	3.27	2.39
64	1.33	1.01

6   Discussion

Reflecting on our simulation results, it seems clear that approximation quality improves markedly as an algorithm pays closer attention to the *geometry* of the shape. The wavelet algorithm uses a "generic" form of compression to reduce the representation size, which tends to treat all vertices the same. This has the virtue of simplicity, but often leads to poor approximation.

We also considered the Douglas–Peucker polygon simplification algorithm [4], which is popular in geographical information systems (GIS), computational geometry, and computer graphics. This is a greedy scheme that starts with a coarse representation (say, the four extreme vertices) then successively refines it by adding a new line segment at each step. At each step, the algorithm adds segments to the vertex that is farthest from the current approximation. In typical

GIS applications, the refinement continues until the maximum distance between the approximation and the input polyline drops below a specified threshold. In our setting, the termination occurs when the approximation reaches k vertices.

By design, the Douglas-Peucker scheme has a centralized flavor: at each step, it requires global computation to determine the vertex of the contour that is *farthest* from the current approximation. We have developed *distributed* variants of Douglas-Peucker, but decided to emphasize ADAPTIVE-GROUP-MERGE (AGM) for several reasons. First, much of the simplicity and computational efficiency of Douglas-Peucker is lost in adapting it to a distributed environment. Second, while it generally yields good approximations in practice, Douglas-Peucker does not have a good worst-case theoretical guarantee, while AGM does. And, finally, our experiments showed that distributed AGM produces approximations at least as good as the *centralized* versions of Douglas-Peucker, and hence we expect AGM to be the algorithm of choice in distributed settings.

By design, AGM is well-tailored for distributed environments. The localized nature of AGM allows the algorithm to carry out contour data reduction *independently* at nodes. In particular, if consecutive portions of the contour are available at m different nodes, then each node can reduce the contour size to $O(k)$ through entirely *local processing*, without risking global approximation quality. Thus, AGM may be especially well-suited for heterogeneous multi-tiered architectures like Tenet [7] where clusterhead nodes will aggregate data from nearby motes, and the application software will run only on clusterheads.

AGM guides its approximation by discarding those vertices whose removal leads to least increase in the error, and thus pays close attention to local features of the input—a long sequence of nearly collinear points may get replaced by just the endpoints, while peaks are preserved. We find it encouraging that (1) such a *locally adaptive* scheme yields a worst-case approximation guarantee, which others including wavelets and Douglas–Peucker do not; (2) even though AGM is limited by being tailored to the distributed architecture of sensor networks, it outperforms both wavelets and Douglas–Peucker in the quality of its approximation; and (3) in most cases, AGM performs almost as well as the optimal dynamic programming scheme (which is both centralized and slow).

References

1. N. Bulusu, J. Heidemann, and D. Estrin. GPS-less low cost outdoor optimization for very small devices. *IEEE Personal Communications Magazine*, 7(5):28–34, 2000.
2. T. Chan. Faster core-set constructions and data stream algorithms in fixed dimensions. In *Proc. 20th Annu. ACM Sympos. Comput. Geom.*, pages 152–159, 2004.
3. K. K. Chintalapudi and R. Govindan. Localized edge detection in sensor fields. In *IEEE Intl. Workshop on Sensor Network Protocols and Applications*, pages 59–70, 2003.
4. D. Douglas and T. Peucker. Algorithms for the reduction of the number of points required to represent a digitized line or its caricature. *Canadian Cartographer*, 10(2):112–122, December 1973.

5. R. M. Dudley. Metric entropy of some classes of sets with differentiable boundaries. *J. Approx. Theory*, 10:227–236, 1974.

6. A. C. Gilbert, Y. Kotidis, S. Muthukrishnan, and M. J. Strauss. One-pass wavelet decompositions of data streams. *IEEE Trans. on Knowledge and Data Engineering*, 15(3):541–554, 2003.

7. R. Govindan, E. Kohler, D. Estrin, F. Bian, K. Chintalapudi, O. Gnawali, S. Rangwala, R. Gummadi, and T. Stathopoulos. Tenet: An architecture for tiered embedded networks. Technical report, University of California, Los Angeles, November 10 2005.

8. J. M. Hellerstein, W. Hong, S. Madden, and K. Stanek. Beyond average: Toward sophisticated sensing with queries. In *Information Processing in Sensor Networks: 2nd Intl. Workshop*, pages 63–79. Springer-Verlag, 2003. LNCS 2634.

9. J. Hershberger and J. Snoeyink. Speeding up the Douglas-Peucker line simplification algorithm. In *Proc. 5th Intl. Symp. on Spatial Data Handling*, pages 134–143, 1992.

10. A. Kolesnikov. *Efficient Algorithms for Vectorization and Polygonal Approximation*. PhD thesis, Department of Computer Science, University of Joensuu, 2003.

11. D. Kotz, C. Newport, and C. Elliott. The mistaken axioms of wireless-network research. Technical Report TR2003-467, Dept. of Computer Science, Dartmouth College, July 2003.

12. S. Madden, M.J. Franklin, J. Hellerstein, and W. Hong. Tag: a tiny aggregation service for ad-hoc sensor networks. In *Proc. of OSDI '02*, 2002.

13. S. Mallat. *A Wavelet Tour of Signal Processing*. Academic Press, 1998.

14. R. Moses, D. Krishnamurthy, and R. Patterson. A self-localization method for wireless sensor networks. *EURASIP J. Applied Signal Processing*, 2003(4):348–358, 2003.

15. R. Nowak and U. Mitra. Boundary estimation in sensor networks: Theory and methods. In *Information Processing in Sensor Networks: 2nd Intl. Workshop*, pages 80–95. Springer-Verlag, 2003. LNCS 2634.

16. N. Shrivastava, C. Buragohain, D. Agrawal, and S. Suri. Medians and beyond: New aggregation techniques for sensor networks. In *Proc. of SenSys'04*, 2004.

17. M. Singh, A. Bakshi, and V. K. Prasanna. Constructing topographic maps in networked sensor systems. In *Proc. of Workshop on Algorithms for Wireless and Mobile Networks (ASWAN)*, 2004.

18. G. T. Toussaint. Solving geometric problems with the rotating calipers. In *Proc. IEEE MELECON '83*, pages A10.02/1–4, 1983.

19. M. Yarvis, N. Kushalnagar, H. Singh, A. Rangarajan, Y. Liu, and S. Singh. Exploiting heterogeneity in sensor networks. In *Proc. of IEEE INFOCOM 2005*, 2005.

20. J. Zhao and R. Govindan. Understanding packet delivery performance in dense wireless sensor networks. In *Proc. of the 1st Intl. Conf. on Embedded Networked Sensor Systems, SenSys'03*, pages 1–13, 2003.

A Distortion-Aware Scheduling Approach for Wireless Sensor Networks

Periklis Liaskovitis and Curt Schurgers

University of California San Diego, Electrical and Computer Engineering Department
{pliaskov, curts}@ucsd.edu

Abstract. An important class of applications for wireless sensor networks is to use the sensors to provide samples of a physical phenomenon at discrete locations. Through interpolation-based reconstruction, a continuous map of the monitored environment can be built. In this paper, we leverage the spatial correlation characteristics of the physical phenomenon and find the minimum set of nodes that needs to be active at each point in time for a sufficiently accurate reconstruction. Furthermore, multiple such sets of nodes are found so that a different set can report at each point in time in a rotating fashion. This is crucial in improving network lifetime. To perform all related scheduling tasks we employ a novel approach which does not assume a-priori knowledge of the underlying phenomenon. Instead it jointly estimates process characteristics and performs node selection online. We illustrate that significant gains in network lifetime can be achieved with minimal impact on the overall reconstruction quality, measured in terms of distortion.

Keywords: spatial random process, irregular sampling, distortion, lifetime, energy efficiency.

1 Introduction

Large scale networks of wireless micro-sensors are envisioned to enable the monitoring of physical phenomena without supervision for long periods of time. As sensor nodes often contain a limited energy supply, the individual nodes, and more importantly the network as a whole, have to operate in a highly energy efficient fashion. At the same time, it is crucial that any energy saving mechanism preserves a certain monitoring quality of the network, as this is its primary functionality.

It has been realized that sensors in close physical proximity can generate correlated readings, which can be exploited to increase the overall network lifetime with minimal effect on the monitoring quality. The basic principle is that at each moment in time, only a subset of all available sensor nodes is kept active, while the others are in an energy efficient sleep mode. If the active set of sensors is chosen appropriately, the negative impact on overall sensing quality can be kept to a minimum. It is crucial to note here that such a strategy does not lead to an increase in network lifetime by itself. In fact, only if multiple such sets of sensors, i.e. sets providing adequate coverage, are found and activated sequentially, does the overall network lifetime increase. This process of finding rotating sets is the idea behind sensing topology management.

P. Gibbons et al. (Eds.): DCOSS 2006, LNCS 4026, pp. 372–388, 2006.
© Springer-Verlag Berlin Heidelberg 2006

We can broadly identify two classes of sensing applications, both of which can benefit from such topology management, namely event-driven and continuous-monitoring applications. Event-driven sensor networks focus on detecting single events, such as the presence of an intruder or when the temperature exceeds a certain alarm level. Topology management for these systems has generally been investigated as k-coverage, where each point in space has to be sensed by at least k sensors [6]. Most approaches have assumed a known sensing range or sensing behavior [13].

On the other hand, our work focuses on the second class of applications, namely continuous-monitoring. In this case, the goal is to use the readings from the sensors to reconstruct a spatio-temporal estimate of a physical phenomenon. For example, a sensor network could be deployed to build a temperature map of an area. This essentially boils down to interpolation, as we want to reconstruct a continuous phenomenon from a discrete set of measurement points, i.e. readings at specific sensor node locations. In this case, topology management should try to find the minimum set of sample points needed to generate a sufficiently accurate interpolation, e.g. within a specified distortion bound. Note that this criterion for selection is distinctly different from the k-coverage problem for event-driven applications.

The primary focus of the work presented here is finding an efficient way to reduce the amount of data, i.e. the amount of sample points, output by the network at any point in time, while still being able to accurately reconstruct the monitored phenomenon in the entire spatial domain. We do not yet examine how this data actually reaches the data processing center. Our notion is that a reduced amount of data can benefit any type of aggregation scheme subsequently imposed on the sensors to eventually communicate this data. In this sense, our work most closely compares with Slepian-Wolf encoding of correlated sources as presented for instance in [7]. It is however, to the best of our knowledge, the first to practically address two vital issues: firstly, how the correlation structure of the sensed phenomenon can be discovered by the network on the fly, without a-priori assumptions on process statistics, which is often the case in real deployments; and secondly, how sensing behavior is determined by the physical correlation structure and leads to energy efficient reporting schedules. Our motivation is that the driving element for any real sensor network should be adjusting to the physical phenomenon at hand. For instance, the sensing range is not necessarily circular and the statistical behavior is not always Gaussian.

We will tackle the sensing topology management problem for continuous-monitoring applications assuming one-dimensional scenarios. This setup corresponds to applications such as monitoring the light intensity in a hallway or the humidity along a coast line and there exists a mobile access point (i.e. a robot or aircraft) for the sensor nodes, as discussed in [4] and references thereof. It also provides a useful first step into the more general, but more complex, problem in multiple dimensions.

2 Problem Overview

Consider N monitoring sensor nodes deployed over a one-dimensional spatial observation interval of length L. At distinct time instants t_i a data processing center seeks to obtain a reconstruction of the monitored phenomenon, within a distortion bound D_0. The reconstruction will be an interpolation of the measured values, as the phenomenon is a continuous function of space.

If N is large, correlation in the phenomenon translates to redundancy in measurements, so that, ideally, not all N readings are necessary to produce a sufficiently accurate interpolation. For energy efficient operation, the center should try to identify the minimum set of readings it needs to provide a sufficiently accurate interpolation, and request that only the respective sensor nodes report in the future. This set inevitably depends on the correlation characteristics of the phenomenon. In addition, the set of nodes should not remain fixed, but rotate among the available nodes to increase the lifetime of the monitoring system as a whole. The data center must thus devise a sequence of selections over time, i.e. impose a reporting schedule on the sensors. This operation can be succinctly described by the following optimization problem:

$$\text{Minimize:} \quad \sum_{m=1}^{N} \sum_{i=1}^{k} I(M_i, m) \tag{1}$$

$$\text{Subject to:} \quad \sum_{i=1}^{k} I(M_i, m) \le E_0 \quad m = 1, 2, ..., N \tag{2}$$

$$\text{and} \quad E[D] \le D_0 \tag{3}$$

M_i is a vector of length N where each element corresponds to one sensor. The element is equal to one if that sensor reports at t_i and zero otherwise. $I(M_i, m)$ stands for a function indicating whether the m-th position of vector M_i is unity or, equivalently, if sensor m is a member of the set reporting at t_i. $E[D]$ is an average measure of distortion and t_k is the last time instant of interest. The initial energy of the nodes is denoted as E_0. Minimization problem (1) is over the total number of messages, i.e. samples, sent from sensors to the data center. Constraint (2) states that the total number of messages sent by any sensor m cannot exceed its initial energy, assuming unit energy cost for a message. This simple model is adequate for our purposes of describing the energy consumption of the network especially since direct communication capability of the sensors to a mobile access point has been assumed.

There are two further observations to be made on our formulation:

- When a schedule is constructed, its performance, i.e. if it actually meets the distortion bound, can only be evaluated in an average sense as per (3), because the monitored phenomenon is a random process. The processing center must therefore possess statistics for the process.
- The most information the processing center can obtain about the monitored phenomenon at any single time instant is N sensor measurements. It has to rely on those N discrete values, perhaps over multiple time instants, to estimate the necessary statistics of the process it requires.

Building on these two observations, we propose a two-phase approach. In the first phase, the learning phase, all N nodes report their readings to the data processing center. This phase lasts for K time instants, i.e. K periods of periodic reporting. Based on the information gathered this way about the underlying physical phenomenon, the reporting sets are computed and the nodes are notified of their membership to these sets. Only designated nodes will report their measurements in future time instants.

This whole procedure assumes that the statistics of the underlying physical phenomenon do not change significantly during the lifetime of the network, i.e. that the phenomenon is stationary in time.

During the first phase, we use a procedure with two logical steps. In the first step, the available readings are used to distill useful statistical information about the underlying physical phenomenon. This is used in the second step to devise the rotating sets that satisfy the desired distortion bound. The architecture is graphically depicted in Figure 1.

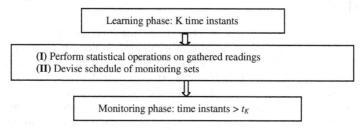

Fig. 1. Monitoring Architecture

The learning phase is potentially costly in terms of energy consumption, as all nodes need to report their measurements for K time instants. Note however, we will evaluate different alternative schemes to devise the reporting sets, which might require different levels of statistical knowledge of the process and hence different values of K. Before introducing our schemes, we will first give a more detailed mathematical description of our setup in the next section. This includes the modeling of the physical process and a discussion of the basic interpolation at the data processing center.

3 Physical Process

At each point of continuous space, the measurable value of the phenomenon is a random variable of unknown distribution. Values at points in close proximity are, in general, correlated with one another, thus forming a spatial random process $S(x)$. At each point in time t_i the values of the process in space constitute a sample function or realization of the process $S_i(x)$.

We assume that $S(x)$ is real, wide sense stationary (WSS) and ergodic in mean and correlation over the entire observation interval. This is a mild assumption since it holds for many real world processes as well as most correlation models appearing in current literature [1][4][7]. Again, without compromising the applicability of our methodology on real-world random processes we assume that $S(x)$ has a continuous covariance function $R_S(x-y)$ and power spectral density $\phi(\omega)$ given by:

$$R_S(x-y) = R_S(\tau) = E_S[S(x) \cdot S(y)] = \int_{-\infty}^{\infty} \phi(\omega) \cdot e^{j\omega\tau} d\omega \qquad (4)$$

The power spectral density function is also assumed to satisfy the condition:

$$\phi(\omega) = 0 \qquad \omega \notin (-B, B) \qquad B < \pi \qquad (5)$$

This condition means that the process can effectively be represented by its equidistant samples without loss in spectral information, i.e. that it is band-limited. The band-limited assumption has been extensively employed in recent research work to describe smoothly varying physical phenomena such as temperature and humidity [2]. Note that   refers to cyclic frequency normalized by the rate with which these equidistant samples are obtained, F_s.

4 Interpolation

Sensor nodes (we also refer to them as 'sensors' for short) are distributed over the spatial observation interval L in a uniformly random fashion. Their exact locations $\{x_n\}$, where $\{x_n\}$ is an ordered set, are assumed to be known and fixed over the whole lifetime of the network. This is achievable by running a localization service in the network. In addition, sensors are indexed by increasing order of their positions. The messages sent to the data center, are assumed to contain exactly one time-stamped value $S_i(x_n)$ and impose a standard unit energy cost. When it is not reporting, a sensor is in sleep mode. In this mode, the node saves energy by not sensing, not processing data and not sending it, where the last contribution is often dominant. If direct transmission to the data processing center is not feasible, nodes might have to resort to multi-hop communication. This imposes an additional constraint on when nodes can turn their radio off, and which is governed by communication topology management. In this paper, as previously discussed, we solely focus on the sensing topology problem, postponing the interaction with multi-hop communication to future work.

With the assumption of no measurement and no communication noise, at each time instant of interest t_i the data processing center possesses the values $S_i(x_i)$. The vector x_i contains only the positions of those sensors that were scheduled and reported and has as many elements as are the non-zero entries of M_i. The operation performed on this data to reconstruct the process is of the form:

$$\hat{S}(M_i, x) = \sum_{k=1}^{|x_i|} g_k(x, x_i) \cdot U_k \qquad (6)$$

where $U_k = S_i(x_k)$. The processing center basically treats the realization $S_i(x)$ as a deterministic function of space and interpolates its reported values. The term 'interpolation' stresses the requirement that the resulting reconstructed version of the realization should, at least, induce zero error at the sampling points, i.e. $\hat{S}(M_i, x_k) = S_i(x_k)$, $k = 1, ..., |x_i|$. Note that (6) is a linear operator with respect to U_k. Distortion can be readily defined as:

$$E[D_{M_i}] = E[\int_0^L (\hat{S}(M_i, x) - S(x))^2 dx] \qquad (7)$$

Equation (7) indicates the average reconstruction performance of a given set of reporting sensors M_i over all possible realizations of the process in time.

The general interpolation equation (6) utilizes all sampled values and all sampling positions to compute the estimate at a point x of the observation interval. Specific interpolation schemes should define the form of the weighting functions $g_k(x, x_i)$, usually referred to as interpolation kernels [8]. The characteristic of interest for an interpolation kernel is the support it requires on the observation space, i.e. how many sampled values need to be taken into account for the computation of the estimate at a point x. A kernel with large sample support results in good reconstruction for intermediate points of the observation interval but enhances edge effects in a real system and is computationally burdensome.

A generic class of interpolators with finite sample support is Lagrange interpolators defined by:

$$\hat{S}(M_i, x) = \sum_{k=m+1}^{m+p+q} (\prod_{\substack{n=m+1 \\ k \ne n}}^{m+p+q} \frac{x - x_n}{x_k - x_n}) \cdot U_k \tag{8}$$

where $x_{m+1} < \ldots < x_{m+p} < x < x_{m+p+1} < \ldots < x_{m+p+q}$ are the positions of $p+q$ sampled values, of which p immediately precede x and q immediately follow x, as indicated by the vector x_i. For $p = q = 1$ the scheme corresponds to linear interpolation.

For reasons that will be clarified later on, we have primarily used the following interpolation scheme in our system:

$$\hat{S}(M_i, x) = \sum_{k=\left\lfloor \frac{h_i(x)}{\Delta} \right\rfloor - 1}^{\left\lfloor \frac{h_i(x)}{\Delta} \right\rfloor + 2} r(h_i(x) - k \cdot \Delta) \cdot U_k \tag{9}$$

with r(x) being the cubic four-point interpolation kernel [8] with $\square$ = -0.5, which we do not reproduce here due to lack of space and $h_i(x)$ is a nonlinear space transformation such that:

$$v_i(x) = \frac{dh_i(x)}{d(x)} > 0 \text{ in } (0,L) \quad (10) \quad \text{and} \quad \int_{x_k}^{x_{k+1}} v_i(x)dx = h_i(x_{k+1}) - h_i(x_k) = \Delta \tag{11}$$

The basic idea behind the scheme (9) is to 'stretch' space so that consecutive sampling positions become effectively equidistant in the transformed space, with $\square$ being their resulting distance. The function $h_i(x)$ defines the one-to-one mapping between a point in the original space and a point in the transformed space. Equations (10) and (11) can for instance be satisfied by a piecewise-linear mapping function.

Next, we present the main contributions of this paper: how to construct generic monitoring schedules that meet pre-specified distortion bounds through online estimation of process statistics, and how to best exploit available statistics to devise highly energy efficient schedules.

5 Monitoring Schemes

A monitoring schedule, as previously mentioned, is a sequence of sensor set selections over time. Selected sensors have to be able to provide reconstruction within some distortion bound D_0. In addition, selected sets should be as disjoint from each other as possible. To that end, the first scheduling scheme we examine is uniformly random selection.

Random Selection: At each time instant choose M out of N sensors uniformly at random to report.

A conceptually similar scheduling scheme is:

Random Binning: Divide the observation interval in A_1 bins, each of length $L /_1$. In each bin choose A_2 sensors uniformly at random to report. To reach the desired number of active sensors M, the data center may have to sequentially choose one sensor at random to report from the bin that currently contains the most (not previously selected) sensors. The intuition behind this scheme is to have at least one sensor active and reporting in each bin at all times. Therefore, if the choices of parameters are such that $M < L / A_1 * A_2$ the value of A_2 is forced to unity.

The next scheduling scheme stems from the familiar notion of equidistant sampling of a signal:

Semi-Equidistant Sampling: At time instant t_i choose the M sensors that are closer to positions

$$x_k = Offset_i + k \cdot \frac{L}{M} \qquad k = 0,...., \frac{L - Offset_i}{\left\lfloor \frac{L}{M} \right\rfloor}$$

to report. What this scheme basically does is, starting off at a different initial point for each time instant, find positions that are spaced $\frac{L}{M}$ apart and then schedule the sensors that are closest to those positions to report.

The remaining question is how to choose M so that the M selected sensors give distortion D_0 on the average. At system initialization, the processing center has no information about what M should be, because it has no knowledge about the process it is monitoring. Even for these simple monitoring schedules there is a need for distortion prediction ahead of time. Specifically, the data processing center needs a way to be able to evaluate the average performance of a configuration of sensors M, without actually activating that configuration.

This operation corresponds to step 1 of the scheduling algorithm, as was shown in Figure 1. Overall, monitoring with the above schemes is performed as follows:

- **Step 1:** Construct a distortion prediction curve i.e. a function $\hat{E}[D_M]$ for all vectors M. From the curve find the number of sensors M that gives average distortion D_0.
- **Step 2:** At each time instant select M out of N sensors in the way defined by the corresponding scheduling scheme.

We next elaborate on an algorithm that, given a systematic way to select M out of N sensors, produces a distortion prediction curve tailored to this selection scheme. The method requires reporting from all available sensors for K initial time instants.

5.1   Distortion Prediction

Introducing the discrete process $Q(x)$: $Q_{i,M_i}(x) = S_i(x) - \hat{S}(M_i, x)$ (12)

equation (7) can be re-written as: $E[D_{M_i}] = E[\int_0^L Q_{M_i}^2(x)dx]$ (13)

This new process evolves in time and space according to how close the estimate of $S(x)$ being constructed at the data processing center is to the actual process.

---

1. Begin operation at $t = t_0$.

2. Activate all N sensors for the initial K time instants. Record K sets of reported values.

3. Perform interpolation for each set separately according to (9). Call the interpolated sequences $\hat{S}_n(M_0, x)$, $n = 1, \ldots K$.

4. To predict distortion for a target sensor set M_i, interpolate the values of the sensors, comprising set M_i to produce $\hat{S}_n(M_i, x)$, $n = 1, \ldots K$. Then compute the quantity:

$$\hat{E}[D_{M_i}] = \frac{1}{K} \cdot \sum_{n=1}^{K} \int_0^L (\hat{S}_n(M_0, x) - \hat{S}_n(M_i, x))^2$$

Algorithm 1. Distortion Prediction

To predict distortion, we need an estimate of the second moment of the process $Q_{Mi}(x)$ as per (13); this can be achieved if we consider that the realizations of $S(x)$ and hence their reconstructed versions $\hat{S}(M_i, x)$ are ergodic. The natural estimator

$$\hat{E}[D_{M_i}] = \hat{E}[\int_0^L Q^2_{M_i}(x)dx] = \frac{1}{K} \cdot \sum_{n=1}^{K} \int_0^L Q^2_{n,M_i}(x)dx \tag{14}$$

is in this case the optimal one. In order to use (14), however, we still need the exact values of the realization $S_n(x)$. We propose to approximate $S_n(x)$ by $\hat{S}_n(M_0, x)$, where M_0 is the set of all available sensors. To obtain a sufficient number of values of $Q(x)$ all sensors have to be active and reporting for K time instants. Algorithm 1 systematically presents these operations.

5.2 Simulations

Reliable distortion prediction is an integral part of scheduling. We present simulation results on the quality of our distortion prediction algorithm for the three scheduling schemes presented in section 5. First, we elaborate on our simulation setup, which is the same throughout the paper. We generate one realization of a spatially correlated process by feeding zero mean uniformly random white noise of unit variance into a low-pass filter, referred to as the spatial filter. The two filters we have experimented with have transfer functions as shown in Figure 2.

As can be seen, the process generated by filter H_1 has negligible energy outside the interval (-0.0314, 0.0314) rads/sample resulting in bandwidth $B_1 = 0.0314$ rads/sample while $B_2 = 0.157$ rads/sample. The specific choice of filters was made because H_1 imposes the same gain to all frequencies of interest, thus generating process realizations with an accurately defined bandwidth, whereas H_2 is frequency selective.

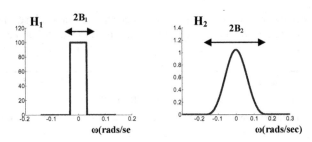

Fig. 2. Filters for process generation

We choose a one dimensional field of length $L = 10$ m. To mimic a continuous process in our calculations, we have divided the field in tiny slots of length $\Box x = 0.001$ m. Furthermore:

$$L = \Theta \cdot \Delta x \qquad (15)$$

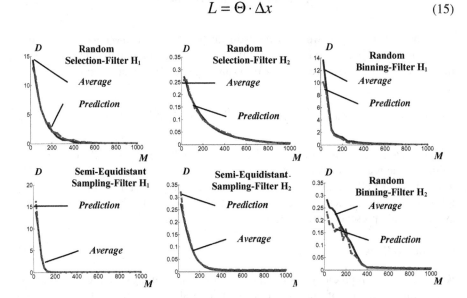

Fig. 3. Predicted and true average distortion for three scheduling schemes

In this case, $\Box$ is 10000. The N sensors are deployed uniformly random over this field. We chose $N = 1000$ for the results presented here, but also experimented with other values. We evaluated the quality of distortion prediction as described by Algorithm 1. Always starting from the same random initial deployment of N sensors, we removed $N - M$ sensors according to each of the scheduling schemes described in section 5 and logged the true distortion suffered by the resulting set of M sensors, averaged over a hundred process realizations. For distortion prediction, we generated K realizations of the process, performed interpolation of all N initial sensor values and then for each reduced set of M sensors estimated distortion by regarding these 'best' interpolated versions as estimates of the true signal. We experimented with different values of K. These results are not included due to space limitations. A good compromise between prediction accuracy and overhead was found to be $K = 7$. The resulting true vs. predicted distortion curves are shown in Figure 3.

The parameters chosen for the random binning scheme were $A_2 = 2$ for both filters and $A_1 = 0.1$ for filter H_1 and $A_1 = 0.025$ for filter H_2. We observe that, in all cases, prediction closely matches the average up to the knee of the curve, which is the interesting region for most practical scenarios, i.e. those requiring relatively low distortion.

The overall topology management procedure is thus as follows. In step 1 (see also section 5), these prediction curves of distortion versus M (the number of sensors that remains active in each set) are generated. Given a predefined desired level of distortion, the appropriate value of M is found. This M is the only input needed for the random selection, random binning and semi-equidistant sampling procedures in order to devise the reporting sets of sensors.

5.3   Minimal Semi-equidistant Sampling

Two key observations can be made on the monitoring schemes discussed so far:

- They can suffer in terms of energy consumption from a potentially high value of K. K defines the duration of the energy-expensive learning phase where all sensors have to report. Its value essentially depends on the temporal correlation characteristics of the process and may greatly fluctuate for different scenarios.
- The curves shown in Figure 3 represent a continuous tradeoff of energy spent for sensing and reporting versus distortion induced when reconstructing at the data center. But we also notice that the curves corresponding to Semi-Equidistant Sampling have a relatively sharp knee, specifically at $M \cong 120$ sensors for filter H_1 and at $M \cong 400$ sensors for filter H_2.

The latter observation is related to the fact that we are performing equidistant sampling on a band-limited process. It would be beneficial therefore to discover the minimum possible M that can give us arbitrarily low distortion in the Semi-Equidistant Sampling scheme. According to the Nyquist criterion, a band-limited continuous signal with W as its highest cyclic frequency component can be reconstructed perfectly from its equidistant samples, if they are obtained at a cyclic sampling frequency $F_{s,min}$ at least equal to $W / $. Equivalently the sampling period $T_{s,max}$ can be at most $ / W$.

The maximum sampling period is roughly the threshold distance after which equidistant samples of the signal become uncorrelated and therefore cannot provide any useful information about the signal. Thus, we will hereafter refer to the maximum sampling period as the correlation distance C_d of the signal (this is not exactly accurate nomenclature, but is done to make the discussion more intuitive).

Our goal now is to estimate the bandwidth of the process or, equivalently, its correlation distance. We can then obtain the fewest possible number of equidistant sensors necessary to reconstruct the process with arbitrarily low distortion. However, we are constraint to the actual (random) positions of the N deployed sensors, and therefore typically cannot achieve this ideal equidistant sampling. However, as we will show, we can still utilize this information to generate a minimal semi-equidistant sampling monitoring scheme.

For true equidistant sampling, the minimum number of sensor nodes needed is:

$$M_{min} = \frac{L}{C_d} \tag{16}$$

This is essentially the Nyquist rate of the process. Note that the discretization of the space in slots of ▫x, imposes the following relationship between the bandwidth of the continuous space process, W, and the normalized bandwidth, B [12]:

$$B = W \cdot \Delta x \Rightarrow T_{s,\,max} = \frac{\pi}{B} \cdot \Delta x = C_d \tag{17}$$

A substitution in (17) readily gives us the Nyquist rates for the processes generated by the filters H_1 and H_2 as 100 sensors and 500 sensors respectively, which are close to the knees of the curves in Figure 3, as cited earlier.

With the assumption of possessing an algorithm (presented in the next subsection) that estimates the bandwidth B, the minimal semi-equidistant monitoring policy will be:

Minimal Semi-Equidistant Sampling:

Step 1: Estimate B, C_d and M_{min} from (17) and (16).
Step 2: At time instant t_i choose the M_{min} sensors that are closer to positions

$$x_k = Offset_i + k \cdot \frac{L}{M_{min}} \qquad k = 0,...,\ \frac{L - Offset_i}{\left\lfloor \frac{L}{M_{min}} \right\rfloor}$$

to report.

One can observe that the sensor selection taking place in step 2 is identical to that of Semi-Equidistant Sampling. Minimal Equidistant Sampling is a special case of Semi-Equidistant Sampling for $M = M_{min}$. Because it uses the minimum possible M that can achieve arbitrarily low distortion, it is also the best possible semi-equidistant sampling scheme in terms of producing disjoint sets of sensors. The maximum number of disjoint sets thus produced is N / M_{min}. We refer to this ratio as the oversampling factor of the network. Of course, this policy would be most beneficial, if also bandwidth estimation in step 1 needed a low value for K. In the following section we present such a bandwidth estimation algorithm.

5.4 Bandwidth Estimation

The problem of bandwidth estimation readily translates to estimating the spectrum of the process, i.e. discovering which frequencies hold the bulk of its power. This needs to be done at the data center from a limited number of process samples $S_i(x_n)$ that are irregular in space, because the sensors have been deployed at random. To achieve this we take advantage of the extensive research that has been conducted on the topic of spectrum estimation i.e. estimating the power spectral density of the process as defined by (4).

Two categories of estimators have emerged as prevalent in this area. The first category consists of estimators that directly operate on the irregularly sampled data, such as the Lomb-Scargle estimator [11]. The second one includes estimators that first perform resampling, i.e. interpolation of the data on a regular grid, and then utilize one of the numerous techniques for spectrum estimation of regularly sampled data, such as the smoothed periodogram or auto-regressive modeling [9]. For reasons of simplicity, we have chosen to use an estimator of the second category.

Resampling, essentially means interpolation on the regular grid defined in (15). Initially, values from all available sensors are used for resampling and eventually we obtain $\hat{S}(M_0, x)$ computed at a set of $\square$ discrete positions of the observation interval $\{y_k\}$. To leverage accuracy in resampling, we demand that all sensors report for the initial K time instants and concatenate the resampled sequences. The assumption here is that the initial set of sensors is numerous enough to provide us with a near errorless representation of the process realization so that the bandwidth estimate will be close to the true bandwidth. Recall that a similar assumption holds for distortion prediction.

To get a reliable resampled sequence for our bandwidth estimation procedure, we choose to interpolate the initially reported values with the Lagrange method (8) instead of (9). Our initial experiments indicated that Lagrange interpolation results in very low distortion for $p + q > 4$ (Lagrange parameters), but only if the used number of sensor values well exceeds the Nyquist rate. Otherwise, distortion can be very high, which is undesirable and therefore we use (9) in all other cases.

The method we have used to estimate the normalized spectrum B from the $\square$ resampled values is the smoothed periodogram or Blackman-Tukey estimator [12]:

$$\hat{\phi}(\omega) = \frac{1}{2 \cdot \pi} \sum_{n=-\Theta_s}^{\Theta_s} e^{-j\omega n} \cdot w\left(\frac{n}{\Theta_s}\right) \cdot \hat{R}_n \qquad (18)$$

where $w(x)$ is Parzen's spectral window [10] used as a smoothing kernel, and $\hat{R}_n$ is an estimate of the covariance of the process, computed as:

$$\hat{R}_n = \frac{1}{\Theta} \sum_{k=1}^{\Theta-|n|} \hat{S}(M_0, y_k) \cdot \hat{S}(M_0, y_{k+|n|}) \qquad (19)$$

$\hat{S}(M_0, x)$ in (20) is assumed to have been normalized in the mean. $\square_s$ is a parameter of the spectral window satisfying $\Theta_s \to \infty$ and $\Theta_s / \Theta \to 0$ as $\Theta \to \infty$. Eventually the bandwidth estimate is obtained as per equation (20) where $\square$ is chosen close to unity. The algorithm is depicted in the corresponding box.

$$\hat{B} = \arg\min_A \left\{ \int_A^A \hat{\phi}(\omega)d\omega = \varepsilon \cdot \int_{-\pi}^{\pi} \hat{\phi}(\omega)d\omega \right\} \qquad (20)$$

As stressed in the previous section, it is desirable that the number K of time instants that all sensors have to report for bandwidth estimation is low. To verify this

---

- Begin operation at t = t_0.

- Activate all N sensors for the initial K time instants. Record K sets of reported values.

- Perform interpolation for each set according to (8). Then concatenate the interpolated sequences. Call the resulting sequence $\hat{S}(M_0, x)$.

- Estimate the autocorrelation sequence according to (19), the power spectral density according to (18), the bandwidth according to (20).

Algorithm 2. Bandwidth Estimation

we performed simulations over a hundred independently generated initial node deployments and process realizations. The results, translated into Nyquist rates by way of (17) and (16) are shown in Figure 4.

We also experimented with values of K higher than two but obtained very similar results, enabling us to infer that low values are sufficient for practical purposes. The Nyquist rate estimate for filter H_1 is quite close to its true value for initial numbers of sensors down to the true Nyquist rate of 100 sensors. It can be observed that raising K only improves the estimate when the initial number of sensors is lower than this threshold. For the case of H_2 there is a slight bias towards lower Nyquist rates, i.e. lower bandwidths. This can be attributed to the specific shape of the spatial filter and the type of spectrum estimation employed in (18) that tends to underestimate low energy spectral content as in the tails of H_2. This bias however does not practically affect us because the knee in the Semi-Equidistant Sampling distortion curve for filter H2 (Figure 3) is itself somewhat ill-defined.

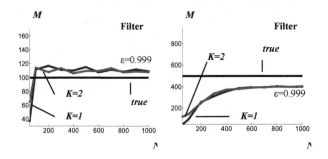

Fig. 4. Nyquist rate estimates for filters H_1 and H_2

The schemes presented in section 5 allow a direct tradeoff between distortion and energy by way of adjusting the number of reporting sensors M. Minimal semi-equidistant sampling allows such a tradeoff indirectly, through adaptation of the parameter □ (see equation (20)). This parameter steers the accuracy of spectral estimation and therefore the resulting energy consumption. In fact, the tradeoff of distortion versus □ can be quantified and predicted in much the same way as for all the schemes of section 5.

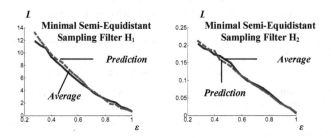

Fig. 5. Predicted and true average distortion for the Minimal Semi-Equidistant Sampling Scheme

The related curves for $K = 7$ are shown in Figure 5. The magnitude of distortion is comparable with that of the diagrams in Figure 3. However, such distortion prediction necessitates a long learning phase, and thus relinquishes the advantage Minimal Semi-Equidistant Sampling has over the schemes of Section 5.

6 Comparative Results

We have thus far presented four monitoring policies. Three of them, namely Random Selection (RS), Random Binning (RB) and Semi-Equidistant Sampling (SES) require the construction of a distortion prediction curve when the network starts operating. This translates to all sensors having to report for $K = K_1$ initial time instants. Figure 6 comparatively shows average distortion results for these three schemes.

The fourth scheme presented, namely Minimal Semi-Equidistant Sampling (MSES) requires all sensors to report for $K = K_2$ initial time instants, where, in general, $K_2 < K_1$. This scheme in essence tries to directly obtain M at the knee of the SES curve in Figure 6 without going through the expensive distortion prediction procedure.

The last step in our simulation study is to actually impose a schedule on the network and track reconstruction distortion as well as the status of sensors over time. We assumed all sensors start out with the same amount of a thousand energy units and that they spend twenty energy units each time they send a message to the data processing center.

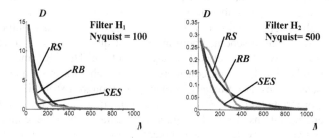

Fig. 6. Average distortion for three heuristic monitoring schemes

Keeping the same initial random sensor deployment as in the distortion prediction scenarios of Figure 3, we first ran the MSES monitoring scheme with $\mathbf{a} = 0.999$. The initial average distortion observed for this scheme was 1.1194 for filter H_1 and 0.0078443 for filter H_2. For a fair comparison with monitoring based on the two random schemes (in this case, SES is equivalent to MSES) we have to specify the number of sensors M, so that both of them achieve the same average distortion as MSES. This can be readily acquired from the distortion prediction curves of Figure 3 resulting in $M = 290$ and 794 for scheme RS and filters H_1 and H_2 respectively and $M = 184$ and 678 for scheme RB and filters H_1 and H_2 respectively. The plotted results include the overhead of the learning phase of K_1 and K_2 time instants. The final results are shown in Figure 7, where we have also added a black solid line to indicate when the network would completely die out if no schedule was imposed.

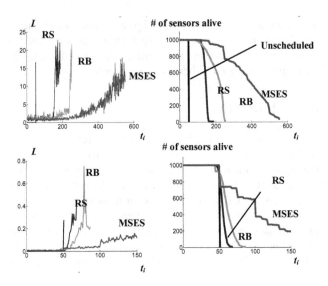

Fig. 7. Distortion and number of alive sensors over time for (a)Filter H_1 (b)Filter H_2

The distortion remains low as long as the sensors remain alive, and only increases once nodes start dying. Note that lifetime is much higher for filter H_1 than for filter H_2. This is due to the fact that the oversampling factor N / M_{min} is much higher for the process generated by H_1 than that by H_2, i.e. 1000 / 100 vs. 1000 / 500. The oversampling factor is essentially an upper bound to the lifetime gain we can obtain through scheduling sensor reports. The MSES policy significantly outperforms the distortion prediction based schemes both in terms of quality of reconstruction and of number of sensors alive in the network at any given time. The RS (random selection) scheme on the other hand, can be considered as a benchmark to compare the others against, as it selects sensors uniformly at random (although it still utilizes knowledge of the most appropriate value of M). It is interesting to point out that the benchmark scheduling scheme attains more than three-fold gain in lifetime compared to an unscheduled network, provided there is potential for such a gain in terms of initial oversampling.

All four schemes allow a tradeoff between distortion and energy when a long learning phase can be tolerated. The MSES scheme however, has the ability to inexpensively predict the knee of the curves shown in Figure 6. For lower numbers on M beyond this knee point, distortion increases rapidly while only resulting in marginal lifetime improvements. In most situations therefore, operating at the knee point is probably desirable.

7   Related Work

In [1], a spatial correlation based Medium Access Control (MAC) protocol, CMAC, is presented, assuming Gaussian statistics. The described system however does not aim at providing a representation of the process over the whole observation area but rather computing the best estimated value of the point source, i.e. it addresses a more basic monitoring problem. Recently, [2] proposed heuristics based on mutual information criteria for optimally placing a given set of sensors on an area. Sensor readings can effectively be modeled as a multivariate Gaussian process. This approach however finds the best achievable distortion performance given a target number of sensors, whereas we are interested in finding the fewest sensors that result in a target distortion. A related approach is presented in [6], where mutually exclusive sets of nodes are selected, so that each of the sets completely covers the observation area. Coverage in this case is based on the notion of a circular sensing radius, which is not clearly related to the underlying spatial phenomenon. The work in [4] analyzes noisy sampling of a one-dimensional Gaussian-Markov random field, not however targeting lifetime improvements. In [3], a distributed approach for in-network spatial data modeling is presented. The model computes weights of local basis functions when sensor measurements have been partitioned on the basis of kernel functions. Kernel functions model the correlation among various positions in the field and therefore assume prior knowledge of this correlation by the query center. In our work, we do not make a-priori assumptions neither on the distribution of the process nor on the form of the dependence of its correlation function on distance.

8 Conclusion

In this paper we have presented a two-phase approach tackling energy-efficient, distortion-driven monitoring of a spatial random process. Based on the premise that none of the process statistics are a-priori known but have to be estimated at network initialization, we proposed four distinct monitoring schemes. Three of the proposed schemes rely on distortion prediction as a first step while the fourth one bypasses this need by exploiting the notion of bandwidth as a measure of spatial correlation. We showed that considerable improvements in lifetime can be achieved this way. Our current work focuses on extending these results to the two dimensional case.

References

[1] M. C. Vuran and I. F. Akyildiz, "Spatial Correlation-based Collaborative Medium Access Control in Wireless Sensor Networks", to appear in IEEE/ACM Transactions on Networking, June 2006.
[2] C. Guestrin, A. Krause and A. P. Singh, "Near Optimal Sensor Placements in Gaussian Processes", in Proceedings of the 22nd International Conference on Machine Learning, 2005.
[3] C. Guestrin, P. Bodik, R. Thibaux, M. Paskin and S. Madden, "Distributed Regression: An Efficient Framework for Modeling Sensor Network Data", in IPSN 2004.
[4] M. Dong, L. Tong and B. M. Sadler, "Effect of MAC Design on Source Estimation in Dense Sensor Networks", in ICASSP 2004.
[5] Y. Yu, D. Ganesan, L. Girod, D. Estrin and R. Govindan, "Synthetic Data Generation to Support Irregular Sampling in Sensor Networks", in Geo Sensor Networks 2003, October 2003.

[6] S. Slijepcevic, M. Potkonjak, "Power Efficient Organization of Wireless Sensor Networks", in ICC 2001.

[7] R. Cristescu, B. Beferull-Lozano and M. Vetterli, "On Network Correlated Data Gathering", in INFOCOM 2004.

[8] T. M. Lehmann, C. Gönner and Klaus Spitzer, "Survey: Interpolation Methods in Medical Image Processing", in IEEE Transactions on Medical Imaging, Vol. 18, no. 11, November 1999.

[9] S. de Waele and P. M. T. Broersen, "Reliable LDA Spectra by Resampling and ARMA-Modeling", in IEEE Transactions on Instrumentation and Measurement, Vol. 48, No. 6, December 1999.

[10] E. Masry, D. Klamer and C. Mirabile, "Spectral Estimation of Continuous Time Processes: Performance Comparison between Periodic and Poisson Sampling Schemes", in IEEE Transactions on Automatic Control, Vol. 23, no. 4, August 1978.

[11] J. D. Scargle, "Studies in Astronomical Time Series Analysis. II. Statistical Aspects of Spectral Analysis of Unevenly Spaced Data", in The Astrophysical Journal, 263:835-853, December 1982.

[12] J. G. Proakis, D. G. Manolakis, "Digital Signal Processing Principles, Algorithms and Applications", Third Edition, Prentice Hall.

[13] B.Cărbunar, A. Grama, J. Vitek and O. Cărbunar, "Coverage Preserving Redundancy Elimination in Sensor Networks".

Optimal Placement and Selection of Camera Network Nodes for Target Localization

Ali O. Ercan[1], Danny B. Yang[2], Abbas El Gamal[1], and Leonidas J. Guibas[2]

[1] Dept. of Electrical Engineering, Stanford University, Stanford, CA 94305, USA
aliercan@stanford.edu, abbas@ee.stanford.edu
[2] Dept. of Computer Science, Stanford University, Stanford, CA 94305, USA
danny@riya.com, guibas@cs.stanford.edu

Abstract. The paper studies the optimal placement of multiple cameras and the selection of the best subset of cameras for single target localization in the framework of sensor networks. The cameras are assumed to be aimed horizontally around a room. To conserve both computation and communication energy, each camera reduces its image to a binary "scan-line" by performing simple background subtraction followed by vertical summing and thresholding, and communicates only the center of the detected foreground object. Assuming noisy camera measurements and an object prior, the minimum mean squared error of the best linear estimate of the object location in 2-D is used as a metric for placement and selection. The placement problem is shown to be equivalent to a classical inverse kinematics robotics problem, which can be solved efficiently using gradient descent techniques. The selection problem on the other hand is a combinatorial optimization problem and finding the optimal solution can be too costly to implement in an energy-constrained wireless camera network. A semi-definite programming approximation for the problem is shown to achieve close to optimal solutions with much lower computational burden. Simulation and experimental results are presented.

1 Introduction

A wireless sensor network (WSN) comprises a collection of many low cost, low-power nodes each with sensing, processing and communication capabilities. WSNs have many advantages over traditional sensing modalities including wide coverage, robustness, scalability, and the ability to observe large scale phenomena distributed over space and time [1, 2].

The scarcest resource in a WSN is energy, as typically nodes operate untethered. The limited battery life of a node imposes severe constraints on its communication and computation capabilities. Consequently, recent work on WSNs has focused mainly on very low data rate sensor nodes [3]. In many applications, however, high data rate sensors are needed to perform the desired tasks. The most notable example is video cameras, which are widely used for surveillance and monitoring. Current surveillance camera installations are expensive and use outdated infrastructure. All captured video data is shipped to a central station for human operators to watch which makes the system non-scalable. As a result, there is a growing need to develop less costly wireless networks of cameras with automated task-driven capabilities. Such development faces many challenges.

P. Gibbons et al. (Eds.): DCOSS 2006, LNCS 4026, pp. 389–404, 2006.
© Springer-Verlag Berlin Heidelberg 2006

First, current video cameras are expensive and have high power consumption. Second, video cameras are high data rate devices, so transmitting all the data is costly in terms of energy. Third, video processing algorithms are in general computationally expensive, require floating point arithmetic, and are costly to implement locally.

The camera cost and power problems can be addressed by recent advances in CMOS technology, which enable the integration of sensing, processing and communication [4]. It is currently feasible to design very low cost and power camera systems suitable for deployment in a wireless network. To address the communication and computation challenges facing the development of wireless camera networks, simple local processing algorithms that produce only the essential information needed for the network to collaboratively perform a task or answer a query are needed (e.g., see [5]).

Energy consumption can also be minimized by reducing the number of cameras used to answer a query. This can be achieved by judicious placement of the cameras with respect to the objects and selection of the best subset of cameras to collaboratively answer the query. A proper placement of cameras increases the accuracy of sensing, while selecting a good subset allows for efficient sensing with little performance degradation relative to using all the cameras. Selection also allows the network to scale to large numbers of nodes because of the savings in communication, computation, and sensing.

In this paper, we investigate the problems of placement and selection of camera nodes in order to minimize the localization error for a single object. Localizing an object is important in many applications such as tracking, surveillance, and human computer interaction. For example, if we could localize an object at every time step, the tracking and correspondence problems become trivial. Very accurate localization is also important in several robotics applications, such as navigation through a complex environment or controlling an end effector to perform a delicate task. Specifically, we focus on 2-D object localization, i.e., location on the ground plane, because this is the most relevant information for many real world applications. We assume that the cameras are placed horizontally around a room. The local processing framework in [5] is used to reduce the image to a scan-line and only the center of the detected object from each camera is communicated to the central processor. Given these noisy measurements and the object prior distribution, the minimum mean squared error (MSE) of the best linear estimate of the object location in 2-D is used as a metric for placement and selection. To find the best camera placement, we optimize this metric with respect to the camera positions. For a circularly symmetric object prior distribution and sensors with equal noise, we show that a uniform sensor arrangement is optimal. Somewhat surprisingly, we establish that the general problem is equivalent to solving the inverse kinematics of a planar robotic arm which can be solved efficiently using gradient descent techniques. We then devise a semi-definite programming approximation of the optimal solution for the selection problem. We show that this method performs close to optimal and outperforms naive heuristics (e.g., picking greedily or the closest or uniform sensors).

The rest of the paper is organized as follows: In Sect. 2, we review related sensor networks, computer graphics and computer vision work. In Sect. 3, we introduce the camera model and define the placement and selection problems. In Sect. 4, we derive the optimal solution for the placement problem. In Sect. 5, we develop an approximate solution to the selection problem and compare our approximation method to other heuristics and to the optimal solution, both in simulation and experimentally. Section 6 discusses how to handle some non-idealities. Finally, in Sect. 7, we conclude.

2 Related Work

Sensor placement and selection have been addressed in the sensor networks, computer vision, and computer graphics literature. Selection has been studied in wireless sensor networks with the goal of decreasing energy cost and increasing scalability. Viewpoint selection, or the next best view, has been studied in computer graphics and vision for picking the most informative views of a scene. We summarize the work related to this paper in this section.

Sensor placement: Camera placement has been studied in computer vision and graphics. In photogrammetry [6, 7, 8], the goal is to place the cameras so as to minimize the 3D measurement error. The error propagation is analyzed to derive an error metric that is used to rank camera placements. The best camera placement is then solved numerically. The computational complexity of this approach only allows solutions involving a few cameras. In our approach we simplify the camera model, derive the localization error analytically as a function of camera places and minimize it to find the best placement. This is computationally lighter compared to above numerical methods. In [9] the problem of how to position (general) sensors with 2-D measurement noise to minimize the overall error is investigated. The paper also presents an algorithm to compute the optimal sensor placement. Our measurements are 1-D after local processing and we pose the placement problem as a special case of classical inverse kinematics problem. The emphasis of our work is also on selection, which due to its combinatorial nature is different from placement.

Sensor Selection: In [10] a technique referred to as IDSQ is developed to select the next best sensor node to query in a sensor network. The technique is distributed and uses a utility measure based on the expected posterior distribution. However, expected posterior distribution is expensive to compute because it involves integrating over all possible measurements. In [11] the mutual information metric is used to select sensors. This is shown to be equivalent to minimizing the expected posterior uncertainty, but with significantly less computation. The work in [12] expands on [11] and shows how to select the sensor with the highest information gain. An entropy-based heuristic that approximates the mutual information and is computationally cheaper is used. All these methods greedily select the next best sensor based on an entropy metric. In our work, we show how to select the next best *group* of sensors via combinatorial optimization. Also, in our approach the utility function is an analytical expression, which makes it much faster to evaluate than methods requiring numerical integration.

Camera Selection: Sensor selection has also been studied for camera sensors. In [13, 14, 15], a metric is defined for the next best view based on most faces seen (given a 3-D geometric model of the scene), most voxels seen, or overall coverage. The solution requires searching through all camera positions to find the highest scoring viewpoints. In [5], a subset of horizontal camera sensors are selected to minimize the visual hull of all objects in the scene. This problem is solved using heuristics. These works use numerical techniques or heuristics to compute the viewpoint scores. We investigate a simpler problem, devise an analytical metric for it and find the optimal solution using combinatorial optimization techniques.

3   Problem Formulation

Given a number of noisy camera sensors, our goal is to localize an object as accurately as possible in the ground plane. We first describe the camera model and then formulate the utility metric used in determining the best placement and selection.

We assume that the cameras are aimed roughly horizontally. An overhead camera may have a less occluded view, and may allow better localization. But overhead cameras are often impractical to deploy and can only observe a small area limited by the field of view. Horizontal cameras are often more practical to install. They can also observe a larger area. Additionally targets may be easier to identify in a horizontal view.

As discussed earlier, the camera nodes in a WSN must perform cheap local image processing to reduce the video data. Following a similar approach to [5], we limit this processing to background subtraction. The background subtracted images are vertically summed and thresholded, as the horizontal location of the object in the camera plane is most relevant to the 2-D localization (see Fig. 1). We refer to the resulting linear bitmap as a "scan-line". For the localization method that we use, only the center of the detected foreground object in the scan-line is communicated to the central processor. This reduces the data per frame from a 2-D array of pixel values to a single integer. This approach requires very little computation and communication and is thus compatible with a resource constrained WSN framework.

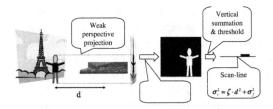

Fig. 1. Local processing at each camera

We assume that the cameras are far enough from the object that they can be modeled by weak perspective projections. We assume that the measurement error variance is of the form $\sigma_v^2 = \zeta d^2 + \sigma_f^2$, where d is the distance from the camera to the object (see Fig. 1). It can be shown that making camera noise variance dependent on d effectively models the weak perspective projection while allowing the usage of projective model in the equations. Our noise model also accounts for errors in the calibration of the cameras. Errors in the 2-D camera locations can be accounted for in σ_f and errors in the orientation can be accounted for in ζ.

Our specific problem is to localize one point object in a room with N cameras placed around its perimeter (See Fig. 2). As there is only one object to localize, we do not need to consider occlusions from other objects. For now, we also assume that there is no static object that occludes the view of the cameras. We discuss how static occlusions can be handled later in Sect. 6. The orientations of the cameras with respect to the abscissa are given by θ_i, $i = 1, 2, \ldots, N$. For simplicity, it is assumed that the point to be localized is in the FOV of all cameras. The consequences of limited FOVs is also addressed in Sect. 6. We assume that the mean μ_x and covariance Σ_x of the prior distribution of the object location are known.

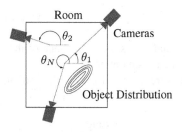

Fig. 2. Illustration of the problem definition

We formulate the problem of camera placement for target localization in the framework of linear estimation. Given the first and second order statistics of the object prior and the camera noise parameters, we use the minimum MSE of the best linear estimate of the object location, which is a function of the camera orientations θ_i, as a measure for localization error. The best placement is then obtained by finding the camera orientations that minimize this metric. As explained in Sect. 4, the best orientations give the best positions of the cameras uniquely based on our assumptions for the placement problem. The same formulation is then used to investigate the camera selection problem, with some of the simplifying assumptions removed.

The measurement model is illustrated in Fig. 3. As explained before, we use the projective model in our equations. The weak perspective dependence of the measurements is hidden in the error model for the cameras. Let the object location be $x = [x_1, x_2]^T$, and the measurements from the N cameras be $z = [z_1 \ldots z_N]^T$. The relationship between the measurements and the object location is then given by

$$z = A\,x + v \ ,$$

where A is given by

$$A = \begin{pmatrix} -\sin\theta_1 & \cos\theta_1 \\ -\sin\theta_2 & \cos\theta_2 \\ \vdots & \vdots \\ -\sin\theta_N & \cos\theta_N \end{pmatrix} \ ,$$

and v is the measurement error.

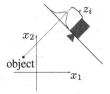

Fig. 3. The measurement model

The best linear unbiased estimator for x is given by

$$\hat{x} = \mu_x + \Sigma_x A^{\mathrm{T}} (A \Sigma_x A^{\mathrm{T}} + \Sigma_v)^{-1}(z - A\mu_x) \ , \tag{1}$$

where μ_x is the mean and Σ_x is the covariance of the object location prior, and Σ_v is the covariance of the measurement noise.

Assuming the measurement noise is independent for different cameras, its covariance is

$$\Sigma_v = \mathrm{diag}(\sigma_{v_1}^2, \cdots, \sigma_{v_N}^2) \ .$$

The object prior can be assumed to be diagonal with a horizontal major axis without loss of generality because one can rotate everything, solve the problem and rotate everything back, so

$$\Sigma_x = \sigma_x^2 \mathrm{diag}(1, 1/\alpha) \ ,$$

where $\alpha \geq 1$. It can be shown that the MSE of the best linear estimate in (1) is

$$\mathrm{MSE} = \frac{4 \left(\frac{\alpha+1}{\sigma_x^2} + \sum_{i=1}^{N} \frac{1}{\sigma_{v_i}^2} \right)}{\left(\frac{\alpha+1}{\sigma_x^2} + \sum_{i=1}^{N} \frac{1}{\sigma_{v_i}^2} \right)^2 - \left(\frac{\alpha-1}{\sigma_x^2} + \sum_{i=1}^{N} \frac{\cos 2\theta_i}{\sigma_{v_i}^2} \right)^2 - \left(\sum_{i=1}^{N} \frac{\sin 2\theta_i}{\sigma_{v_i}^2} \right)^2} \ . \tag{2}$$

Note that only the latter two terms in the denominator of (2) are functions of the camera orientations. Given fixed σ_{v_i}s, the MSE is minimized by setting these two squared terms as close to zero as possible. If these two terms are set to zero, we obtain a lower bound for MSE:

$$\mathrm{MSE} \geq \frac{4}{\frac{\alpha+1}{\sigma_x^2} + \sum_i \frac{1}{\sigma_{v_i}^2}} \ . \tag{3}$$

The cameras with lower noise achieve a smaller lower bound. In our noise model, closer cameras have less noise, so this means closer cameras achieve a smaller lower bound. The two additional terms in the denominator of MSE (2) balance this low noise criterion with the direction criterion. Given σ_{v_i}s, (3) can be utilized to estimate the number of cameras that must be used to achieve a certain allowable error.

Given the above formulation, we define the following two problems:

Placement: Minimize (2) over $\theta_1, \theta_2, \ldots, \theta_N$. The resulting θ_is provide the locations of the cameras that result in the best localization error.

Selection: N cameras are previously placed. This means that the θ_is and σ_{v_i}s are fixed. We need to select the best subset of k cameras. As we are using a subset of the cameras, the summations in (2) also need to run over the selected set of cameras. The metric to minimize then becomes

$$\frac{4\left(\frac{\alpha+1}{\sigma_x^2} + \sum_{i \in S} \frac{1}{\sigma_{v_i}^2}\right)}{\left(\frac{\alpha+1}{\sigma_x^2} + \sum_{i \in S} \frac{1}{\sigma_{v_i}^2}\right)^2 - \left(\frac{\alpha-1}{\sigma_x^2} + \sum_{i \in S} \frac{\cos 2\theta_i}{\sigma_{v_i}^2}\right)^2 - \left(\sum_{i \in S} \frac{\sin 2\theta_i}{\sigma_{v_i}^2}\right)^2}. \tag{4}$$

where S is the set of selected cameras. Formally defined, the selection problem becomes: Minimize (4) over the set S such that $S \subseteq \{1, \ldots, N\}$ and $|S| = k$. The resulting set S gives the best selection.

In the following section we show that the optimal placement problem can be solved efficiently. An approximate solution to the selection problem is presented in Sect. 5.

4   Placement

We wish to find θ_is that minimize (2), given the object location prior statistics and the σ_{v_i}s for each camera.

The placement of the cameras is usually done before there is actually any object in the room. Therefore, it is natural to assume in this problem that the object prior is centered in the room. However, it does not have to be circularly symmetric, as people might tend to walk along certain directions more than others. For example, in a hallway, the major axis of the prior distribution would be aligned with the hallway.

Cameras are usually placed on the walls of the room, so the distance of the camera to the object prior cannot vary by much and can be approximated by a constant. Therefore, for the placement problem specifically, we are trying to come up with the best *orientations* of the cameras, and we assume a circular room for simplicity. As the cameras are fixed to the periphery and oriented towards the center, the orientations of the cameras (θ_is) uniquely determine their placement.

Note that all of the above assumptions are specific to the placement problem. In the selection problem (see Sect. 5), the circular room assumption and the centered object prior assumption are removed and our approach is applicable to any room shape, camera configuration and object prior. In Sect. 6 we give a recipe on how to handle general case for the placement problem with non-circular room and non-centered prior.

Given these assumptions, the camera noise parameters are constant. From (2) we note that only the last two terms of the denominator depend on the θ_is. Thus minimizing (2) is the same as minimizing

$$\left(\frac{\alpha-1}{\sigma_x^2} + \sum_{i=1}^{N} \frac{\cos 2\theta_i}{\sigma_{v_i}^2}\right)^2 + \left(\sum_{i=1}^{N} \frac{\sin 2\theta_i}{\sigma_{v_i}^2}\right)^2. \tag{5}$$

It is clear that (5) is bounded below by 0. The following subsections show when this can be achieved for the optimal camera orientations.

4.1 Symmetric Case

When the cameras have the same error variance, $\sigma_{v_i} = \sigma_v$ for all $1 \leq i \leq N$, and the object prior is circularly symmetric ($\alpha = 1$), the problem of minimizing (5) reduces to minimizing

$$\left(\sum_{i=1}^{N} \cos 2\theta_i \right)^2 + \left(\sum_{i=1}^{N} \sin 2\theta_i \right)^2 . \tag{6}$$

This is equivalent to the norm-squared of the sum of N unit vectors with angles $2\theta_i$. Thus (6) is equal to zero when the θ_is are chosen uniformly between 0 and π. This leads to the intuitive conclusion that when the object prior is circularly symmetric and the cameras have the same amount of noise, uniform placement of cameras is optimal. An illustration of this result for 6 cameras is depicted in Fig. 4(a). The angles of the vectors are twice the orientation angles of the cameras. For example, for two cameras, an orthogonal placement of the cameras is optimal, so that the unit vectors are 180 degrees apart.

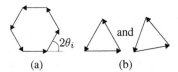

(a)   (b)

Fig. 4. (a) Uniform placement of 6 cameras that minimizes (6). (b) Locally optimal clusters are globally optimal. Relative orientations of clusters do not matter.

Uniform placement, however, is not the only optimal way to place the cameras. If we partition the vectors into subgroups and all subgroups of vectors sum to zero, then the combination of all the vectors also sums to zero. This means that we can cluster the cameras into (local) groups (with at least 2 cameras in each group) and solve the problem distributedly in each cluster. If each group finds a locally optimal solution, then the combined solution is globally optimal (see Fig. 4(b)). This is true no matter what the relative orientations between the groups of cameras are.

4.2 General Case

We now discuss the general placement problem, i.e., when $\alpha \neq 1$ and the σ_{v_i}s are not all equal. The problem corresponds to minimizing (5). Again this is the sum of N vectors, but the vectors can have different lengths $1/\sigma_{v_i}^2$. The MSE is minimized when the sum equals $-\frac{\alpha-1}{\sigma_x^2}$ (offset from zero) on the abscissa. Again, the resulting angles of the vectors are twice the optimal θ_i of the cameras (see Fig. 5).

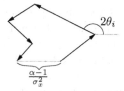

Fig. 5. An optimal solution

This problem can be thought of as an inverse kinematics robotics problem. Our vectors describe a planar revolute robot arm with N linkages. The base of the robot arm is at the origin and it is trying to reach a point $-\frac{\alpha-1}{\sigma_x^2}$ on the abscissa with its end effector. If the σ_{v_i}s are ordered such that

$$\sigma_{v_N} \geq \sigma_{v_{N-1}} \geq \ldots \geq \sigma_{v_1} \;,$$

then any point in an annulus with inner and outer radii

$$r_{\text{out}} = \sum_i 1/\sigma_{v_i}^2 \;,$$

$$r_{\text{in}} = \max\left(0, 1/\sigma_{v_1}^2 - \sum_{i \neq 1} 1/\sigma_{v_i}^2\right)$$

is achievable. If the point the robot is trying to reach is inside the annulus, we use gradient descent algorithms to find an optimum solution that minimizes (5) by setting it to zero [16]. If the point is outside the annulus, we minimize the distance to the point the robot arm is trying to reach by lining up all the vectors along the abscissa such that the tip of the arm touches the outer or inner radius of the annulus. This configuration does not zero out (5) but gives the minimum achievable error. Figure 6 illustrates these two cases.

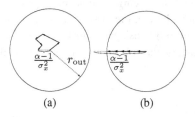

(a)   (b)

Fig. 6. Inverse kinematics can solve for the best θ_i: The case when the point to reach is (a) inside the annulus, and (b) outside of the annulus. Note that r_{in} is 0 for this example.

The gradient descent technique might give different solutions for different starting arm configurations. However, under the assumptions made earlier, all such solutions yield the same MSE on average. This leaves room for further relaxations of these assumptions. Some of these generalizations will be discussed in Sec. 6.

In Fig. 6(b), note that all the vectors point in the same direction. The best placement for this scenario is putting all cameras orthogonal to the object prior's major axis (such that twice the angles are $180°$). This seems counterintuitive since we expect an orthogonal placement to be better for triangulation. However in this case, the prior uncertainty along the minor axis is small enough $((\alpha-1)/\sigma_x^2 > \sum_i 1/\sigma_{v_i})$ that the optimal solution is to place all cameras to minimize the uncertainty along the major axis.

An example placement for $N = 4$, $\alpha = 5$, $\sigma_x = 4$ and $\sigma_{v_i}^2 = 5, 10, 15$, and 20 is given in Fig. 7(a). The room, object prior and resulting camera placements are shown. Note that the three higher noise cameras are placed close to each other, while the first camera is placed separately. The interpretation here is that the similar views from these bad cameras are averaged by the linear estimator to provide one good measurement. This is verified by the example in Fig. 7(b). Here, an optimal placement for two high quality cameras ($\sigma_{v_i}^2 = 5$ for both) is shown. The first camera is placed roughly at the same position as before, while the second camera is placed in the middle of the three bad camera positions. Note that the cameras are placed in $[0, \pi)$, as the corresponding vectors have angles in $[0, 2\pi)$ and they are twice the angles of the cameras. However, one can flip any camera to the opposite side of the room without changing its measurement.

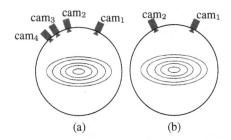

$$(a) \qquad\qquad (b)$$

Fig. 7. Two optimal placements for the given object prior. (a) One good camera and 3 worse ones. (b) Two good cameras.

Note that for the general case, clustering the cameras into multiple groups can still achieve global optimality while solving the placement problem for each cluster as long as the clusters zero out their share of the offset. Suppose N cameras are clustered into c groups. Then one algorithm might ask each cluster's "arm" to reach $-\frac{\alpha-1}{c\sigma_x^2}$. If this can be achieved by all the clusters, the solution is globally optimal.

5   Selection

For the selection problem defined in Sect. 3, the camera locations are fixed and we wish to select the best set S of size k that minimizes (4). In this problem, the object prior is not necessarily assumed to be at the center of the room. Also, the room does not have to be circular, and the cameras neither have to be placed at the periphery nor be oriented towards the center. This problem is difficult to solve because:

- All the terms in (4) change with different selections since the summation involving the σ_{v_i}s is a function of the selected set S.
- A naive search for the global optimum requires a combinatorial search among all possible sets S, which is $O(N^k)$. This is too costly.

To overcome these difficulties, we drop the numerator of (4) and focus only on optimizing the denominator. This is reasonable because it is equivalent to maximizing the mutual information between the measurements and the object location assuming Gaussian distributions, which is another good metric for camera selection ([11, 12]). Simulations also show that this modification does not introduce much performance degradation. We added the weights w_i inside the sums of (4), instead of running them over the subset S. The weights can be either 0 or 1. That is, they have to satisfy $w_i^2 - w_i = 0$. With the above, the problem can be formulated as follows:

$$
\text{Maximize} \left[\left(\frac{\alpha+1}{\sigma_x^2} + \sum_{i=1}^{N} \frac{w_i}{\sigma_{v_i}^2} \right)^2 - \left(\frac{\alpha-1}{\sigma_x^2} + \sum_{i=1}^{N} \frac{w_i \cos 2\theta_i}{\sigma_{v_i}^2} \right)^2 - \left(\sum_{i=1}^{N} \frac{w_i \sin 2\theta_i}{\sigma_{v_i}^2} \right)^2 \right]
$$

$$
\text{Subject to} \qquad \sum_{i=1}^{N} w_i = k,
$$

$$
w_i^2 - w_i = 0, \forall i \ .
$$

This problem is not convex – neither the objective function nor the feasible set is necessarily convex. We use the following heuristic to find a good solution. We form the dual of the problem and use semi-definite programming (SDP) to find the dual optimal variables [17]. We then plug the dual optimal variables in the Lagrangian and solve for the w_is. We select the k cameras with the highest weights.

In practice, the selection technique we described might be performed distributedly as follows. The user asks for the location of an object with a desired accuracy. The query is passed to a cluster head near the object prior. The cluster head knows the locations of the other cameras near him. Using (3) with the lowest noise cameras, it computes a lower bound of the number of required cameras. It can also compute an upper bound assuming uniformly placed cameras and that the final selection will do better than the uniform selection. Using this lower and upper bounds, the cluster head decides on k. It then can compute the optimal selection of k cameras. If the predicted MSE from (4) for this selection is above the desired accuracy, k is incremented and the selection algorithm is repeated. Computing the optimal selection is feasible on a sensor node because it is computationally inexpensive. Finally, the selected cameras are queried for a measurement and the result is sent back to the cluster head to compute the measured localization. The relatively cheap processing is only done on the cluster head, and only the minimum number of camera nodes are queried. This limits the amount of networking and processing required of each node and is expected to greatly extend the overall network lifetime.

5.1   Simulation Results

We performed Monte-Carlo simulations to compare the performance of the above SDP approach to the optimal using brute-force enumeration as well as to the following other heuristics:

- *Uniform:* Pick uniformly placed cameras.
- *Closest:* Pick the closest cameras to the object mean.

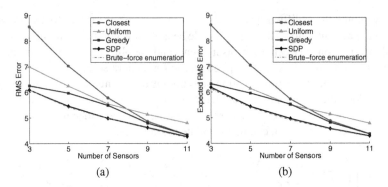

Fig. 8. (a) Localization performance for different selection heuristics. (b) Expected RMS localization error.

– *Greedy:* Pick one camera at a time, using the *expected* posterior after each selected camera's measurement (w/o actually making a measurement) as the prior for the next camera.

In Fig. 8(a) we show a typical simulation run for $k = 3$ to 11 cameras out of 30 uniformly placed cameras on a circle of radius 100 units. The camera noise parameters used were

$$\sigma_{v_i}^2 = (0.1 \times d_i)^2 + 4 \ ,$$

where d_i is the distance from the ith camera to the object mean. For this run we chose $\alpha = 5$ and $\sigma_x^2 = 80$. As seen in the figure, the SDP approach achieves very close to optimal and clearly outperforms the other heuristic approaches. We can also predict the expected RMS of the localization using (2), for the selected cameras. Figure 8(b) shows the predicted RMS values, which are close to the RMS errors in Fig. 8(a).

5.2 Experimental Results

We tested our selection algorithm in an experimental setup consisting of 12 web cameras placed around a $22' \times 19'$ room. The horizontal FOV of the cameras used is $49°$, and they all look toward the center of the room. The relative positions of the cameras in the room can be seen in Fig. 9(a). The cameras are hooked up to a PC via an IEEE 1394 (FireWire) interface and can provide 8-bit 3-channel (RGB) raw video at 15 Frames/s. The PC connected to a camera models a sensor node with processing and communication capabilities. Each PC is connected to 2 cameras, but the data from each camera is processed independently. The data is then sent to a central PC, where further processing is performed.

A process was run for each camera to perform background subtraction and generate the scan-lines (as described in Sect. 3). Only the selected cameras need to take a measurement and send the scan-line data over the network to an aggregating node where the localization is performed. The actual object that is localized is a point light source (see Fig. 9(b)).

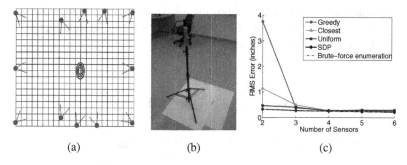

(a)   (b)   (c)

Fig. 9. (a) The object prior and positions of the cameras in the real setup. Cones show FOV of cameras and grid spacing is 1'. (b) Object to be localized. (c) Experimental Results.

The object was randomly placed 100 times according to the prior:

$$\Sigma_x = (6'')^2 \times \begin{bmatrix} 1 & 0 \\ 0 & \frac{1}{4} \end{bmatrix} \ ,$$

and localized using the selection algorithm. The noise parameters for the cameras were measured separately. The selection algorithm was applied for 2 to 6 cameras using the object prior and noise statistics. Figure 9(c) shows the localization error of the selection heuristics. For $k = 2$ and 3, the SDP and brute-force enumeration heuristics perform the best. While uniform selection scheme is also much better compared to greedy and closest schemes, it performs about 40% worse than SDP or brute-force. For this specific localization problem, around 4 cameras is enough to localize the object to the accuracy of the ground truth, so for $k \geq 4$, the localization error of all the heuristics levels off to a similar value. From the figure we see that a good selection scheme can allow us to task a very small number of cameras (2 or 3 out of 12) and localize with an accuracy that is very close to optimum. On the other hand, if we have the luxury of tasking a large number of the cameras (e.g., 6 out of 12), then a simple scheme like uniform selection works just as well.

6   Discussion

In this section, we discuss how some of the non-idealities can be handled in our framework.

Static Occlusions: As we are localizing a single object, there is no occlusion from other moving objects. But there might be occlusions due to static objects such as partitions, tables, etc. For the case of selection, handling these is simple. If a camera cannot see a considerable portion of the object prior (if the prior probability of the object being in the area that a camera cannot see is bigger than a user defined threshold), we simply discard that camera from the feasible set of cameras and limit the search to the remaining set. For the placement, the following approach could be used: As mentioned in Sect. 4, flipping the camera to the other side of the room does not affect the result. So, the type of occlusions that do not have any occluding object at the other side of the room can easily be handled. If the solution obtained in Sect. 4 places the camera behind such an occluding object, one can still achieve the same result by flipping the cameras to the other side of the room. (See Fig. 10(a)).

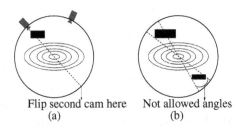

Flip second cam here   Not allowed angles
        (a)   (b)

Fig. 10. (a) Static occlusions can be avoided if there is no other occluding object on the other side of the room. (b)We cannot place the cameras at some angles due to occlusion.

If it is the case that for some region of angles there are occluding objects at both sides of the room, then we cannot place the cameras at those angles (Fig. 10(b)). This places regions of angles that are not allowed (and regions that are allowed) for our inverse kinematics solution of Sect. 4. For the case that the number of allowed regions is one, a solution is given in [18]. Note that the situation illustrated in Fig. 10(b) is an example of one region of allowed angles (not two), as two flip sides of the room are equivalent. However, for the case of more than one such allowed regions, there is no general inverse kinematics solution. For this case, we can restrict each joint angle to be in a specific region and try all possible combinations until a feasible solution is found. Although the complexity of this search is exponential in the number of allowed regions, in practice we do not expect the number of occluding objects to be too high to make the computation infeasible.

Limited FOV: We can also handle limited FOV of cameras. For placement, one can place the cameras such that the object prior mean is mostly in the FOV of the camera. Of course there is still a possibility that the actual object position is too far away from the prior mean and it is out of the FOV of some cameras but this is very unlikely. For selection, again one can discard the cameras that cannot see a considerable portion of the object prior beforehand and restrict the selection only to the remaining cameras.

General Case for Placement: In Sect. 4 we assumed the object prior is centered in a circular room. Any general room shape and non-centered prior can be also handled as follows. We discretize all possible locations that a camera can be placed and assume there exists a camera at all of the locations. Then using the SDP heuristic described above, we select best k cameras. The actual placement is then done at the locations of these k selected cameras. Note that the selection heuristic we used did not require a circular room or centered object prior. This placement method can also handle the non-idealities such as limited FOV of the cameras and static occlusions, using the extensions described above.

7   Conclusion

With the goal of accurately localizing an object, we have shown how to efficiently compute the best sensor placement and selection in an idealized setting. Our results are analytical, and yield algorithms that are more computationally efficient than the numerical utility maximization techniques for sensor placement/selection [10, 11, 12]. For the placement problem, the globally optimal solution is found. For selection, an efficient approach using SDP is proposed. We demonstrated using simulation and experimentally that our selection algorithm performs as well as the exhaustive search and outperforms other heuristics.

Our algorithm for selection are performed assuming a static object. It can be easily extended to a moving object using a Kalman filter approach. Initially, the prior can be assumed circular. Using the measured data, the posterior of the object location can be computed and used as a prior for the next iteration.

In order to make the analysis tractable, we made several simplifying assumptions that resulted in some performance degradation in experiments. However, as the optimization criterion tries to distribute the viewing directions of the cameras, the results were still promising. When the noisy weak perspective camera assumption is valid (when the object is far enough from the cameras), our method is expected to work even better, as the results of our simulations showed.

Acknowledgments

The work in this paper was supported by Stanford SNRC Consortium, Stanford Media-X Consortium, Max Planck Center for Visual Computing and under NSF NeTS NOSS grant 0535111. We wish to thank Prof. John T. Gill III for his help in setting up the experimental lab, and to Prof. Jack Wenstrand, Prof. Balaji Prabhakar, Helmy Eltouhkhy, Sam Kavusi and James Mammen for their comments.

References

1. Pottie, G.J., Kaiser, W.J., Clare, L., Marcy, H.: Wireless integrated network sensors. Communications of the ACM **43** (2000) 51–58
2. Akyildiz, I.F., Su, W., Sankarasubramaniam, Y., Cayirci, E.: Wireless sensor networks: A survey. Computer Networks **38** (2002) 393–422
3. Mainwaring, A., Polastre, J., Szewczyk, R., Culler, D., Anderson, J.: Wireless sensor networks for habitat monitoring. In: Proceedings of First International Workshop on Sensor Networks and Applications. (2002)
4. Yazawa, Y., Oonishi, T., Watanabe, K., Nemoto, R., Kamahori, M., Hasebe, T., Akamatsu, Y.: A wireless biosensing chip for DNA detection. In: Proceedings of ISSCC'05. (2005)
5. Yang, D.B.R., Shin, J.W., Ercan, A.O., Guibas, L.J.: Sensor tasking for occupancy reasoning in a network of cameras. In: Proceedings of BASENETS'04. (2004)
6. Chen, X., Davis, J.: Camera placement considering occlusion for robust motion capture. Stanford University Computer Science Technical Report, CS-TR-2000-07 (2000)
7. Olague, G., Mohr, R.: Optimal camera placement for accurate reconstruction. Pattern Recognition **35** (2002) 927–944

8. Wu, J., Sharma, R., Huang, T.: Analysis of uncertainty bounds due to quantization for three-dimensional position estimation using multiple cameras. Optical Engineering **37** (1998) 280–292

9. Zhang, H.: Two-dimensional optimal sensor placement. IEEE Transactions on Systems, Man, and Cybernetics **25** (1995)

10. Chu, M., Haussecker, H., Zhao, F.: Scalable information-driven sensor querying and routing for ad hoc heterogeneous sensor networks. The International Journal of High Performance Computing Applications **16** (2002) 293–313

11. Ertin, E., Fisher III, J.W., Potter, L.C.: Maximum mutual information principle for dynamic sensor query problems. In: Proceedings of IPSN '03. (2003)

12. Wang, H., Yao, K., Pottie, G., Estrin, D.: Entropy-based sensor selection heuristic for localization. In: Proceedings of IPSN '04. (2004)

13. Vazquez, P.P., Feixas, M., Sbert, M., Heidrich, W.: Viewpoint selection using viewpoint entropy. In: Proceedings of the Vision Modeling and Visualization'01. (2001)

14. Wong, L., Dumont, C., Abidi, M.: Next best view system in a 3d object modeling task. In: Proceedings of Computational Intelligence in Robotics and Automation. (1999)

15. Roberts, D., Marshall, A.: Viewpoint selection for complete surface coverage of three dimensional objects. In: Proceedings of the British Machine Vision Conference. (1998)

16. Welman, C.: Inverse kinematics and geometric constraints for articulated figuremanipulation. Master's Thesis, Simon Fraser University (1993)

17. Poljak, S., Rendl, F., Wolkowicz, H.: A recipe for semidefinite relaxation for (0,1)-quadratic programming. Journal of Global Optimization **7** (1995) 51–73

18. Goldenberg, A.A., Benhabib, B., Fenton, R.G.: A complete generalized solution to the inverse kinematics of robots. IEEE Journal of Robotics and Automation **RA-1** (1985) 14–20

An Optimal Data Propagation Algorithm for Maximizing the Lifespan of Sensor Networks

Aubin Jarry, Pierre Leone*, Olivier Powell, and José Rolim**

Department of Informatics,
University of Geneva, Switzerland
{jarry, leone, powell, rolim}@cui.unige.ch

Abstract. We consider the problem of *data propagation* in wireless sensor networks and revisit the family of mixed strategy routing schemes. We show that maximizing the lifespan, balancing the energy among individual sensors and maximizing the message flow in the network are equivalent. We propose a distributed and adaptive data propagation algorithm for balancing the energy among sensors in the network. The mixed routing algorithm we propose allows each sensor node to either send a message to one of its immediate neighbors, or to send it directly to the base station, the decision being based on a potential function depending on its remaining energy. By considering a simple model of the network and using a linear programming description of the message flow, we prove the strong result that *an energy-balanced mixed strategy beats every other possible routing strategy* in terms of lifespan maximization. Moreover, we provide sufficient conditions for ensuring the dynamic stability of the algorithm. The algorithm is inspired by the gradient-based routing scheme but by allowing to send messages directly to the base station we improve considerably the lifespan of the network. As a matter of fact, we show experimentally that our algorithm is close to optimal and that it even *beats the best centralized multi-hop routing strategy*.

1 Introduction

Recently advances in micro-electro-mechanical systems (MEMS) have enabled the development of very small sensing devices called sensor nodes [1]. These sensor nodes are smart devices with sensing, data processing and transmission (typically radio) capabilities. A typical application of wireless sensor networks (WSN) is *area monitoring*, where sensors are dispersed over a region and monitor some event (heat increase, pressure variation, intrusion, etc...). When a sensor detects an event it needs to report to (one of) the base station(s), which

* Swiss SER Contract No. C05.0030.
** Research partially funded by the *Swiss National Research Foundation* (SNRF) and FP6-015964 AEOLUS.

P. Gibbons et al. (Eds.): DCOSS 2006, LNCS 4026, pp. 405–421, 2006.
© Springer-Verlag Berlin Heidelberg 2006

has much more resources than sensor nodes and will be able to take appropriate action (such as sending a report on the Internet, to a satellite, etc...). We consider the problem of maximizing the lifespan of a wireless sensor network that carries out data propagation duties, collecting information from the monitored area [2, 3, 4, 5]. The lifespan of such a network is limited by the available energy in its nodes, and thus, in order to maximize its lifespan, the network (and each of its nodes) needs to save energy [6]. We assume that the most energy expensive operation for sensor nodes is radio transmission and since the energy cost of sending a message from a node to another grows in proportion to the the square of the distance [2, 3], it makes sense to prefer multi-hop data propagation algorithms to single-hop algorithms, (unless the distance from nodes to the base station is very small [7]). However, multi-hop routing algorithms tend to overuse a few bottleneck nodes (typically the nodes close to the base station), making them run out of energy and eventually putting the whole network down (when too many routes are broken) despite the fact that plenty of energy may still be available in other regions of the network. To overcome this unbalanced energy consumption, it was proposed in [8] to use *mixed routing strategies*. The idea of mixed strategies is to let sensors make a choice: they can either send messages to a neighbor, or send them directly to the base station (which may cost a lot more energy when the base station is far away). In [8], it was shown that a randomized mixed strategy can be used to balance the energy consumption among nodes and substantially increase the lifespan of the network. However, the probabilistic decisions made by each sensor (either to send a message directly to the sink or to a neighbor) are computed offline by a centralized algorithm. In [9], the possibility of computing solutions online with an adaptive algorithm is considered. However, the solutions are still computed by a centralized algorithm, which need to be broadcast in the network from time to time. In [10], the question of the existence of energy-balanced mixed strategies is first addressed, and several properties (such as theorem 2 which is presented in section 3.2) are discovered. An offline centralized polynomial time algorithm is also given, which finds a lifespan maximizing mixed routing strategy, even when an energy-balanced solution does not exist.

We investigate the properties of mixed strategies by introducing a constrained-flow linear program description of data propagation (having some similarities with [11]). We answer the important question of the appropriateness of mixed strategies by proving in theorem 1 that *mixed strategies beat every other possible strategy in terms of lifespan maximization of the network* under the weak condition that an energy-balanced mixed strategy exists. We also show that lifespan maximization, flow maximization and energy-balancing are equivalent problems (section 3.2). Improving on [12], we propose an algorithm which is both adaptive (i.e. online) *and* distributed and which aims at balancing the energy consumption and thus the lifespan of the network. We prove that the algorithm is stable under realistic conditions, and show through simulations that it approaches the centralized offline optimal solution.

2   Model

2.1   Data Propagation

We consider a Wireless Sensor Network (WSN) with sensor nodes scattered randomly over a region where some phenomenon is being monitored. When an event is detected by one of the nodes, it needs to be reported to the closest base station. In order to report such an event, a message is generated and routed from node to node towards (one of) the base stations. We allow only two ways of transmitting messages: from a node to one of its neighbors, (where neighborhood means being neighbor in the *unit disc graph* built upon the nodes of the WSN), or alternatively, directly from a node to a base station in a single-hop long range transmission. We simplify the cost of communications as follows: when a node sends a message to one of its neighbors, it spends x energy unit, when it sends a message directly to one of the base stations, it spends $h^2 \times x$ energy units, where h is the hop distance (in the unit disc graph) from the node to the closest base station and x is a constant. To simplify the model further, we will suppose throughout the paper that all nodes have the same amount of available energy: 1. While this latter simplification (of normalized and uniform batteries) enables us to unclutter our equations, it is worth noting that the proofs and experiments presented in this paper would still be valid without it.

2.2   The Slice Model

In our analytic study, we partition the monitored area into slices (following [8, 9, 10]). Each slice contains all the nodes which are at the same hop distance (in the unit disc graph) from the base station, as illustrated on the left hand side of figure 1. We consider that each sensor has 1 energy unit available, and thus the amount of energy b_i available at each slice is equal to the number of nodes it contains. For the example illustrated in figure 1 we have $b_1 = 1$, $b_2 = 2$, $b_3 = 2$, $b_4 = 4$ and $b_5 = 3$. We also need to model the detection of external events (i.e. the phenomenon which is being monitored by the WSN). We define g_i to be

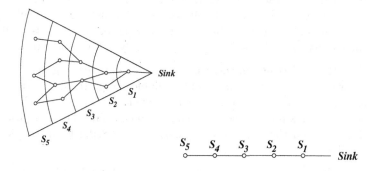

Fig. 1. Slices and simplified model

the number of events detected by the ith slice over a given period of time. If, for example, one of the nodes of S_5 detects 4 events, the second node 2 events and the third node of S_5 detects 0 events, we would set $g_5 = 6$. The g_i's can be seen as relative *input rates* between slices (where an *input* is the detection of an event).

3   Sensor Network Flows

Consider a slice model of a WSN with N slices $S_1, \ldots, S_N$ such that slice S_i contains b_i sensors and such that the input rate for slice S_i is g_i. We model the routing of messages with a linear program (LP) (definition 1), which is very similar to a flow problem. For each slice S_i, we represent the input rate by the parameter $f_{0,i}$ and the number of messages sent from S_i directly to the base station by the parameter $f_{i,0}$ [1]. We represent the number of messages sent from S_i to S_j by the parameter $f_{i,j}$. We need to ensure that the input rates are respected (equations 1), that the flow is conserved (equations 2), that no slice spends more energy than available (equations 3), and that messages are only sent to a 1-hop neighbor or directly to the base station (equations 4).

Definition 1 (mixed-flow maximization problem). *Let* $\{(g_i, b_i)\}_{1 \leq i \leq N}$ *be the description of the slice model of a WSN. A mixed-flow for this WSN is* $F = \{f_{i,j}\}_{0 \leq i,j \leq N}$ *satisfying equations 1 to 4 in the LP hereunder for some positive real* T. *The* mixed-flow maximization problem *is to find a mixed-flow* F *maximizing* T.

$$f_{0,i} = T g_i \qquad 1 \leq i \leq N \tag{1}$$

$$\sum_{j=0}^{N} f_{i,j} = \sum_{j=0}^{N} f_{j,i} \quad 1 \leq i \leq N \tag{2}$$

$$\sum_{i=1}^{N} f_{i,j}(i-j)^2 \leq b_i \quad 1 \leq i \leq N \tag{3}$$

$$f_{i,j} = 0 \qquad\qquad 1 \leq i \leq N, j \notin \{0, j-1\} \tag{4}$$

Since equations 4 allow a routing algorithm which is a *mix* of single-hops and multi-hops, we call this LP the *mixed-flow* maximization problem.

Remark 1. The LP is not a network flow problem in a strict sense (because of inequalities 3), so we should probably follow [11] which calls a similar problem an *energy constrained* flow problem. For the sake of conciseness, and since it brings no confusion in this article, we shall simply call it a flow problem.

A solution to the LP maximizes the amount of data propagated by a *mixed-strategy* for given input rates (g_i's) and available energy (b_i's) (i.e. the algorithm is offline), c.f. [11] for a similar approach. If one looks carefully at the LP above, it appears that maximizing the flow reduces to choosing the appropriate ratios, for each i, between the $f_{i,i-1}$'s and the $f_{i,0}$'s, c.f. [8, 9, 10] and section 3.2 for more details.

---

[1] One can think of S_0 as a "virtual" slice which plays the role of both the source and the sink.

3.1   Optimality of Mixed-Flows

The LP of the previous section can be solved by an offline[2] and centralized LP solver, thus obtaining for each slice the optimal ratio of messages which should be sent to the next slice and the ratio which should be sent directly to the base station. As we shall see in section 4, there is a fairly simple *totally distributed* and *adaptive* algorithm which can be implemented in each node of a WSN and which approaches this solution. However one may ask if mixed-flows are too restrictive: nodes (or slices) are only allowed to send messages to the sink and to neighbors which are one hop away in the unit disc graph. It is natural to ask oneself how much the message flow can be increased if the nodes are allowed to send messages not only to the base station and to 1-hop neighbors, but also to 2-hop neighbors, 3-hop neighbors, etc... [3]. There is one a priori objection to this approach: implementing this idea in a distributed algorithm such as the one we propose in section 4 seems much more complicated, and there is one good reason to avoid this extra complication, which is the important finding (theorem 1) that *generalized flows* (i.e. flows allowing messages to be passed 1, 2, 3, etc.. hops away or directly to the base station) do not improve the max flow reached by an energy-balanced mixed strategy (i.e. where message are passed either 1 hop away or directly to the base station). Details follow. First of all, the LP of definition 2 can be rewritten to allow 1, 2, 3, etc... hops as follows.

Definition 2 (Generalized flow maximization problem). *The* generalized flow maximization problem *is, on input* $\{(g_i, b_i)\}_{1 \leq i \leq N}$, *to find a flow* $F = \{f_{i,j}\}_{0 \leq i \leq N}$ *maximizing* T *in the LP program hereunder.*

$$f_{0,i} = Tg_i \qquad 1 \leq i \leq N \tag{5}$$

$$\sum_{j=0}^{N} f_{j,i} = \sum_{j=0}^{N} f_{i,j} \quad 1 \leq i \leq N \tag{6}$$

$$\sum_{j=0}^{N} f_{i,j}(i-j)^2 \leq b_i \quad 1 \leq i \leq N \tag{7}$$

The main theoretical result of this paper is that an energy-balanced solution to the mixed-flow maximization problem (definition 1) is also a solution to the general flow maximization problem of definition 2.

Remark 2. It was shown in [10] that solutions to the mixed-flow maximization problem are always reached by an energy-balanced solution, when an energy-balanced flow exists, c.f. theorem 2).

Definition 3 (energy-balanced flow). *Let* F *be a flow (either a mixed or a generalized flow). The energy spent by slice* S_i *is defined as* $e_i := \sum_{j=0}^{N} f_{i,j}(i - j)^2$. *A flow is called* energy-balanced *if there is a constant* c *such that for each* i *with* $1 \leq i \leq N$: $\frac{e_i}{b_i} = c$

---

[2] Because the input rates are known a priori.

[3] We are thankful to the research community for asking this question at conferences.

Our main theoretical result (theorem 1) shows that the *maximum generalized flow* is reached by an *energy-balanced mixed-flow*, when such a flow exists. This finding is important, since it implies that there is no need to search for generalized flows when one can find an energy-balanced mixed-flow. The proof follows.

Lemma 1. *The following two equations are true.*

$$\forall\, n > 0, \quad \frac{1}{n(n+1)} + \frac{(n-1)^2}{n(n-1)} = \frac{n^2}{n(n+1)} \tag{8}$$

$$\forall\, n > i > 0, \quad \frac{(n-i)^2}{n(n+1)} + \frac{i^2}{i(i+1)} \geq \frac{n^2}{n(n+1)} \tag{9}$$

Proof. The proof of equation 8 is straightforward. Equation 9 is a consequence of equation 8 when $i = n - 1$. By multiplying both terms by $n(n+1)(i+1)$ (a strictly positive number), we see that equation 9 is equivalent to

$$(n-i)^2(i+1) + in(n+1) - n^2(i+1) \geq 0$$
$$(i+1)(i^2 - 2ni) + i(n-i-1+i+1)(n+1) \geq 0$$
$$i((i+1)(i - 2n + n + 1) + (n-i-1)(n+1)) \geq 0$$
$$i(n-i-1)((n+1) - (i+1)) \geq 0$$

Theorem 1. *An energy-balanced solution to the mixed-flow maximization problem is also a solution to the generalized flow maximization problem.*

Proof. Let $(T, \{f_{i,j}\}_{i,j\in[0,N]})$ be a solution to the mixed-flow maximization problem. We will first prove by induction on N that for any path $\mu = (s_{i_0}, s_{i_1})$ $(s_{i_1}, s_{i_2})..(s_{i_{l-1}}, s_l)$ from a vertex s_i to the base station (i.e. $i_0 = i$ and $i_l = 0$) such that $f_{x,y} > 0$ if (s_x, s_y) appears in μ^4 , we have

$$\sum_{0 \leq j < l} \frac{(i_j - i_{j+1})^2}{i_j(i_j + 1)} = \frac{i^2}{i(i+1)} \tag{10}$$

This is obviously true for $N = 1$. Supposing that the property is true for $N - 1$, we can observe that a path from s_N to the base station is either composed of a single edge (s_N, s_0) (in which case equation 10 holds), or of an edge (s_N, s_{N-1}) and of a path μ' from s_{N-1} to the base station, in which case we have

$$\sum_{0 \leq j < l} \frac{(i_j - i_{j+1})^2}{i_j(i_j + 1)} = \frac{1}{N(N+1)} + \sum_{1 \leq j < l} \frac{(i_j - i_{j+1})^2}{i_j(i_j + 1)} = \frac{1}{N(N+1)} + \frac{(N-1)^2}{N(N-1)} \tag{11}$$

Equations 8 and 11 imply that equation 10 holds.

Now let $(\overline{T}, \{\overline{f_{i,j}}\}_{i,j\in[0,N]})$ be a solution to the generalized flow maximization problem. We will prove by induction that for any simple[5] path $\overline{\mu} = (s_{i_0}, s_{i_1})$

---

[4] That is, μ is a component path of the flow $\{f_{i,j}\}_{i,j\in[0,N]}$.
[5] i.e. a path without loops.

$(s_{i_1}, s_{i_2})..(s_{i_{l-1}}, s_l)$ from a vertex s_i to the base station, if $\overline{\mu}$ is a *component path*[6] of $\{\overline{f}_{i,j}\}_{i,j \in [0,N]}$ we have

$$\sum_{0 \leq j < l} \frac{(i_j - i_{j+1})^2}{i_j(i_j + 1)} \geq \frac{i^2}{i(i + 1)} \tag{12}$$

This is obviously true for $N = 1$. Supposing that the property is true for $N-1$, we can observe that a simple path from s_N to the base station is either composed of a single edge (s_N, s_0) (in which case equation 12 holds), or of an edge (s_N, s_{N-i}) and of a simple path μ' from s_i to the base station which does not go back through s_N, in which case we have

$$\sum_{0 \leq j < l} \frac{(i_j - i_{j+1})^2}{i_j(i_j + 1)} = \frac{1}{N(N + 1)} + \sum_{1 \leq j < l} \frac{(i_j - i_{j+1})^2}{i_j(i_j + 1)} \geq \frac{(N - i)^2}{N(N + 1)} + \frac{i^2}{i(i + 1)}$$

Recalling equation 9, equation 12 holds for all simple paths from s_N to the base station. Simple paths going from s_i to the base station through s_N will give us

$$\sum_{0 \leq j < l} \frac{(i_j - i_{j+1})^2}{i_j(i_j + 1)} \geq \frac{N^2}{N(N + 1)} \geq \frac{i^2}{i(i + 1)}$$

Equation 12 is thus proved.

Decomposing flows into paths enables us to deduce from equations 10 and 12 the following:

$$\sum_{0 < i \leq N, 0 \leq j \leq N} \frac{(i - j)^2 f_{i,j}}{i(i + 1)} = \sum_{0 < i \leq N} \frac{i^2 f_{0,i}}{i(i + 1)}$$

$$\sum_{0 < i \leq N, 0 \leq j \leq N} \frac{(i - j)^2 \overline{f}_{i,j}}{i(i + 1)} \geq \sum_{0 < i \leq N} \frac{i^2 \overline{f}_{0,i}}{i(i + 1)}$$

If $(T, \{f_{i,j}\}_{i,j \in [0,N]})$ is an energy-balanced solution to the mixed-flow maximization problem and if $(\overline{T}, \{\overline{f}_{i,j}\}_{i,j \in [0,N]})$ is a solution to the generalized flow maximization problem, we have

$$\sum_{0 < i \leq N} \frac{b_i}{i(i + 1)} = \sum_{0 < i \leq N, 0 \leq j \leq N} \frac{(i - j)^2 f_{i,j}}{i(i + 1)} = \sum_{0 < i \leq N} \frac{i^2 f_{0,i}}{i(i + 1)}$$

$$\sum_{0 < i \leq N} \frac{b_i}{i(i + 1)} \geq \sum_{0 < i \leq N, 0 \leq j \leq N} \frac{(i - j)^2 \overline{f}_{i,j}}{i(i + 1)} \geq \sum_{0 < i \leq N} \frac{i^2 \overline{f}_{0,i}}{i(i + 1)}$$

and since $\forall\, 0 < i \leq N$, $f_{0,i} \leq \overline{f}_{0,i}$, it follows that $\forall\, 0 < i \leq N$, $f_{0,i} = \overline{f}_{0,i}$, so $T = \overline{T}$: an energy-balanced solution to the mixed-flow maximization problem is also a solution to the generalized flow maximization problem.

---

[6] c.f. definition of μ above.

3.2   Lifespan, Flow and Energy-Balance

In this section we revisit the *lifespan maximization problem* introduced and studied in [10], which builds upon previous results: [8, 9]. As we show in lemma 2, the *lifespan maximization problem* from [10] is equivalent, but in another formalism, to the mixed-flow maximization problem from definition 2. This lemma, together with theorem 2, shows that *maximizing the flow, maximizing the lifespan* and *balancing the energy are equivalent*. The WSN model considered in [10] is also the slice model, but a routing strategy (equivalent to a mixed-flow) was described by values p_i (for $1 \leq i \leq N$) which represent the fraction of messages sent from slice S_i to slice S_{i-1}; $1 - p_i$ being the fraction sent from S_i directly to the base station. In this formalism, the energy e_i spent by slice S_i is computed in the following way: for each slice S_i, let m_i be the sum of g_i, the number of events detected by S_i and the number of messages that slice S_i receives from S_{i+1}, thus $m_i = g_i + p_{i+1}m_{i+1}$. More precisely:

$$m_N = \qquad g_N \tag{13}$$
$$m_i = \quad m_{i+1}p_{i+1} + g_i \quad 1 \leq i \leq N - 1 \tag{14}$$
$$e_i = m_i(p_i + i^2(1 - p_i)) \; 1 \leq i \leq N - 1 \tag{15}$$

In [12], the following definitions were given

Definition 4. *Let $\{(g_i, b_i)\}_{1 \leq i \leq N}$ be the description of a WSN.*

- *A mixed strategy is the choice of a probability p_i for $1 \leq i \leq N$*
- *A mixed strategy is called* energy-balanced *if $\frac{e_i}{b_i} = \frac{e_{i+1}}{b_{i+1}}$ for $1 \leq i \leq N - 1$*
- *The lifespan of the network is defined as $\min\{\frac{b_i}{e_i}\}_{1 \leq i \leq N}$*

Definition 5 (Mixed-strategy lifespan maximization). *The mixed-strategy lifespan maximization problem is, on input $\{(g_i, b_i)\}_{1 \leq i \leq N}$ (which is the description of a WSN), to find a mixed strategy which maximizes the lifespan of the network.*

Lemma 2. *Solving the mixed-strategy lifespan maximization problem is equivalent to solving the mixed-flow maximization problem.*

Proof. By definition, the input of both problems is the same. It thus suffices to show that any solution to one of the problems can be (efficiently) transformed into a solution to the other problem. The fundamental observation is that the mixed-flow T grows with the lifespan, since the total amount of events reported by the WSN before it goes down is the time during which the WSN is active multiplied by the input rate:

$$T = \text{input rate} \cdot \text{lifespan} = \sum_1^N g_i \cdot \text{lifespan} = \text{cst} \cdot \text{lifespan}$$

If $\{f_{i,j}\}_{0 \leq i,j \leq N}$ is a solution to the maximization of T, then $p_i = \frac{f_{i,i-1}}{f_{i,i-1} + f_{i,0}}$ is a solution to the mixed-strategy lifespan maximization problem. Reciprocally, if $\{p_i\}_{1 \leq i \leq N}$ is a solution to the mixed-flow maximization problem, then the energy e_i spent and the received messages m_i can be computed (as in equations 13, 14 and 15). A solution to the mixed-flow maximization problem is then given by the following equations:

$$1 \leq i \leq N \quad f_{0,i} = g_i \min \left\{ \frac{b_j}{e_j} \right\}_{1 \leq j \leq n} \tag{16}$$

$$1 \leq i \leq N \quad f_{i,0} = m_i(1 - p_i) \min \left\{ \frac{b_j}{e_j} \right\}_{1 \leq j \leq n} \tag{17}$$

$$2 \leq i \leq N \quad f_{i,i-1} = m_i p_i \min \left\{ \frac{b_j}{e_j} \right\}_{1 \leq j \leq n} \tag{18}$$

Details are left to the reader.

In [10], the following result (restated in our notation) is shown:

Theorem 2. *Given* $\{(g_i, b_i)\}_{1 \leq i \leq N}$ *a WSN, if an energy-balanced mixed strategy exists, it is the unique solution to the mixed-strategy lifespan maximization problem.*

Using lemma 2, theorem 1, and theorem 2 we deduce that if an energy-balanced mixed strategy exists, it is the unique solution to the *general* lifespan maximization problem:

Theorem 3. *Maximizing the flow, the lifespan and balancing the energy are equivalent.*

4   The Algorithm

In this section, we propose a simple blind and online distributed algorithm for lifespan maximization of a WSN in a data propagation scenario. Following theorem 3, lifespan maximization can be achieved by reaching energy balance. The algorithm is fairly simple: each node has the choice between sending messages to one of its neighbors or directly to the base station. It makes its decision using a potential function. We show in section 4.1 that the algorithm is stable, and we show experimentally in section 4.2 that the energy consumption is almost energy balanced (and thus close to maximal lifespan).

Definition 6 (Algorithm). *We consider a WSN scattered on a surface with one or more base stations. We consider the unit disc graph with a vertex for each sensor and a vertex for each base station, and edges between vertexes at a maximum distance of 1 from each other. Let n be a sensor. V_n is the neighborhood of n, i.e. all sensors linked to n in the unit disc graph. Each sensor n has an associated potential $pot(n)$, and each sensor knows the value of the potential of each of its neighbors. When n detects an event or receives a message which it must pass on to another sensor, it makes the following decision: let m be the sensor of V_n with the lower potential value: $pot(m) \leq pot(m')$ for all m' in V_n.*

- If $pot(n) > pot(m)$ then n sends the message to m (spending one energy unit).
- Otherwise, n sends the message directly to the closest base station, spending h^2 energy units, where h is the length of the shortest path from n to a base station in the unit disc graph.

The potential function $pot(n)$ we use can be separated in two parts: $pot_s(n)$, a static component which does not evolve over time and $pot_d(n)$ a dynamical component which evolves over time. $pot_s(n)$ is an estimation of the distance from n to the closest base station, so for example $pot_s(n)$ could be equal to h, the length of the shortest path from n to the closest base station. $pot_d(n)$ is the energy spent by sensor n (thus it evolves over time as the sensor n consumes its energy). The assumption that each sensor is aware of the potential of its neighbors can be implemented in a real WSN by making each sensor include its current potential in the header of each message it sends.

4.1   Stability of the Algorithm

In this subsection we restrict the analysis to the case where we have only one base station (or sink) and the topology of the slice model is the one depicted in figure 1. The potential function at node n is given by the current energy spent by the node $pot_d(n)$. If node n has to handle a message the two options are to send it to the sink (if $pot_d(n) < pot_d(n-1)$), spending n^2 units of energy or to the next sensor $n-1$, spending 1 unit of energy (if $pot_d(n) \geq pot_d(n-1)$). Actually, because of the particular form of the potential energy and since the time dependence becomes important, we introduce the notation $x_n(t)$ to denote the energy spent by the sensor n at time t. The time takes discrete values $t = 0, 1, 2, \ldots$ since it only refers to the occurrence of a message to be handled. This amounts to considering only the Markov Chain embedded in the time continuous dynamical process, and has no relationship with considering synchronism in the network. The network states are described by vectors whose entries are the total energy consumed by slice n, and the discrete dynamics is expressed as a map

$$
\begin{pmatrix} x_N(t) \\ x_{N-1}(t) \\ \vdots \\ x_1(t) \end{pmatrix} \Longrightarrow \begin{pmatrix} x_N(t+1) \\ x_{N-1}(t+1) \\ \vdots \\ x_1(t+1) \end{pmatrix}
$$

Actually, since our main concern is with balancing the energy it is more convenient to deal with the reduced state vector $X(t) \in \mathcal{R}^{N-1}$ which must ideally vanish if the energy is balanced

$$
X(t) = \begin{pmatrix} x_N(t) - x_{N-1}(t) \\ x_{N-1}(t) - x_{N-2}(t) \\ \vdots \\ x_2(t) - x_1(t) \end{pmatrix} \Longrightarrow X(t+1) = \begin{pmatrix} x_N(t+1) - x_{N-1}(t+1) \\ x_{N-1}(t+1) - x_{N-2}(t+1) \\ \vdots \\ x_2(t+1) - x_1(t+1) \end{pmatrix} \approx \begin{pmatrix} 0 \\ 0 \\ \vdots \\ 0 \end{pmatrix}
$$

The occurrences of the events in the network leading to the generation of messages are random and we denote by λ_n, $n = 1, \ldots, N$ the probability that the event occurs in slice n. To get back to the previous notation, one has $\lambda_n \approx g_n / \sum_j g_j$. The evolution of the dynamics $X(t) \Longrightarrow X(t+1)$ is then described by a Markov chain.

Theorem 4. *The sequence of random vectors* $\{X(t)\}_{t \geq 1}$ *is a Markov chain.*

The proof is clear and omitted. Moreover, we have

Theorem 5. *If* $\lambda_i > 0$ *the Markov chain is irreducible.*

Proof. The initial state vector is the null vector $X(0) = (0, 0, \ldots, 0)$ and to show the irreducibility of the chain it is sufficient to show that given any state vector $X(t)$ one can with strictly positive probability get back to the initial state. Let us denote $\widetilde{X}(t) = (x_N(t), x_{N-1}(t), \ldots, x_1(t))^T$ the given state vector. We consider the following sequence of events which occurs with a positive probability and which leads to perfect energy-balance, $X(t+u) = (0, 0, \ldots, 0)^T$, with $u \geq 0$. We first assume that a sequence of events is detected in the N-th slice. The generated messages are first ejected by slice N until the consumed energy is larger that the consumed energy in slice $N - 1$ and then forwarded to slice $N - 1$. If there are enough events occurring in the last slice N it will, in the end, happen that

$$x_N(t_1) \geq x_{N-1}(t_1) \geq \ldots \geq x_1(t_1), \quad t_1 \geq t$$

This happens with positive probability since $\lambda_n > 0$. Now, assume that $x_N(t_1) - x_{N-1}(t_1)$ events occur in slice $N - 1$. Each generated message is forwarded to slice $N - 2$, leading to energy-balance between slice $N - 1$ and N. The event occurs with probability $\lambda_{N-1}^{x_N(t_1) - x_{N-1}(t_1)} > 0$. To conclude the proof, we consider the occurrence of events in slice $N - 2$ leading to energy-balance $x_{N-2}(t_2) = x_{N-1}(t_2)$, and so on up to the first slice.

The restriction $\lambda_i > 0$ is necessary to ensure irreducibility but is not always realistic. Indeed, one can assume that some nodes are only responsible for handling the messages and not for sensing the environment. In this case, the previous result cannot be applied directly. To get some sufficient conditions, we investigate the stability of the algorithm.

Definition 7. *The Markov chain* $\{X(t)\}_{t \geq 1}$ *is stable at the origin* o *if there exists a neighborhood* $\mathcal{D}$ *of the origin which is positive recurrent, i.e. such that given any* $X \in \mathcal{R}^{N-1}$ *the Markov chain with initial condition* $X(0) = X$ *satisfies*

$$Prob(X(t) \in \mathcal{D}, \ t < \infty) = 1 \tag{19}$$

The stability of the Markov chain means that the chain can leave the neighborhood of the origin but that it will always go back to it.

The next result provides sufficient conditions for stability. We state the result by assuming $\lambda_i > 0$ since the proof is neater and we discuss how these conditions can be relaxed later.

Theorem 6. *If we assume that* $\lambda_i > 0$, $i = 1, \ldots, N$, *then the set*

$$\mathcal{A} = \left\{ X \in \mathcal{R}^{N-1} \ : \ \mid x_i(t) - x_{i-1}(t) \mid \leq \frac{i^2}{2}, \ i = 2, \ldots, N \right\}$$

is positive recurrent if

$$i^2 \lambda_i > (i-1)^2 \lambda_{i-1}, \quad i = 2, \ldots, N \tag{20}$$

Proof. The demonstration of the sufficient conditions is based on proving that the function

$$f\big(X(t)\big) = \sum_{i=2}^{N} \mid x_i(t) - x_{i-1}(t) \mid$$

satisfies $\forall X(t) \notin \mathcal{A}$

$$E\left(f\big(X(t+1)\big) - f\big(X(t)\big) \Big| X(t) \right) < -\epsilon, \quad \epsilon > 0, \tag{21}$$

where $E(\mid)$ is the conditional expectation. The function $f(X)$ is a Lyapunov function for the Markov chain and (21) proves that $\mathcal{A}$ is positive recurrent since it shows that $f(X(t))$ is a convergent super-martingale and because $\epsilon > 0$ it cannot converge to a point not belonging to $\mathcal{A}$ and hence will reach this set of points (full details can be found in [13], chapter 2). However, once we have $X(t) \in \mathcal{A}$, the super-martingale property (21) no longer holds and the Markov chain will eventually leave out of the set and oscillate between inside and outside of $\mathcal{A}$. To prove the existence of the bound (21) we consider different cases.

Case 1: $x_{i+j}(t) < x_{i+j-1}(t) > \ldots > x_i(t) > x_{i-1}(t) > x_{i-2}(t)$

This corresponds to the situation where messages generated in slices $i + j - 1, \ldots, i$ are forwarded to the next slice without changing the value of $x_i(t) - x_{i-1}(t)$. Messages generated in slice $i - 1$ are forwarded to the next slice and it leads to $x_i(t+1) - x_{i-1}(t+1) = x_i(t) - x_{i-1}(t) - 1$. The messages generated in the other slices are not handled by slice i nor $i - 1$. Hence, we have

$$E\big(x_i(t+1) - x_{i-1}(t+1)\big) = \big(x_i(t) - x_{i-1}(t)\big) \sum_{j=1, \ j\neq i-1}^{N} \lambda_j + \lambda_{i-1}\big(x_i(t) - x_{i-1}(t) - 1\big)$$

$$= \underbrace{x_i(t) - x_{i-1}(t)}_{>0} - \lambda_{i-1}$$

Case 2: $x_{i+j}(t) < x_{i+j-1}(t) > \ldots > x_i(t) < x_{i-1}(t) > x_{i-2}(t)$

A similar reasoning leads to

$$E\big(x_i(t+1) - x_{i-1}(t+1)\big) = (\lambda_{i+j-1} + \ldots + \lambda_i)(x_i(t) - x_{i-1}(t) + i^2) +$$
$$+ \lambda_{i-1}(x_i(t) - x_{i-1}(t) - 1) + \ldots$$
$$= \underbrace{x_i(t) - x_{i-1}(t)}_{<0} + i^2(\lambda_{i+j-1} + \ldots + \lambda_i) - \lambda_{i-1}$$

where the dots ... in the expression on the second line from the bottom stand for messages generated in slices which do not change the value of $x_i(t) - x_{i-1}(t)$. We use the same form below to denote such events.

Case 3: $x_{i+j}(t) < x_{i+j-1}(t) > \ldots > x_i(t) > x_{i-1}(t) < x_{i-2}(t)$
leads to

$$E\big(x_i(t+1) - x_{i-1}(t+1)\big) = (\lambda_{i+j-1} + \ldots + \lambda_i)(x_i(t) - x_{i-1}(t) + 1 - (i-1)^2) +$$
$$+ \lambda_{i-1}(x_i(t) - x_{i-1}(t) - (i-1)^2) + \ldots$$
$$= \underbrace{x_i(t) - x_{i-1}(t)}_{>0} - (\lambda_{i+j-1} + \ldots + \lambda_{i-1})(i-1)^2 + (\lambda_{i+j-1} + \ldots + \lambda_i)$$

Case 4: $x_{i+j}(t) < x_{i+j-1}(t) > \ldots > x_i(t) < x_{i-1}(t) < x_{i-2}(t)$
finally, leads to

$$E\big(x_i(t+1) - x_{i-1}(t+1)\big) = (\lambda_{i+j-1} + \ldots + \lambda_i)(x_i(t) - x_{i-1}(t) + (i)^2) +$$
$$+ \lambda_{i-1}(x_i(t) - x_{i-1}(t) - (i-1)^2) + \ldots$$
$$= \underbrace{x_i(t) - x_{i-1}(t)}_{<0} + (\lambda_{i+j-1} + \ldots + \lambda_i)i^2 - \lambda_{i-1}(i-1)^2$$

In order to ensure that there exists $\epsilon_i > 0$ such that

$$E\big(|\, x_i(t+1) - x_{i-1}(t+1)\,|\big) - |\, x_i(t) - x_{i-1}(t)\,| < -\epsilon_i,$$

in the four cases listed above, one can check that (20) is a sufficient condition to ensure that the expected value of $|\, x_i(t) - x_{i-1}(t)\,|$ decreases if $x(t) \notin \mathcal{A}$. To conclude, notice that $\sum_i \epsilon_i = \epsilon$ satisfies the condition in (21).

Before ending this section one has to discuss the case where some λ_i vanish. We first notice that if $\lambda_i \neq 0$ and $\lambda_{i-1} \neq 0$ then $x_i(t) - x_{i-1}(t)$ is bounded if the sufficient conditions (20) hold for i, i.e. the (20) conditions are local. When $\lambda_{i-1} = 0$ then $|\, x_i - x_{i-1}\,| \leq i^2$ since messages handled by the $(i-1)$–th slice are forwarded by slice i. The only case where the energy consumption difference can be unbounded is $\lambda_i = 0$, $\lambda_{i-1} \neq 0$, since the energy in the $(i-1)$–th slice can increase independently of the energy level in the i–th slice. Let us assume $\lambda_{i+j} \neq 0$, $\lambda_{i+j-1} = 0, \ldots,$ $\lambda_i = 0$, $\lambda_{i-1} \neq 0$ and $x_{i+j} = x_{i+j-1} = \ldots = x_i < x_{i-1}$. The first message handled by the $(i+j)$–th slice will be forwarded up to the i–th slice which will eject the message since $x_i < x_{i-1}$. The next one will be ejected by the $(i+1)$–th slice since we now have $x_{i+1} < x_i$, and so on up to $x_{i+j} > x_{i+j-1} > \ldots > x_i$. In this configuration, the next message handled by the $(i+j)$–th slice will be handled by the i–th slice and forwarded or ejected according to $x_i < x_{i-1}$ or $x_i \geq x_{i-1}$. This simple reasoning shows the proportion of messages reaching the $(i-1)$–th slice approaches $\frac{1}{j}$. The other messages are ejected in an intermediate slice for which $\lambda_k = 0$. In order to balance the energy the sufficient condition (20) becomes

$$(i+j)^2 \frac{\lambda_{i+j}}{j} > (i-1)^2 \lambda_{i-1} \tag{22}$$

This brief analysis is suitably completed with the following observation. If the sufficient condition (20) are violated with a strict equality, i.e. if $i^2\lambda_i = (i-1)^2\lambda_{i-1}$, then the network is in the situation where all sensors have to eject the data in order to balance the energy. The dynamic is not stable since, with probability one, it is impossible to recover from an energy difference between sensors. However, condition (22) means that the number of messages generated in the $(i+j)$–th slice must be multiplied by j to take into account messages which are ejected by sensors in the slices $i+j-1,\dots,i$.

4.2   Simulations

We randomly disperse sensors in a circle or a sector according to a uniform distribution, and add one or more base stations. Simulations are then made by discretizing the time in rounds.

– *Event generation:* During each round, one (or more) event is generated at a random location (using the uniform distribution). An event can be detected by any sensor close enough: at most at distance 1, and we make the assumption that only the closest sensor to an event detects it (i.e. each event is detected by only one sensor).
– *Message transmission:* During each round, each sensor can *send one message* either to a neighbor (spending 1 energy unit), or directly to the base station, spending h^2 energy units. Message transmission is made synchronously. Collisions between messages are not taken into account

If we let simulations run for a period of time t, we can observe the energy spent by each sensor. We can also observe the input rates for each sensor: the number of events detected by each sensor, and the total number of messages $m(t)$ which have been reported to one of the base stations (which is equal to the total number of messages detected by a sensor minus the number of messages which are still inside the WSN). Making the assumption that the input rates are fixed, we can compute the flow $F(t)$ of the algorithm: $F(t) = \frac{m(t)}{e(t)}$ where $e(t)$ is the maximum energy consumption over all sensors at time t. The flow $F(t)$ allows the comparison of the performance of our algorithm with other routing schemes. In particular, given the input rates for each sensor we can write an LP program similar to the LP of definition 1 but including each node: flow must be conserved, input rates respected, and energy constraint satisfied. Solving this LP permits to find an upper bound $U(t)$ on the flow $F(t)$. $U(t)$ is the maximum flow which can be reached using a mixed strategy. Of course, $U(t)$ will be greater than $F(t)$ since our algorithm is handicapped by the fact that it is blind (or online) and distributed. We also compute L (using a similar LP) which is the maximum flow that can be achieved using a multi-hop routing strategy, but with no direct transmission from a sensor to a base station (unless the sensor and the base station are at distance less than 1). It is trivial that $L \le U$. We show experimentally that under the configurations considered (uniform distribution of sensors and uniform distribution of events) it holds that $L < F < U$, and we conjecture those inequalities hold for any reasonable distribution of sensors and events.

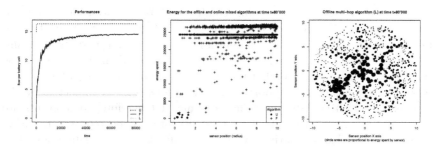

Fig. 2. First simulation: flows

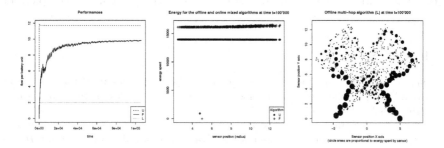

Fig. 3. Second simulation

First Simulation. The first simulation is done by scattering 1000 sensors in a 10 meter diameter disc with a single base station at the center of the disc. Five events are generated during each round. The potential functions we use are $pot_s(n) = hop(n)$ and $pot_d(n) = energy(n)$, and we use the following ordering: $pot(n) < pot(m)$ if and only if $[pot_s(n) < pot_s(m)$ and $pot_d(n) < pot_d(m)]$ (thus the hop distance from a message to a base station always diminishes). Flows are plotted in figure 2. We can also see on the central plot of figure 2 that the offline ideal flow U balances energy consumption amongst all the nodes (which is in accordance with theorem 1), and that our online algorithm approaches energy-balance between nodes. The right-hand side of figure 2 shows a circle for each of the 1000 sensors, and the area of the circle is proportional to the energy spent by the sensor under the offline multi-hop flow L. As we can see, it does not balance energy well: nodes close to the base station spend more energy than others (they are bottleneck nodes), and only a few privileged routes are used.

Other Simulations. We have conducted many simulations similar to the one above, and results are comparable if the distribution of sensors is reasonable. Events can be generated in a less uniform way (e.g. only certain nodes are sensors, the others are just used as data transmission gateways), and still results are comparable, as is consistent with the stability results of section 4.1. Of course, it can be imposed that $L = U$ by a deliberate but unrealistic dispersion of sensors (typically braking conditions 20), and thus it would not be true that $L < F$. We present in figure 3 similar plots to those of figure 2, but for a simulation where 600 sensors are dispersed randomly and uniformly in two sectors of 30 degrees angle and 10 meter diameter, with a base station at the narrow end of both sectors (c.f. rightmost plot of figure 3). Events are then generated randomly and uniformly, and are detected by the closest sensor which is at no more than distance 1 (therefore, some events are missed), at the rate of 3 events per round. Events can be reported to either base station. For this simulation, we used a different potential function than for the previous simulation, which is the following: $pot(n) = hop(n) + pot_d(n)$, with $pot_d(n)$ being the energy used by sensor n so far. The simulation results (which are similar to the previous simulation) are shown in the plots of figure 3.

References

1. Jan M. Rabaey, M. Josie Ammer, Julio L. da Silva, Danny Patel, and Shad Roundy. Picoradio supports ad hoc ultra-low power wireless networking. *Computer*, 33(7):42–48, 2000.
2. Jamal N. Al-Karaki and Ahmed E. Kamal. A taxonomy of routing techniques in wireless sensor networks. In Mohammad Ilyas and Imad Mahgoub, editors, *Handbook of Sensor Networks: Compact Wireless and Wired Sensing Systems*, pages 6.1–6.24. CRC Press, 2005.
3. Kemal Akkaya and Mohamed Younis. A survey on routing protocols for wireless sensor networks. *Ad Hoc Network Journal*, 3/3:325–349, 2005.
4. I. Chatzigiannakis, T. Dimitriou, S. Nikoletseas, and P. Spirakis. A probabilistic algorithm for efficient and robust data propagation in smart dust networks. In *5th European Wireless Conference on Mobile and Wireless Systems beyond 3G (EW 2004)*, pages 344–350, 2004.
5. I. Chatzigiannakis, S. Nikoletseas, and P. Spirakis. Smart dust protocols for local detection and propagation. In *2nd Workshop on Principles of Mobile Computing (POMC)*, pages 9–16. ACM, ACM Press, 2002.
6. I. F. Akyildiz, W. Su, Y .Sankarasubramaniam, and E. Cayirci. Wireless sensor networks: a survey. *Computer Networks*, (38):393–422, 2002.
7. W.R. Heinzelman, A. Chandrakasan, and H. Balakrishnan. Energy efficient communication protocol for wireless microsensor networks. In *Hawaii International Conference on Sytem Sciences (HICSS)*, number 33, 2000.
8. C. Efthymiou, S. Nikoletseas, and J. Rolim. Energy balanced data propagation in wireless sensor networks. Invited paper in the Wireless Networks (WINET, Kluwer Academic Publishers) Journal, Special Issue on "Best papers of the 4th Workshop on Algorithms for Wireless, Mobile, Ad Hoc and Sensor Networks (WMAN 2004)" to appear in 2005.
9. P. Leone, S. Nikoletseas, and J. Rolim. An adaptive blind algorithm for energy balanced data propagation in wireless sensor networks. In *The First International Conference on Distributed Computing in Sensor Systems (DCOSS)*, number 3560 in Lecture Notes in Computer Science. Springer Verlag, June/July 2005.
10. Olivier Powell, Pierre Leone, and Jose Rolim. Energy optimal data propagation in wireless sensor networks. *arXiv.org automated e-print archives*, 2005. Report CS-0508052. Journal version submitted for publication.

11. Bo Hong and Viktor K. Prasanna. Maximum data gathering in networked sensor systems. *International Journal of Distributed Sensor Networks*, 2005.
12. Olivier Powell, Aubin Jarry, Pierre Leone, and Jose Rolim. Gradient based routing in wireless sensor networks: a mixed strategy. *arXiv.org automated e-print archives*, 2005. Report CS-0511083.
13. G. Fayolle, V.A. Malyshev, and M.V. Menshikov. *Topics in the Constructive Theory of Countable Markov Chains*. Cambridge University Press, 1995.

Lifetime Maximization of Sensor Networks Under Connectivity and k-Coverage Constraints⋆

Wei Mo, Daji Qiao, and Zhengdao Wang

Iowa State University, Ames, IA 50011, USA
{mowei, daji, zhengdao}@iastate.edu

Abstract. In this paper, we study the fundamental limits of a wireless sensor network's lifetime under connectivity and k-coverage constraints. We consider a wireless sensor network with n sensors deployed independently and uniformly in a square field of unit area. Each sensor is active with probability p, independently from others, and each active sensor can sense a disc area with radius r_s. Moreover, considering the inherent irregularity of a sensor's sensing range caused by time-varying environments, we model the sensing radius r_s as a random variable with mean r_0 and variance $r_0^2\sigma_s^2$. Two active sensors can communicate with each other if and only if the distance between them is smaller than or equal to the communication radius r_c.

The key contributions of this paper are: (1) we introduce a new definition of a wireless sensor network's lifetime from a novel probabilistic perspective, called ω-lifetime $(0 < \omega < 1)$. It is defined as the expectation of the time interval during which the probability of guaranteeing connectivity and k-coverage simultaneously is at least ω; and (2) based on the analysis results, we propose a near-optimal scheduling algorithm, called PIS (Pre-planned Independent Sleeping), to achieve the network's maximum ω-lifetime, which is validated by simulation results, and present a possible implementation of the PIS scheme in the distributed manner.

1   Introduction

Energy conservation is perhaps the most important issue in wireless sensor networks [1, 2]. Most sensor devices are battery-powered and hence have a very limited amount of energy. It is, therefore, very important to extend the battery operation time of individual sensors and, consequently, the network's lifetime. Operating each sensor device in a low duty-cycle has been recognized as an effective way to achieve this goal, where *duty-cycle* is defined as the fraction of

---

⋆ The research reported in this paper was supported in part by the Information Infrastructure Institute (I-Cube) of Iowa State University. The authors would also like to acknowledge the support from the National Science Foundation under Grants No. ECS 0428040, CCF 0431092, and CNS 0520102. Preliminary results of this research were presented at *the 43-rd Annual Allerton Conference on Communication, Control, and Computing*, Monticello, IL, Sept. 2005.

P. Gibbons et al. (Eds.): DCOSS 2006, LNCS 4026, pp. 422–442, 2006.
© Springer-Verlag Berlin Heidelberg 2006

time that a sensor device is active. On the other hand, a wireless sensor network typically has two major tasks: *sensing* and *communication*. It is always desirable to have all active sensors connected and, at the same time, to have the entire sensing field k-covered. The connectivity among active sensors is required in order for an active sensor to report its sensing results back to the user, and the reason for requiring k-coverage rather than just 1-coverage is to increase the detection probability and accuracy of tracking. Obviously, the lower the duty-cycle of individual sensors, the longer the wireless sensor network's lifetime, but at the same time, there are a smaller number of active sensors at a given time and, hence, more likely either active sensors are not connected or the k-coverage of the sensing field cannot be guaranteed. So, there are inherent tradeoffs, and the key contribution of this paper is to present an integrated study on connectivity, k-coverage, and lifetime of a large-scale wireless sensor network.

1.1   Related Work

Several researchers [3, 4, 5, 6, 7, 8] have addressed the coverage and connectivity issues in wireless sensor/ad hoc networks. Gupta et al. [3] studied scaling laws for asymptotic connectivity of sensors placed at random over a unit area, and provided bounds on connectivity probability for finite-size networks. In [4], the authors presented an analytical procedure to compute the node isolation probability in an ad hoc network in the presence of channel randomness, and showed that, under the assumption that sensing relies on the same wave propagation laws that also guide signal propagation, the coverage probability coincides with the complement of the node isolation probability. In [5], the authors studied the relation between k-coverage and k-connectivity when the communication radius is at least twice the sensing radius, where the sensing radius is deterministic. However, no statistical properties of either k-coverage or k-connectivity were given. In [6], three fundamental coverage measures of large-scale sensor networks were studied: area coverage, node coverage, and detectability. In [7] and [8], the asymptotic coverage problem was addressed for mostly-sleeping (unreliable) wireless sensor networks, where 1-coverage was studied in [7] and k-coverage in [8], but neither one provided the sufficient-and-necessary condition for asymptotic coverage. None of the above work studied network's lifetime under connectivity and coverage constraints.

Recently, research efforts [9, 10] have been made to analyze the lifetime of a wireless sensor network with coverage requirements. The definitions of network's lifetime in these literature are different from ours. In [9], the lifetime refers to the time it takes for the coverage — defined as the fraction of the area covered by working sensors — to drop below a pre-defined threshold. In [10], the α-lifetime of a wireless sensor network is defined as the interval during which at least α portion of the sensing region is covered by at least one sensor node. These lifetime definitions are all from the deterministic point of view, while in this paper, considering the fact that the deployment and dynamics of wireless sensor networks are random and, hence, the coverage of the sensing field and the connectivity among active sensors are also random variables, we study network's lifetime from a (different) probabilistic perspective. Moreover, neither [9] nor [10] studied the effect of the communication radius on the network's lifetime.

1.2   Key Contributions

This paper explores the fundamental limits of a wireless sensor network's lifetime under connectivity and k-coverage constraints, and the contributions are twofold. First, we introduce a new definition of network's lifetime from a probabilistic perspective, namely ω-lifetime, which is defined as the expectation of the time interval during which the probability of guaranteeing connectivity and k-coverage simultaneously is at least ω. By solving two convex optimization problems, we obtain a lower bound and an upper bound on the network's maximum ω-lifetime. Second, based on the obtained lower bound, we propose a near-optimal scheduling scheme, called PIS (Pre-planned Independent Sleeping), to maximize the network's ω-lifetime, and describe a possible distributed implementation of the PIS scheme.

1.3   Organization

The rest of this paper is organized as follows. Section 2 describes our network model and gives the problem statement. In Section 3, we derive the sufficient-and-necessary condition for maintaining k-coverage with probability one as the number of sensors goes to infinity. Section 4 describes the details of the proposed ω-lifetime and PIS scheduling scheme. Section 5 presents and evaluates the simulation results and, finally, the paper concludes in Section 6.

2   Network Model and Problem Statement

2.1   Network Model

Consider a wireless sensor network with n sensors deployed independently and uniformly in a square sensing field $\mathcal{D}$ of unit area. In order to extend network's lifetime, an appropriate duty cycle and a well-designed sleeping schedule are required, and we propose the following Pre-planned Independent Sleeping (PIS) scheme for this purpose: *time is divided into rounds, and at the beginning of a round, each alive sensor becomes active with probability p or inactive (sleeping) with probability (1−p), independently from others; the value of p and active sensors' communication radius may vary with the round, and their variation patterns are pre-determined by the performance metric to be optimized.* Here, alive sensors refer to the sensors with enough energy to operate. The PIS scheme is based on the Randomized Independent Sleeping (RIS) scheme proposed in [8] and the details of PIS will be discussed in Section 4. Note that, in general, RIS-like schemes are energy-efficient, lightweight, and easy to implement because each sensor determines its own sleeping schedule independently without interacting with others. In comparison, the Neighborhood Cooperative Sleeping (NCS) schemes [11, 9, 12, 13] allow neighbor sensors to collaborate with each other to determine their sleeping schedules, hence improving the coverage performance further but with increased complexity. Design and analysis of NCS schemes are out of the scope of this paper.

Sensing model. To consider the sensing radii irregularity caused by time-varying environments, we assume a random sensing radius model where (1) each active sensor has a *sensing radius* of r_s; (2) any object within a disc of radius r_s centered at an active sensor can be reliably-detected by the sensor; and (3) r_s's are independently identically distributed (i.i.d) random variables with mean r_0 and variance $r_0^2\sigma_s^2$, and the underlying distribution is assumed unknown. A point in the sensing field $\mathcal{D}$ is said to be *k-covered* if it is within the sensing radius of at least k active sensors. The field $\mathcal{D}$ is said to be *k-covered* if every point in $\mathcal{D}$ is k-covered.

Communication model. Two active sensors can communicate directly with each other if and only if the distance between them is less than r_c. The radius r_c is referred to as the *communication radius* and may vary from round to round in the PIS scheme. For the purpose of simplicity, we assume that, at each round, all active sensors have the same and deterministic communication radii. The network is said to be connected if the underlying graph of active sensors is connected. Moreover, we assume torus convention (also known as the toroidal model) [14], i.e., each disc (communication or sensing) that protrudes one side of the field $\mathcal{D}$ enters $\mathcal{D}$ again from the opposite side. This eliminates the edge effects and simplifies the problem.

ω-lifetime. Due to the randomness in sensor deployment and sleeping schedule, it is impossible to guarantee connectivity and k-coverage with probability one with finite number of sensors, unless the communication disc and the sensing disc of each active sensor can cover the entire field. However, the physical limitations prohibit such large communication radius and sensing radius. In other words, there is no deterministic guarantee of connectivity or k-coverage for randomly-deployed wireless sensor networks in practice. Such facts motivate us to study the network's lifetime from a probabilistic perspective. More specifically, we define the *ω-lifetime* of a randomly-deployed wireless sensor network as the expectation of the time interval during which the probability of guaranteeing k-coverage of field $\mathcal{D}$ and the connectivity of the network simultaneously is at least ω, where $0 < \omega < 1$. For example, suppose that the PIS scheduling scheme is employed, then the network's ω-lifetime is $T_\omega = \mathrm{E}\left[\sum_{i=1}^{M} T_i\right]$, where T_i is the duration of the i-th round, and M is the maximum number of rounds during which the network can function properly. In other words, for any round i $(i \leqslant M)$, the probability of guaranteeing both connectivity and k-coverage simultaneously, denoted by $P_{c\&c}$, is at least ω, but for round $(M+1)$, $P_{c\&c}$ is smaller than ω.

2.2   Problem Statement

The kernel problem we study in this paper is:

- For a finite-size wireless sensor network, how to find the optimal parameters (p and r_c) for the PIS scheme to maximize the ω-lifetime of the network?

This is an interesting problem and the results may serve as good guidelines in deploying finite-size wireless sensor networks. In order to address this problem, we first study the following companion problem, which is referred to as the critical condition for asymptotic k-coverage:

– What relation among n, p, r_0, and σ_s^2 would be the sufficient-and-necessary condition to guarantee that the probability of the entire field $\mathcal{D}$ being k-covered approaches 1 as n goes to infinity?

3   Critical Condition for Asymptotic k-Coverage

In this section, we derive the sufficient-and-necessary condition, under our random sensing radius model described in Section 2.1, to guarantee that the entire sensing field $\mathcal{D}$ is k-covered with probability one as the total number of deployed sensors n goes to infinity. Similar to [10], we apply the coverage process techniques introduced in [14] to solve the problem.

Lemma 1. *Let n points distributed independently and uniformly in a square field $\mathcal{D}$ of unit area within $\mathbb{R}^2$, then for sufficiently large n, these points form a stationary Poisson point process with density n.*

Lemma 1 is a well-known result and its proof is given by Hall in [14]. Let $\mathcal{P} \equiv \{\xi_i, i \geqslant 1\}$ denote the set of active sensors. It is shown in Lemma 2 that $\mathcal{P}$ is also a stationary Poisson point process with density np for sufficiently large n.

Lemma 2. *Let n points distributed independently and uniformly in a square field $\mathcal{D}$ of unit area within $\mathbb{R}^2$. Each point is marked independently as an active point with probability p, where $0 < p \leqslant 1$. Then the set of active points, $\mathcal{P} = \{\xi_i, i \geqslant 1\}$, is a stationary Poisson point process with density np for sufficiently large n.*

Let S_i denote a random disc with radius $r_{s,i}$ centered at the origin of $\mathbb{R}^2$, which is defined as $S_i \equiv \{x \in \mathbb{R}^2 : |x| \leqslant r_{s,i}\}$, where $r_{s,i}$ is the sensing radius of the i-th active sensor ξ_i. Here, we assume that all sensing radii are i.i.d random variables following an *unknown* distribution $F(r)$, with *known* mean r_0 and variance $r_0^2 \sigma_s^2$, i.e., all S_i's are distributed as S:

$$S \equiv \{x \in \mathbb{R}^2 : |x| \leqslant r, \ r \sim F(r)\}. \tag{1}$$

Then, the sensing disc (abbreviated as disc) centered at active sensor ξ_i can be defined as $D_i \equiv \xi_i + S_i = \{\xi_i + y : y \in S_i\}$. The set of $\{D_i, i \geqslant 1\}$ forms a stationary coverage process. For such a coverage process, Lemma 3 gives the distribution of the number of discs with certain properties.

Lemma 3. *Let $\mathcal{Q} = \{\xi_i + S_i, i \geqslant 1\}$ denote a stationary coverage process, where $\{\xi_i\}$ is a stationary Poisson point process with density λ within $\mathcal{D}$, and S_i's are distributed as S defined in (1). For a given deterministic condition C, let*

Y denote the number of discs in Q that satisfy the condition C. Then, Y is Poisson-distributed with mean

$$\mu = \lambda \cdot \mathrm{E}\Big[\|\{x : I_C(x + S) = 1\}\|\Big],$$

where $I_C(\cdot)$ is the indicator function of whether a disc satisfies the condition C or not, and $\|\cdot\|$ denotes the area.

The proofs of Lemma 2 and Lemma 3 are omitted due to space limitation. Interested readers can refer to the full version of this paper [15].

Let $Y(x)$ denote the number of active sensors that cover a point x, and $I_k(x)$ denote the indicator function of whether the point x is covered by at most $(k-1)$ active sensors, i.e.,

$$I_k(x) = \begin{cases} 1, & \text{if } Y(x) < k, \\[2mm] 0, & \text{otherwise.} \end{cases}$$

Then, the expectation of Bernoulli random variable $I_k(x)$ is

$$\mathrm{E}[I_k(x)] = P(x \text{ is at most } (k-1)\text{-covered}) = P(Y(x) < k).$$

By Lemma 3, we know that $Y(x)$ is Poisson-distributed with mean

$$\mu = np \cdot \mathrm{E}\Big[\|\{x : (x + S) \cap \{x\} \neq \emptyset\}\|\Big] = np \cdot \mathrm{E}\big[\|x - S\|\big] = npa_s,$$

where $a_s \equiv \mathrm{E}\big[\|S\|\big] = \pi r_0^2(1 + \sigma_s^2)$. Therefore,

$$\mathrm{E}[I_k(x)] = e^{-npa_s} \sum_{j=0}^{k-1} \frac{(npa_s)^j}{j!}. \tag{2}$$

Let the k-vacancy V_k denote the area within $\mathcal{D}$ that is covered by at most $(k-1)$ active sensors, then the random variable V_k can be expressed as $V_k = \int_{\mathcal{D}} I_k(x)\mathrm{d}x$. Using Fubini's theorem [16] and exchanging the order of integral and expectation, we obtain the expected value of the k-vacancy as:

$$\mathrm{E}[V_k] = \int_{\mathcal{D}} \mathrm{E}[I_k(x)]\mathrm{d}x = e^{-npa_s} \sum_{j=0}^{k-1} \frac{(npa_s)^j}{j!}. \tag{3}$$

K-coverage of the sensing field $\mathcal{D}$ means that each point in $\mathcal{D}$ should be covered by at least k active sensors, which implies $V_k = 0$. Because sensors are deployed independently and uniformly within $\mathcal{D}$, it cannot guarantee $P(V_k = 0) = 1$ with finite n for $a_s < 1$ regardless of the value of n. However, if $np \to \infty$ as $n \to \infty$, it is possible that $P(V_k = 0) \to 1$ as $n \to \infty$. Before studying the asymptotic behavior of $P(V_k = 0)$, we first give an upper bound and a lower bound on $P(V_k = 0)$ for finite n. Similar bounds have been proved in [10] for the case of deterministic sensing radius model and non-sleeping sensor networks. Theorem 1 is a generalization of the results in [10] for the random sensing radius model.

Theorem 1. *For $n > 1$, $0 < p \leqslant 1$, and $a_s < 1$,*

$$P_l < P(V_k = 0) < P_u, \tag{4}$$

in which

$$P_u = \frac{4(k+1)!(1+\sigma_s^2)(np)^{-1}(npa_s)^{-k} \cdot e^{npa_s}}{1 + 4(k+1)!(1+\sigma_s^2)(np)^{-1}(npa_s)^{-k} \cdot e^{npa_s}}, \tag{5}$$

and

$$P_l = 1 - 2e^{-npa_s}\left(1 + (n^2p^2a_s' + 2npr_0)\sum_{i=0}^{k-1}\frac{(npa_s)^i}{i!}\right), \tag{6}$$

where $a_s' \equiv \pi r_0^2(1 + \sigma_s^2/2)$.

Proof: (i) Upper bound.
    By the Cauchy-Schwartz inequality [14],

$$E[V_k] = E[V_k \cdot I(V_k > 0)] \leqslant \{E[V_k^2]P(V_k > 0)\}^{1/2},$$

where $I(\cdot)$ denotes the indicator function, thus

$$P(V_k > 0) \geqslant \frac{(E[V_k])^2}{E[V_k^2]}, \tag{7}$$

where $E[V_k^2] = E\left[\int\int_{\mathcal{D}^2} I_k(x_1)I_k(x_2)dx_1dx_2\right] = \int\int_{\mathcal{D}^2} E[I_k(x_1)I_k(x_2)]dx_1dx_2$. Let Y_1 denote the number of active sensors that cover x_1, Y_2 the number of active sensors that cover x_2, and Y_3 the number of active sensors that cover x_2 but not x_1, then

$$E[I_k(x_1)I_k(x_2)] = P(Y_1 < k, Y_2 < k) \leqslant P(Y_1 < k, Y_3 < k). \tag{8}$$

Lemma 4. *For the random variables Y_1 and Y_3 defined above, we have the following results:*

- *Y_1 is Poisson-distributed with mean npa_s,*
- *Y_3 is Poisson-distributed with mean npb_s,*
- *Y_1 and Y_3 are independent,*

where $b_s \equiv E\left[\|\{x : (x + S) \cap \{x_1\} = \emptyset, (x + S) \cap \{x_2\} \neq \emptyset\}\|\right]$.

    The proof of Lemma 4 is omitted here due to space limitation. Interested readers can refer to [15]. Using Lemma 4 and (8), we have

$$E[I_k(x_1)I_k(x_2)] \leqslant P(Y_1 < k) \cdot P(Y_3 < k) = E[I_k(x_1)] \cdot P(Y_3 < k)$$

$$= E[I_k(x_1)] \cdot \left(e^{-npb_s}\sum_{j=0}^{k-1}\frac{(npb_s)^j}{j!}\right). \tag{9}$$

Let $z = x_1 - x_2$, then

$$b_s = \mathrm{E}\Big[\|\{x : (x + S) \cap \{x_1\} = \emptyset, (x + S) \cap \{x_2\} \neq \emptyset\}\|\Big]$$
$$= \mathrm{E}\Big[\|\{x : (x + S) \cap \{z\} = \emptyset, (x + S) \cap \{0\} \neq \emptyset\}\|\Big] = a_s - \rho(z),$$

where

$$\rho(z) = \mathrm{E}\Big[\|\{x : (x + S) \cap \{z\} \neq \emptyset, (x + S) \cap \{0\} \neq \emptyset\}\|\Big] = \int_0^\infty r^2 B(|z|/2r) dF(r),$$
and

$$B(x) = \begin{cases} 4 \int_x^1 \sqrt{(1 - y^2)} dy & \text{if } 0 \leqslant x \leqslant 1, \\ \\ 0 & \text{otherwise} \end{cases}$$

is the area of the lens of intersection of two unit discs centered $2x$ apart, and $F(r)$ is the distribution of sensing radius r_s.

It is shown in [15] that $B(x) \leqslant \pi(1 - x)$ for $0 \leqslant x \leqslant 1$, then using the fact that $\rho(z) \geqslant 0$ and after some algebraic manipulation, we can bound $\rho(z)$ as

$$\begin{cases} \rho(z) \leqslant a_s - \pi r_0 |z|/2 & \text{if } |z| < 2r_0(1 + \sigma_s^2), \\ \\ \rho(z) = 0 & \text{if } |z| \geqslant 2r_0(1 + \sigma_s^2). \end{cases}$$

If $|z| \geqslant 2r_0(1 + \sigma_s^2)$, then $b_s = a_s$. Using (9), we have

$$\mathrm{E}[I_k(x_1)I_k(x_2)] \leqslant \mathrm{E}[I_k(x_1)] \cdot \mathrm{E}[I_k(x_2)].$$

Therefore,

$$I_1 \equiv \int\int_{\mathcal{D}^2 \cap \{|x_1 - x_2| \geqslant 2r_0(1 + \sigma_s^2)\}} \mathrm{E}[I_k(x_1)I_k(x_2)] dx_1 dx_2 \tag{10}$$
$$\leqslant \int\int_{\mathcal{D}^2} \mathrm{E}[I_k(x_1)] \cdot \mathrm{E}[I(x_2)] dx_1 dx_2 = (\mathrm{E}[V_k])^2.$$

Similarly, if $|z| < 2r_0(1 + \sigma_s^2)$, then $b_s \geqslant \pi r_0 |z|/2$. Using (9), we have

$$\mathrm{E}[I_k(x_1)I_k(x_2)] \leqslant \mathrm{E}[I_k(x_1)] \cdot \left(e^{-np\frac{\pi r_0}{2}|z|} \sum_{j=0}^{k-1} \frac{(np\pi r_0|z|)^j}{2^j \cdot j!} \right).$$

Therefore,

$$I_2 \equiv \int\int_{\mathcal{D}^2 \cap \{|x_1 - x_2| < 2r_0(1 + \sigma_s^2)\}} \mathrm{E}[I_k(x_1)I_k(x_2)] dx_1 dx_2$$
$$\leqslant \int_{\mathcal{D}} \mathrm{E}[I_k(x_1)] dx_1 \int_0^{2r_0(1 + \sigma_s^2)} e^{-np\pi r_0 z/2} \sum_{i=0}^{k-1} \frac{(np\pi r_0 z)^i}{2^i \cdot i!} 2\pi z dz$$

$$= \mathrm{E}[V_k] \cdot \left(\int_0^1 e^{-\lambda u} \sum_{i=0}^{k-1} \frac{(\lambda u)^i}{i!} 8\pi r_0^2 \left(1 + \sigma_s^2\right)^2 u\,du \right)$$

$$< 4a_s(1 + \sigma_s^2)k(k+1)\lambda^{-2},$$

where $\lambda = npa_s$. The proof of the last inequality above can be found in [15]. Hence, we have

$$I_2 < 4a_s(1 + \sigma_s^2)k(k+1)(npa_s)^{-2} \cdot \left(e^{-npa_s} \sum_{i=0}^{k-1} \frac{(npa_s)^i}{i!} \right). \qquad (11)$$

Since $\mathrm{E}[V_k^2] = I_1 + I_2$, combining (7), (3), (10), and (11), we can upper-bound $P(V_k = 0)$ as follows:

$$P(V_k = 0) = 1 - P(V_k > 0) \leqslant 1 - \frac{(\mathrm{E}[V_k])^2}{\mathrm{E}[V_k^2]} < \frac{\beta}{1 + \beta},$$

where

$$\beta = \frac{4(1 + \sigma_s^2)a_s k(k+1)(npa_s)^{-2}}{e^{-npa_s} \sum_{i=0}^{k-1} (npa_s)^i / i!} \leqslant 4(1 + \sigma_s^2)(k+1)!(np)^{-1}(npa_s)^{-k} \cdot e^{npa_s}.$$

Therefore, we obtain the upper bound on $P(V_k = 0)$ as

$$P(V_k = 0) < \frac{4(k+1)!(1 + \sigma_s^2)(np)^{-1}(npa_s)^{-k} \cdot e^{npa_s}}{1 + 4(k+1)!(1 + \sigma_s^2)(np)^{-1}(npa_s)^{-k} \cdot e^{npa_s}}.$$

(ii) *Lower bound.*
Observe that

$$p(V_k = 0) = 1 - p_1 - p_2 - p_3,$$

where

$$p_1 = P(\text{no active sensors centered within } \mathcal{D}) = e^{-np} < e^{-npa_s}.$$

Here, we assume $a_s < 1$, meaning that, even for the random sensing radius model, the expected sensing area of one sensor will not cover the entire field $\mathcal{D}$.

$p_2 = P(\text{at least one disc centered within } \mathcal{D}, \text{ but none of the discs intersects with}$
$\quad\quad \text{any other disc, and none of the discs intersect the boundary of } \mathcal{D})$
$\quad \leqslant P(\text{at least one disc centered within } \mathcal{D}) \cdot P(\text{a given disc intersects with no other discs})$
$\quad = (1 - e^{-np}) \cdot e^{-np\pi \mathrm{E}[\pi(r_{s,1} + r_{s,2})^2]} = (1 - e^{-np}) \cdot e^{-2np\pi r_0^2(2 + \sigma_s^2)} < e^{-npa_s},$

where $r_{s,1}$ and $r_{s,2}$ are sensing radii of two active sensors, which are i.i.d with mean r_0 and variance $r_0^2 \cdot \sigma_s^2$, and the second equality is due to Lemma 3.

$p_3 = P(\mathcal{D} \text{ is not } k\text{-covered, at least one disc centered within } \mathcal{D}, \text{ and at least}$
$\quad\quad \text{one disc intersects with another disc or the boundary of } \mathcal{D}).$

Therefore

$$p(V_k = 0) > 1 - 2e^{-npa_s} - p_3. \tag{12}$$

Our next task is to derive an upper bound on p_3.

Define a *crossing* to be either an intersection point of the boundaries of two discs or an intersection point of the boundary of an disc and the boundary of the field $\mathcal{D}$. A crossing is said to be k-covered if it is within at least k discs. It is proved in [5] that, field $\mathcal{D}$ is k-covered if there exist crossings and every crossing is k-covered. Therefore, if $\mathcal{D}$ is not k-covered, if one or more discs are centered within $\mathcal{D}$, and if there exist crossings in $\mathcal{D}$, then at least one of the discs has two or more crossings that are not k-covered. Thus

$$p_3 \leqslant P(M_k \geqslant 2) \leqslant \mathrm{E}[M_k]/2, \tag{13}$$

where M_k denotes the number of crossings that are not k-covered.

Define L_1 and L_2 as the number of crossings created by two discs intersecting with each other, and the ones created by a disc intersecting the boundary of field $\mathcal{D}$. We first study the expected value of L_1. The expected number of crossings created by a given active sensor ξ_1 with other active sensors is

$$\mathrm{E}[2np \cdot \pi(r_{s,1} + r_{s,2})^2] = 8npa'_s,$$

where $a'_s \equiv \pi r_0^2(1 + \sigma_s^2/2)$, and the expected number of discs centered within $\mathcal{D}$ is np. Therefore,

$$\mathrm{E}[L_1] = np \cdot 8npa'_s/2 = 4n^2p^2a'_s.$$

If a disc intersects the edge of field $\mathcal{D}$, at most two crossings will be created; if a disc intersects the corner of field $\mathcal{D}$, at most four crossings will be created (due to the toroidal model assumption). Thus the expected value of L_2 is bounded by

$$\mathrm{E}[L_2] \leqslant 8npr_0.$$

The probability that a given crossing is not k-covered is given by (2). Therefore,

$$\mathrm{E}[M_k] = (\mathrm{E}[L_1] + \mathrm{E}[L_2])e^{-npa_s} \sum_{j=0}^{k-1} \frac{(npa_s)^j}{j!} \leqslant 4(n^2p^2a'_s + 2npr_0)e^{-npa_s} \sum_{j=0}^{k-1} \frac{(npa_s)^j}{j!}. \tag{14}$$

By (12), (13), and (14), we have

$$P(V_k = 0) > 1 - 2e^{-npa_s}\left(1 + (n^2p^2a'_s + 2npr_0)\sum_{i=0}^{k-1} \frac{(npa_s)^i}{i!}\right).$$

This completes the proof.   □

In what follows, we establish the sufficient-and-necessary condition for asymptotic k-coverage.

Corollary 1. *Assume $np \to \infty$ as $n \to \infty$, and let*

$$\pi r_0^2 \left(1 + \sigma_s^2\right) = \frac{\ln(np) + k \ln\ln(np) + c_1(np)}{np}, \tag{15}$$

then the entire unit square field $\mathcal{D}$ is k-covered with probability one as $n \to \infty$, if and only if $c_1(np) \to \infty$ as $n \to \infty$.

Proof: The entire unit square field $\mathcal{D}$ is k-covered with probability one means that $P(V_k = 0) \to 1$ as $n \to \infty$. First, we prove if $c_1(np) \to \infty$ as $n \to \infty$, $P(V_k = 0) \to 1$.

By (4) and (6) in Theorem 1, we have

$$P(V_k = 0) > 1 - 2e^{-npa_s} - (b_1 + b_2) \cdot (np)(npa_s)^k e^{-npa_s},$$

where $b_1 \equiv 2k\frac{1+\sigma_s^2/2}{1+\sigma_s^2} > 0$ is independent of n, and $b_2 \equiv \frac{4k}{\pi r_0(1+\sigma_s^2)np}$. Let $npa_s = \ln(np) + k \ln\ln(np) + c_1(np)$, then $npa_s \to \infty$, $e^{-npa_s} \to 0$, and $b_2 \to 0$, as $n \to \infty$. Therefore, when $c_1(np) \to \infty$,

$$\begin{aligned}
\ln\left((b_1 + b_2) \cdot (np)(npa_s)^k e^{-npa_s}\right) = {} & \ln(b_1 + b_2) - k\ln\ln(np) - c_1(np) \\
& + k \cdot \ln\left(\ln(np) + k\ln\ln(np) + c_1(np)\right) \\
& \to -\infty,
\end{aligned}$$

and consequently, $P(V_k = 0) \to 1$. The first part is proved.

If $c_1(np) \leqslant C_1$ for some finite $C_1 > 0$ as $n \to \infty$, then for sufficiently large n,

$$\begin{aligned}
4(k+1)!(1+\sigma_s^2)(np)^{-1}(npa_s)^{-k}e^{npa_s} = {} & 4(k+1)!(1+\sigma_s^2)e^{c_1(np)} \\
& \leqslant 4e^{C_1}(k+1)!(1+\sigma_s^2).
\end{aligned}$$

Therefore, by (4) and (5), we have

$$P(V_k = 0) < \frac{4e^{C_1}(k+1)!(1+\sigma_s^2)}{1 + 4e^{C_1}(k+1)!(1+\sigma_s^2)} < 1.$$

It means that $P(V_k = 0) \to 1$ only if $c_1(np) \to \infty$ as $n \to \infty$. This completes the proof. $\qquad\square$

Remark: The bounds obtained in Theorem 1 is valid for finite n. Therefore, they can be used as performance criteria for designing finite-size sensor networks, as will be shown in the next section.

4 ω-lifetime of Finite-Size Wireless Sensor Networks

In this section, we address the problem of finding optimal parameters for the PIS scheme to maximize the ω-lifetime of a finite-size wireless sensor network.

Let A denote the event of the sensing field $\mathcal{D}$ being k-covered, and B denote the event of the sensor network being connected. The probability of guaranteeing simultaneously k-coverage of field $\mathcal{D}$ and connectivity of the network is $P_{c\&c} \equiv P(A \cap B)$.

Definition 1. *ω-lifetime, denoted by T_ω, of a sensor network is defined as the expectation of the time interval during which the probability of guaranteeing simultaneously k-coverage of field $\mathcal{D}$ and the connectivity of the network is no less than ω, i.e., $P_{c\&c} \geqslant \omega$, where $0 < \omega < 1$.*

In order to study the ω-lifetime, we first introduce the energy consumption model of each wireless sensor. We assume that inactive sensors do not consume energy and the communication traffic is evenly distributed across the network. The energy consumption of an active sensor consists of two parts: *communication* and *sensing*. Thus, the power consumption P_0 of each active sensor can be modeled as

$$P_0 = Q \cdot \frac{1}{r_c} \cdot r_c^\beta + \Delta, \tag{16}$$

where

- r_c^β is proportional to the communication energy consumption per bit, and the typical values of β range from 3 to 4 for different propagation models [17];
- $1/r_c$ is proportional to the average traffic rate of active sensors (we assume that all active sensors have the same traffic rate, following the assumption of evenly distributed traffic.);
- Δ is the power consumption for continuous sensing and listening;
- $Q > 0$ is a constant.

As the communication radius r_c decreases, the average number of hops required for packets transmitted from one point to another increases inversely. For this reason, we incorporate the factor of $1/r_c$ into the average traffic rate expression. We further assume that all active sensors have the same communication radius r_c, which results in the same individual lifetime:

$$T_0(r_c) = \frac{E_0'}{P_0} = \frac{E_0}{r_c^{\beta-1} + \eta}, \tag{17}$$

where E_0' is the initial energy of each active sensor, $E_0 = \frac{E_0'}{Q}$, and $\eta = \frac{\Delta}{Q}$, respectively. This assumption is typical when analyzing the network's lifetime, e.g., in [10] and [18].

Next, we formally define the PIS scheme which can extend the ω-lifetime of a wireless sensor network. Suppose that time is divided into rounds. At the beginning of round i, there are $n^{(i)}$ alive sensors, and each alive sensor decides independently whether to remain sleeping (with probability $1 - p^{(i)}$), or become active (with probability $p^{(i)}$). All active sensors choose the same communication radius of $r_c^{(i)}$. Both $p^{(i)}$ and $r_c^{(i)}$ are chosen such that $P_{c\&c} \geqslant \omega$. Next, all active sensors will operate continuously until batteries die out. Since we assume that all active sensors have the same individual lifetime, they will die out at the same time instant, which is defined as the end of this round. The same procedure is repeated for the next rounds until there are not enough alive sensors to satisfy the "$P_{c\&c} \geqslant \omega$" requirement, regardless of the choices of p and r_c.

The major differences between PIS and RIS in [8] are as follows. In PIS, p and r_c are chosen for each round to satisfy both connectivity and k-coverage requirements, and they may vary from round to round. The round duration is the same as an individual sensor's lifetime, i.e., within each round, all active sensors operate continuously until batteries die out. In comparison, the round duration of the RIS scheme is selected to be sufficiently-small, and the values of p and r_c in RIS are fixed throughout the network operation, where p is chosen to satisfy the k-coverage requirement but with no optimization on r_c. This way, batteries of all sensors die out at approximately the same time around the end of the network's lifetime.

In the rest of this section, we study the ω-lifetime with the proposed PIS scheme and try to find the optimal parameters to maximize the ω-lifetime of the network.

4.1 ω-Lifetime Study

Suppose that n sensors are deployed independently and uniformly within a unit-area square field $\mathcal{D}$, and the network can operate M rounds following the PIS scheduling scheme. Then, the ω-lifetime of the wireless sensor network is

$$T_\omega = \mathrm{E}\left[\sum_{i=1}^{M} T_0(r_c^{(i)})\right] = \mathrm{E}\left[\sum_{i=1}^{M} \frac{E_0}{\left(r_c^{(i)}\right)^{\beta-1} + \eta}\right], \tag{18}$$

subject to both connectivity and k-coverage requirements, and the expectation is with respect to M. Define $n_{\mathrm{eff}}^{(i)} = n^{(i)}p^{(i)}$, which is the expected number of active sensors in round i. It is easy to verify that the probability mass function (pmf) of M is

$$P(M = m) = \sum_{\substack{n=n^{(1)} \geqslant n^{(2)} \geqslant \cdots \geqslant n^{(m)} \\ n^{(i)} \geqslant n_{\mathrm{eff}}^{(i)},\ i=1,\ldots,m}} \cdots \sum_{n^{(m+1)}=0}^{n_{\mathrm{eff}}^{(m+1)}-1} \prod_{i=1}^{m} \binom{n^{(i)}}{n^{(i+1)}} \left(1 - p^{(i)}\right)^{n^{(i+1)}} \left(p^{(i)}\right)^{n^{(i)}-n^{(i+1)}}.$$

Thus, the problem of maximizing the ω-lifetime of the network can be expressed as

$$T_\omega^{\mathrm{max}} = \max_{r_c^{(i)}, n_{\mathrm{eff}}^{(i)}} T_\omega = \max_{r_c^{(i)}, n_{\mathrm{eff}}^{(i)}} \mathrm{E}\left[\sum_{i=1}^{M} \frac{E_0}{\left(r_c^{(i)}\right)^{\beta-1} + \eta}\right], \tag{19}$$

subject to $\quad P_{\mathrm{c\&c}} = P(A \cap B) \geqslant \omega \quad$ for each round. $\tag{20}$

Using the union bound, we have

$$\min\{P(A), P(B)\} \geqslant P_{\mathrm{c\&c}} \geqslant P(A) + P(B) - 1. \tag{21}$$

Since it is hard to analyze $P_{\mathrm{c\&c}}$ directly, we next focus on finding a lower bound and an upper bound on the optimal ω-lifetime, T_ω^{max}.

Lower bound. Restricting the constraint in (20) by replacing it with the lower bound in (21), and assuming that all $n_{\text{eff}}^{(i)}$ and $r_c^{(i)}$'s are the same for each round, we can obtain a lower bound on $T_w^{\max}$ by solving the following optimization problem:

$$\max_{n_{\text{eff}}, r_c, \epsilon} \; E[M] \cdot \frac{E_0}{r_c^{\beta-1} + \eta}, \tag{22}$$

$$\text{subject to} \quad P(A) \geqslant \omega + \epsilon, \; P(B) \geqslant 1 - \epsilon \quad \text{for } 0 < \epsilon < 1 - \omega. \tag{23}$$

Using the result $P(A) > P_l$ in Theorem 1, and the following result in [3]:

$$P(B) \approx 1 - P(\exists \text{ isolated active sensors}) > 1 - n_{\text{eff}} e^{-n_{\text{eff}} \pi r_c^2},$$

where the edge effects are avoided by the toroidal model assumption, we can restrict the constraints in (23) as

$$P_l \geqslant \omega + \epsilon, \; r_c \geqslant \sqrt{[\ln(n_{\text{eff}}/\epsilon)]/(\pi n_{\text{eff}})} \quad \text{for } 0 < \epsilon < 1 - \omega. \tag{24}$$

Notice that the value of ω is usually larger than 90% in practice, then the P_l defined in (6) can be approximated as

$$P_l \approx 1 - g(n_{\text{eff}}) \equiv 1 - 2n_{\text{eff}}^2 a_s' e^{-a_s n_{\text{eff}}} \sum_{i=0}^{k-1} \frac{(a_s n_{\text{eff}})^i}{i!}. \tag{25}$$

Let X_i denote the number of active sensors in round i, then $n^{(m)} = n - \sum_{i=1}^{m-1} X_i$, and conditional on $n^{(i)}$, X_i is Binomial-distributed as $\text{BIN}(n^{(i)}, p^{(i)})$. Next, we use the expectation of $n^{(i)}$ to obtain an approximation of $p^{(i)}$ as

$$p^{(i)} = \frac{n_{\text{eff}}}{n^{(i)}} \approx \frac{n_{\text{eff}}}{n - (i-1)n_{\text{eff}}} = \frac{1}{M_0 + 1 - i}, \tag{26}$$

where $M_0 \equiv n/n_{\text{eff}}$. Using (26) and the central limit theorem, we can approximate $n^{(m)}$ as a Gaussian random variable with mean $n - (m-1)n_{\text{eff}}$ and variance $A(m)n_{\text{eff}}$, where $A(m) = \sum_{i=1}^{m-1}(1 - p^{(i)})$. Then, we have

$$P(M \leqslant m) = P(n^{(m+1)} < n_{\text{eff}}) = Q\left(\frac{n - (m+1)n_{\text{eff}}}{\sqrt{A(m+1)n_{\text{eff}}}}\right),$$

$$P(M \geqslant m) = P(n^{(m)} \geqslant n_{\text{eff}}) = Q\left(\frac{mn_{\text{eff}} - n}{\sqrt{A(m)n_{\text{eff}}}}\right),$$

where $Q(\cdot)$ is complementary cumulative distribution function (CCDF) of Gaussian distribution. Therefore,

$$P(M \leqslant \lfloor M_0 \rfloor - 2) = Q\left(\frac{n - (\lfloor M_0 \rfloor - 1)n_{\text{eff}}}{\sqrt{A(\lfloor M_0 \rfloor - 1)n_{\text{eff}}}}\right) \leqslant Q\left(\sqrt{\frac{n_{\text{eff}}}{A(M_0 - 1)}}\right),$$

and

$$P(M \geqslant \lfloor M_0 \rfloor + 2) = Q\left(\frac{(\lfloor M_0 \rfloor + 2)n_{\mathrm{eff}} - n}{\sqrt{A(\lfloor M_0 \rfloor + 2)n_{\mathrm{eff}}}}\right) \leqslant Q\left(\sqrt{\frac{n_{\mathrm{eff}}}{A(M_0 + 1)}}\right),$$

where the floor function $\lfloor x \rfloor$ denotes the largest integer that is not greater than x. For $m < M_0 + 2$, $A(m)$ can be upper-bounded as

$$A(m) \leqslant (m - 1) - \int_{M_0+2-m}^{M_0} \frac{1}{x}\mathrm{d}x = (m - 1) - \ln\frac{M_0}{M_0+2-m}.$$

Then, for n and n_{eff} in the range of our interests, we have

$$P(M \geqslant \lfloor M_0 \rfloor + 2) \leqslant Q\left(\sqrt{\frac{n_{\mathrm{eff}}}{A(M_0+1)}}\right) \leqslant Q\left(\sqrt{\frac{n_{\mathrm{eff}}}{M_0-\ln M_0}}\right) \approx 0.$$

Similarly, we have $P(M \leqslant \lfloor M_0 \rfloor - 2) \approx 0$. Thus, the pmf of M are mostly concentrated at 3 points: $\lfloor \frac{n}{n_{\mathrm{eff}}} \rfloor - 1$, $\lfloor \frac{n}{n_{\mathrm{eff}}} \rfloor$, and $\lfloor \frac{n}{n_{\mathrm{eff}}} \rfloor + 1$. Monte Carlo simulation results also verify this conclusion. Therefore, we have the lower bound on $\mathrm{E}[M]$ as

$$\mathrm{E}[M] \geqslant \left\lfloor \frac{n - n_{\mathrm{eff}}}{n_{\mathrm{eff}}} \right\rfloor. \tag{27}$$

Since $E_0/(r_c^{\beta-1} + \eta)$ is a decreasing function in r_c, using (24), (25) and (27), we obtain a new lower bound on T_ω^{max} as

$$T_\omega^L = \max_{n_{\mathrm{eff}}} T_1(n_{\mathrm{eff}}) \equiv \max_{n_{\mathrm{eff}}} \left\lfloor \frac{n - n_{\mathrm{eff}}}{n_{\mathrm{eff}}} \right\rfloor \cdot \frac{E_0}{\left(\frac{1}{\pi n_{\mathrm{eff}}} \ln \frac{n_{\mathrm{eff}}}{1-\omega-g(n_{\mathrm{eff}})}\right)^{(\beta-1)/2} + \eta},$$

subject to $n_{\mathrm{eff}} > g^{-1}(1 - \omega)$,

where $g^{-1}(\cdot)$ is the inverse function of $g(n_{\mathrm{eff}})$. By temporarily removing the floor function $\lfloor \cdot \rfloor$, we have the following convex optimization problem (given $\beta > 3$):

$$\max_{n_{\mathrm{eff}}} \frac{E_0(n - n_{\mathrm{eff}})}{n_{\mathrm{eff}}\left(\frac{1}{\pi n_{\mathrm{eff}}} \ln \frac{n_{\mathrm{eff}}}{1-\omega-g(n_{\mathrm{eff}})}\right)^{(\beta-1)/2} + \eta \cdot n_{\mathrm{eff}}}, \tag{28}$$

subject to $n_{\mathrm{eff}} > g^{-1}(1 - \omega)$.

The verification of the concavity of the objective function is omitted due to space limitation.

The convex optimization problem defined in (28) can be solved easily by numerical methods. Suppose that the solution of such problem is $\bar{n}_{\mathrm{eff}}$, then

$$T_\omega^L = \max\{T_1(n_{\mathrm{eff}}^1), T_1(n_{\mathrm{eff}}^2)\},$$

where $n_{\text{eff}}^1 = n \big/ \left\lfloor \frac{n}{\tilde{n}_{\text{eff}}} \right\rfloor$, $n_{\text{eff}}^2 = n \big/ \left\lceil \frac{n}{\tilde{n}_{\text{eff}}} \right\rceil$, and $\lceil x \rceil$ denotes the smallest integer that is equal to or greater than x. We can also obtain the corresponding n_{eff}^L and r_c^L as

$$n_{\text{eff}}^L = \arg \max_{n_{\text{eff}}^1, n_{\text{eff}}^2} T_1(n_{\text{eff}}), \quad r_c^L = \sqrt{[\ln(n_{\text{eff}}^L/(1 - \omega - g(n_{\text{eff}}^L)))]/(\pi n_{\text{eff}}^L)}. \quad (29)$$

Upper bound. Next, we present an approximate upper bound on $T_\omega^{\max}$. Relaxing the constraint in (20) with the upper bound in (21), we obtain the relaxed constraints as

$$P(A) \geqslant \omega, \quad P(B) \geqslant \omega. \quad (30)$$

Then, we use the lower bounds to approximate $P(A)$ and $P(B)$ as

$$P(A) \approx P_l \approx 1 - g(n_{\text{eff}}^{(i)}), \quad P(B) \approx 1 - n_{\text{eff}}^{(i)} e^{-n_{\text{eff}}^{(i)} \pi \left(r_c^{(i)}\right)^2}. \quad (31)$$

Next, we assume that the number of active sensors in round i is approximately equal to $n_{\text{eff}}^{(i)}$. Then the maximum number of rounds, M, is a deterministic quantity, and satisfies the constraint $\sum_{i=1}^M n_{\text{eff}}^{(i)} \leqslant n$. Using (30) and (31), we obtain an approximate upper bound on $T_\omega^{\max}$ by solving the following optimization problem:

$$\max_{n_{\text{eff}}^{(i)}} \sum_{i=1}^M \frac{E_0}{\left(\frac{1}{\pi n_{\text{eff}}^{(i)}} \ln \frac{n_{\text{eff}}^{(i)}}{\omega}\right)^{(\beta-1)/2} + \eta},$$

$$\text{subject to} \quad n_{\text{eff}}^{(i)} \geqslant g^{-1}(1 - \omega), \quad \sum_{i=1}^M n_{\text{eff}}^{(i)} \leqslant n. \quad (32)$$

It is easy to verify that, given M, (32) is a convex optimization problem. By Lagrange multiplier, we obtain a new upper bound on $T_\omega^{\max}$ as

$$T_\omega^U = \max_{n_{\text{eff}}} T_2(n_{\text{eff}}) \equiv \max_{n_{\text{eff}}} \left\lfloor \frac{n}{n_{\text{eff}}} \right\rfloor \cdot \frac{E_0}{\left(\frac{1}{\pi n_{\text{eff}}} \ln \frac{n_{\text{eff}}}{1-\omega}\right)^{(\beta-1)/2} + \eta}, \quad (33)$$

$$\text{subject to} \quad n_{\text{eff}}^{(i)} \geqslant g^{-1}(1 - \omega).$$

Similarly, we temporarily remove the floor function $\lfloor \cdot \rfloor$. It is easy to verify that the resulting optimization problem is a convex problem. Suppose that the solution of such problem is $\tilde{n}_{\text{eff}}$, then

$$T_\omega^U = \max\{T_2(n_{\text{eff}}^1), T_2(n_{\text{eff}}^2)\}, \quad n_{\text{eff}}^U = \arg \max_{n_{\text{eff}}^1, n_{\text{eff}}^2} T_2(n_{\text{eff}}), \quad \text{and}$$

$$r_c^U = \sqrt{[\ln(n_{\text{eff}}^U/(1 - \omega))]/(\pi n_{\text{eff}}^U)},$$

where $n_{\text{eff}}^1 = n \big/ \left\lfloor \frac{n}{\tilde{n}_{\text{eff}}} \right\rfloor$ and $n_{\text{eff}}^2 = n \big/ \left\lceil \frac{n}{\tilde{n}_{\text{eff}}} \right\rceil$.

As an example, we let $E_0 = 1$, $\beta = 3.5$, $\eta = 0.001$, $\omega = 0.92$, and $k = 1$. Numerical results show that the relative difference between the lower bound (T_ω^L) and the upper bound (T_ω^U) is at the level of 10% for n from 10000 to 40000, which suggests that the derived lower bound is a good approximation of the optimal ω-lifetime of the sensor network.

4.2   PIS Scheme Design

We propose to choose the operational parameters for the PIS scheme according to the derived lower bound on the optimal ω-lifetime, i.e., choosing $p^{(i)}$ and $r_c^{(i)}$ for round i as

$$p^{(i)} = \min\left\{n_{\text{eff}}^L/n^{(i)},\ 1\right\},\quad r_c^{(i)} = r_c^L, \tag{34}$$

where $n^{(i)}$ is the number of alive sensors at the beginning of round i ($i \geqslant 1$), and n_{eff}^L and r_c^L are given in (29). Obviously, (34) provides a centralized solution, since $n^{(i)}$ is a global information. At the beginning of each round, such information is required for each alive sensor to calculate $p^{(i)}$ online.

In resource-constrained wireless sensor networks, we always prefer distributed solutions. In our case, distributed solutions mean that the choices of $p^{(i)}$'s should be independent of $n^{(i)}$. As shown in Section 4.1, the expected number of active sensors in each round, $n^{(i)}p^{(i)}$, is the key parameter to determine whether the network satisfies the "$P_{\text{c\&c}} \geqslant \omega$" requirement or not. According to the lower bound on the optimal ω-lifetime, we define *outage* of round i as the event that $n^{(i)}p^{(i)} < n_{\text{eff}}^L$, which means that the "$P_{\text{c\&c}} \geqslant \omega$" requirement can not be satisfied at round i. The probability that an outage occurs at round i is denoted by $P_{\text{out}}^{(i)}$. For the centralized solution in (34), $P_{\text{out}}^{(i)}$ is always 0 for the rounds that $n^{(i)} \geqslant n_{\text{eff}}^L$.

As an approximation to (34), we propose a distributed solution as follows:

$$p^{(i)} = \begin{cases} \dfrac{n_{\text{eff}}^L\left(1+\epsilon^{(i)}\right)}{n - n_{\text{eff}}^L \sum_{l=1}^{i-1}\left(1+\epsilon^{(l)}\right)} & 1 \leqslant i < M \\[2em] 1 & i = M \end{cases},\quad r_c^{(i)} = r_c^L, \tag{35}$$

where M is the maximum number of rounds, $\epsilon^{(1)} = 0$, and for $1 < i < M$, $\epsilon^{(i)}$'s are chosen such that

$$P_{\text{out}}^{(i)} = P\left(n^{(i)}p^{(i)} < n_{\text{eff}}^L\right) = \delta, \tag{36}$$

where $\delta > 0$ is a pre-defined small quantity.

With the choice of $p^{(i)}$ in (35), where $1 < i < M$, we can approximate $n^{(i)}$ as a Gaussian random variable by the central limit theorem:

$$n^{(i)} \sim \mathcal{N}\left(n - n_{\text{eff}}^L \sum_{l=1}^{i-1}\left(1+\epsilon^{(l)}\right),\ n_{\text{eff}}^L \sum_{l=1}^{i-1}\left(1+\epsilon^{(l)}\right)\left(1-p^{(l)}\right)\right).$$

Then, $\epsilon^{(i)}$'s in (35) can be calculated recursively according to

$$\epsilon^{(i)} = \begin{cases} 0 & i = 1 \\ \dfrac{Q^{-1}(\delta)}{a^{(i)} - Q^{-1}(\delta)} & 1 < i < M \end{cases}, \tag{37}$$

and

$$a^{(i)} = \frac{n - n_{\text{eff}}^{L} \sum_{l=1}^{i-1} \left(1 + \epsilon^{(l)}\right)}{\sqrt{n_{\text{eff}}^{L} \sum_{l=1}^{i-1} \left(1 + \epsilon^{(l)}\right) \left(1 - p^{(l)}\right)}}, \tag{38}$$

where $Q^{-1}(\cdot)$ is the inverse function of $Q(\cdot)$. The maximum number of rounds (M) is defined as

$$M = \arg\max_{i>1} \left\{ a^{(i)} > 0 \right\}.$$

The idea of this distributed solution is to use the expected number of alive sensors to replace $n^{(i)}$ in (34), and increase the expected number of active sensors slightly by $n_{\text{eff}}^{L} \epsilon^{(i)}$ such that the outage probability $(P_{\text{out}}^{(i)})$ can be controlled at a given level (δ). In fact, this algorithm sacrifices the total number of rounds, equivalently network's lifetime, to achieve the distributed property.

5 Simulation Results

In this section, we use simulation results to demonstrate the performance of the proposed PIS scheduling schemes. The performance criterion is the ω-lifetime of the network. As a comparison, we include the results of a PIS-like scheme that simply fixes the communication range to be twice the mean of the sensing radius $(r_c = 2r_0)$, and $n_{\text{eff}}^{(i)}$ to be n_{eff}^{A}, regardless of i. Here, n_{eff}^{A} is obtained by solving the following equation: $P(A) = \omega$, where $P(A)$ is given in (31). We call this scheme the PIS-naive scheme.

We simulate a square sensing field $\mathcal{D}$ of unit area in which n sensors are deployed independently and uniformly. The sensing radius r_s is assumed to be a uniformly distributed random variable on $[0.0384, 0.1216]$, which corresponds to $r_0 = 0.08$ and $\sigma_s = 0.3$. Let $E_0 = 1$, $\beta = 3.5$, $\eta = 0.001$, $\omega = 0.92$, and $k = 1$, i.e, we considerer 1-coverage as an example. With this network setup, the centralized and distributed PIS schemes select $p^{(i)}$ and $r_c^{(i)}$ according to (34) and (35), respectively. For the distributed PIS scheme, the outage probability threshold (δ) is set to 10^{-2}. The PIS-naive scheme selects $p^{(i)}$ according to (34) with n_{eff}^{L} replaced by n_{eff}^{A}.

First, we simulate the operation of a network with $n = 10000$ using different scheduling schemes. We divide the field $\mathcal{D}$ into a grid of size 62×62, and approximate that the field $\mathcal{D}$ is k-covered if all grid points are k-covered. For the connectivity, we approximate that the network is connected if there is no isolated active sensors. The torus convention is also employed for simulations to avoid

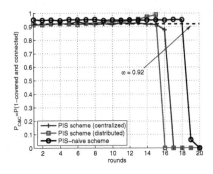
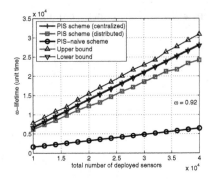

Fig. 1. Three snapshots of the network operation

Fig. 2. ω-lifetime comparison for different scheduling schemes

edge effects. Then, $P_{c\&c}$ at each round of the network operation is estimated as follows: given a deployment, the network is operated according to the particular scheduling scheme until the batteries of all sensors die out. Repeat this experiment 2500 times with the same deployment. For round i of experiment j, define $\delta_j^i = 1$ if the field $\mathcal{D}$ is k-covered and active sensors are connected, 0 otherwise. Then, $P_{c\&c}$ of round i can be estimated as $P_{c\&c}^i = \frac{1}{2500} \sum_{j=1}^{2500} \delta_j^i$.

Fig. 1 shows three snapshots of the network operation using PIS-naive scheme, centralized and distributed PIS scheduling schemes, respectively. It is seen that all scheduling schemes can guarantee that the network satisfies the connectivity and k-coverage requirements as long as the expected number of active sensors is no less than n_{eff}^L. Therefore, in the simulation of the network's ω-lifetime, we only need to simulate how many rounds a network can operate properly following a particular scheduling scheme. Notice that the PIS-naive scheme can operate more rounds than the PIS schemes. However, each round is shorter in the PIS-naive scheme, since r_c is not optimally selected. As seen in the next simulation, the PIS schemes have longer ω-lifetime than the PIS-naive scheme.

Second, we compare the ω-lifetime of a network using different scheduling schemes with n from 10000 to 40000, and the results are plotted in Fig. 2. The derived lower bound and upper bound for the PIS scheme are also shown in the figure. The estimate of the ω-lifetime is calculated as:

$$\widehat{T}_{\text{net}} = \frac{1}{N} \sum_{j=1}^{N} M_j \cdot T_0(r_c), \qquad (39)$$

where N is the number of Monte Carlo realizations (we set N to 1000 in this simulation), $T_0(r_c)$ is the duration of each round defined in (17), and M_j is the number of rounds the network can operate properly at the j-th Monte Carlo realization. At each Monte Carlo realization, the network is said to operate properly at round i if the expected number of active sensors at round i is at least n_{eff}^L, i.e.,

$$n^{(i)} p^{(i)} \geqslant n_{\text{eff}}^L.$$

We observe that for the centralized PIS scheme, the simulation result is very close to the theoretical lower bound, T_ω^L, which was derived in Section 4. By comparing the PIS schemes and the PIS-naive scheme, we clearly see that the ω-lifetime's of both centralized and distributed PIS schemes are much longer than that of the PIS-naive scheme, and the differences become larger with more deployed sensors. Such fact demonstrates the importance of joint optimization of lifetime, connectivity, and coverage. We also see that the ω-lifetime of the distributed PIS scheme is close to that of the centralized one, which suggests that the distributed PIS scheme is a good choice for real applications.

6   Conclusions and Future Work

In this paper, we investigate the fundamental limits of a wireless sensor network's lifetime under connectivity and k-coverage constraints. The contributions of the paper are twofold. First, we study the lifetime of a wireless sensor network from a novel probabilistic perspective and introduce a new concept, called network's ω-lifetime, which is defined as the expectation of the time interval during which the probability of guaranteeing connectivity and k-coverage simultaneously is at least ω. Second, we propose PIS (Pre-planned Independent Sleeping) as a near-optimal scheduling scheme to maximize the ω-lifetime of a finite-size wireless sensor network, describe a possible distributed implementation of the PIS scheme, and demonstrate the PIS performance by simulation results.

Future work includes extending the analysis to more generic and realistic scenarios such as when only a portion of the sensing field needs to be k-covered, or when the sensing field is of irregular shape, or when the communication radius is also a random variable.

References

1. I. Akyildiz, W. Su, Y. Sankarasubramaniam, and E. Cayirci, "A survey on sensor networks," *IEEE Communication Magazine*, vol. 40, no. 8, pp. 102–114, Aug. 2002.
2. D. Estrin, R. Govindan, J. Heidemann, and S. Kumar, "Next century challenges: scalable coordination in sensor networks," in *Proc. ACM MobiCom'99*, Seattle, WA, 1999, pp. 263–270.
3. P. Gupta and P. Kumar, "Critical power for asymptotic connectivity," in *Proc. the 37th IEEE Conference on Decision and Control*, vol. 1, 1998, pp. 1106–1110.
4. D. Miorandi and E. Altman, "Coverage and connectivity of ad hoc networks in presence of channel randomness," in *Proc. IEEE INFOCOM'05*, Miami, FL, Mar. 2005, pp. 491–502.
5. X. Wang, G. Xing, Y. Zhang, C. Lu, R. Pless, and C. Gill, "Integrated coverage and connectivity configuration in wireless sensor networks," in *Proc. ACM SenSys'03*, Los Angeles, CA, 2003, pp. 28–39.
6. B. Liu and D. Towsley, "A study of the coverage of large-scale sensor networks," in *Proc. IEEE MASS'04*, Fort Lauderdale, FL, Oct. 2004, pp. 475–483.
7. S. Shakkottai, R. Srikant, and N. Shroff, "Unreliable sensor grids: coverage, connectivity and diameter," in *Proc. IEEE INFOCOM'03*, vol. 2, 2003, pp. 1073–1083.

8. S. Kumar, T. Lai, and J. Balogh, "On k-coverage in a mostly sleeping sensor network," in *Proc. ACM MobiCom'04*, Philadelphia, PA, 2004, pp. 144–158.

9. F. Ye, G. Zhong, J. Cheng, S. Lu, and L. Zhang, "PEAS: a robust energy conserving protocol for long-lived sensor networks," in *Proc. IEEE ICDCS'03*, Providence, RI, May 2003, pp. 28–37.

10. H. Zhang and J. Hou, "On deriving the upper bound of α-lifetime for large sensor networks," in *Proc. ACM MobiHoc'04*, Roppongi Hills, Tokyo, Japan, 2004, pp. 121–132.

11. Y. Xu, J. Heidemann, and D. Estrin, "Geography-informed energy conservation for ad hoc routing," in *Proc. ACM MobiCom'01*, Rome, Italy, July 2001.

12. C. Hsin and M. Liu, "Network coverage using low duty-cycled sensors: random and coordinated sleep algorithms," in *Proc. IEEE IPSN'04*, Berkeley, CA, Apr. 2004, pp. 433–442.

13. C. Gui and P. Mohapatra, "Power convervation and quality of surveillance in target tracking sensor networks," in *Proc. ACM MobiCom'04*, Philadelphia, PA, Sept. 2004.

14. P. Hall, *Introduction to the Theory of Coverage Process*. John Wiley and Sons, 1988.

15. W. Mo, D. Qiao, and Z. Wang, "Mostly sleeping wireless sensor networks: connectivity, k-coverage, and lifetime," Electrical and Computer Engineering Department, Iowa State University, Tech. Rep., 2005.

16. P. Billingsley, *Probability and Measure*. New York: Wiley, 1979.

17. T. Rappaport, *Wireless Communications: Principles and Practice*. Prentice Hall, 2001.

18. D. Blough and P. Santi, "Investigating upper bounds on network lifetime extension for cell-based energy conservation techniques in stationary ad hoc networks," in *Proc. MobiCom'02*, 2002, pp. 183–192.

Network Power Scheduling for TinyOS Applications

Barbara Hohlt and Eric Brewer

Electrical Engineering and Computer Sciences Department
University of California at Berkeley
Berkeley, CA USA
{hohltb, brewer}@eecs.berkeley.edu

Abstract. This paper presents a study of the Flexible Power Scheduling protocol and evaluates its use for real-world sensor network applications and their platforms. FPS uses dynamically created schedules to reserve network flows in sensor networks allowing nodes to turn off their radio during idle times. We show that network power scheduling has high end-to-end packet reception and can achieve power savings of 2-5x for two well-known TinyOS applications over their existing power-management schemes, and over 150x compared with no power management. Twinkle is our second-generation implementation of FPS and provides additional application support.

1 Introduction

Power is one of the dominant problems in wireless sensor networks. Constraints imposed by the limited energy stores on individual nodes require planned use of resources, particularly the radio. Sensor network energy use tends to be particularly acute as deployments are left unattended for long periods of time, perhaps months or years. Communication is the most costly task in terms of energy [2,9,27,21]. At the communication distances typical in sensor networks, listening for information on the radio channel is of a cost similar to transmission of data [23]. Worse, the energy cost for a node in idle mode is approximately the same as in receive mode. Therefore, protocols that assume receive and idle power are of little consequence are not suitable for sensor networks. Idle listening, the time spent listening while waiting to receive packets, comprises the most significant cost of radio communication. Even for hand-held devices Stemm et al. observed that idle listening dominated the energy costs [30]. Thus, the biggest single action to save power is to turn the radio off during idle times.

Unfortunately, turning the radio off implies that you must know that the radio will be idle in advance, and the easiest way to do this is to have a schedule. An obvious approach is to use TDMA to turn the radio off at the MAC layer during idle slots. However, this requires tight time synchronization and typically hardware support. Scheduling network *flows* helps for multi-hop topologies, which play a significant role in wireless sensor networks. Pottie and Kaiser [21] cover the many advantages of multi-hop topologies, including reduced energy use and routing around obstructions. In multi-hop networks the farthest nodes have more chances to drop packets, and thus using only hop-by-hop decisions (rather than flows), as with any MAC-layer approach, tend to achieve lower bandwidth and less fairness.

P. Gibbons et al. (Eds.): DCOSS 2006, LNCS 4026, pp. 443–462, 2006.
© Springer-Verlag Berlin Heidelberg 2006

Flexible Power Scheduling (FPS) [13] introduced the approach of scheduling the network for power savings in sensor networks and proposed a two-level architecture that combines coarse-grain dynamic scheduling at the network layer to plan radio on-off times, and simple CSMA to handle channel access at the MAC-layer. The FPS paper presented the distributed scheduling algorithm details and microbenchmarks, but no performance evaluation with real applications.

In this paper we present a study of the FPS protocol and evaluate its use for real-world sensor network applications with studies of two well-known sensornetwork applications, GDI and TinyDB, on three mote platforms, `mica`, `mica2dot`, and `mica2`. Our second-generation implementation of FPS, named Twinkle[1], is used in these studies. We compare the power savings of GDI and TinyDB running their default radio power management against these two applications running Twinkle radio power management.

The main contribution of this paper is the implementation and evaluations from two real applications using Twinkle, our second-generation implementation of FPS. In particular, we provide an application-level evaluation of the power savings using two well-known and deployed TinyOS applications [11]: the Great Duck Island [18,31] deployment and a TinyDB application that collects data on Redwood trees [17, 29]. We also compare Twinkle with low-power listening, an alternative proposal for power savings.

The contributions of this paper include:

- An implementation and evaluation of network power scheduling with two well-known TinyOS applications accross three platforms
- A 4x power savings for the Great Duck Island application.
- A 4.3x power savings for a 35-mote sensor network using TinyDB, compared with the default "duty cycling" power management scheme, and 150X versus no power management.
- A detailed comparison between Twinkle and Low-Power Listening with measured power data from real motes. This reveals a 2x or more power savings due to Twinkle.

Section 2 presents an overview of the basic FPS scheduling approach to provide background for these studies. Section 3 and Section 4 present evaluations using two real applications. Finally, Section 5 covers related work, and we conclude in Section 6.

2   Background

Flexible Power Scheduling (FPS) [13] introduced the approach of scheduling the network for power savings in sensor networks and proposed a two-level architecture that combines coarse-grain dynamic scheduling at the network layer to plan radio on-off times, and simple CSMA to handle channel access at the MAC-layer. The original protocol only supported communication in one direction, from the network to the gateway. Twinkle is our second-generation implementation of the FPS protocol and adds

---

[1] The name "Twinkle" comes from observing the network: scheduling avoids collisions and thus the network twinkles if you turn on an LED every time a node transmits.

broadcast capability to enable communication from the gateway to the network while running the FPS protocol.

In this section we give a general description of the FPS protocol with an overview of the new broadcast support to provide context for the studies that follow. The focus and scope of this paper is to provide real-world experiences and evaluations of FPS with TinyOS applications and their platforms.

2.1   Power Scheduling

Power scheduling is primarily useful for low-bandwidth long-lived data-gathering applications such as GDI and TinyDB. The FPS scheme exploits the structure of a tree to build the schedule, which makes it useful primarily for data collection applications, rather than those with any-to-any communication patterns. A large class of TinyOS applications fit this model, including equipment tracking, building-wide energy monitoring, habitat monitoring [31, 29], conference-room reservations [5], art museum monitoring [26], and automatic lawn sprinklers [8]. The basic approach is to use a schedule that tells every node when to listen and when to transmit. As the bandwidth needs are low, most nodes are idle most of the time, and the radio can be turned off during these periods.

FPS scheduling is receiver initiated. In particular, the schedule spreads from the root of the tree down to the leaves based on the required bandwidth: parents advertise available slots and children that need more bandwidth request a slot. Applied recursively, this allows bandwidth allocation for all of the nodes in the network. Although this schedule ensures that parents and their children are contention free, there may still be contention due to other nodes in the network or poor time synchronization; however, this contention is rare and can be handled by a normal CSMA MAC layer.

FPS reservations correspond to a unit flow from source-node to root, and thus the schedule is really a schedule of flows. Scheduling flows reduces contention and increases fairness, and form one reason why higher-level scheduling has more value than traditional TDMA. To allow adaptive schedules, advertising continues after the initial schedule is built. If new nodes arrive, or bandwidth demands change, children can request more bandwidth or release some.

2.2   Making Reservations

Time is divided into cycles and cycles are divided into slots. Each node maintains a local schedule that indicates in what slot it transmitts, receives, or idles. The main operation is as follows:

1. Parent selects an idle slot S and advertises the slot
2. Child hears the advertisement and sends a request for slot S
3. Parent receives the request and sends an acknowledgement.

Here the parent node is the route-through node, closest to the base station. In Step 1, the parent node selects an idle slot S at random from its list of idle slots and advertises slot S during slot C (a specific slot known to its children). In Step 2, a child hears the advertisement and subsequently sends a request for slot S during slot S. In Step 3, the parent hears and acknowledges requests during time slot S. Thereafter the child transmits during slot S and the parent receives during slot S. The reservation does not need to be renegotiated and remains in effect until the child cancels the reservation or the parent times out the reservation because no receptions occur after some number of cycles. No acknowledgement implies a request was denied, and the child must petition for the next advertised reservation slot. A parent may additionally advertise slots at random times i.e. not in the C slot.

A node keeps its radio off during idle time slots. The one exception is when a node joins the network or switches parents. In this case it must leave its radio on until it makes an initial reservation and learns the slot C specific to its parent. Although made locally, these reservations represent bandwidth allocation for entire traffic flows from source to sink. This is because all nodes preallocate some amount of flow in advance. Generally speaking, local nodes observe a rule that the amount of transmission slots in their schedule must be kept greater than the amount of receive slots.

2.3   Partial Flows and Broadcast

The original FPS protocol reserves entire flows from source to sink. Twinkle introduces a new reservation type called *partial flows*. A partial flow is one that terminates at some node other than the root, i.e. the reservation is not from source to sink. Partial flows can be used for various operations such as data aggregation and compression. For example, partial flows can be used to enable in-network data aggregation, in which the flow terminates at the node that does the aggregation.

Broadcast is essential for systems like TinyDB that need to inject queries or commands into the network. In Twinkle, a broadcast channel is an instance of a partial flow. In this case the partial flows are used in the reverse direction: each node reserves a partial flow with its parent that it will use as a broadcast channel for its children. Upon joining the network, each node acquires at least one partial flow reservation that terminates at its parent. This is called the *Comm* channel (slot C) and is used by the node as a broadcast channel for sending synchronization packets, advertisements, and forwarding messages injected from the base station. Twinkle protocol messages always include the slot number of the Comm channel (slot C). In this way, children nodes know in which slot to listen for broadcasts from their parent.

Twinkle maintains two forwarding queues: one used for broadcasting or forwarding commands away from the base station, and one used for forwarding packets toward the base station. When a node receives a command message it invokes the appropriate command message handler and places the message on the command queue for forwarding. The Comm channel is shared; both injected commands and time sync packets (with slot advertisements) use the same channel. The convention is if there is a command to be forwarded that is sent first followed by the time sync packet.

```
if current slot == Comm slot
   if command in command queue
      broadcast command message
   endif
   broadcast sync packet
endif
```

The GDI application in Section 3 uses the Comm channel for time sync packets and injecting commands to start and stop the experiments. The TinyDB application in Section 4 uses the Comm channel for time sync packet and injecting TinyDB queries.

3   Application: Great Duck Island

Our first target application, GDI [18, 31], is a habitat monitoring application deployed on Great Duck Island, Maine. GDI is a sense-to-gateway application that sends periodic readings to a remote base station, which then logs the data to an Internet-accessible database. The architecture is tiered, consisting of two sensor patches, a transit network, and a remote base station. The transit network consists of three gateways and connects the two sensor patches to the remote base station. There are two classes of mica2dot hardware: the *burrow* mote and the *weather* mote. The burrow motes monitor the occupancy of birds in their underground burrows and the weather motes monitor the climate above the ground surface. In this section, we will draw on information about the *weather* motes provided by the study of the Great Duck Island deployment [31].

Of the two weather mote sensor patches, one is a singlehop network and the other is a multihop network. The singlehop patch is deployed in an ellipse of length 57 meters and has 21 weather motes. Data is sampled and sent every 5 minutes. The multihop network is deployed in a 221 x 71 meter area and has 36 weather motes. Data is sampled and sent every 20 minutes.

In this section we compare the end-to-end packet reception, or *yield*, and power consumption of Twinkle/FPS with the low-power listening technique [12] used at Great Duck Island. Both schemes will be running the GDI application on a 30 node laboratory testbed. We will additionally investigate the phenomena of *overhearing* in the low-power listening case.

3.1   GDI with Low-Power Listening

The GDI application uses low-power listening to reduce radio power consumption. In low-power listening, the radio periodically samples the wireless channel for incoming packets. If there is nothing to receive at each sample, the radio powers off, otherwise it wakes up from low-power listening mode to receive the incoming packet. Messages include very long preambles, so they are at least as long as the radio channel sampling interval. The advantages of low-power listening are that it reduces the cost of idle listening, integrates easily, and is complementary with other protocols. It is characterized by high end-to-end packet reception, or *yield*. This is due to the long packet preamble acting as an in-band busy-tone.

Density and multihop also impact power consumption. The GDI study [31] reports a much higher power consumption in the multihop patch than the single hop patch which resulted in a shortened network lifetime — 63 of the 90 expected days — for the multihop patch. Two causes are attributed. First, messages have a higher transmission and reception cost due to their long preambles. Second, nodes wake up from low-power listening mode not only to receive their own packets, but anytime a packet is heard, regardless of the destination. *Overhearing* is the main contributor to the higher power consumption in the multihop patch.

We also observe that although low-power listening reduces the *cost* of idle listening it does not reduce the *amount* of idle listening, so that at very low data-sampling intervals its advantage declines because the radio must continue to turn on to check for incoming packets although there are none to receive. For very low data rates, we will show that scheduling such as Twinkle becomes more attractive because the radio (and potentially other subsystems) can be deterministically powered down until it is time to be used.

3.2   GDI with Twinkle

We implemented a version of GDI in TinyOS that uses Twinkle for its radio power management. This was a rather straight forward integration that consisted of wiring the GDI application component to the Twinkle component and disabling low-power listening. The Vanderbilt TimeSync, SysTime, and SysAlarm [19] components are used for time synchronization and timers. At the time of this work, TimeSync only supported the use of SysTime, which uses the CPU clock. The implication being, that for these experiments, GDI was not able to power manage the CPU. In all of our data presented here, we subtracted the draw of the CPU as if we had used a low-power Timer implementation. A version of TimeSync using the external crystal will become available shortly.

3.3   GDI Experiments

We conducted a total of 12 experiments on two versions of the GDI application. GDI-lpl uses low-power listening for radio power management and GDI-Twinkle uses Twinkle for radio power management The experiments were run on a 30-node in-lab multihop sensor network of `mica2dot` motes.

Twinkle supports data-gathering type applications like GDI where the majority of traffic is assumed to be low-rate, periodic, and traveling toward a base station. We ran a simple routing tree algorithm provided by Twinkle based on grid locations to obtain a realistic multihop tree topology and then used the same tree topology for the 12 experiments. As is done in the Great Duck Island deployment, no retransmissions are used in these experiments.

In each experiment we varied the data sample rate: 30 seconds, 1 minute, 5 minute, and 20 minutes. For experiments with 30 second and 1 minute sample rates, 100 messages per node were transmitted. For experiments with 5 minute and 20 minute sample rates, 48 and 12 messages were transmitted per node respectively. In the GDI-lpl experiments we varied the channel sampling interval: 485 ms and 100 ms. All experiments collected node id, sequence number, routing tree parent, routing tree depth, node temperature, and node voltage. The GDI-Twinkle experiments additionally collected the number of children, number of reserved slots, current transmission slot, current cycle, and number of radio-on slots per sample period.

3.4   Measuring Power Consumption

During the experiments, we measure the power consumption directly, using an oscilloscope, of two nodes located in two separate places of interest in the network. One node, we call the *inner node*, is located one hop from the base station and has a heavy amount of route-through traffic that is similar to its routing one-hop siblings. This should give us an estimate of the maximum lifetime of the network. This is a common method, documented by several researchers, for example [29]. In addition we measure the current at a second node. The second node is a *leaf node* that is one-hop from the base station as well. As it does not route-through any traffic, we should be able to see the effect of overhearing on power consumption at a node in a busy part of the network. If the measured current of the *inner node* and *leaf node* are similar in their active cycles, then we know the inner node is experiencing overhearing since all other factors remain the same. This is an important aspect of evaluating low-power listening.

At the lower sample rates, it is not feasible to take a measurement over the entire sample period, so we design our experiments so that we take some direct measurements and extrapolate others. For GDI-Twinkle, we define a cycle to be 30 seconds. Thus, full sample periods for the 30-second, 1-minute, 5-minute, and 20-minute sample rates are 1, 2, 10, and 40 cycles respectively. We schedule all data traffic during one cycle of each sample period called the *active cycle*. The unscheduled cycles are called *passive cycles*. Both active and passive cycles include protocol traffic (i.e. sending advertisements and listening for requests). We then measure the current at the two motes capturing data from both active and passive cycles during the 1 minute sample rate experiment. Then we take a running windowed average over a full 1-minute period, which gives us the power draw for both an active and passive cycle. Table 1 presents these direct power measurements.

For GDI-lpl we follow a similar method. We measure current at the two motes capturing data from both active and passive periods during the 1-minute sample period experiment. To represent an active period, we take a running average over the full 1-minute period. This also captures all the overhearing that occurs at the mote during a full period of any given sample rate. To represent a passive period, we took the longest chain of data from the measurements in which only idle channel sampling occurred. From this information we calculate the power consumption for the 5-minute and 20-

Table 1. Power Measurement (mW)

Power Management	Period (Sec)	Inner (mW)	Leaf (mW)
Twinkle active	30	2.18	0.69
Twinkle passive	30	0.33	0.33
Lpl-485 active	60	16.5	16.0
Lpl-485 passive	60	0.99	0.99
Lpl-100 active	60	8.20	7.60
Lpl-100 passive	60	3.90	3.90

minute sample rate experiments. The 30-second sample rate was measured separately (not calculated) and is shown in Table 1.

3.5   Evaluation

In this section we discuss the results of the data from all 12 experiments, and we also compare with the actual GDI deployment data.

Power Comparison with Low-Power Listening. Given the direct power measurements from Table 1, we can estimate the power consumption for the 5-minute and 20-minute sample rate experiments. For example, for Twinkle, we read off the following: an active cycle at the inner mote consumes 2.18 mW and a passive cycle consumes 0.33 mW. Given these numbers, for a 20-minute sampling rate we expect 1 active cycle and 39 passive cycles, for a weighted average of 0.38 mW. For the leaf mote, an active cycle consumes 0.69 mW and a passive cycle consumes 0.33 mW, giving a weighted average of 0.34 mW.

Similarly, to compute the GDI-lpl power consumption at a 20-minute sample rate we assume that for one minute the application consumes the energy of the active period and for the remaining 19 minutes the application consumes the energy of the passive period. Using the values from Table 1, the inner mote during the 20-minute sample rate Lpl-100 experiment, would consume an average of 4.12 mW ((8.2+19*3.9)/20 = 4.12mW).

Figure 1 shows all four sample periods: the 30-second and 1-minute rates are measured, and the 5-minute and 20-minute periods are estimated as above. For Twinkle, the inner node consistently has a greater draw than the leaf node. In contrast, for LPL, the inner and leaf nodes consistently have almost the same draw. This indicates that Twinkle's main power draw depends on the routed traffic, and in most of the cases LPL's main power draw depends on the overheard traffic. However, from Table 1 we see the passive power draw for LPL-100 is 3.9 mW, which forms an asymptote as the

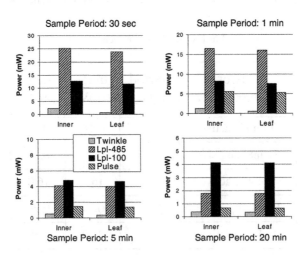

Fig. 1. Relative power consumption of Twinkle and LPL for four different sample periods. Pulse is a newer version of LPL discussed below.

sample period increases. Overall, as the sample rate gets lower and the preambles get shorter, overhearing does not play as big role.

The next thing to notice is at the higher sample rates, LPL-485 has a higher power consumption than LPL-100, but at the lower sample rates, LPL-485 has a lower power consumption than LPL-100. This reveals a relationship within LPL where as the cost of transmitting increases with longer preambles, the cost of channel sampling decreases with longer sampling intervals.

Finally, we added a newer variation of LPL to the figure, called Pulse. Pulse was developed as part of BMAC [20], and optimizes the power consumption of LPL by listening for energy in the channel rather than the decoded preamble. This reduces the cost of listening substantially. Because it has much stricter timing constraints, Pulse cannot run on the `mica2dot` platform. However, we can compute the active and passive estimates for Pulse as if it were running on the `mica2dot` given our power traces and Table 2 from the BMAC paper, which provides the raw listening cost. Although Pulse does perform better than LPL, it is still 2x to 5x higher power consumption than Twinkle. Across the board, Twinkle has better power consumption than LPL, with improvements that range from 2x (over Pulse for low rates) to 10x (in cases where the listening interval is poorly chosen).

Table 2. Yield and Fairness Comparison

Power Scheme	Sample Period	Yield	Max/Min
Twinkle	0.5	0.80	2.11
Twinkle	1	0.90	1.74
Twinkle	5	0.84	1.92
Twinkle	20	0.83	2.4
Lpl-485	0.5	0.40	15.6
Lpl-485	1	0.68	94.0
Lpl-485	5	0.72	11.8
Lpl-485	20	0.69	12.0
Lpl-100	0.5	0.85	3.45
Lpl-100	1	0.83	2.23
Lpl-100	5	0.78	2.76
Lpl-100	20	0.77	4.00

Yield and Fairness. Table 2 shows the average yield (end-to-end packet reception) for all 12 experiments, and the ratio of the best and worst throughputs (Max/Min). This ratio indicates fairness: lower ratios are more fair. At 30 seconds, the LPL-485 network is saturated due to the long preambles and this accounts for its low yield. Overall, both Twinkle and LPL-100 are significantly better than LPL-485. Twinkle shows better fairness than LPL-100 and, other than the 30 second sample rate, Twinkle has higher yield than LPL-100.

Understanding the GDI Field Study. Viewing the data in comparison to the data provided by the GDI study [31], we find the results in the laboratory are remarkably close to the results in the field. The Great Duck Island deployment used a low-power listening channel sampling interval of 485 ms, a data sample period of 20 minutes in the multihop patch, and a data sample period of 5 minutes in the singlehop patch.

Table 3 presents results taken from the GDI field study, labeled GDI-485, and ;ncludes data from four of our in-lab experiments, labeled LPL-485 and Twinkle. For each row, we report the sample period, average yield, inner and leaf power consumption, and the number of nodes in the experiment. *For GDI-485, the yield figure represents the average yield from the first day of deployment.*

A close comparison can be drawn between LPL-485 and GDI-485 at the 20 minute sample rate. LPL-485 has a power draw of ~1.76 mW while GDI-485 has a power draw of 1.6 mW. The GDI-485 figure is expected to be lower for two reasons: in the laboratory, the two measured nodes are from the busier section of the testbed, and the testbed has a constant load rather than a decreasing one. In the GDI deployment, some multihop motes died and stopped sourcing traffic, which is why we report yield from the first day of deployment.

The yield data is extremely close as well. All yields for LPL-485 and GDI-485 are ~70%. The only large difference between the two data sets is the power consumption

Table 3. Comparison of our lab data with the actual GDI field study

Power Mgnt	Sample Period	Yield	Inner (mW)	Leaf (mW)	#
GDI-485 (single)	5	0.70	n/a	0.71	21
GDI-485 (multi)	20	0.70	1.60	n/a	36
Lpl-485	5	0.72	4.09	3.99	30
Lpl-485	20	0.69	1.77	1.74	30
Twinkle	5	0.84	0.52	0.36	30
Twinkle	20	0.83	0.38	0.34	30

at the 5-minute sample period. This is easily explained by recalling that at the 5-minute sample period, GDI-485 is singlehop while LPL-485 is multihop, and the LPL-485 measurements include a large amount of overhearing.

The closeness of the LPL-485 and GDI-485 data gives us high confidence in the corrrectness of our methodology and the results of our laboratory experiments. We expect the Twinkle numbers are a good estimate of how Twinkle would do were we to have access to a field deployment. Our laboratory experiments show that Twinkle consumes at least 4x less power and provides about 14% better yield.

4   Application: Redwoods with TinyDB

Our second target application, TinyDB [17], is a distributed query processor for TinyOS motes. TinyDB consists of a declarative SQL-like query language, a virtual database table, and a Java API for issuing queries and collecting results. Conceptually the entire network is viewed as a single table called *sensors* where the attributes are inputs of the motes (e.g. temperature, light) and queries are issued against the *sensors* table. The SQL language is extended to include an "EPOCH DURATION" clause that specifies the sample rate.

A typical query looks like this:

```
SELECT nodeid, temperature
FROM sensors
EPOCH DURATION 3 min
```

TinyDB allows up to two queries running concurrently: one for sensor readings and one for network monitoring. In this section we compare the power savings of TinyDB using Twinkle versus TinyDB using application-level duty cycling — the power management scheme currently used in TinyDB. We estimate the power savings of the two approaches using the TinyDB Redwood deployment in the Berkeley Botanical Garden [14] as our topology and traffic model.

4.1   Estimating Power Consumption

Determing the power consumption of TinyDB with application-level duty cycling is straight forward. For this analysis we will estimate the power consumption of both the mica and mica2 platforms and take an in-depth look at a radio trace generated by TinyDB with Twinkle. We use the following three-part methodology:

1. Estimate the amount of time the radio is on and off for each scheme. Our metric for this will be radio on time per hour, measured in seconds.
2. For Twinkle, we validate this estimate in Section 4.5 by looking in detail at one of the motes. The radio on time for application-level duty cycling is easy to estimate.
3. We use actual measured current we obtained from mica and mica2 motes using an oscilloscope (Table 6) to estimate power consumption for radio on/off times. (In the GDI application we measured the current directly during the experiment.)

Listening for information on the radio channel is of a cost similar to transmission of data [23,24,4], so this combination provides a reasonably accurate overall view of power consumption, which although not perfect, is certainly very accurate relative to the 4.3X advantage in power shown by Twinkle in Section 4.6.

4.2   Topology and Traffic Model

The Redwood deployment has 35 mica2dot motes dispersed across two trees reporting to one base station in the Berkeley Botanical Gardens. Each tree has 3 tiers of 5 nodes each and 2 nodes placed at each crest. One tree has 1 additional node at a bottom branch. Every 2.5 minutes each mote transmits its query results, which are multi-hopped and logged at the base station.

By examining the records in the redwood database, we can derive the actual topology information, and from this construct a general topology that reflects its state the majority of the time.

Out of 35 nodes, generally 2/3 of the nodes are one hop from the base station and 1/3 of the nodes are two hops from the base station at any given time. We start by computing the radio on time per hour for the case with no power management:

$$60 \text{ sec/min} * 60 \text{ min/hour} = 3600 \text{ sec/hour}$$
$$\text{No power management} = 3600 \text{ sec/hour}$$

This number is the average amount of time each radio is on per hour throughout the deployment. We next estimate this metric for duty cycling followed by an estimate for FPS.

4.3   Duty Cycling

In TinyDB duty cycling, the default power management scheme, all nodes wake up at the same time for a fixed waking period every EPOCH. During the waking period nodes exchange messages and take sensor readings. Outside the waking period the processor, radio, and sensors are powered down. Estimating the radio-on time is thus straightforward: all 35 nodes wake up at the same time every 2.5 minutes for 4 seconds and exchange messages. The sample rate is thus 24 samples per hour. Each node is on for 96 sec/hour.

$$24 \text{ samples/hour} * 4 \text{ sec/sample} = 96 \text{ sec/hour}$$
$$\text{Duty Cycling} = 96 \text{ sec/hour}$$

As expected, this approach is subject to very high packet losses due to the contention produced by exchanging packets at nearly the same time. A recent TinyDB empirical study [29] shows high losses, between 43% and 50%, and high variance using duty cycling. Although we do not test it explicitly, there is no reason to expect the yield for Twinkle (or low-power listening) would deviate from the 80% shown in Section 3.

4.4   Twinkle

Topology, time-slot duration, protocol traffic, and data traffic are factors in estimating the radio-on time for Twinkle. We will use the same topology as above for estimating

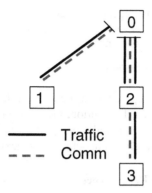

Fig. 2. Topology and Traffic for Estimates

the radio-on time of the 35 nodes. Time-slot duration and number of slots per cycle are configuration parameters in Twinkle. For this example, the time slot duration is 128 ms and for simplicity, the number of slots per cycle is 1172, which is roughly 2.5 minutes. Because of the long cycle length, we will add an extra advertisement per cycle.

Figure 2 depicts our subtree topology and traffic model. Solid lines represent data traffic (T/R) that is forwarded from the network to the base station every cycle. Dashed lines represent a Broadcast channel used for protocol traffic (B/RB). The Broadcast channel is used for TinyDB queries, network protocol messages, and advertisements.

Given the topology in Figure 2 and traffic in Table 4 we can now calculate the radio-on time for each node. Node 0 is the base station and has no cost. There is a cost of 3 time slots for advertisements (A): one advertisement, one receive pending, and one receive pending for the advertisement sent during the Broadcast.

Thus, this model captures data traffic as well as protocol traffic (i.e. sending advertisements and listening for requests).

Table 4. Traffic per Cycle (number of time slots)

Node	T	R	B	RB	A
1	1	0	1	0	3
2	2	1	1	1	3
3	1	0	1	0	3

For each node the cost is 0.767 seconds per cycle:

```
5(T/R) + 4(B/RB) + 9(A)
   = 18 * 128ms
   = 2.3 sec/cycle per 3 nodes
   = 0.767 sec/cycle (per node)
```

At 24 samples per hour, on average, each node is on 18.4 sec/hour:

```
24 samples/hour * 0.767 sec/cycle
     = 18.4 sec/hour
```

Twinkle = 18.4sec/hour

This is a savings of 5.2x compared with the duty cycle approach and 196x compared with no power management. In addition, the radio-on time is actually overestimated. Transmit slots do not leave the radio on for the whole slot since they can stop once their message is sent; this is shown in detail in the next section.

4.5   Twinkle Validation and Radio Trace

We implemented a prototype of TinyDB that uses Twinkle for radio power management. To validate our prototype and the radio on/off times, we ran the following experiment on three mica2dot motes and one mica2 mote as base station arranged in a topology shown in Figure 2. We monitored intermediate node 2 while it forwarded packets and sent advertisements. There are 64 slots of 128 ms each per cycle. We instrumented TinyDB-Twinkle to record the time of each call to turn the radio on and radio off, the beginning time of each time slot, and the state of each slot.

From the TinyDB Java tool we issue the query:

```
SELECT nodeid
FROM sensors
EPOCH DURATION 8192 ms
```

The intermediate mote is connected to an Ethernet device, and the debug records are logged over the network to a file on the PC. The regular query results are multi-hopped to the base station and displayed by the Java tool. In this experiment, we expect to have 1 advertisement, 2 receive pending slots, 3 transmit slots (one is a broadcast), 2 receive slots, and 56 idle slots per 64-slot cycle. We validate both the count of idle slots against the radio off time shown in Table 5.

Table 5. Validating Idle Slots

Metric	Slots	Idle %
Predicted Idle Slots	56/64	89.1%
Measured Idle Slots	56/64	89.1%
Measured Radio Off Time	—	91%

Note that the radio off time is higher than the percentage of idle slots because Transmit slots turn the radio off early — as soon as their messages have been sent.

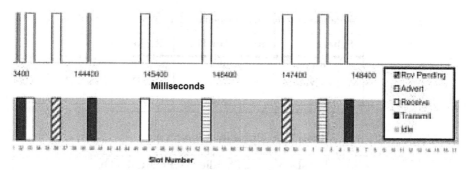

Fig. 3. A subsection of the validation experiment. The top graph shows actual radio on/ off times in milliseconds. The bottom graph shows the measured Twinkle state versus slot numbers aligned with time Note that the radio is always off for Idle cycles and that for Transmit cycles the on time is just long enough to transmit the queued messages.

Figure 3 shows a subsection of the validation experiment. The top graph shows actual radio on/off times (milliseconds). The bottom graph shows the measured Twinkle state versus slot numbers aligned with time; this subsection shows the active portion of a cycle (slots not shown are idle). Note that the radio is always off for Idle slots and that for Transmit slots the on time is just long enough to transmit the queued messages. In this experiment, the time slot duration is 128 ms, there are 64 slots per cycle, and the advertising frequency is once per cycle. This cut shows two advertisement slots, which is fine given that they are actually in two different cycles.

This experiment validates our methodology and shows that the power estimate for Twinkle in the previous section is actually conservative (since we count all of the Transmit slot time).

4.6   Power Savings

Finally, given the validated radio on times, we can estimate the power savings. First, however we need to know the current draw for a mote depending on whether or not the radio is on, and/or the CPU is on. With an oscilloscope, we measured the current of the mica and mica2 motes in three states: asleep, cpu idle, and both cpu and radio on. The results are shown in Table 6.[2]

Given these current draws, we estimate power consumption as:

```
Power (mAh) =
(On time)*(On draw) + (Off time)*(Off draw)
```

Using this equation and the radio-on times summarized in Table 7, we estimate the power consumption depicted in Figure 4. In all cases, both Duty Cycling and Twinkle perform substantially better (lower power) than no power management, so we focus on the difference between Twinkle and Duty Cycling.

---

[2] Mica2 radio power varies from 7.4 to 15.8 mA depending on transmit power, plus 7.8 mA for the *active* CPU draw for a total of 15.2 to 23.6 mA. We use 20mA as an overall estimate.

Table 6. Power Consumption of Motes (mA)

Scheme	Radio On Time	Ratio
None	3600	196
Duty Cycling	96	5.2
Twinkle	18.4	1

Table 7. On Times (seconds per hour)

Mote	Asleep	CPU Idle	CPU+Radio On
Mica1	0.01	0.4	8.0
Mica2	0.03	3.9	20

The biggest issue for estimating the power savings is whether or not the CPU is asleep when the radio is off. Neither system needs the CPU per se during idle times, but some sensors may require CPU power. Thus we expect for both the mica1 and mica2 the "CPU asleep" numbers are more realistic and we will quote these in our overall conclusions. However, we include the "CPU on" case for completeness. Note that even for cases where the CPU is needed for sensor sampling, the "CPU asleep" graph is more accurate, since the CPU would be asleep most of the time.

For the CPU on case, Twinkle outperforms Duty Cycling by 37% on the mica1 and 8% on the mica2, which has a higher CPU current draw. Compared to no power management, the advantage for Twinkle is 18X and 5X respectively.

For the more realistic "CPU asleep" case, i.e. the CPU is asleep during Idle slots, Twinkle outperforms Duty Cycling by 4.4X on the Mica1 and 4.3X on the Mica2. Note that this is consistent with the 5.2X reduction in radio on time. Compared to no power management, the advantage for Twinkle is 160X and 150X respectively.

Thus to summarize, for the TinyDB application with the Redwood study workload, we see a power savings of about 4.3X over Duty Cycling and 150X over no power management.

5   Related Work

Power consumption is an important issue in wireless sensor networks and energy optimizations are considered at all layers of the hardware and software platform. Many researchers have investigated energy efficient protocols in software to reduce communication costs.

In the area of energy-efficient MAC layers, there are two broad classes of approaches: contention based [22,33,7] and TDMA based [28,1,6]. PAMAS [22] enhances the MACA protocol with the addition of a signaling channel. It powers down the radio when it hears transmissions over the data channel or receptions over the signaling channel. S-MAC [33] incorporates periodic listen/sleep windows of fixed sizes similar to 802.11 PS mode [16]. In order to communicate, neighboring nodes periodically exchange their listen schedules. In the listen phase nodes transmit RTS/CTS packets and in the sleep phase nodes either transmit data or sleep if there is no data to

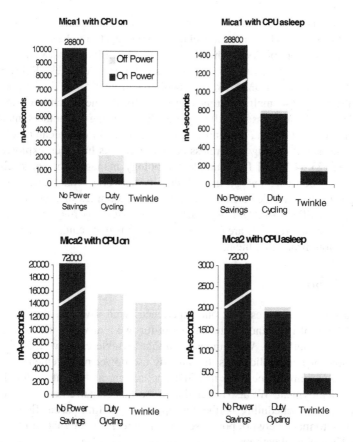

Fig. 4. Estimating power savings for two families of motes (Mica1, top, and Mica2, bottom), with the CPU on or asleep when the radio is off. Each vertical axis has a different scale, and in all cases the "No power savings" column goes off the top (Mica1 28800, Mica2 72000 mA-secsonds). Light gray is the radio-off power consumed (per hour), while dark gray is the radio-on power consumed.

send. T-MAC [7] is a variation on S-MAC. Instead of using a fixed listen window size, it transmits all messages in bursts of variable length, and sleeps between bursts.

TDMA-based protocols have natural idle times built into their schedules where the radio can be powered down. Additionally they do not have to keep the radio on to detect contention and avoid collisions. Centralized energy management [1] uses cluster-heads to manage CPU and radio consumption within a cluster. Centralized solutions generally do not scale well because inter-cluster communication and interference is hard to manage. Self organization [28] does not use clusters or hierarchies. It has a notion of super frames similar to TDMA frames for time schedules and requires a radio with multiple frequencies. It assumes a stationary network and generates static schedules. This scheme has less than optimal bandwidth allocation. Slot reservations can only be used by the node that has the reservation. Other nodes cannot reuse the slot reservation.

ReOrgReSync[6] uses a combination of topology management (ReOrg) and channel access (ReSync) and relies on a backbone for connectivity. Relay Organization is a topology management protocol which systematically shifts the network's routing burden to energy-rich nodes (wall powered and battery powered nodes). Relay Synchronization (ReSync), is a TDMA-like protocol that divides time into epochs. Nodes periodically broadcast small intent messages at a fixed time which indicate when they will send the next data message. All neighbors listen during each others intent message times. It assumes a low data rate and only one message per epoch can be sent.

Energy-efficient routing in wireless ad-hoc networks has been explored by many authors, see [25,34,15,10] for examples. Topology management approaches exploit redundancy to conserve energy in high-density networks. Redundant nodes from a routing perspective are detected and deactivated. Examples of these approaches are GAF [32] and SPAN [3]. Our approach does not seek to find minimum routes or redundancy. These protocols are designed for systems that require much more general communication throughout the network.

6   Conclusion

In this paper we have presented our experiences with Twinkle, the next-generation implementation of FPS, and evaluated its use for two real-world TinyOS applications and three mote platforms. We demonstrated that Twinkle can save 2-5x of the power consumption for real applications that already use power management of some kind. We saw a 2-4x improvement for the GDI application, and about 4x for the TinyDB Redwoods deployment. We also covered an important enhancement to the idea of network-layer power scheduling — the concept of scheduling partial flows that enable broadcast — to make network power scheduling a realistic alternative for real deployments of TinyOS applications.

Acknowledgments. We are much indebted to several individuals of the TinyOS community for their collaborations, suggestions, and support. Rob Szewczyk for his continuous help and advice on power management, TinyOS, and GDI. Brano Kusy and Miklos Maroti for their support on timers and time synchronization. Sam Madden for his support on TinyDB and Gilman Tolle for his work on the Redwoods database. This work was supported, in part, by the Defence Department Advanced Research Projects Agency (grants F33615-01-C-1895 and N6601-99-2-8913), the National Science Foundation (grant NSF IIS-033017), and Intel Corporation. Research infrastructure was provided by the National Science Foundation (grant IEA-9802069).

References

[1]   K.A. Arisha, M.A. Youssef, M.F. Younis,"Energy-aware TDMA based MAC for sensor networks," IEEE IMPACCT 2002, New York City, NY, USA, May 2002.

[2]   G. Asada, M. Dong, T. S. Lin, F. Newberg, G. Pottie, W. J. Kaiser, H. O. Marcy,"Wireless integrated network sensors: low power systems on a chip," ESSCIRC '98. Proceedings of the 24th European Solid-State Circuits Conference, The Hague, Netherlands, September 1998.

[3]   B. Chen, K. Jamieson, H. Balakrishnan, and R. Morris, "Span: an energy-efficient coordination algorithm for topology maintenance in ad hoc wireless networks,"MobiCom 2001, Rome Italy, July 2001.

[4]   Chipcon. http://www.chipcon.com/files/CC1000_Data_Sheet_2_3.pdf

[5]   W.S. Conner, L. Krishnamurthy, and R. Want, "Making everyday life a little easier using dense sensor networks," Proceeding of ACM Ubicomp 2001, Atlanta, GA, Oct. 2001.

[6]   W.S. Conner, J.Chhabra, M. Yarvis, L.Krishnamurthy, "Experimental Evaluation of Topology Control and Synchronization for In-building Sensor Network Applications," ACM Workshop on Wireless Sensor Networks and Applications, September 2003.

[7]   T.van Dam, K. Langendoen, "An Adaptive Energy-Efficient MAC Protocol for Wireless Sensor Networks," SENSYS 2003, Los Angeles, CA, USA, November 2003.

[8]   Digital Sun, Inc.: http://digitalsun.com

[9]   L. Doherty, B.A. Warneke, B.E. Boser, K.S.J. Pister, "Energy and Performance Considerations for Smart Dust," International Journal of Parallel Distributed Systems and Networks, Volume 4, Number 3, 2001, pp. 121-133.

[10] Z. Haas, J. Halpern, and L. Li, "Gossip-based ad-hoc routing," IEEE INFOCOM 2002, New York, NY, USA, June 2002.

[11] J. Hill, R. Szewczyk, A. Woo, S. Hollar, D. Culler, and K.S.J. Pister, "System architecture directions for networked sensors," ASPLOS 2000, Cambridge, MA, USA, November 2000.

[12] J. Hill, D. Culler, "Mica: a wireless platform for deeply embedded networks," IEEE Micro, 22(6):12-24, November/December 2002.

[13] B. Hohlt, L. Doherty, E. Brewer, "Flexible Power Scheduling for Sensor Networks,"IPSN 2004, Berkeley, CA, USA, April 2004.

[14] W. Hong,"TASK In Redwood Trees", http://today.cs.berkeley.edu/retreat-1-04/weihong-task-redwood-talk.pdf, NEST Retreat, Jan 2004.

[15] B. Karp and H.T. Kung, "GPSR: Greedy Perimeter Stateless Routing for wireless networks," MobiCom 2000, Boston, MA, USA, August 2000.

[16] LAN MAN Standards Committee of the IEEE Computer Society,"IEEE Standard 802.11, Wireless LAN Medium Access Control (MAC) and Physical Layer (PHY) specifications," IEEE, August 1999.

[17] S.R. Madden, M.J. Franklin, J.M. Hellerstein, and W. Hong, "TAG: a tiny aggregation service for ad-hoc sensor networks,"5th Symposium on Operating Systems Design and Implementation, Boston, MA, USA, December 2002.

[18] A. Mainwaring, J. Polastre, R. Szewczyk, D. Culler, J. Anderson,"Wireless sensor networks for habitat monitoring," WSNA 2002, Atlanta, GA, USA, September 2002.

[19] M. Maroti, B. Kusy, G. Simon, A. Ledeczi, "The Flooding Time Synchronization Protocol," SenSys 2004, Baltimore, MD, USA, November 2004.

[20] J.Polastre,J.Hill,D.Culler,"Versatile Low Power Media Access for Wireless Sensor Networks", SenSys 2004, Baltimore, ML,USA.

[21] G.J. Pottie, W.J. Kaiser, "Wireless Integrated Network Sensors,"Communications of the ACM, vol. 4, no. 5, May 2000.

[22] C.S. Raghavendra and S. Singh, "PAMAS - Power aware multi-access protocol with signaling for ad hoc networks,"ACM Communications Review, vol. 28, no. 33, July 1998.

[23] V. Raghunathan, C. Schurgers, S. Park, and M.B. Srivastava, "Energy-aware wireless microsensor networks," IEEE Signal Processing Magazine, vol. 19, no. 2, March 2002.

[24] RFM Monolithics. http://www.rfm.com/products/data/tr1000.pdf.

[25] E. M. Royer and C-K. Toh. "A review of current routing protocols for ad-hoc mobile wireless networks," IEEE Personal Communications, April 1999.

[26]   Sensicast Systems: http://www.sensicast.com.

[27]   K. Sohrabi, J. Gao, V. Ailawadhi, and G.J. Pottie, "Protocols for self-organization of a wireless sensor network," IEEE Personal Communications, Oct. 2000.

[28]   K. Sohrabi and G.J. Pottie, "Performance of a novel self-organization for wireless ad-hoc sensor networks," IEEE Vehicular Technology Conference, 1999, Houston, TX, May 1999.

[29]   P. Buonadonna, J. Hellerstein, W. Hong, D. Gay, S. Madden, "TASK: Sensor Network in a Box", European Workshop on Wireless Sensor Networks 2005, Istanbul, Turkey, February 2005.

[30]   M. Stemm and R. Katz, "Measuring and reducing energy consumption of network interfaces in hand-held devices," IEICE Trans. on Communications, vol. E80-B, no. 8, pp. 1125-1131, August 1997.

[31]   R. Szewczyk, A. Mainwaring, J. Polastre, J. Anderson, D. Culler, "An Analysis of a Large Scale Habitat Monitoring Application", SenSys 2004, Baltimore, ML, USA, November 2004.

[32]   Y. Xu, J. Heidemann, D. Estrin, "Geography-informed energy conservation for ad hoc routing," MobiCom 2001, Rome, Italy, July 2001.

[33]   W. Ye, J. Heidemann, D. Estrin, "An energy-efficient MAC protocol for wireless sensor networks," IEEE INFOCOM 2002, New York City, NY, USA, June 2002.

[34]   Y. Yu, R. Govindan, and D. Estrin. "Geographical and Energy Aware Routing: a recursive data dissemination protocol for wireless sensor networks," UCLA Computer Science Department Technical Report UCLA/CSD-TR-01-0023, May 2001.

Approximation Algorithms for Power-Aware Scheduling of Wireless Sensor Networks with Rate and Duty-Cycle Constraints[*]

Rajgopal Kannan[1] and Shuangqing Wei[2]

[1] Department of Computer Science
[2] Department of Electrical and Computer Engineering
Louisiana State University, Baton Rouge, LA 70803, USA
{rkanna1, swei}@lsu.edu
www.csc.lsu.edu/~rkannan, www.ece.lsu.edu/~swei

Abstract. We develop algorithms for finding the minimum energy transmission schedule for duty-cycle and rate constrained wireless sensor nodes transmitting over an interference channel. Since traditional optimization methods using Lagrange multipliers do not work well and are computationally expensive given the non-convex constraints, we develop fully polynomial approximation schemes (FPAS) for finding optimal schedules by considering restricted versions of the problem using multiple discrete power levels. We first show a simple dynamic programming solution that optimally solves the restricted problem. For two fixed transmit power levels (0 and P), we then develop a 2-factor approximation for finding the optimal fixed transmission power level per time slot, P_{opt}, that generates the optimal (minimum) energy schedule. This can then be used to develop a $(2, 1 + \epsilon)$-FPAS that approximates the optimal power consumption and rate constraints to within factors of 2 and arbitrarily small $\epsilon > 0$, respectively. Finally, we develop an algorithm for computing the optimal number of discrete power levels per time slot $(O(1/\epsilon))$, and use this to design a $(1, 1 + \epsilon)$-FPAS that consumes less energy than the optimal while violating each rate constraint by at most a $1 + \epsilon$ factor.

1 Introduction

Energy-efficiency is a critical concern in many wireless networks, such as cellular networks, ad-hoc networks or wireless sensor networks (WSNs) that consist of large number of sensor nodes equipped with unreplenishable and limited power resources. Since wireless communication accounts for a significant portion of node energy consumption, network lifetime and utility are dependent on the design of energy-efficient communication schemes including low-power signaling and energy-efficient multiple access protocols.

Delay is also an important constraint in many wireless network applications, for example battlefield surveillance or target tracking in which data with finite

---

[*] This work was supported by NSF grants IIS-0329738, ITR-0312632 and by AFRL.

© Springer-Verlag Berlin Heidelberg 2006

lifetime-information must be delivered before a deadline. Delay constraints in wireless networks can also be examined in terms of node operation under periodic duty cycles, in which time is divided into active (awake) and inactive (asleep) periods. [1], [2, 3] establish the idea of duty cycles in WSNs as a practical means of conserving node energy. Minimizing transmission energy subject to latency constraints has been studied [4, 5]. Several approaches for maximizing information transmission over a shared channel subject to average power constraints have been proposed [6, 7, 8, 9, 10]. [11] addresses the issue of minimizing transmission power, subject to a given amount of information being successfully transmitted and derives power control multiple access (PCMA) algorithms for autonomous channel access.

We consider N sensor nodes transmitting to their destinations over a typical AWGN interference channel over a time period T. These nodes could represent reasonably close neighbors communicating as part of some MAC protocol. We assume that time T is divided into M slots of equal duration. Let P_{it} be the transmit power used by node i during time slot t, $1 \leq t \leq M$. Let R_{it} represent the achievable transmission rate for node i during time slot t over this N-node interference channel. Single user decoding is assumed at each receiver to decode the information from its own transmitter while treating the remaining information as Gaussian interference. Thus we have,

$$R_{it} = \frac{1}{2} \log_2 \left(1 + \frac{\alpha_{ii}^t P_{it}}{\mathcal{N}_i^t + \sum_{j \neq i} \alpha_{ji}^t P_{jt}} \right), \quad 1 \leq i \leq N, \quad 1 \leq t \leq M \qquad (1)$$

where α_{ji}^t represent the channel attenuation at i's receiver due to transmitter j, which captures the effects of path-loss, shadowing and frequency nonselective fading, and $\mathcal{N}_i^t$ represents the background interference (usually $\mathcal{N}_i^t = \mathcal{N}_0$), during time slot t. We assume these parameters remain fixed over a (short) time slot of duration T/M but can vary from slot to slot.

We are interested in the following scheduling and energy minimization problem (labeled MESP: minimum energy scheduling problem)

$$\min f : \sum_{i=1}^{N} \sum_{t=1}^{M} P_{it}$$

$$\text{s.t} \quad g : \sum_{t=1}^{M} A_{it} R_{it} \geq \tilde{R}_i \quad i = 1, 2, \ldots, N$$

$$A_{it} = \begin{cases} 0 \text{ if node } i \text{ is idle} \\ 1 \text{ otherwise} \end{cases}$$

$$\sum_{t=1}^{M} A_{it} \leq \mu_i \quad i = 1, 2, \ldots, N$$

$$(2)$$

The objective function in MESP is to determine the schedule which minimizes the total energy. Since all slots are assumed to be of fixed duration, this is equivalent to minimizing the total transmitted power. Each node must maintain an average rate constraint $\tilde{R}_i$ over the M slots. Further, we assume that nodes operate under duty-cycles where time T is divided into active and idle time slots, wireless sensor networks for example, operate under such constraints [2, 1]. The duty-cycle constraint of node i is given by μ_i: the maximum number of time slots it can remain active, $1 \leq \mu_i \leq M$, $i = 1, 2, \ldots N$. $A_{it} \in \{0, 1\}$ depending on whether the node is idle or active during slot t, $1 \leq t \leq M$. Note that in this formulation of MESP, we do not have any overall power budget constraint (only duty-cycle constraints for limiting node activity) and we are looking to minimize the total power/energy over the universe of available power values. Individual/overall power budget constraints can be incorporated in our algorithm, if desired.

It can be seen that the rate constraints above are non-convex in the power variables P_{it}, even for the restricted version of MESP with two users ($N = 2$). Unfortunately this implies that traditional analytical optimization methods such as Lagrange multipliers [12] will not work well, since convexity of the constraints is a necessary condition for obtaining the global minimum using the Lagrangean $H = f + \lambda_k g_k$ (where g_k are the constraints), and computing $\nabla_{P_{it}, \lambda_k} = 0$. Moreover finding the global minimum through exhaustive search of all possible solutions of $\partial h / \partial P_{it} = 0$ is likely to be computationally expensive. Alternately computing the optimal dual $\max_\lambda \min_x h()$ introduces a duality gap which vanishes only under certain conditions on the number of constraints and parameters N and M [12, 13].

In this paper, we develop approximation algorithms for finding the optimal rate and duty-cycle constrained energy schedule by considering restricted versions of the problem using discrete power levels. From the algorithmic perspective, the MESP problem is NP-hard and related to the generalized assignment problem [14]. We develop fully polynomial approximation schemes (FPAS) for MESP using ideas related to bin-packing and the knapsack problem [14, 15]. We first show a simple dynamic programming solution (of exponential complexity in M) that optimally solves the restricted problem. For two fixed transmit power levels (0 and P), we then develop a 2-factor approximation for finding the optimal fixed transmit power level per time slot, P_{opt}, that generates the optimal (minimum) energy schedule. This can then be used to develop a $(2, 1 + \epsilon)$-FPAS that approximates the optimal power consumption and rate constraints to within factors of 2 and arbitrarily small $\epsilon > 0$, respectively. Finally, we develop an algorithm for computing the optimal number of discrete power levels per time slot $(O(1/\epsilon))$, and use this to design a $(1, 1 + \epsilon)$-FPAS that consumes less energy than the optimal while violating each rate constraint by at most a $1 + \epsilon$ factor.

2   Basic Dynamic Programming Solution

First, we consider a simple relaxation of the minimum energy scheduling problem using two discrete transmit power levels. In this restricted version of the problem, a node is allowed to be either idle or transmit with a given (fixed) power P during its active slot. We illustrate our schemes using two nodes ($N = 2$) over M time slots. As mentioned above, even the restricted two node case is not amenable to traditional optimization methods. Later in section 6, we extend the dynamic program and approximations are extended to the N-node, M time slot case.

The restricted optimization problem is described by:

$$\min \sum_{i=1}^{2} \sum_{t=1}^{M} P_{it}$$

$$\text{s.t} \sum_{t=1}^{M} R_{it} \geq \tilde{R}_i, \quad i = 1, 2$$

$$P_{it} \in \{0, P\}, \quad i = 1, 2; \quad t = 1, \ldots, M \tag{3}$$

$$A_{it} = \begin{cases} 0 \text{ if } P_{it} = 0 \\ 1 \text{ otherwise} \end{cases}$$

$$\sum_{t=1}^{M} A_{it} \leq \mu_i, \quad i = 1, 2$$

$$\tag{4}$$

We assume that $\mu_1 + \mu_2 \geq M$, i.e the two nodes have to interleave during some of the slots. A more restricted version of 4 with $\alpha_{ji}^t = \alpha_{ji}$ independent of t is analyzed in [16].

Let $\bar{R}_{i,j}^{kP,a,b} = \{<R_1, R_2>\}$ represent the set of rate vector (rate pairs) corresponding to cumulative transmission rates for user 1 and user 2 from time slots i through j, $1 \leq i \leq j \leq M$, while using a total power (node 1 + node 2) of kP. For notational simplicity, if $i = j$, we drop one of the redundant subscripts in the rate vector. In the above definition, $R_l = \sum_{t=i}^{j} R_{lt}$, where R_{lt}, $l = 1, 2$, is the achievable rate for node l during time slot t, depending on the actions of the other node i.e active/asleep. The number of active slots for user 1 and 2 in this period is denoted by a and b, respectively, where $0 \leq a, b \leq j-i+1$. Since a node uses fixed power P during an active slot, $a+b = k$, in this case. Thus for a given time slot t, we have four different rate vectors specified by,

$$\bar{R}_t^{0,0,0} = <0, 0>$$

$$\bar{R}_t^{P,0,1} = <0, \frac{1}{2} \log_2 \left(1 + \alpha_{22}^t P/\mathcal{N}_2^t\right)>$$

$$\bar{R}_t^{P,1,0} = <\frac{1}{2} \log_2 \left(1 + \alpha_{11}^t P/\mathcal{N}_1^t\right), 0>$$

$$\bar{R}_t^{2P,1,1} = <\frac{1}{2} \log_2 \left(1 + \frac{\alpha_{11}^t P}{\mathcal{N}_1^t + \alpha_{21}^t P}\right), \frac{1}{2} \log_2 \left(1 + \frac{\alpha_{22}^t P}{\mathcal{N}_2^t + \alpha_{12}^t P}\right)>$$

$$\tag{5}$$

The restricted version of the problem consists of finding a transmission schedule of minimum total energy in which active nodes transmit at a fixed power during each active time slot while also satisfying the given duty-cycle and rate constraints. For fixed power level P, the optimal schedule is easily specified by the following dynamic program which maintains the current best-solution for each total power level and duty-cycle value. The boundary conditions are given by the rate vectors in Eq. 5. The recursive formula for each power level kP and duty-cycles a, b, $1 \leq k \leq 2M$, $0 \leq a \leq \mu_1$, $0 \leq b \leq \mu_2$ is

$$\bar{R}_{i,j}^{kP,a,b} = \text{vectormax}\Big\{ \bar{R}_{i,j-1}^{kP,a,b} \bigcup \Big(\bar{R}_{i,j-1}^{(k-1)P,a-1,b} + \bar{R}_j^{P,1,0} \Big) \bigcup \Big(\bar{R}_{i,j-1}^{(k-1)P,a,b-1} + \bar{R}_j^{P,0,1} \Big)$$
$$\bigcup \Big(\bar{R}_{i,j-1}^{(k-2)P,a-1,b-1} + \bar{R}_j^{2P,1,1} \Big) \Big\} \tag{6}$$

where the rate vectors in each union operation above are computed using pairwise addition of the individual vectors. The vectormax operation eliminates all dominated rate pairs from a set of rate pairs, i.e. $\forall \{ <R_1, R_2>, <R_3, R_4> \} \in \bar{R}_{i,j}^{kP,a,b}$ either $R_1 > R_3$ and $R_2 \leq R_4$ or vice versa. Using the recursive function, the table of rate vector values is evaluated in increasing order of time slots from 1 to M. There are $O(MP\mu_1\mu_2)$ rate vectors and the set of feasible schedules correspond to those rate vectors $\geq$ $<\tilde{R}_1, \tilde{R}_2>$ under the usual meaning of vector comparison. The optimal schedule for a given transmit power level P is the one whose rate vector satisfies

$$\bar{R}_{opt}^P = \underset{k=1,2\ldots,2M}{\text{argmin}} \Big\{ \exists \ <R_1, R_2> \in \bar{R}_{1,M}^{kP,\mu_1,\mu_2} \mid <R_1, R_2> \ \geq \ <\tilde{R}_1, \tilde{R}_2> \Big\}$$
$$\tag{7}$$

In practice, it is likely that many of the vectors in $\bar{R}_{i,j}^{kP,a,b}$ would be dominated and hence eliminated by the vectormax operation. However in the worst-case, even after the vectormax operation, the size of $\bar{R}_{i,j}^{kP,a,b}$ can quadruple with each additional slot. Thus the above dynamic program is clearly exponential in terms of the slot parameter M, even though each slot contains only four rate vectors. This motivates us to consider a $(1+\epsilon, 1+\epsilon)$ FPAS for the problem, as described in Section 5.

3   2-Approximate Minimum Energy Schedule

Let $\mathcal{A}^P$ denote the (exponential time) dynamic programming algorithm for finding the optimal schedule under duty-cycle constraints and using only two fixed transmit power levels of 0 or P per slot. We note it is possible under $\mathcal{A}^P$ that $\forall k$, $\bar{R}_{1,M}^{kP,\mu_1,\mu_2}$ $<$ $<\tilde{R}_1, \tilde{R}_2>$. Thus $\bar{R}_{opt}^P = \phi$ and no feasible schedule exists for the given transmit power value P. In this case, we wish to find the optimal feasible transmit power $P = P_{opt}$ for which a feasible schedule exists under $\mathcal{A}^P$ and that uses minimum possible energy $E_{\mathcal{A}}^{P_{opt}}$ among all such feasible powers. In this section, we describe a 2-approximation for finding $E_{\mathcal{A}}^{P_{opt}}$. Subsequently (in Section 5), we develop an FPAS using $O(1/\epsilon)$ power levels, that approximates P_{opt} and the corresponding minimal energy schedule to within an ϵ-factor.

Let $E_{\mathcal{A}}^P$ denote the total energy of the schedule produced by $\mathcal{A}^P$. Let P_a and P_b, where $P_a > P_b$, represent two different transmit power levels. Consider two instances of the scheduling problem. In the first instance, each node can either transmit at power P_a or be idle during each slot. Likewise, with power P_b in the second instance.

Claim. For each $<R_1, R_2> \in \bar{R}_{i,j}^{kP_b,a,b}$ *there is a rate pair* $<R_3, R_4> \in \bar{R}_{i,j}^{kP_a,a,b}$ *such that* $<R_1, R_2> \; < \; <R_3, R_4>$.

Proof. From Eq. 5 it can be seen that for any slot t, we have $\bar{R}_t^{kP_a,a,b} > \bar{R}_t^{kP_b,a,b}$, $k = 1, 2, a = 0, 1, b = 0, 1$. The proof follows in a straightforward manner by induction.

Let P_{min} be the minimum (fixed) transmit power level per active slot for which a feasible schedule exists. Without loss of generality, we assume $P_{min} \geq 1$.

Theorem 1. $\lceil P_{min} \rceil$ *can be found in* $O(\lceil \log_2 P_{min} \rceil)$ *calls to the dynamic programming algorithm* $\mathcal{A}^P$.

Proof. Initialize $P = 1^1$. While $\bar{R}_{opt}^P = \phi$, set $P = 2P$ and run algorithm $\mathcal{A}^P$. By Claim 1, the values of the rate vectors increase with P and hence the process will terminate with $\bar{R}_{opt}^P \neq \phi$. Let P_m be the terminating value of P which is found in $\lceil \log_2 P_{min} \rceil$ calls. $\lceil P_{min} \rceil$ can then be obtained through binary search in the interval $[P_m/2, P_m]$ with $O(\log_2(P_m/2))$ further calls to $\mathcal{A}^P$.

Note that Claim 1 for rate vectors cannot be translated to total energy values i.e $P_a > P_b$ does not imply $E_{\mathcal{A}}^{P_a} > E_{\mathcal{A}}^{P_b}$. $E_{\mathcal{A}}^P$ is not convex and can have multiple local minima for $P_a > P_{min}$. Thus to obtain a 2-approximation of the global minimum energy schedule, we first need to restrict the space of feasible transmit powers by finding an upper bound P_{max} such that $E_{\mathcal{A}}^{P_{opt}} < E_{\mathcal{A}}^P$ for all $P > P_{max}$.

A simple upper bound is $P_{max} = \left(\frac{\mu_1 + \mu_2}{2}\right) P_{min} \leq M P_{min}$. Note that $E_{\mathcal{A}}^{P_{opt}} \leq E_{\mathcal{A}}^{P_{min}} \leq P_{min}(\mu_1 + \mu_2)$. Since each node is active during at least one slot, $E_{\mathcal{A}}^P > E_{\mathcal{A}}^{P_{min}}$ for all $P > P_{max}$. Further, since $P_{opt} \in [P_{min}, P_{max}]$, we note that P_{opt} can be found by searching in an interval of size bounded by $O(M P_{min})$.

We can obtain a smaller bound on P_{max} (and hence the search space for P_{opt}) by using the following lemma: Let S_1^P, S_2^P and S_3^P be the set of time slots occupied by node 1 only, node 2 only and both nodes, under the schedule created by $\mathcal{A}^P$. Let $R_{i,S_i^P}^P$ denote the total rate obtained by node i over S_i^P, $i = 1, 2$. Let $S_{i,s}^P \subset S_i^P$ represent the set of $\lfloor |S_i^P|/2 \rfloor$ time slots with the *smallest* rates $\log_2(1 + \alpha_{ii}^t P/\mathcal{N}_i^t)/2$ among the slots in S_i^P. Similarly, let $S_{3,s}^P(i) \subset S_3^P$ denote the set of $\lfloor |S_3^P|/4 \rfloor$ slots with the smallest rates calculated as $\log_2(1 + \alpha_{ii}^t P/\mathcal{N}_i^t)/2$ among the slots in S_3^P and let $R_{i,S_{3,s}^P(i)}^P$ denote the corresponding total rate over these slots. A sufficient condition for finding P_{max} is then given by:

---

[1] Note that a better initial value can be obtained by using $P = \min(P_1', P_2')/M$ from Eq 10 in the next section.

Lemma 1. $P \leq P_{max} < 2P$ if $S_3^P \cap S_3^{2P} \neq \emptyset$, $R_{i,S_{3,s}^P}^P(i) \geq (\lceil |S_3^P|/4 \rceil)/2$ and $R_{i,S_{i,s}^P} \geq (\lceil |S_i^P|/2 \rceil)/2$, $i = 1, 2$.

For a detailed proof, please refer to [17]. The last rate condition of the lemma is derived from the fact that doubling the power over any set S of solo slots can increase the achieved rate by less than $|S|/2$. Thus if the worst half-set of slots $(S_{i,s}^P)$ has a total rate at least $|S_i^P|/4$, $i = 1, 2$, then doubling the power over the best half-set of slots (thereby expending the same energy) cannot achieve the same rate as before. The second rate condition is derived using the fact that doubling the power still leads to overlapping slots. The first condition states that if overlapping slots persist even after doubling the transmit power, and simultaneously the second rate condition is also satisfied with respect to the worst $S_{3,s}^P(i)/4$ slots (pretending that each node i is transmitting without interference from the other in these slots), then no amount of further increases in transmit power can decrease the overall energy. Thus $P_{max} < 2P$.

We use the above bound on P_{max} to obtain a 2-approximation for $E_{\mathcal{A}}^{P_{opt}}$, the energy of the optimal (minimum energy) schedule as follows:

Theorem 2. *Let*

$$P^* = \operatorname*{argmin}_{P=2^t P_{min}, \ t=0,1..., \lceil \log_2 \frac{P_{max}}{P_{min}} \rceil} E_{\mathcal{A}}^P.$$

Then $E_{\mathcal{A}}^{P^*}$ is a 2-approximation to $E_{\mathcal{A}}^{P_{opt}}$, the minimum energy schedule generated by the optimal transmit power P_{opt}. The algorithm for finding $E_{\mathcal{A}}^{P^*}$ uses $\lceil \log_2 \frac{P_{max}}{P_{min}} \rceil = o(\log_2 M)$ calls to $\mathcal{A}^P$.

Proof. We run the $\mathcal{A}^P$ algorithm starting with $P = P_{min}$ and doubling P with each iteration until we reach a P_{max} as defined by lemma 1. The total energy can oscillate between $E_{\mathcal{A}}^{P_{min}}$ and $E_{\mathcal{A}}^{P_{max}}$ as we sequentially double the power. For any solution using power P_a, $P < P_a < 2P$, the number of active slots $t_{P_a} = |S_1^{P_a}| + |S_2^{P_a}| + 2|S_3^{P_a}|$ cannot increase between t_P and t_{2P} i.e $t_P \geq t_{P_a} \geq t_{2P}$ (using claim 1). Thus $E_{\mathcal{A}}^{P_a} \geq (1/2) \min E_{\mathcal{A}}^P, E_{\mathcal{A}}^{2P}$. Let P^* be the power yielding the minimum energy among the iterations and choose $E_{\mathcal{A}}^{P^*}$ as the output of our algorithm. By the above arguments, $E_{\mathcal{A}}^{P^*} \leq 2E_{\mathcal{A}}^{P_{opt}}$ and therefore this algorithm is a 2-approximation. Since $P_{max} = o(MP_{min}$, the number of iterations is $o(\log_2 M)$.

4 Minimum Energy Schedule with Multiple Power Levels

We now consider the scheduling problem with multiple discretized power levels, where each node can choose from a set of power levels per time slot. As shown below, if the power levels are chosen appropriately, the cost of the resulting minimum energy schedule approximates the cost of the optimal schedule to within an ϵ-factor.

For the optimization problem with multiple power levels, let P and L_t denote the maximum allowable transmit power and the number of discrete power levels available per time slot, respectively, with values as defined below. For this problem, the constraint 3 of Eq. 4 is replaced with

$$P_{it} \in \{P_l\}, \ l = 0, 1, \ldots L_t; \ \ 0 = P_0 \le P_l \le P_{L_t} = P; \ \ i = 1, 2; \ \ t = 1, \ldots, M. \quad (8)$$

Note that the corresponding optimal version of the minimum energy scheduling problem contains the constraint

$$0 \le P_{it} \le P, \qquad\qquad i = 1, 2; \ \ t = 1, \ldots, M \qquad (9)$$

Let $\mathcal{A}^{P^*}$ denote the optimal algorithm for the above restricted version of MESP with per slot maximum power constraints (Eq. 9), i.e nodes select an optimal power value $0 \le P_{it}^* \le P$ in each slot, to satisfy their rate and duty-cycle constraints. Let R_{it}^* denote the corresponding optimal rate achieved per time slot, $i = 1, 2, \ t = 1, 2, \ldots M$. Finally, let $P^* = \sum \sum P_{it}^*$ and $R_i^* = \sum_t R_{it}^*$ denote the overall optimal power and rate allocations. In general, an (α, β) approximation of the optimal minimum energy scheduling problem is one which provides a feasible schedule with total power $\hat{P} \le \alpha P^*$ and each rate constraint violated by at most a β-factor i.e $\beta \hat{R}_i \ge R_i^*$, for each node i. Note that $R_i^* \ge \tilde{R}_i$ and hence $\beta \hat{R}_i \ge \tilde{R}_i$. Given some $\epsilon > 0$, we first show the construction of a more computationally expensive $(1 + \epsilon, 1 + \epsilon)$-approximation in order to illustrate our approach and then describe a more efficient $(1, 1 + \epsilon)$-approximation to the optimal.

Let $P' = P_1' + P_2'$, where P_i' is the solution to the problem

$$\min P_i' = \sum_{t=1}^{M} P_{it}, \ \ i = 1, 2$$

$$\text{s.t} \sum_{t=1}^{M} \frac{1}{2} \log_2 \left(1 + \frac{\alpha_{ii}^t P_{it}}{\mathcal{N}_i^t} \right) \ge \tilde{R}_i, \ \ i = 1, 2$$

$$P_{it} \ge 0 \ \ i = 1, 2; t = 1, .., M$$

$$\sum_{t=1}^{M} A_{it} \le \mu_i, \ \ i = 1, 2$$

$$A_{it} = \begin{cases} 0 \ \text{if} \ P_{it} = 0 \\ 1 \ \text{otherwise} \end{cases}$$

$$(10)$$

P_j' is the solution to the problem of zero-interference scheduling of node j with variable (non-discrete) power levels and can be found using standard Lagrange multiplier techniques [12]. Thus P' is a lower bound for the minimum energy scheduling problem using discrete power levels. Now define

$q = \min_{i,t} \left\{ \frac{P'}{M}, \frac{\alpha_{ji}^t}{\alpha_{ji}^t} \left(2^{\epsilon \tilde{R}_i / M} - 1 \right) \right\}, i, j = 1, 2, \ 1 \le t \le M$. Let k be the largest solution to the equation $kq = 2 \ln kP$ such that

$$e/P < k \le \frac{2(2^{\epsilon \tilde{R}_i / M} - 1)}{q \left(1 + \epsilon - 2^{\epsilon \tilde{R}_i / M} \right)} \tag{11}$$

else set $k = 0$. For the given $\epsilon > 0$, choose $\delta_1 = \frac{\epsilon q}{2 + kq}$. If $k = 0$, let $r_0 = \lfloor \frac{2P}{q\epsilon} \rfloor$, otherwise $r_0 = \lceil \frac{2 + kq}{\epsilon kq} \rceil$. Let $s_0 = \lfloor \ln_{1 + k\delta_1} P / r_0 \delta_1 \rfloor$.

Allocate power to nodes in each time slot by dividing the total available power P into the following $L_t = r_0 + s_0 + 2$ discrete power levels.

$$P_r = \begin{cases} r\delta_1, & 0 \le r \le r_0 \\ (1 + k\delta_1)^{r - r_0} P_{r_0}, & r_0 + 1 \le r \le r_0 + s_0 \end{cases}$$
$$P_{r_0 + s_0 + 1} = P$$

$$\tag{12}$$

Lemma 2. *For given max power level P and constraints $\tilde{R}_i$, the number of discrete power levels per slot L_t is $O(\frac{1}{q\epsilon})$.*

Proof. Note that we are allocating power levels by dividing the range of available power into two types of intervals: the first r_0 intervals of fixed size δ_1 and remaining intervals of geometrically increasing size. Since geometric intervals are small in the beginning, the total number of power levels would be much larger using only geometrically increasing intervals. Therefore we use intervals of fixed size initially and choose integer r_0 such that the size of the first geometric interval, $k\delta_1^2 r_0$ is the same as the size of the previous fixed interval δ_1. The overall objective is to find optimal values of k and δ_1 that minimize the total number of power levels, yet allow us to closely approximate the overall energy consumption and rate constraints. From the energy approximation requirements (as shown below), we will get the constraint $\delta_1 = q\epsilon/(2 + kq)$. Hence $k\delta_1 < \epsilon$ and thus for small ϵ, the total number of levels $L_t = r_0 + s_0 = 1/(k\delta_1) + \ln_{1 + k\delta_1} kP$ can be approximated by $\frac{1 + \ln kP}{k\delta_1} = (1/\epsilon)(1 + \ln kP)(1 + 2/(kq))$. Thus the objective is to find k that minimizes L_t. The solution to this minimization is $\ln kP = kq/2$ subject to $\ln kP > 1$. If k does not satisfy these conditions then $\delta_1 = q\epsilon/2$ and the number of power levels is $\lceil \frac{2P}{q\epsilon} \rceil$.

The remaining constraints on k as specified in Eq. 11, are obtained from the rate approximation requirements shown below.

Theorem 3. *For small $\epsilon > 0$, let $\mathcal{A}^{\hat{P}}$ denote the modified version of the (exponential) dynamic programming algorithm $\mathcal{A}^P$ in which each node can select from discrete power levels per time slot as specified by Eq. 12, subject to overall duty-cycle and rate constraints $\tilde{R}_i(1 - \epsilon)$. Then $\mathcal{A}^{\hat{P}}$ is a $(1 + \epsilon, 1 + \epsilon)$-approximation of $\mathcal{A}^{P^*}$.*

Proof. Divide the set of time slots $T = \{1, 2, \ldots, M\}$ into disjoint sets T_{11} and T_{12} (resp. T_{21} and T_{22}) such that

$$t \in T_{11}(\text{resp. } T_{21}) \text{ if } P_{1t}^*(\text{resp. } P_{2t}^*) \in [0, r_0\delta_1]$$
$$t \in T_{12}(\text{resp. } T_{22}) \text{ if } P_{1t}^*(\text{resp. } P_{2t}^*) \in (r_0\delta_1, P]$$

(13)

Let $\hat{P}_{it}$ and $\hat{R}_{it}$ denote the (discrete) power levels and rate allocations per node per time slot under $\mathcal{A}^{\hat{P}}$. Since $\mathcal{A}^{\hat{P}}$ considers combinations of power levels over M slots, the errors in power levels and rate allocations per slot (either absolute or relative) must be bounded from above. Consider the solution in $\mathcal{A}^{\hat{P}}$ that simply rounds up the optimal power level in each slot to the nearest (larger) discrete power level. For this solution, the absolute error is bounded by $\hat{P}_{it} - P_{it}^* < \delta_1, \quad t \in T_{i1}$, and the relative error by $\hat{P}_{it} < (1 + k\delta_1)P_{it}^*, \quad t \in T_{i2}$, $i = 1, 2$. Therefore we have

$$\hat{P} = \sum_i \sum_{t \in T_{i1}} \hat{P}_{it} + \sum_i \sum_{t \in T_{i2}} \hat{P}_{it}$$
$$\leq P^* + \frac{q\epsilon \left(|T_{11}| + |T_{21}|\right)}{2 + kq} + \frac{kq\epsilon}{2 + kq} \sum_i \sum_{t \in T_{i2}} P_{it}^*$$
$$\leq P^* + \frac{2Mq\epsilon}{2 + kq} + \frac{\epsilon kq}{2 + kq} P^*$$

(14)

The overall relative error in energy P_{err}, of this solution $\hat{P}$ is defined as

$$P_{err} = \frac{\hat{P} - P^*}{P^*}$$

(15)

Therefore we can bound the relative error as

$$P_{err} = \frac{2\epsilon}{2 + kq} \cdot \frac{Mq}{P^*} + \frac{\epsilon kq}{kq + 2} \leq \epsilon$$

(16)

since $q \leq P'/M \leq P^*/M$ as P' is a lower bound for the optimal energy value P^*. Hence this particular solution of algorithm $\mathcal{A}^{\hat{P}}$ approximates the optimal energy value of the minimum energy schedule to within an ϵ factor.

To complete the proof, we just need to show that the above power allocation is also a feasible solution in terms of the rate constraints i.e the overall rates achieved by $\mathcal{A}^{\hat{P}}$ also approximate each rate constraint to within an ϵ factor. First consider the achieved rate $\hat{R}_{1t}$, for the case $t \in T_{21}$.

$$\hat{R}_{1t} \geq \frac{1}{2} \log_2 \left(1 + \frac{\alpha_{11}^t P_{1t}^*}{N_1^t + \alpha_{21}^t (P_{2t}^* + \delta_1)}\right)$$
$$\geq \frac{1}{2} \log_2 \left(1 + \frac{\alpha_{11}^t P_{1t}^*}{N_1^t + \alpha_{21}^t P_{2t}^*} \cdot \frac{1}{1 + \frac{\alpha_{21}^t \delta_1}{N_1^t + \alpha_{21}^t P_{2t}^*}}\right)$$

$$\geq R_{1t}^* - \frac{1}{2}\log_2\left(1 + \frac{\delta_1}{P_{2t}^* + \frac{\mathcal{N}_1^t}{\alpha_{11}^t}\cdot\frac{\alpha_{11}^t}{\alpha_{21}^t}}\alpha_{21}^t\right) \tag{17}$$

Using the fact that $P_{2t}^* \geq 0$, and the background noise $\mathcal{N}_1^t/\alpha_{11}^t \geq 1$ for each time slot $t \in T_{11}$, we can bound the absolute R_1 rate error $= R_1^* - \hat{R}_1$ over all such time slots by

$$\frac{M}{2}\log_2\left(1 + \max_t\left(\frac{\alpha_{21}^t}{\alpha_{11}^t}\right)\delta_1\right) \leq \frac{\epsilon\tilde{R}_1}{2}$$

by using the fact that $\delta_1 \leq \epsilon q \leq \min_t\left(\frac{\alpha_{11}^t}{\alpha_{21}^t}\right)\epsilon\left(2^{2\epsilon\tilde{R}_1/M} - 1\right)$.

Next, for $t \in T_{22}$ (when $k > 0$), we get

$$\hat{R}_{1t} = \frac{1}{2}\log_2\left(1 + \frac{\alpha_{11}^t\hat{P}_{1t}}{\mathcal{N}_1^t + \alpha_{21}^t\hat{P}_{2t}}\right)$$

$$\geq \frac{1}{2}\log_2\left(1 + \frac{\alpha_{11}^t P_{1t}^*}{\mathcal{N}_1^t + \alpha_{21}^t P_{2t}^*(1 + k\delta_1)}\right)$$

$$\geq \frac{1}{2}\log_2\left(1 + \frac{1}{1+k\delta_1}\cdot\frac{\alpha_{11}^t P_{1t}^*}{\frac{\mathcal{N}_1^t}{1+k\delta_1} + \alpha_{21}^t P_{2t}^*}\right)$$

Since $k\delta_1 \geq 0$, this implies

$$\hat{R}_{1t} \geq \frac{1}{2}\log_2\left(1 + \frac{\alpha_{11}^t P_{1t}^*}{\mathcal{N}_1^t + \alpha_{21}^t P_{2t}^*}\right) - \frac{1}{2}\log_2(1 + k\delta_1)$$

$$= R_{1t}^* - \frac{1}{2}\log_2(1 + k\delta_1) \tag{18}$$

Hence the total error in R_1 over all the time slots when $t \in T_{22}$ is at most $(M/2)\log_2(1+k\delta_1) \leq \epsilon\tilde{R}_1/2$ using the upper bound on k as specified in Eq. 11. Combining the two cases, the total absolute error in $R_1 = \tilde{R}_1 - \hat{R}_1 \leq \epsilon\tilde{R}_1$ and thus the relative error in R_1 is bounded by ϵ i.e. $\hat{R}_1 \geq \tilde{R}_1(1 - \epsilon)$. The analysis is identical for rate R_2. Since algorithm $\mathcal{A}^{\hat{P}}$ uses $\tilde{R}_i(1 - \epsilon)$ as the rate constraint for user i, therefore the choice of power levels described above is a feasible choice and hence the algorithm is a $(1 + \epsilon, 1 + \epsilon)$ approximation.

For the algorithm above, note that the number of discrete power levels per slot L_t, is a function of the channel quality parameters $\alpha_{ji}^t/\alpha_{ii}^t$. While the α's are exponentially distributed random variables with typically small means [18], the ratios can still be quite large, thereby increasing the number of levels. Therefore we consider a more optimal scheme where the rate and energy approximations are obtained independent of channel quality parameters.

Let $\tilde{R}_m = \min(\tilde{R}_1, \tilde{R}_2)$ and $k_1 = (M\log_2(1+P) - 2\tilde{R}_m)/\log_2\left(\frac{1+P}{1+1/k}\right)$. Define $\delta_1 > 0$ and $k > 0$ as the solutions to

$$\min \frac{1}{k\delta_1} + \ln_{1+k\delta_1} kP$$

$$\text{s.t } k_1\delta_1 + M\log_2(1 + k\delta_1) = 2\epsilon\tilde{R}_m$$

$$k > \frac{1}{2^{2\tilde{R}_m/M} - 1} \tag{19}$$

δ_1 and k can be obtained using standard constrained minimization techniques such as Lagrange multipliers [12]. However if no solution exists above, then δ_1 and k are the solutions obtained by replacing the constraints in Eq. 19 above by the constraint

$$\delta_1 + \log_2(1 + k\delta_1) = \frac{2\epsilon\tilde{R}_m}{M} \tag{20}$$

If no solution still exists, then $\delta_1 = \epsilon\tilde{R}_m/M$ and $k = (2^{\epsilon\tilde{R}_m/M} - 1)/\delta_1$. Now divide the available power per time slot into discrete power levels as specified by Eq. 12 using the δ_1 and k values above.

Theorem 4. *For $\epsilon > 0$, let $\mathcal{A}^{\overline{P}}$ denote the (exponential) dynamic programming algorithm for finding a minimal energy schedule using the discrete power levels defined above, subject to overall duty-cycle and rate constraints $\tilde{R}_i(1-\epsilon)$. Then $\mathcal{A}^{\overline{P}}$ is a $(1, 1+\epsilon)$-approximation of $\mathcal{A}^{P^*}$.*

Proof. For each slot t, round down the optimal power level choice P_{it}^* to the nearest discrete power level, represented by $\overline{P}_{it}$ and let $\overline{R}_{it}$ denote the corresponding achieved rate per slot. As before, divide the M time slots into sets T_{ij}, $i, j = 1, 2$, based on the value of P_{it}^*. We show below that $\overline{P}_{it}$ represents a feasible allocation of power levels under the rate constraints $\tilde{R}_i/(1-\epsilon)$. Hence $\mathcal{A}^{\overline{P}}$ is a $(1, 1+\epsilon)$-approximation since the total energy consumption of $\mathcal{A}^{\overline{P}}$ is at most $\sum\sum\overline{P}_{it} \leq \sum\sum P_{it}^*$.

First, for $t \in T_{12}$, using $\overline{P}_{1t} \geq P_{1t}^*/(1 + k\delta_1)$ and $\overline{P}_{2t} \leq P_{2t}^*$, we get

$$\overline{R}_{1t} \geq \frac{1}{2}\log_2\left(1 + \frac{\alpha_{11}^t P_{1t}^*}{(1 + k\delta_1)(\mathcal{N}_1^t + \alpha_{21}^t \overline{P}_{2t})}\right)$$

$$\geq R_{1t}^* - \frac{1}{2}\log_2(1 + k\delta_1) \tag{21}$$

Thus the absolute error in R_{1t} per time slot for this case is $\leq \frac{1}{2}\log_2(1 + k\delta_1)$.

Next, for $t \in T_{11}$, define the total interference, $\overline{I}_{1t} = (\mathcal{N}_1^t + \alpha_{21}^t \overline{P}_{2t})/\alpha_{11}^t$, and likewise I_{1t}^*, where $I_{1t}^* \geq \overline{I}_{1t} \geq 1$ (minimum total interference ≥ 1). Therefore we have,

$$R_{1t}^* - \overline{R}_{1t} \leq \frac{1}{2}\log_2\left(1 + \frac{P_{1t}^*}{\overline{I}_{1t}}\right) - \frac{1}{2}\log_2\left(1 + \frac{\overline{P}_{1t}}{\overline{I}_{1t}}\right)$$

Using the fact that $\ln x - \ln y < x - y$ for $x > y > 1$, we get $R_{1t}^* - \overline{R}_{1t} < (P_{1t}^* - \overline{P}_{1t})/2 \leq \delta_1/2$. Thus the absolute error in R_{1t} per time slot for this case is $\leq \delta_1/2$.

Combining the two cases, we can bound the overall rate error over M time slots as

$$T_{err} = \frac{|T_{11}|\delta_1}{2} + \frac{|T_{12}|\log_2(1 + k\delta_1)}{2} \tag{22}$$

For $\mathcal{A}^{\overline{P}}$ to be a $(1, 1 + \epsilon)$ algorithm, we must have $T_{err} \leq \epsilon\tilde{R}_1$. To finish the proof, note that the maximum R_1 rate we can obtain under this algorithm in any $t \in T_{12}$ is $\frac{1}{2}\log_2(1+P)$ and $\frac{1}{2}\log_2(1+r_0\delta_1) = \frac{1}{2}\log_2(1+1/k)$ in any $t \in T_{11}$. The maximum value of $|T_{12}|$ is M. (Clearly $\log_2(1+P)$ should be $\geq 2\tilde{R}_1(1-\epsilon)/M$, otherwise $\mathcal{A}^{\overline{P}}$ does not have a solution). However the maximum value of $|T_{11}|$ is $|T_{11}| \leq (M\log_2(1 + P) - 2\tilde{R}_1)/\log_2\left(\frac{1+P}{1+1/k}\right)$ if $\log_2(1 + 1/k) < 2\tilde{R}_1/M$ else $|T_{11}| \leq M$. When $|T_{11}|$ takes the first value, the total number of power levels per slot is minimized by choosing δ_1 and k as in Eq. 19, whereas in the second case it is minimized by Eq. 20. If both cases do not yield a solution then we set the two error components $\delta_1 = \log_2(1 + k\delta_1) = \epsilon\tilde{R}_m/M$ which makes the relative error over M slots $\leq \epsilon$ as desired.

Finally, we note that the worst-case values of k and $k\delta_1$ are $O(\epsilon\tilde{R}_m/M)$ and therefore

Theorem 5. *Given rate constraints $\tilde{R}_i$ and max power P, the number of discrete power levels per slot is $O(\frac{1}{\epsilon})$.*

Note that the time complexity of $\mathcal{A}^{\overline{P}}$ is still exponential. Using the fact that the number of power levels per slot required to closely approximate rate and energy constraints is $O(\frac{1}{\epsilon})$, we develop an FPAS in the next Section.

5   An FPAS for Rate Constraints

We now describe a simple Fully Polynomial Approximation Scheme that solves the minimum energy scheduling problem by using a β-relaxation on the rate constraints for some arbitrary constant $\beta > 0$. For clarity, we describe the FPAS using two power levels 0 and P per time slot. The algorithm for the multiple power level case is a simple extension as described later.

The FPAS solves the same restricted problem of Eq. 4 with only each rate constraint replaced by

$$\sum_{t=1}^{M} R_{it} \geq (1 - \beta)\tilde{R}_i \quad i = 1, 2 \tag{23}$$

For any $\delta > 0$, define the following

Definition 1. *A rate vector $<R_1, R_2>$ δ-dominates another vector $<R_3, R_4>$ iff either $R_3(1-\delta) \leq R_1 \leq R_3$ and $R_2 \geq R_4$ or $R_3 \leq R_1(1-\delta)$ and $R_4(1-\delta) \leq R_2$. For $R_1 \geq \tilde{R}_1$, the δ-dominant vector is the one with $\max R_2$ among all such vectors.*

Note that dominance (under standard vector comparison) implies δ-dominance but not vice-versa.

Definition 2. *Let $\bar{R}$ be a set of rate vectors. Define the operation vector-maxdelta($\bar{R}$) as one that eliminates all δ-dominated and dominated vectors from $\bar{R}$.*

Operation vectormaxdelta is equivalent to dividing the two-dimensional vector space into horizontal and vertical strips, each of whose left endpoint is $(1-\delta)$ times its right endpoint and choosing at most one vector per strip. A simple algorithm for implementing vectormaxdelta($\bar{R}$) is as follows. Assume $\bar{R}$ has been sorted by R_1 values. First obtain the δ-dominant vector for $R_1 \geq \tilde{R}_1$ if such R_1's exist. Then find the δ-dominant vectors successively in the strips defined by R_1 intervals $(\tilde{R}_1(1-\delta), \tilde{R}_1]$, $(\tilde{R}_1(1-\delta)^2, \tilde{R}_1(1-\delta)]$ $(\tilde{R}_1(1-\delta)^3, \tilde{R}_1(1-\delta)^2]$ and so on. Dominated vectors are eliminated simultaneously. Since $\bar{R}$ has been sorted by R_1, this can be done in one pass through $\bar{R}$, in decreasing order of R_1 values.

Choose $\delta = \frac{\beta}{2M}$. Let $\mathcal{A}_\beta^P$ denote the following dynamic programming algorithm for the fixed power minimum energy scheduling problem. The boundary conditions (i.e rate vectors for each slot t) are the same as before in Eq. 5. The main recursive step in the algorithm is derived by replacing the vectormax operation with vectormaxdelta. Let $\hat{R}_{i,j}^{kP,a,b}$ represent the set of δ-dominating rate pairs corresponding to cumulative transmission rates for user 1 and user 2 from time slots i through j, $1 \leq i \leq j \leq M$, while using a total power of kP, $1 \leq k \leq 2M$.

$$\hat{R}_{i,j}^{kP,a,b} = \text{vectormaxdelta}\Big\{ \hat{R}_{i,j-1}^{kP,a,b} \bigcup \Big(\hat{R}_{i,j-1}^{(k-1)P,a-1,b} + \hat{R}_j^{P,1,0} \Big)$$
$$\bigcup \Big(\hat{R}_{i,j-1}^{(k-1)P,a,b-1} + \hat{R}_j^{P,0,1} \Big) \bigcup \Big(\hat{R}_{i,j-1}^{(k-2)P,a-1,b-1} + \hat{R}_j^{2P,1,1} \Big) \Big\} \quad (24)$$

The terminating condition for the algorithm occurs when the rate vectors are $\geq \tilde{R}_i(1 - \beta)$, $i = 1, 2$. The optimal schedule corresponds to the minimum total power rate vector that satisfies the terminating condition.

Theorem 6. *$\mathcal{A}_\beta^P$ is a FPAS for the minimum energy scheduling problem with two fixed transmit power choices 0 or P per slot.*

Proof. First we show that the running time of $\mathcal{A}_\beta^P$ is polynomial in $1/\beta$. The number of δ-dominant vectors in $\hat{R}_{i,j-1}^{kP,a,b}$ is bounded by

$$1 + \ln_{1+\delta} \tilde{R}_1 = 1 + \frac{\ln \tilde{R}_1}{\ln(1 + \delta)} = O\left(\frac{M}{\beta} \cdot \ln \tilde{R}_1 \right)$$

since we keep only one vector for each $1-\delta$-factor interval. and using $1/(1 - \delta) = 1 + \delta$. The running time for the creation of each $\hat{R}_{i,j}^{kP,a,b}$ is also polynomial since it includes sorting followed by the vectormaxdelta operation. There are $O(MP\mu_1\mu_2)$ such rate vector sets, each of size polynomial in $1/\beta$ and hence the overall running time is also polynomial in $1/\beta$.

Next we need to show that algorithm $\mathcal{A}_\beta^P$ provides a β-approximation of the rate constraints. Let $<R_1, R_2> \in \bar{R}_{1,j}^{kP,a,b}$ be an arbitrary non-dominated vector from the exponential time algorithm $\mathcal{A}^P$ up to time slot j. We can show by induction that $\exists <R_3, R_4> \in \hat{R}_{1,j}^{kP,a,b}$ such that $R_3 \geq R_1(1-\delta)^j$ and $R_4 \geq R_2(1-\delta)^j$. The 'parent' of $<R_1, R_2>$ (the vector that produced $<R_1, R_2>$ in stage $j-1$) is approximated within $(1-\delta)^{j-1}$ by the induction hypothesis. After combining with the vectors of stage j and implementing vectormaxdelta, at most a further $(1-\delta)$-factor error in R_1 and R_2 is introduced. Thus the total error in each dimension is bounded by $(1-\delta)^j$ after j slots. Therefore every rate vector in $\bar{R}_{1,M}^{kP,\mu_1,\mu_2}$ is approximated to within $(1-\delta)^M$ by a rate vector from algorithm $\mathcal{A}_\beta^P$. Using $\delta = \beta/2M$, we can see that there exist 'approximate' rate vectors $<R_3, R_4> \in \hat{R}_{1,M}^{kP,\mu_1,\mu_2}$ such that $R_3 \geq R_1(1-\beta)$ and $R_4 \geq R_2(1-\beta)$ for all 'actual' rate vectors $<R_1, R_2> \in \bar{R}_{1,M}^{kP,\mu_1,\mu_2}$. Hence $\mathcal{A}_\beta^P$ is a β-approximation.

Algorithm $\mathcal{A}_\beta^P$ above can be easily modified to incorporate multiple power levels per slot. For any small $\alpha > 0$, choose $\epsilon = \beta = \alpha/2$ and then set δ_1 and k as per Eq. 19 with L_t power levels per user per slot. Eq. 5 is modified to reflect $(L_t)^2 = O(1/\alpha^2)$ (from Theorem 5) total rate vectors per time slot t, corresponding to all combinations of power levels. Define a new algorithm $\mathcal{A}_\beta^{P_{L_t}}$ in which the vectormaxdelta operation applies to combinations of these $(L_t)^2$ rate vectors. The total number of table entries (for rate vectors) in the modified dynamic program is now increased to $(L_t)^2 M P \mu_1 \mu_2$. However by applying the vectormaxdelta operation, the size of each rate vector set remains the same size, $O(1/\beta)$, as before.

Theorem 7. *For any $\alpha > 0$ and $\epsilon = \beta = \alpha/2$, $\mathcal{A}_\beta^{P_{L_t}}$ is a $(1, 1+\alpha)$-Fully Polynomial Approximation Scheme for the minimum energy scheduling problem with L_t power levels per slot.*

Proof. By choosing multiple power levels as defined above, each rate vector is no more than a $1 - \epsilon = (1 - \alpha/2)$-factor away from the ideal rate vector for that stage. For each such vector, the vectormax operation selects another which is at most another $1 - \alpha/2$-factor away. Thus at the end of algorithm $\mathcal{A}_\beta^{P_{L_t}}$, the rate constraints are violated by at most a factor of $(1 - \alpha/2)^2 < (1 - \alpha)$. For given M, P, μ_1 and μ_2, the total number of table entries and related operations is $O(1/\alpha^2)$ and hence $\mathcal{A}_\beta^{P_{L_t}}$ is a $(1, 1+\alpha)$ FPAS.

Finally, we note that the 2-factor approximation of Section 3 that finds a minimal energy schedule corresponding to optimal transmit power P_{opt} can be improved by using $\mathcal{A}_\beta^{P_{L_t}}$ instead of the exponential $\mathcal{A}^P$. We increase P by a factor of $1 + k\delta_1 = 1 + \alpha$ in each iteration rather than doubling as in Theorem 2. Unlike the two fixed transmit powers case, $E_{\mathcal{A}_\beta^{P_{L_t}}}^{P(1+k\delta_1)} \leq E_{\mathcal{A}_\beta^{P_{L_t}}}^P$ since the former contains the schedule of the latter as a subset. The other arguments of Theorem 2 remain valid and by outputting the lowest energy value from the iterations, we obtain a $(1 + \alpha, 1 + \alpha)$-approximation algorithm that finds the optimal maximum transmit power level P_{opt} and the corresponding minimum energy schedule in $O(\log_{1+k\delta_1} P_{opt}/P_{min}) = O(\frac{1}{\alpha} \cdot \frac{P_{opt}}{P_{min}}) = O(\frac{M}{\alpha})$ iterations (since $P_{opt} \leq P_{max} \leq M P_{min}$), where each iteration takes time $O(1/\alpha^2)$. Since this is the solution to the unrestricted MESP problem of Eq. 2, we have

Theorem 8. *There is a $(1 + \alpha, 1 + \alpha)$ FPAS for solving the unrestricted MESP problem.*

6 Multiple Node Case

Even with N nodes, the number of discrete power levels L_t, required to approximate each nodes rate and overall energy within a $(1+\alpha)$-factor, remains the same as defined by Eq. 12 and Eq. 19 since the arguments of Theorem 4 apply even with interference from multiple nodes. Hence each node can select from $O(1/\alpha)$ power levels per slot. Even with only 2 power levels, the number of rate vectors per slot is 2^N, and in general $O(1/\alpha)^N$. However, we can extend the preceding algorithm to the multiple node case by defining δ-dominance for N-tuple rate vectors. If the number of users is treated as a fixed constant N, this extended algorithm is still an FPAS since 1) the number of rate vectors per slot t is polynomial in $1/\alpha$ and 2) the size of each table entry (corresponding to the rate vector set upto the jth slot) is $O\left(\left(\frac{M \ln \tilde{R}_m}{\alpha}\right)^{N-1}\right)$, where $\tilde{R}_m = \min_i\{\tilde{R}_i\}$, since the number of δ-dominant vectors in the smallest dimension is $O((M \ln \tilde{R}_m)/\beta)$ and we are considering dominant vectors over an N-dimensional hypercube of vector elements.

7 Conclusions

We have considered the problem of finding a minimum energy transmission schedule for duty-cycle and rate constrained wireless sensor nodes. Since traditional optimization methods using Lagrange multipliers are computationally expensive given the non-convex constraints, we develop fully polynomial time approximation schemes by considering restricted versions of the problem using discrete power levels. We derive an $(1+\epsilon, 1+\epsilon)$-FPAS for MESP that approximates the optimal energy consumption and rate constraints to within an $1 + \epsilon$-factor.

References

1. Singh, S., Raghavendra, C.: Pamas: Power aware multi-access protocol with signalling for ad hoc networks (1999)
2. Ye, W., Heidemann, J., Estrin, D.: An energy-efficient mac protocol for wireless sensor networks (2002)
3. Ye, W., Heidemann, J., Estrin, D.: Medium access control with coordinated, adaptive sleeping for wireless sensor networks (2003)

4. Uysal-Biyikoglu, E., Prabhakar, B., El Gamal, A.: Energy-efficient packet transmission over a wireless link. IEEE/ACM Transactions on Networking **10** (2002) 487– 499
5. Uysal-Biyikoglu, E., Gamal, A.E.: On adaptive transmission for energy efficiency in wireless data networks. IEEE Trans. Inform. Theory (2004)
6. Hanly, S., Tse, D.: Power control and capacity of spread-spectrum wireless networks. Automat. **35**(12) (1999) 1987–2012
7. Wang, K., Chiasserini, C., Rao, R., Proakis, J.: A distributed joint scheduling and power control algorithm for multicasting in wireless ad hoc networks. In: IEEE International Conference on Communications, 2003. ICC '03. Volume 1. (2003) 725–731
8. ElBatt, T., Ephremides, A.: Joint scheduling and power control for wireless ad hoc networks. IEEE Transactions on Wireless Communications **3** (2004) 74–85
9. Foschini, G.J., Miljanic, Z.: A simple distributed autonomous power control algorithm and its convergence. IEEE Transactions on Vehicular Technology (1993) 641–646
10. Bambos, N.: Toward power-sensitive network architectures in wireless communications: concepts, issues, and design concepts. IEEE Personal Communications (1998) 50–59
11. Bambos, N., Kandukuri, S.: Power control multiple access (pcma). Wireless Networks (1999)
12. Bertsekas, D.P.: Nonlinear Programming. Second edn. Athena Scientific, Belmont, Massachusetts (1999)
13. Bertsekas, D., Lauer, G., Sandell, N., Posbergh, T.: Optimal short-term scheduling of large-scale power systems. IEEE Transactions on Automatic Control (1983) 1–11
14. Dorit Hochbaum, E.: Approximation Algorithms for NP-Hard Problems. First edn. PWS Publishing Company, Boston, MA (1997)
15. Martello, S., Toth, P.: Knapsack Problems. First edn. J. Wiley and Sons, Chichester (1990)
16. Kannan, R., Wei, S., Chakravarthi, V., Seetharaman, G.: Using misbehavior to analyze strategic versus aggregate energy minimization in wireless sensor networks (2006)
17. Kannan, R., Wei, S.: Lsu-cs-tr-06-3. Technical report, LSU (2006)
18. Cover, T.M., Thomas, J.A.: Elements of Information Theory. First edn. Wiley, New York (1991)

MobiRoute: Routing Towards a Mobile Sink for Improving Lifetime in Sensor Networks⋆

Jun Luo, Jacques Panchard, Michał Piórkowski, Matthias Grossglauser, and Jean-Pierre Hubaux

School of Computer and Communication Sciences
Ecole Polytechnique Fédérale de Lausanne (EPFL)
CH-1015 Lausanne, Switzerland
`firstname.lastname@epfl.ch`

Abstract. Improving network lifetime is a fundamental challenge of wireless sensor networks. One possible solution consists in making use of mobile sinks. Whereas theoretical analysis shows that this approach does indeed benefit network lifetime, practical routing protocols that support sink mobility are still missing. In this paper, in line with our previous efforts, we investigate the approach that makes use of a mobile sink for balancing the traffic load and in turn improving network lifetime. We engineer a routing protocol, MobiRoute, that effectively supports sink mobility. Through intensive simulations in TOSSIM with a mobile sink and an implementation of MobiRoute, we prove the feasibility of the mobile sink approach by demonstrating the improved network lifetime in several deployment scenarios.

1 Introduction

Many proposals on using mobile sinks to improve the lifetime of *wireless sensor networks* (WSNs) have appeared recently [1, 2, 3, 4, 5, 6, 7, 8, 9]. However, the doubt that moving sinks is practical persists in the research community (e.g., [10]). One of the major concerns behind this doubt is that mobility inevitably incurs additional overhead in data communication protocols and the overhead can potentially offset the benefit brought by mobility. In this paper, we intend to dismiss the doubt.

We focus on a scenario where all nodes are fixed and have limited energy reserves and where a mobile sink endowed with significantly more resources serves as the data collector. In this scenario, the sink mobility can increase network lifetime through two different methods, depending on the relationship between the sink moving *speed* and the tolerable *delay* of the data delivery.

In the *fast mobility* regime, the speed produces tolerable data delivery delay. The WSNs may then take advantage of *mobility capacity* [11]. This *mobile relay*

⋆ The work presented in this paper was supported (in part) by the National Competence Center in Research on Mobile Information and Communication Systems (NCCR-MICS), a center supported by the Swiss National Science Foundation under grant number 5005-67322. (`http://www.terminodes.org`)

P. Gibbons et al. (Eds.): DCOSS 2006, LNCS 4026, pp. 480–497, 2006.
© Springer-Verlag Berlin Heidelberg 2006

approach [1, 2, 3] uses the mobile sink to transport data with its mechanical movements. It trades data delivery latency for the reduction of node energy consumption. We refer to [3] and [4] for simulations and field studies in this regime. In the *slow mobility* regime, the sink mobility takes a **discrete** form: the movement trace consists of several *anchor points* between which the sink moves and at which it pauses. Consequently, the network cannot benefit from mobility capacity. However, it has recently been observed [5, 6, 7, 8, 9] that sink mobility can still improve network lifetime. The reason is that the typical many-to-one traffic pattern in WSNs imposes a heavy forwarding load on the nodes close to sinks. While no energy conserving protocol alleviates such a load, moving the sink (even very **infrequently**) can distribute over time the role of bottleneck nodes and thus even out the load. Unfortunately, theoretical analysis [5, 6, 7, 8, 9] may produce misleading results due to its simplified system model (an example is given in Section 5: footnote 4); simulations involving a detailed protocol implementation are necessary to fully understand the benefit of using mobile sinks.

We argue that the slow mobility conditions exist in many realistic applications of WSNs. For example, suppose that a WSN is equipped with batteries that cannot be replaced, e.g., because the sensor nodes are not accessible, or because changing batteries would be hazardous or costly. This may be the case for sensors in smart buildings, where batteries might be designed to last for decades, and for environmental or military sensing under hostile or dangerous conditions (e.g., avalanche monitoring). In this case, it may be desirable and comparatively simple to move a sink infrequently (e.g., once a day or a week) by a human or by a robot. For example, in the avalanche monitoring scenario, a sink may be deployed at the periphery[1] of the monitored area and then moved once in a while by a helicopter. In the building scenario, the sink can be "virtually" moved: computers in different offices serve as the sink in shifts. In the military surveillance scenario, moving the sink may require some effort, but it can be acceptable if done infrequently.

Numerous routing protocols have been proposed in the last decade to support data communications in either mobile ad hoc networks (MANETs) or WSNs. On one hand, the protocols for MANETs (e.g., [12, 13]) are definitely an overkill for supporting mobile sinks because the basic assumption in MANETs is that every node moves in an unpredictable way. On the other hand, the protocols for WSNs usually take static sinks for granted (e.g., [14, 15], although some exceptions exist [16, 17, 18, 3]). It is true that a routing protocol that supports mobile sinks to collect data in WSNs, compared with existing protocols for static WSNs, will have higher protocol complexity and overhead. However, the following favorable features of the sink mobility (in the low mobility regime) and of the existing routing protocols help to limit the side effects:

- The mobility is controllable and thus predictable,
- The pause time of a sink along its moving trace is much longer than the actual moving time,

---

[1] As we have shown in [8], the optimum trace (in terms of network lifetime) for a mobile sink is the network periphery.

– Existing routing protocols usually possess proactive features to cope with
  the dynamics in link quality; this can be exploited to support sink mobility.

In the spirit of [8] (where we theoretically prove the superiority of mobile
sinks over static ones), we further investigate in this paper the performance,
with respect to both lifetime and reliability (measured by packet delivery ratio),
of WSNs with a mobile sink. We consider a scenario where nodes periodically
transfer data through multi-hop routes towards the sink, while the sink intermittently changes its position according to certain predefined traces. We propose a
routing protocol, MobiRoute, dedicated to support sink mobility. It takes into
account the favorable features we mentioned above and thus only marginally
increases the protocol complexity and overhead. We use TOSSIM [19] as the
simulator. Our simulation results demonstrate the efficiency of MobiRoute, in
terms of both an improved network lifetime and an undegraded reliability, in
several deployment scenarios. Our contribution with respect to [5, 6, 7, 8, 9] are:

– Our investigation is based on a practical routing protocol. Consequently, we
  take into account realistic conditions such as control overhead with a routing
  protocol and collision/overhearing [20, 21] at the MAC layer.
– We also look at the reliability issue incurred by sink mobility, which is not
  considered by the numerical simulations of the previous work.

The rest of this paper is organized as follows: Section 2 surveys related work.
Section 3 clarifies our metrics and methodologies. Section 4 presents our MobiRoute protocol. Section 5 describes the algorithm that controls the sink mobility in an adaptive way. Simulation results are reported in Section 6. Finally,
Section 7 concludes the paper.

2   Related Work

In this section, we briefly survey the existing routing protocols that are designed
to support sink mobility. Whereas the previous proposals [5, 6, 7, 8, 9] serve as
the theoretical basis of this paper, we will not discuss them further because this
paper focuses on the practicality of sink mobility instead of theoretical analysis.
We refer to Section 1 for the principle of these proposals.

There are a few proposals for data collection with mobile sinks [16, 17, 18, 3].
While most of them [16, 17, 18] consider sink mobility as an inherent behavior of
WSNs and try to **cope with** it, only [3] share the same opinion of **exploiting**
controllable sink mobility as ours.

The *two-tier data dissemination* (TTDD) approach [16] is actually based on
the existing idea of virtual backbone. The backbone (or grid in TTDD's terminology) is formed proactively upon detection of a stimulus. The mobile sinks
send queries to the nearest grid points with a flooding. Queries are routed along
the grid and data trace the reverse path back to the sinks. As a consequence,
the control overhead introduced by sink mobility is limited to the grid cell where
a sink is located. Aiming at further limiting the widespread diffusion of control

messages introduced by sink mobility, the *scalable energy-efficient asynchronous dissemination* (SEAD) protocol [17] assigns particular nodes as the *access nodes* and relies on such fixed anchors to limit the control traffic. Mobile sinks only need to select one of their neighbors as an access node and maintain the link with it. SEAD constructs an energy-efficient dissemination tree from a source to difference access nodes. Data are routed along the tree and then unicast from the access nodes to their sinks. To handle mobility, sink may handoff from one access node to the other if the tradeoff between the energy consumed to build another tree and the data delivery latency exceeds a given threshold. Both TTDD and SEAD heavily rely on the assumption of location-aware sensor nodes, which are not required in our case. Moreover, the data generation model is also different from ours. Under our data generation model (where every node produces data with the same rate), the virtual backbone used in TTDD to carry most traffic would become fixed and the delayed handoff in SEAD could lead to suboptimal routing trees for a substantial amount of time; they both offset the load balancing effect resulting from sink mobility.

The *hybrid learning-enforced time domain routing* (HLETDR) proposed in [18] comes a bit closer to our situation. They assume a somewhat fixed sink trace and the same data generation model as ours. However, instead of requiring the sink to inform other nodes about its location changes, HLETDR lets sensor nodes to figure out the sink trace through a learning-based approach. This learning process is done through positive and negative reinforcements according to the probability density function indicating how far the sink is from the nodes close to the sink trace. Tour period of the sink is divided into m domains. When a node has data to be forwarded, the probability of selecting next hop is determined according to its possibility being on a shortest path to the sink in that time domain. HLETDR has a high complexity of the routing table, i.e., $O(m)$, when a fine time granularity is required. In addition, the relatively slow learning procedure limits the adaptability of the sink mobility (which is very crucial for load balancing in our case).

The routing protocol described in [3] is based on *directed diffusion* [14], a routing protocol dedicated to *data-centric*[2] communications. [3] extends directed diffusion by adding mainly three components: 1) a pre-move phase for the nodes to learn the sink trace, 2) an acknowledgement-retransmit scheme to handle packets loss during handoff, and 3) a pre-fetch mechanism to improve the data delivery. The data generation model assumed in [3] is quite different from ours: nodes send data only when the data are queried whereas nodes proactively push data to the sink in our model. In addition, the sink moves continuously in [3] because their protocols work in the fast mobility regime (see Section 1 for details). Therefore, the control objective is the moving speed for [3], whereas we consider the adjustment of pause times. In this paper, we take an approach similar to that of [3] in that, instead of developing a routing protocol from scratch, we extend an existing routing protocol with the ability of handling sink mobility.

---

[2] Data generated by sensor nodes are named by attribute-value pairs. A node requests data by sending interests for named data rather than for named nodes.

3   Problem, Metrics and Methodology

We define network lifetime as the time period for the first node to run out of
its energy reserve [22]. When evaluating this quantity, we convert the problem
of maximizing network lifetime to a min-max problem in terms of the **radio**
energy consumption of individual nodes. Another performance index we want to
evaluate is the packet delivery ratio (or *reliability*). In fact, a possible side-effect
brought by sink mobility could be an increase in packet loss due to occasional
topology changes; the lifetime elongation resulting from sink mobility is justifi-
able only if the increase in packet loss is tolerable.

We assume that nodes generate data and send them to the sink with the same
rate. In our approach, the mobility pattern of a sink takes a *discrete* form [6]: the
moving trace consists of several *anchor points* between which the sink moves and
at which the sink pauses. We require each *epoch* (the time during which the sink
pauses) to be much longer than the moving time, such that the routing overhead
introduced by sink mobility becomes negligible due to its amortization across a
long epoch. Imposing these anchor points simplifies the design of the mobile sink[3]
and limits the extra overhead introduced to the routing protocol (see Section 4
for details). In addition, a continuous movement is not necessary, as a granularity
of (sink) displacement smaller than the magnitude of the effective radio range
may not lead to any topological change (whereas topological changes are what
we expect from the sink mobility). In order to better adapt to the topology and
dynamics of a given network, we also intend to control the sink mobility *on-line*
(based on the *off-line* optimization described in [8]).

Our experiment methodology involves simulations with TOSSIM [19]. The
main benefit of using the TOSSIM simulator is that the protocol used for simulations can be directly adopted by real sensor nodes. We simulate a set of
networks with nodes on 4×4, 5×5, and 7×7 point lattices; these scenarios represent outdoor WSNs in general. We also simulate a network that we intend to
deploy as an in-building testbed.

4   MobiRoute: Routing Towards a Mobile Sink

According to the definition of discrete mobility pattern described in Section 3,
the sink changes its location from time to time. A routing protocol that transfers
data towards such a sink should perform the following operations that are not
needed for traditional WSNs:

1. Notify a node when its link with the sink gets broken due to mobility.
2. Inform the whole network of the topological changes incurred by mobility.
3. Minimize the packet loss during the sink moving period.

Operation 1 seems to be encompassed by 2, but the level of urgency is different.
Packets forwarded by a last-hop node will get lost if the node does not detect the

---

[3] The mobile sink can simply be a laptop (moved occasionally by a human), rather
than a sophisticate robot as used in [3].

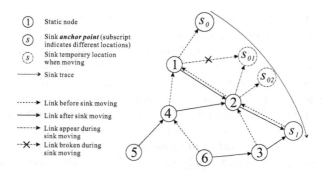

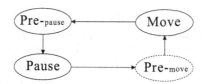

Fig. 1. This example illustrates possible scenarios where additional operations are necessary. Assuming the sink, after its (long) pause at s_0, moves to s_1, (1) the link breakage happening when the sink reaches intermediate location s_{01} (where it loses connectivity with node 1) should be notified to node 1, otherwise the node will have to drop packets sent from other nodes, (2) nodes 3, 4, and 6 should be informed about the topological changes at a proper time, otherwise, for example, 6 might take the following sub-optimal routing path: $6 \rightarrow 4 \rightarrow 1 \rightarrow 2 \rightarrow s$.

Fig. 2. States and transitions involved in MobiRoute. Note that only the protocol running at the sink side has the *pre-move* state.

link breakage, while a remote node can still send its data to the sink successfully without knowing the topological changes. However, the routing optimality is compromised without operation 2. It is not possible to avoid packet loss, because a realistic failure detector (which usually relies on a timer) always has some delay. Therefore, the goal of operation 3 is to minimize rather than eliminate packet loss. Possible scenarios related to these operations are illustrated in Fig. 1.

Our routing protocol, MobiRoute, is a **superset** of Berkeley MintRoute [15]. MobiRoute extends MintRoute by adding functions that perform the aforementioned operations. We first introduce MintRoute briefly in Section 4.1, then we describe the extended functions of MobiRoute separately in Sections 4.2 to 4.4. The state diagram shown in Fig. 2 is used when we present MobiRoute.

4.1 MintRoute

Berkeley MintRoute [15] is a routing protocol designed specifically for the all-to-one data transmission style of WSNs. It takes a distributed distance-vector based approach: route messages (i.e., control packets) are exchanged periodically among neighbor nodes, and the next hop nodes (or *parents* in MintRoute nomenclature) are chosen by evaluating the costs of routing data through different neighbors. The exchanged route messages not only help to measure the distance (in terms of the number of possible transmissions) from the sink but also provide a way to evaluate the link qualities (from both directions) between nodes. As a result, MintRoute applies a **Minimum Transmission** (MT) metric, where the goal is to minimize the total number of transmissions (including retransmissions). Since the data rate in WSNs is low, route messages do not need to be exchanged frequently (the rate is actually a multiple of the data rate in MintRoute). This helps MintRoute to reduce its energy consumption. Although MintRoute does not explicitly apply a metric that considers load balancing, the protocol, according to our experience, balances the traffic load with occasional switches of nodes' parents (which is a direct consequence of the MT metric). This feature makes MintRoute a leading candidate for supporting sink mobility. Finally, MintRoute applies a sequence number for each packet to detect packet loss and thus evaluate link quality; this sequence is shared by both control and data packets.

4.2   Detecting Link Breakage

In order to inform the nodes located close to the sink trace about the state of their links with the sink, MobiRoute applies a beacon mechanism. The sink, during the whole moving period, periodically broadcasts a beacon message (*s-beacon* hereafter). A node, upon receiving a s-beacon, sets (or resets) its **detecting** timer. If the timer times out before receiving the next s-beacon, the failure detector at this node indicates a link breakage and a new parent is chosen (which is taken care by MintRoute). We now discuss several crucial points of this seemingly simple mechanism.

First, we require the sink to transit from the **pause** state to the **pre-move** state before physically beginning to move. The sink begins to broadcast s-beacons under the pre-move state and evolves to the **move** state after a while. The sink moves while broadcasting s-beacons under the move state. A node, after receiving the first s-beacon under its current **pause** state, transits to the **move** state directly. Nevertheless, the pre-move state (of the sink) is necessary: it guarantees the reception of s-beacons at the nodes' side before the link quality changes due to the sink mobility.

Secondly, although only the sink (whose energy reserve is abundant) spends energy to send s-beacons, nodes also spend energy to receive these beacons. Therefore, the frequency of s-beacons should not be too high. On the other hand, low frequency sending retards failure detection, which in turn increases packet loss. We apply a simple heuristic: the frequency is set in the same order as the accumulative packet sending rate. For example, if the sending rate of each node is 1 pkt/min in a 60-node network, the accumulative rate at a last-hop node is at most 1 packet/second, and the beacon frequency is set to 1Hz. A related parameter is the timeout value for the detecting timer. Fortunately, the value can be relatively small, because a node will detect a false-positive when receiving another s-beacon.

Finally, the beacon mechanism is a costly procedure, regardless of which beacon frequency is chosen. Fortunately, since the moving period accounts only for a small fraction of the network lifetime, its costs will be amortized across the lifetime. A continuous sink movement, on the contrary, would incur such costs permanently.

4.3   Conveying Topological Changes

MobiRoute could have relied on MintRoute to propagate the topological changes resulting from sink mobility. However, the rate of route message exchanges in MintRoute is very low. Therefore, it takes a long time to convey the topological changes to the whole network; during this period, many packets are routed through sub-optimal paths, which consumes additional energy and thus offsets the benefit of sink mobility. As a result, MobiRoute needs a speed-up (route message exchange) rate for propagating the topological changes.

Propagating information throughout a network is a costly procedure (message complexity $O(n)$); it cannot be performed frequently. So MobiRoute only performs a propagation upon the sink reaching an anchor, and it tolerates a limited number of sub-optimal routing during the moving period. The sink enters the **pre-pause** state (see Fig. 2) when it stops moving; it then sends route messages with a speed-up rate, which causes their receivers to enter the same state. Nodes that receive messages **directly** from the sink also send speed-up route messages; they re-evaluate the quality of their links with the sink using these exchanges. A node receiving speed-up messages **indirectly** also enters the pre-pause state; it forwards the message only if its distance towards the sink changes significantly (e.g., node 6 in Fig. 1 might not forward messages received from node 3). The energy consumption of the propagation procedure is effectively reduced, because there are nodes that are not affected for a given move of the sink. Every node (including the sink) in the pre-pause state transits to the **pause** state after a short time span controlled by a timer.

4.4   Minimizing Packet Losses

Although packet loss cannot be avoided during the sink moving period due to the lag of the failure detector, there are ways to mitigate the losses. Taking advantage of having a very short moving period (which we would not have if the sink moved continuously), the protocol tries to reduce the sending rate of the last-hop nodes, by asking them to buffer data packets using the interface queue (QueuedSend module) in MintRoute. We also add the following command to QueueControl interface, such that the routing module can access the interface queue to change the next-hop address of the buffered packets upon detecting a link failure.

Nodes can only buffer data packets; control packets should still be sent. However, if we simply picked up control packets from the interface queue and sent them, there would be gaps among the sequence numbers (remember that MintRoute applies the same sequence for both control and data packets). These

```
command void QueueControl.setAddrInQueue(uint16_t parent)
{
    uint16_t i;
    if (!fQueueIdle)
        for (i = dequeue_next; i != enqueue_next;
             i = (i + 1) % MESSAGE_QUEUE_SIZE)
            msgqueue[i].address = parent;
}
```

gaps would mislead a neighbor node about a degradation in the link quality. Two solutions can be applied: 1) using separate sequences and queues for data and control packets or 2) rearranging the sequence number within the queue, such that packets sent have consecutive sequence numbers. In the short term, we adopt the second solution because it is easy to implement, but the first could be desirable in a long-run perspective.

5   Adaptively Controlled Mobility

According to our simulation results in Section 6, a mobility strategy that adapts to the network topology (for which no a priori knowledge exists) performs better than a static schedule. In this section, we describe the adaptive algorithm to control sink mobility. Our algorithm adaptively changes the epoch of the sink at each anchor point, according to the power consumption profile of the network. We derive the algorithm from the following linear program:

$$\text{Maximize lifetime } T = \sum_k t_k \tag{1}$$

$$\text{Constraints} \sum_k t_k \mathbb{P}_k \leq \mathbb{E} \tag{2}$$

where $\mathbb{P}_k$ and $\mathbb{E}$ are vectors that represent the power consumptions of each node (referred to as *P-profile* hereafter) when the sink pauses at a certain anchor point k and the initial energy reserves of all nodes, respectively. This formulation basically means that we weigh, through the epoch t_k, the anchor points based on the corresponding P-profile $\mathbb{P}_k$s, in such a way that the $\mathbb{P}_k$s that complement each other are favored. Although this LP formulation is similar to what is described in [6, 7], our contribution lies in the fact that we define the $\mathbb{P}_k$s by instrumenting the prototype of a routing protocol, while [6, 7] only manipulate flows on a graph[4].

In practice, we propose the following 2-phase algorithm to approximate the above programming problem:

---

[4] In fact, manipulating flow without bearing in mind the behaviors of a realistic routing protocol may leads to a misleading conclusion. For example, the formulation in [6] tries to maximize the lifetime by considering only a subset of flow (which is chosen to simplify the problem). However, this biased choice produces a significant deviation (in terms of the anchor points) from the theoretical optimum [23], on which our empirical settings (see Section 6) are based.

- **Phase I–Initialization:** The mobile sink visits the anchor points one by one and pauses at each point for a short *sampling period*. During each sampling period, the sink collects the power consumption records from all nodes and builds a P-profile for that anchor point. At the end of this phase, the sink performs the programming (1) and drops an anchor point if its weight t_k is extremely low. It is not worth keeping such a point because its corresponding epoch is not long enough to amortize the routing overhead introduced by the sink mobility.

- **Phase II–Operation:** The mobile sink goes through the trajectory repetitively but only pauses at those chosen anchor points. At a given point k, the sink again collects power consumption information and builds a profile $\mathbb{P}_k$. Based on the new profile and previous profiles for other chosen points, the programming (1) is re-solved to deduce t_k. The actual epoch is computed as $\hat{t}_k = t_k/\delta$, where $\delta > 1$ is an integer. Applying the δ makes it possible for the sink to repeat the movement pattern several turns, which allows the sink to be more adaptive to the network dynamics.

We have the following remarks on the algorithm:

- If we make a discrete search over the whole surface covered by the network to obtain the anchor points, the time to finish Phase I could become comparable to the network lifetime; the algorithm would thus lose its adaptability (e.g., the sink might not even get a chance to enter Phase II). Alternatively, we can search over a "good" trace. A candidate of such a trace could be the periphery of the network [8].

- The sink could have directly applied the results (i.e., t_ks) of the first phase to the second phase if the routing topology were fixed. However, according to our experiences with real WSNs, the routing topology keeps evolving even with static nodes. As a result, the P-profiles obtained from the first phase can only be considered as estimations and should be updated if new profiles are available.

6   Simulations

We report two sets of simulations with TOSSIM in this section. In one set, network nodes are located on point lattices; simulation results of this set represent outdoor WSNs in general. In another set, nodes form a ring; the simulations emulate our future field tests with an in-building WSN.

6.1   Grid Networks

We arrange nodes on a point lattice of size 4×4, 5×5, and 7×7. For each network, we either 1) put the sink (node 0) at the network border (the midpoint of one side), or 2) at the center, or 3) let the sink move around the network periphery. There is a constant distance between any two consecutive anchor points; the sink pauses on an anchor point and moves in between two anchors according to

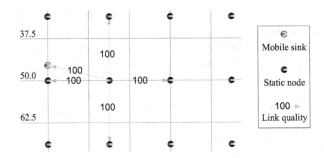

Fig. 3. Neighborhood graph in TinyViz [19]. The number beside a link states the link quality: 100 stands for a perfect link. The numbers on grid lines represent coordinates.

the instruction from a Tython [24] code. The connectivity[5] of a node with other nodes is shown in Fig. 3. The transmission range is set to 1.2 times longer than the distance between two neighbor nodes. Each node generates a data packet every 60 seconds. A control packet (route message) is sent every 120 seconds in the pause state and every 2 seconds (speed-up rate) in the pre-pause state. The s-beacon rate is one per second. The retransmission is disabled for all nodes if not stated otherwise. The epoch of non-adaptive mobility allows each node to send 10 data packets (i.e., 600 seconds)[6], and the moving time is 10 seconds for the 49 nodes network and 20 seconds for the other two. The sink moves at a speed of 1 ft/s in the move state. The full simulation time is just long enough to let the sink go through one round of its trip; the simulation for a given network is repeated 10 times. For the measurement of energy consumptions[7], we use the number of (both control and data) packets that a node is involved to characterize the energy consumption. By doing this, we implicitly assume that 1) radio communication is the dominating energy consumer, 2) sending and receiving a packet consumes the same amount of energy, and 3) control and data packets are of the same size.

---

[5] We take the *fixed radius* model, although it is less realistic than the *empirical* one (we refer to [19] for the definitions of these models). The reason is that, given a set of geo-distributed nodes, applying the empirical model usually leads to small network diameter due to the occasional existence of *shortcuts*. A relatively large network diameter (up to 10 hops) is essential to fully exhibit the benefit of using a mobile sink, but increasing the network size to achieve larger diameter results in a simulation time of unreasonable duration (e.g., 100 hours). By using the fixed radius model, we simply assume that, for a certain node, only nodes within its *effective region* [15] are considered as its neighbors.

[6] This duration is way shorter than what could be in a real deployment, where it might last for days or even weeks. Therefore, the performance of MobiRoute is expected to be better in practice, thanks to a longer amortization period.

[7] Since TOSSIM uses a MAC that never switches off its radio, tools such as Power-TOSSIM [25] always report a flat energy consumption pattern of a network no matter where the sink is located. In reality, motes equipped with B-MAC [21] do switch off their radio when there is no transmission going on.

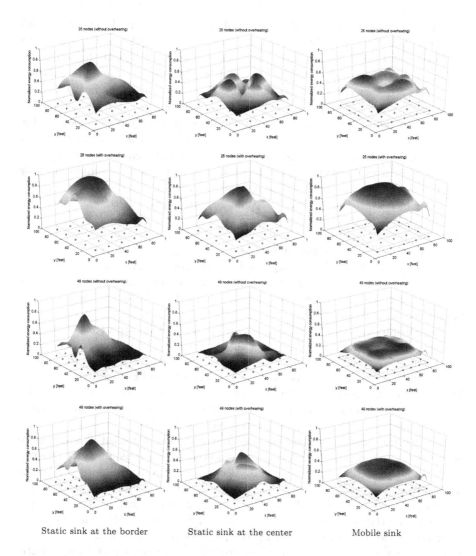

Fig. 4. Energy consumption of WSNs. Two networks with 25 and 49 nodes, respectively, are simulated. For each network, we either put the sink at the network border (the midpoint of one side), or at the center, or let the sink move around the network periphery. For each comparative case (i.e., one row in the figure), the energy consumptions are normalized to a common scale factor.

Non-adaptive Mobility. The spatial distributions of energy consumptions for networks with 25 and 49 nodes are shown in Fig. 4. According to the lifetime definition in Section 3, the smaller the maximum energy consumption in a network, the longer the network lifetime will be. Comparing the two cases with a static sink and the case with a mobile sink, we make the following observations:

- The load-balancing effect of using a mobile sink is evident. The network with a mobile sink always lives longer than the network with a static sink at its border and no shorter than the network with a static sink at the center.
- In the network of 49 nodes, using a mobile sink is the best choice, irrespective of whether the overhearing at the MAC layer exists or not. However, overhearing does offset the benefits of using a mobile sink: the 100% improvement on the lifetime (comparing the network having a mobile sink with the one having a centered static center) is reduced to 50% if overhearing exists.
- In smaller networks of 16 and 25 nodes (only the latter case is shown in Fig. 4 due to their similarity), using a mobile sink is not necessarily helpful, because it does not improve the lifetime compared with using a centered static sink while increasing the accumulative energy consumption of the network.

A straightforward conclusion is that using a mobile sink is more beneficial in large networks. Since the function of the mobile sink is to disperse the traffic flows, the network should be large enough to provide nodes with a sufficient number of alternative routing paths. However, since locating a sink at the network center is not always practical[8], using a mobile sink does help to improve the lifetime in most networks.

Another implication of our observations is that a MAC protocol free of overhearing is very important to improve the effectiveness of using a mobile sink. Unfortunately, the current MAC of motes (i.e., B-MAC [21]) suffers much from overhearing [26], and protocols with the potential to avoid overhearing (e.g., S-MAC [16]) do not necessarily have an overall performance better than B-MAC due to their burdensome synchronization schemes. So, we expect future technology to provide sensor nodes with overhearing-free MACs.

We plot the cumulative distribution functions of the packet delivery ratio in these two networks in Fig. 5. The comparisons are only made between a centered static sink and a mobile sink, because the ratios are quite similar for both networks with a static sink. The figures show that, without retransmission, the packet delivery ratio is always lower in the case of a mobile sink, which is intuitive (see the reasons that we described in Section 4). Also, the difference between the two ratios increases with the network size. The reason is that using a mobile sink increases the worst-case routing path length (actually, a static sink located at one vertex of the network periphery achieves the same ratio). This is not a major problem, because we would expect a much higher reliability in reality, where a node typically sends data only every tens of minutes [26]. Actually, if we enable the retransmission, the packet delivery ratio in the case of a mobile sink can be as high as that in the case of a static sink, but at the cost of increased energy consumption (Fig. 6), whose maximum value is still low enough to justify the benefit of using a mobile sink.

---

[8] For habitat and environment monitoring, unobtrusive observation is key for studying natural phenomena [26]. Although nodes are small enough for this purpose, a sink (especially when it has to transmit the collected data out of the network area) can hardly makes itself invisible in the environment.

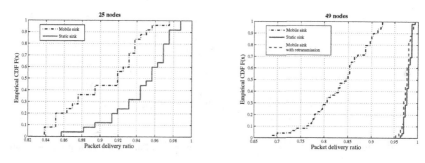

Fig. 5. Comparisons of packet delivery ratio

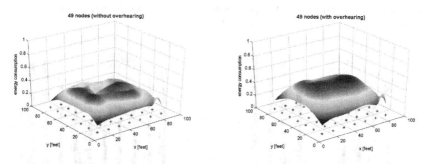

Fig. 6. Energy consumption of a WSN with a mobile sink and retransmission enabled. The scale factors take the same value as used for Fig. 4.

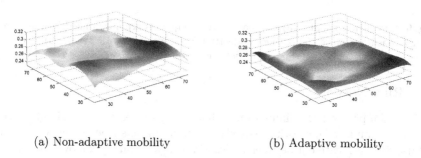

(a) Non-adaptive mobility   (b) Adaptive mobility

Fig. 7. Zooming in the distribution of energy consumption with a mobile sink

Adaptive Mobility. Zooming into the spatial distribution of energy consumption in the network with a mobile sink (as shown in Fig. 7 (a)), we observe that the load taken by nodes near the corner is heavier than that of other nodes. Applying the algorithm described in Section 5, we actually find that the sink should pause less time at those anchors near the corner. The resulting load, shown in Fig. 7 (b), is further balanced; which improves the network lifetime by about 10%. Note that the sink, in our simulations, only circles around the network twice: one in phase I and another in phase II (see Section 5); the network lifetime can be further improved with more rounds in phase II.

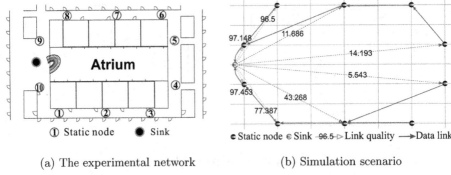

(a) The experimental network   (b) Simulation scenario

Fig. 8. The plan of our network deployment (a) and the simulation scenario (b). Nodes are numbered the same way in (b) as in (a).

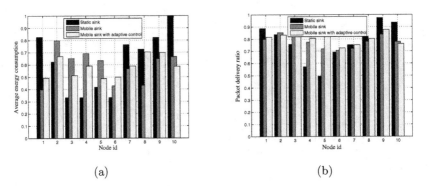

(a)   (b)

Fig. 9. Simulation results. (a) The energy consumptions are normalized by the largest energy consumption observed (i.e., node 10 in the case of a static sink). (b) The averaging effect arises also for the packet deliver ratio.

6.2   Ring Network

This section presents the simulation with a ring network. We use this simulation scenario to emulate a network deployed in our building, as shown in Fig. 8 (a). While a static sink[9] is located in-between nodes 9 and 10, a mobile sink moves around the circle and pauses in between two consecutive nodes. We use the *empirical* model [19] to characterize the connectivity in this set of simulations. As an example, the connectivity graph for the sink (node 0) is shown in Fig. 3 (b). Each node generates a data packet every 30 seconds. A control packet (route message) is sent every 60 seconds in the pause state and every 2 seconds (speed-up rate) in the pre-pause state. The s-beacon rate is one per second. The retransmission is disabled for all nodes. Each of the 10 simulations lasts for 17600 seconds and the epoch of non-adaptive mobility is 1760 seconds. The sink moves at a speed of 1 ft/s in the move

---

[9] The atrium inside of our building prevents us from locating the sink at its optimum position (i.e. the center of the network). This indeed corroborates our claim in Section 6.1 that locating a sink at the network center is not always practical.

state, and the moving time is 25 seconds. The measurement of energy consumptions is the same as for Section 6.1, and the overhearing is not taken into account.

We illustrate the simulation results with bar graphs in Fig. 9. As shown in Fig. 9 (a), the load balancing effect is already very evident by simply moving an uncontrolled sink, which improves the lifetime by 20%. Further improvement is achieved (an additional 15% of improvement on lifetime compared to the non-adaptive mobility) by controlling the mobile sink adaptively. The behavior in packet delivery, plotted in Fig. 9 (b), differs from that shown in Fig. 5; the averaging effect also arises due to the special network topology. In this specific scenario, the averaging effect makes a mobile sink beneficial not only to the network lifetime but also to the reliability, because nodes that are far away from the static sink perform poorly in terms of the reliability of packet delivery.

7   Conclusion

In this paper, we have presented a routing protocol, MobiRoute, to support wireless sensor networks (WSNs) with a mobile sink. This is a follow-up of our previous work [8] where we theoretically proved that moving the sink can improve network lifetime without sacrificing data delivery latency. By intensively simulating MobiRoute with TOSSIM (in which real implementation codes are running), we have demonstrated the benefit of using a mobile sink rather than a static one. We have simulated both general networks with nodes located in point lattices and a special in-building network with nodes forming a ring. The results are very promising: a mobile sink, in most cases, improves the network lifetime with only a modestly degraded reliability in packet delivery.

We are in the process of performing full-scale field tests with the in-building network. We are also considering more comprehensive simulations to evaluate the performance of MobiRoute under diverse conditions (e.g., the number and location of the anchor points, the pause/move ratio, the node number and density, and the protocol parameter such as frequency of s-beacons and δ for adaptive mobility). We will improve MobiRoute based on the experience obtained from our field tests and simulations.

Acknowledgements

The authors would like to thank the anonymous reviewers for their valuable feedback.

References

1. Shah, R., Roy, S., Jain, S., Brunette, W.: Data MULEs: Mobeling a Three-tier Architecutre for Sparse Sensor Networks. In: Proc. of the 1st IEEE SNPA. (2003)
2. Chakrabarti, A., Sabharwal, A., Aazhang, B.: Using Predictable Observer Mobility for Power Efficient Design of Sensor Networks. In: Proc. of the 2nd IEEE IPSN. (2003)

3. Kansal, A., Somasundara, A., Jea, D., Srivastava, M., Estrin, D.: Intelligent Fluid Infrastructure for Embedded Networks. In: Proc. of the 2nd ACM/USENIX MobiSys. (2004)

4. Jea, D., Somasundara, A., Srivastava, M.: Multiple Controlled Mobile Elements (Data Mules) for Data Collection in Sensor Networks. In: Proc. of the 1st IEEE/ACM DCOSS. (2005)

5. Gandham, S., Dawande, M., Prakash, R., Venkatesan, S.: Energy Efficient Schemes for Wireless Sensor Networks with Multiple Mobile Base Stations. In: Proc. of IEEE Globecom. (2003)

6. Wang, Z., Basagni, S., Melachrinoudis, E., Petrioli, C.: Exploiting Sink Mobility for Maximizing Sensor Networks Lifetime. In: Proc. of the 38th HICSS. (2005)

7. Wang, Z., Melachrinoudis, E., Basagni, S.: Voronoi Diagram-Based Linear Programming Modeling of Wireless Sensor Networks with a Mobile Sink. In: Proc. of the IIE Annual Conference and Exposition. (2005)

8. Luo, J., Hubaux, J.P.: Joint Mobility and Routing for Lifetime Elongation in Wireless Sensor Networks. In: Proc. of the 24th IEEE INFOCOM. (2005)

9. Papadimitriou, I., Georgiadis, L.: Maximum Lifetime Routing to Mobile Sink in Wireless Sensor Networks. In: Proc. of the 13th IEEE SoftCom. (2005)

10. Wang, W., Srinivasan, V., Chua, K.C.: Using Mobile Relays to Prolong the Lifetime of Wireless Sensor Networks. In: Proc. of the 11th ACM MobiCom. (2005)

11. Grossglauser, M., Tse, D.: Mobility increases the capacity of ad hoc wireless networks. IEEE/ACM Trans. on Networking 10 (2002) 477–486

12. Johnson, D., Maltz, D., Hu, Y.C.: The Dynamic Source Routing Protocol for Mobile Ad Hoc Networks (DSR). (2004) Internet-Draft, draft-ietf-manet-dsr-10.txt. Work in progress.

13. Perkins, C., Belding-Royer, E., Das, S.: Ad hoc On-Demand Distance Vector (AODV) Routing. (2003) IETF RFC 3561, Network Working Group.

14. Intanagonwiwat, C., Govindan, R., Estrin, D., Heidemann, J., Silva, F.: Directed diffusion for wireless sensor networking. IEEE/ACM Trans. on Networking 11 (2003) 2–16

15. Woo, A., Tong, T., Culler, D.: Taming the Underlying Challenges of Reliable Multihop Routing in Sensor Networks. In: Proc. of the 1st ACM SenSys. (2003)

16. Ye, F., Luo, H., Cheng, J., Lu, S., Zhang, L.: A Two-tier Data Dissemination Model for Large Scale Wireless Sensor Networks. In: Proc. of the 8th ACM MobiCom. (2005)

17. Kim, H., Abdelzaher, T., Kwon, W.: Minimum Energy Asynchronous Dissemination to Mobile Sinks in Wireless Sensor Networks. In: Proc. of the 1st ACM SenSys. (2003)

18. Baruah, P., Urgaonkar, R., Krishnamachari, B.: Learning Enforced Time Domain Routing to Mobile Sinks in Wireless Sensor Fields. In: Proc. of the 1st IEEE EmNets. (2004)

19. Levis, P., Lee, N., Welsh, M., Culler, D.: TOSSIM: Accurate and Scalable Simulation of Entire TinyOS Applications. In: Proc. of the 1st ACM SenSys. (2003)

20. Ye, W., Heidemann, J., Estrin, D.: An Energy-Efficient MAC Protocol for Wireless Sensor Networks. In: Proc. of the 21st IEEE INFOCOM. (2002)

21. Polastre, J., Hill, J., Culler, D.: Versatile Low Power Media Access for Wireless Sensor Networks. In: Proc. of the 2st ACM SenSys. (2004)

22. Chang, J.H., Tassiulas, L.: Energy Conserving Routing in Wireless Ad-hoc Networks. In: Proc. of the 19th IEEE INFOCOM. (2000)

23. Luo, J.: Mobility in Wireless Networks: Friend or Foe – Network Design and Control in the Age of Mobile Computing. PhD thesis, School of Computer and Communication Sciences, EPFL, Switzerland (2006)

24. Demmer, M., Levis, P.: Tython: A Dynamic Simulation Environment for Sensor Networks. (2005) http://www.tinyos.net/tinyos-1.x/doc/tython/tython.html.

25. Shnayder, V., Hempstead, M., Chen, B., Allen, G., Welsh, M.: Simulating the Power Consumption of Large-Scale Sensor Network Applications. In: Proc. of the 2nd ACM SenSys. (2004)

26. Szewczyk, R., Mainwaring, A., Polastre, J., Anderson, J., Culler, D.: An Analysis of a Large Scale Habitat Monitoring Application. In: Proc. of the 2nd ACM SenSys. (2004)

SenCar: An Energy Efficient Data Gathering Mechanism for Large Scale Multihop Sensor Networks*

Ming Ma and Yuanyuan Yang

Dept. of Electrical and Computer Engineering
State University of New York,
Stony Brook, NY 11794, USA
{mingma, yang}@ece.sunysb.edu

Abstract. In this paper, we propose a new data gathering mechanism for large scale multihop sensor networks. A mobile data observer, called *SenCar*, which could be a mobile robot or a vehicle equipped with a powerful transceiver and battery, works like a mobile base station in the network. SenCar starts the data gathering tour periodically from the static data processing center, traverses the entire sensor network, gathers the data from sensors while moving, returns to the starting point, and finally uploads data to the data processing center. Unlike SenCar, sensors in the network are static, and can be made very simple and inexpensive. They upload sensing data to SenCar when SenCar moves close to them. Since sensors can only communicate with others within a very limited range, packets from some sensors may need multihop relays to reach SenCar. We first show that the moving path of SenCar can greatly affect the network lifetime. We then present heuristic algorithms for planning the moving path/circle of SenCar and balancing traffic load in the network. We show that by driving SenCar along a better path and balancing the traffic load from sensors to SenCar, the network lifetime can be prolonged significantly. Our simulation results demonstrate that the proposed data gathering mechanism can greatly prolong the network lifetime compared to a network which has only a static observer, or a network in which mobile observer can only move along straight lines.

Keywords: SenCar, Wireless sensor networks, Data gathering, Load balancing.

1 Introduction and Background

In recent years, wireless sensor networks (WSN) are playing an increasingly important role in a wide-range of applications, such as medical treatment, outer-space exploration,

---

* The research work was supported in part by the U.S. National Science Foundation under grant numbers CCR-0207999 and ECS-0427345 and by the U.S. Army Research Office under grant number W911NF-04-1-0439.

P. Gibbons et al. (Eds.): DCOSS 2006, LNCS 4026, pp. 498–513, 2006.
© Springer-Verlag Berlin Heidelberg 2006

battlefield surveillance, emergency response, etc. [1, 2, 3, 4, 5]. A wireless sensor network is generally composed of hundreds or thousands of sensor nodes, with each sensor capable of sensing the environment and sending data to data observers. Due to the limited battery lifetime and low cost requirement, each sensor node is usually equipped with a simple and low-cost computing module and radio transceiver. Although special attention has been paid to low power consumption when designing sensors, a sensor node can survive only very limited lifetime with current technologies [6, 7]. Therefore, *energy efficiency* is one of the most critical challenges in applications of large scale, resource-limited sensor networks.

Resource-limited sensor nodes are usually thrown into an unknown environment without a pre-configured infrastructure. Before monitoring the environment, sensor nodes must be able to discover nearby sensors and organize themselves into a network. After that, sensor nodes begin to sense the field and forward data to a data observer until they fail or are terminated by the observer. Various types of data gathering mechanisms have been considered and investigated for large scale sensor networks. They can be roughly classified into following categories. First, in a static sensor network, which contains a large number of static sensor nodes and a static observer, the observer must be reachable by all the sensor nodes. Data packets are sent to the observer by one or more hops of forwarding. In such a network, all data traffic flows to the observer. Thus, sensors close to the observer consume much more energy than sensors at the margin of the network. As a result, after these sensors fail, other nodes cannot reach the observer and the network becomes disconnected, even most of the nodes can still survive for a long period. Furthermore, in such a static network, every sensor has to be powerful enough to perform all the functions by itself, such as finding routing paths, obtaining its location information [17, 18], scheduling the packet transmission, and so on. Thus, the network architecture with only one static observer is only suitable for a small network. The second type of architecture introduces a hierarchy to the network. By adding a small number of powerful cluster heads, the network can be divided into clusters. In such a network, sensor nodes are organized into clusters and form the lower layer of the network. At the higher layer, cluster heads collect sensing data from sensors and forward data to outside observers. Such two-layer hybrid networks are more scalable and energy-efficient than homogeneous sensor networks. However, though increasing the number of cluster heads may reduce the burden of sensor nodes, the cost of cluster heads should also be taken into consideration. The third type of sensor networks introduce one or more mobile data observers to collect the data dynamically. A mobile data observer could be a mobile robot or a vehicle equipped with a powerful transceiver, battery and large memory. The mobile data observer starts a tour from the base station, traverses the network, collects sensing data from nearby nodes while moving, returns and uploads data to a remote data processing center. The moving path and the direction of the mobile observer can be random or planned. When the mobile observer moves into the transmission range of some sensors, the sensors send data to the observer directly. Other sensors, which are too far away from the moving path of the mobile observer, can upload data through the relaying of other sensors. The relaying path and transmission time of each packet can be determined by the mobile observer. In addition, a GPS [17] receiver may be optional for sensors, since sensors can estimate the relative location to the mobile observer [18] when the mobile observer moves close to them. By introducing the mobility of the observer, the energy consumption for transmitting packets can be reduced significantly and sensor nodes can be made simpler and less expensive.

In this paper, we consider the problem of planning the moving path of a mobile observer and balancing the traffic load from sensors to the mobile observer to prolong the network lifetime. We define the *network lifetime* as the lifetime of the first failure node in the network. We consider applications, such as environment monitoring for some human-unreachable environment, e.g., outer-space, seabed and so on, where the sensing data is generally collected at a low rate and sensing data is not so delay-sensitive that it can be accumulated into fixed-length data packets and uploaded once a while. For such applications, we have following assumptions. Static sensors are densely deployed onto a two-dimensional working area. Due to the limited transmission power, sensor nodes can only reach nearby nodes within a limited disk-shaped range. The total power consumption for transmitting a packet, including processing power consumption, such as coding/decoding, modulation/demodulation, A/D-D/A, and so on, and radio power both are assumed to be proportional to the size of the packet. For the sake of simplicity, we only consider the major energy consumption for communications and ignore those for sensing and other tasks. The mobile observer can move to anywhere in the working area, and is equipped with a more powerful transceiver and battery with much longer lifetime than sensor nodes. In addition, we assume that the mobile observer has explored the sensing field before collecting data. The location information and the connection patterns of sensor nodes have been obtained by the mobile observer during the exploring phase. In the rest of the paper, we will use **SenCar** to denote the mobile data observer.

The rest of the paper is organized as follows. Section 2 discusses some related work. Section 3 presents our data gathering scheme for a network with an arbitrary topology. Section 4 gives simulation results and some discussions. Finally, Section 5 concludes the paper.

2   Related Work

Mobility of sensor networks has been studied in some literatures recently [3, 4, 5, 8, 9, 10, 11]. In [3] and [4], radio-tagged zebras and whales are used as mobile nodes to collect sensing data in a wild environment. These animal-based nodes wander randomly in the sensing field, and exchange sensing data only when they move close to each other. Thus, sensor nodes in such a network are not necessarily connected all the time. Moreover, the mobility of randomly moving animals is hard to predict and control, thus the maximum packet delay cannot be guaranteed. For sensor networks deployed in an urbane area, where public transportation vehicles, such as buses and trains, always move along the fixed routes. These vehicles can be mounted with transceivers to act as mobile base stations [8, 9]. Compared to the randomly moving animals, the moving path and timing are predictable in this case. However, data exchanging still depends on the existing routes and schedules of the public transportation, and thus is very restrictive. In [10], a number of mobile observers, called *data mules*, traverse the sensing field along parallel straight lines and gather data from sensors. This scheme works well in a large scale, uniformly distributed sensor network. However, in practice, data mules may not always be able to move along straight lines, for example, obstacles or boundaries may block the moving paths of data mules. Moreover, the performance and cost of the data mule scheme depends on the number of data mules and the distribution of sensors. When only a small number of data mules are available and not all sensors are connected, data mules may not cover all the sensors in the network if they only move along straight lines. [5, 11] also consider mobile observers in sensor networks. [5] mainly discussed hardware/software implementation of underwater mobile observers, while [11] proposed an algorithm to schedule the mobile observer, so that there is no data loss due to the buffer overflow. In order to make a data collecting scheme suitable to various network topologies, it is more realistic and efficient to plan the moving path of the mobile observer dynamically based on the distribution of sensors. This is the motivation of our work in this paper. In the following, we will give a data gathering scheme for a network with an arbitrary topology.

3   Data Gathering Scheme

Sensor networks are usually deployed in dangerous or even human-unreachable areas, such as volcano, outer-space, seabed and so on. In such environments, human beings may not move close to the sensing field. A mobile observer, or SenCar, will be sent out to gather data from sensors periodically. Since the network may contain a large number of nodes, each tour may take a long time. In order to save the energy, sensors may turn on their transceivers only when they need to send or relay packets. Except the transmission period, transceivers of sensors could be turned off. The entire sensor network can be divided into several clusters, where sensors in each cluster must be connected to SenCar while it is moving through the cluster. When SenCar moves close to the cluster, all sensors belonging to the cluster will be waken up and prepare to send packets. Sensing data can be collected by SenCar while it is traversing the cluster. To make this scheme work, two issues must be resolved here. The first issue is how to wake up and turn off sensors only when needed. A radio wake-up scheme was proposed in [15], which allows the transceivers of sensors to be deactivated when they are idle. The second issue is how to divide sensors into clusters. As will be described later, a moving path of SenCar consists of a series of connected line segments. Sensors close to each line segment will be organized into a cluster by SenCar, such that the entire network can be divided into a number of clusters. A straightforward way to organize sensor nodes into clusters is to assign each sensor to the "nearest" line segment in the moving path from it. The details of clustering algorithm will be introduced in Section 3.3. While moving, SenCar will poll each sensor one by one to collect data. Relaying path and transmission time of packets are determined by SenCar. Thus, packet collision can be avoided and no routing paths need to be maintained by sensors. In addition, while planning the relaying paths, traffic load needs to be balanced to prolong the lifetime of sensors. Next, we will describe the data gathering scheme in more detail.

3.1   Load Balancing

As discussed above, due to the different amount of traffic each sensor node relays, some nodes may fail sooner than others. In order to maximize the network lifetime, relaying paths must be carefully planned to balance the traffic load. Load balancing problem in static sensor networks has been investigated in some existing work, such as [12, 13, 14]. Next we will describe how to formalize the problem of maximizing network lifetime in our network into a network flow problem.

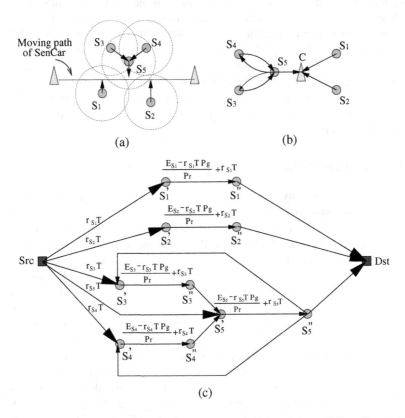

Fig. 1. (a) Connection patterns of a sensor network. (b) Directed graph $G(S, c, A)$ corresponding to the connection patterns of the network. (c) Network flow graph $G'(S', Src, Dst, A')$ for maximizing network lifetime, where the capacities of unmarked arcs are infinity.

Given the connection patterns of the network and the moving path of SenCar, a sensor network can be modeled as a directed graph $G(S, c, A)$, where $S = \{s_1, s_2, \ldots, s_n\}$ is the set of all sensor nodes, c denotes SenCar and A is the set of all directed links $a(i, j)$ where $i \in S, j \in S \bigcup \{c\}$. For each pair of nodes $s_i, s_j \in S$, if s_i can reach s_j in one hop, arc $a(s_i, s_j)$ will be added into A. If the moving path of SenCar traverses the transmission range of s_i, or equivalently s_i can reach SenCar in one hop while SenCar is moving, add arc $a(s_i, c)$ into G. Fig. 1(a) and (b) shows how to construct the directed graph from the connection patterns of a network.

Given the directed graph G of a network, its corresponding flow graph $G'(S', Src, Dst, A')$ can be constructed as follows:

- For each $s_i \in S$, add two vertices s_i' and s_i'' to S', and an arc $a(s_i', s_i'')$ is added into A' with capacity $\frac{Es_i - (r_{s_i} T P_g)}{P_r} + r_{s_i} T$;
- For each arc $a(s_i, s_j) \in A$, where $s_i, s_j \in S$, add an arc $a(s_i'', s_j')$ into A' with infinity capacity;
- A pair of source and destination nodes Src and Dst are added into G', and for each $s_i' \in S'$, connect Src and s_i' by an arc $a(Src, s_i')$ with capacity $r_{s_i} T$;
- For each arc $a(s_i, c) \in A$, where $s_i \in S$, add an arc $a(s_i'', Dst)$ into A' with infinity capacity;

where r_{s_i} and E_{s_i} denote the data generating rate and energy limit of node s_i, P_g and P_r represent the power consumption for generating and relaying a unit of traffic, respectively, and T is the network lifetime. Since SenCar visits sensors periodically, say, every ΔT time. We can set $T = \Delta T$ at the beginning and increase T by ΔT every time. For any given T, this problem is a regular maximum flow problem [16] and can be solved by Ford-Fulkerson algorithm in polynomial time. In this construction, $(r_{s_i} T)$ limits the flow from Src to s_i and represents the flow generated by s_i within time T, which consumes $(r_{s_i} T P_g)$ energy. Due to the energy constraint of node s_i, the maximum flow node s_i can relay within time T is $\frac{Es_i - (r_{s_i} T P_g)}{P_r}$. Thus, the total flow a node s_i can generate and relay in time T is limited by $\frac{Es_i - (r_{s_i} T P_g)}{P_r} + r_{s_i} T$. When the maximum flow equals $\sum_{s_i \in S} r_{s_i} T$, it means until time T, all generated traffic by n sensor nodes is received by SenCar. Thus, all n sensors must be alive until T. We can keep increasing T and running Ford-Fulkerson algorithm to obtain the maximum flow for every T value, until the maximum flow is less than $\sum_{s_i \in S} r_{s_i} T$, which indicates some nodes have failed before time T. Finally, the value of T obtained before the last run of Ford-Fulkerson algorithm is the maximum network lifetime. An example of the construction from the connection patterns of the network to the flow graph is depicted in Fig. 1.

We now analyze the time complexity of this algorithm. Let U denote the maximum units of traffic any sensor node generates and relays within time T^*, where T^* is the maximum network lifetime obtained by the algorithm. Then

$$U = \max_{s_i \in S} \left\{ \frac{Es_i - (r_{s_i} T^* P_g)}{P_r} + r_{s_i} T^* \right\}$$

The running time of this algorithm is $O(Un^2)$, where n is the number of sensor nodes in the network. Therefore, based on the connection patterns of the network and the moving path of SenCar, the optimal traffic relaying paths which maximize the network lifetime can be obtained in polynomial time by running the flow algorithm. Next we will discuss how to determine the moving path of SenCar.

3.2 Determining Turning Points of the Moving Path

Before formally describing the problem we consider, we first give an example to see how the moving path of the SenCar affects the network lifetime. As shown in Fig. 2(a), SenCar traverses the sensing field from A to B, where fifteen nodes are deployed. We assume that each sensor forwards one packet to SenCar, while SenCar moves from

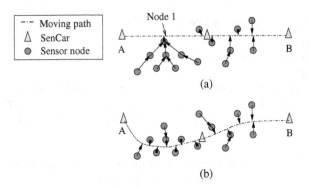

Fig. 2. SenCar moves from A to B and collects data from nearby sensors. (a) SenCar moves along a straight path. (b) SenCar moves along a well-planned path.

A to B. Due to limited transmission power of sensors, packets may need multi-hop relays to reach SenCar. The sensors are organized into spanning trees to forward packets to SenCar. We can see that in Fig. 2(a), node 1 is a bottleneck node, because it has to relay eight packets from itself and its seven child nodes to SenCar. Thus, node 1 consumes energy much faster than other nodes. After node 1 fails, the child nodes of node 1 cannot reach SenCar any more, unless SenCar changes the moving path. Fig. 2(b) shows the relaying paths of sensors when the moving path of SenCar is well planned. We can see that each node has at most one child node and needs to send at most two packets to SenCar. In this example, if we only consider the energy consumption for transmission and roughly measure it by the number of packets transmitted, the well-planned moving path of SenCar can increase the lifetime three times compared to the straight-line moving path. From this simple example, we observe that a well-planned moving path of SenCar may minimize the maximum load of any sensor, save a lot of energy and prolong the network lifetime significantly. In addition to traffic load, the moving path of SenCar can also affect the directions of traffic flow, thereby have a significant impact on the network lifetime. Next we consider the problem of maximizing the lifetime of the network, by carefully planning the moving path of SenCar.

In practice, since it is difficult for vehicles or robots to move along any continuous curve smoothly, we simply assume that the moving path of SenCar consists of $t + 1$ connected straight line segments from the starting point A to the end point B. That means SenCar needs to turn t times before it reaches the end of the path. Let $p_1, p_2, \ldots, p_t$ denote t turning points. Then, the moving path of SenCar can be represented by $A \rightarrow p_1 \rightarrow p_2 \rightarrow \cdots \rightarrow p_t \rightarrow B$. Let (x_A, x_A), (x_B, y_B) and (x_{p_i}, y_{p_i}) denote the coordinates of A, B and p_i, for $i = 1, 2, \ldots, t$. We assume that the x-coordinate of any sensor is between x_A and x_B. We will use the divide and conquer strategy to find t turning points to reduce the maximum traffic load of any sensor needs to send out. Without loss of generality, let $t = 2^k - 1$, where k denotes the rounds of the path planning algorithm and $k = 1, 2, \ldots$. First, given the positions of A and B, we will find the position of the first turning point $p_{\frac{t+1}{2}}$. For the sake of simplicity, we assume that the first turning point can only be chosen from a finite set of points in the bisector of the initial path. Let the x-coordinate of the first turning point $x_{p_{\frac{t+1}{2}}} = \frac{x_A + x_B}{2}$, and

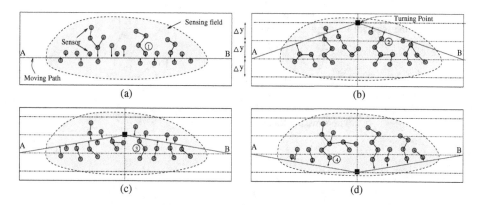

Fig. 3. SenCar moves from A to B and collects data from nearby sensors (a) SenCar moves from A to B through a straight path. (b) SenCar moves from A to B with turning point $(\frac{x_A+x_B}{2}, 2\Delta y)$. (c) SenCar moves from A to B with turning point $(\frac{x_A+x_B}{2}, \Delta y)$. (d) SenCar moves from A to B with turning point $(\frac{x_A+x_B}{2}, -\Delta y)$.

the y-coordinate of the first turning point $y_{p_{\frac{t+1}{2}}} = m \times \Delta y$, where Δy is a fixed grid length and m can be any integer that ensures $(x_{p_{\frac{t+1}{2}}}, y_{p_{\frac{t+1}{2}}})$ to be within the range of the sensing field. After a set of eligible possible locations of the turning point are obtained, we can check each possible turning point and find the one that minimizes the maximum traffic load a sensor has to send out. For example, in Fig. 3(a), the initial path of SenCar begins from A to B. Given the grid length Δy and the range of sensing field, there are three possible locations of the first turning point, located at $(\frac{x_A+x_B}{2}, 2\Delta y)$, $(\frac{x_A+x_B}{2}, \Delta y)$ and $(\frac{x_A+x_B}{2}, -\Delta y)$, as shown in Fig. 3(b), (c) and (d). For each possible turning point, the load balancing algorithm introduced in the previous subsection can be used to obtain the maximum-minimum lifetime of the sensors for its corresponding moving path. Fig. 3(a)-(d) show the connection pattern graph of four different moving paths, where nodes 1, 2, 3 and 4 are the bottleneck nodes in Fig. 3(a)-(d), which need to send four, six, three and nine packets to SenCar, respectively. Thus, the third moving path, turned at $(\frac{x_A+x_B}{2}, \Delta y)$, provides a longer network lifetime than others. In the first step, point $(\frac{x_A+x_B}{2}, \Delta y)$ is chosen as the first turning point of the moving path. Note that sometimes better moving path may not be found by moving the turning point along the bisector of the current path. In this case, the new turning point can be simply set to the mid point between two end points of the current path.

3.3 Clustering the Network Along the Segments of the Moving Path

After the first turning point is obtained, the moving path consists of two connected line segments. Then, sensors will be organized into two clusters, where each cluster corresponds to a line segment. In order to save energy, two clusters of sensors can be waked up sequentially. Sensors in one cluster forward packets to SenCar before it makes the turn, while sensors in the other cluster send data after SenCar turns. A straightforward

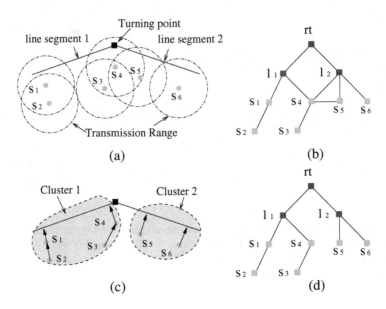

Fig. 4. SenCar moves from A to B and collects data from nearby sensors. (a) Two line segments of the moving path cross transmission ranges of node s_1, s_4, s_5 and s_6. (b) Graph $G(S, L, E)$ of the network. (c) Clustering obtained from the shortest path tree in $G(S, L, E)$. (d) Shortest path tree obtained in $G(S, L, E)$.

way to organize sensor nodes into clusters is to assign each sensor to its "nearest" line segment in the moving path from it. Here, the distance from a sensor to the line segment in the routing path is measured by the hop count. Given a set of sensors S and a set of line segments L, clustering the network can be implemented by running Dijkstra shortest path algorithm in graph $G(S, L, E)$, which can be constructed as follows:

- A root vertex rt is added into V;
- For each line segment, add a vertex l_i into L and an edge $e(rt, l_i)$ into E with weight 1.
- For each sensor $s_j \in S$, add a vertex s_j into V; Connect s_j and l_i by an edge $e(s_j, l_i)$ with weight 1, if and only if sensor s_j can reach line segment l_i in one hop;
- For each pair of nodes $s_j, s_k \in S$, connect s_j and s_k by an edge $e(s_j, s_k)$ with weight 1, if and only if sensor s_j and s_k can reach each other in one hop;

As shown in Fig. 4(a), two line segments of the moving path cross the transmission ranges of node s_1, s_4, s_5 and s_6. The corresponding graph $G(S, L, E)$ of the network is shown in Fig. 4(b). By running Dijkstra algorithm in G, we can find the shortest path from the root vertex to all other vertices, then a shortest path tree can be obtained, which contains $|L|$ first level vertices. Fig. 4(d) and (c) show the shortest path tree of $G(S, L, E)$ and the clustering of the network. Each first level vertex represents a line segment in the moving path. All child vertices of the first level vertex l_i in G represent a cluster of sensors corresponding to line segment l_i in the network.

3.4 Finding the Moving Path: Divide and Conquer

By combining the above algorithms of load balancing, finding turning points and clustering, the moving path planning algorithm can be described as follows: organizing the network into a cluster, determining the turning point from a set of possible locations of turning points, revising the path by adding the new turning point, and then dividing each cluster into two clusters. For each cluster, run the above algorithm recursively. After running k rounds of the moving path planning algorithm, $\sum_{i=1}^{k} 2^{(i-1)}$ turning points are obtained. Fig. 2 gives an example of the moving path planning algorithm. Fig. 2(a)-(d) show the moving paths and network flows of the initial, first, second and fourth round, respectively. We can observe that node 1, 2 and 3 are bottleneck nodes in Fig. 2(a), (b) and (c), which need to relay packets to SenCar from 6, 5 and 2 child nodes, respectively. These bottleneck nodes consume energy much faster than their child nodes. In Fig. 2(d), SenCar traverses through the transmission range of every node. Thus, each node can send data to SenCar directly without relaying from other nodes. The moving path after round 4 increases the lifetime seven times compared to the initial moving path.

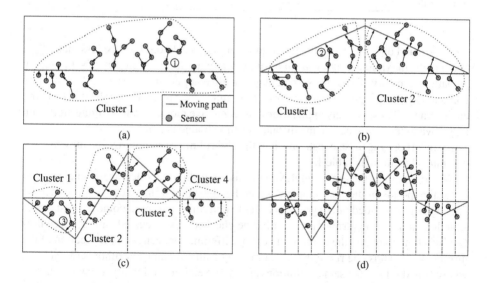

Fig. 5. SenCar moves from A to B and collects data from nearby sensors (a) Initial moving path is a straight line. (b) Moving path contains 1 turning point after round 1. (c) Moving path contains 3 turning points after round 2. (d) Moving path contains 15 turning points after round 4.

In the moving path planning algorithm, we can observe that adding turning points into the moving path will increase the total moving distance of SenCar, according to the triangle inequality rule. However, in practice, the total moving distance or the length of each tour may be restricted by a lot of factors. First, the length of each tour may be determined by the buffer size and data collecting rate of sensors. Sensing data must be gathered by SenCar before the buffer overflows. If all sensors have the same memory

Table 1. Moving Path Planning Algorithm

Moving Path Planning Algorithm

$i = 1$;
flag $= 1$;
while (flag $== 1$)
  Divide the network into $2^{(i-1)}$ clusters;
  for $j = 1$ to $2^{(i-1)}$ **do**
    Find the best turning point in j_{th} cluster from all possible locations of turning points;
    Add the best turning point into the moving path;
    if the total moving distance/time cannot satisfy the constraints after the new
    turning point is added
      flag $= 0$;
      Remove the new turning point from the path;
    end if
  end for
  i^{++};
end while

---

size mem and data rate $rate$, the maximum length of each tour must be less than $\frac{mem}{rate}$. Second, the maximum moving distance of SenCar without recharging may be limited by its battery capacity. Third, for some delay-sensitive applications, sensing data must be uploaded to the data processing center within limited time after being collected from the environment. Thus, in many applications, the recursive moving path planning algorithm may have to terminate before the distance or time bound is reached.

By incorporating these constraints into the algorithm, we summarize the moving path planning algorithm in Table 1.

We now analyze the time complexity of this algorithm. Let t denote the total number of turning points in the moving path, when the algorithm terminates. Suppose that the sensing field is divided into g grids. In order to determine a turning point, at most g possible locations of the turning point would be checked. As discussed earlier, it requires $O(Un^2)$ time to obtain the maximum network lifetime for each possible location of the turning point. Thus, the running time of moving path planning algorithm is $O(tgUn^2)$, where U and n have the same definitions as that in Section 3.1. Finally, we would like to point out that the moving path planning algorithm is executed offline by SenCar before the first data gathering tour. After that, only when some nodes fail or the topology of the network changes, SenCar needs to recalculate the new moving path adaptively.

3.5   Determining the Moving Circle of SenCar

In some applications, SenCar not only needs to traverse the sensing field, but also has to return to the starting point and upload data to the static data processing center. For such applications, moving paths become moving circles. Instead of a one-way straight line, the initial circle becomes a round-trip tour, which consists of two overlapped paths with the same shape but in opposite directions. The initial circle origins from the starting point, traverses the network, turns around and then moves back to the starting point. Both one-way paths of the initial circle pass through the network and divide the network into two parts. Sensors on each side of overlapped paths form a cluster. Each one-way path corresponds to one cluster and can be considered as the initial path of its corresponding cluster. Then, the moving path planning algorithm can be executed recursively in each cluster. Finally, two separate moving paths form a moving circle.

3.6 Avoiding Obstacles in the Sensing Field

We have discussed how to plan the moving path and circle of SenCar in an open sensing field. However, in most real-world applications, the working areas may be partially bounded, or have some irregular-shaped obstacles located within the sensing area. In order to make the moving path planning algorithm feasible in these cases, SenCar has to be able to avoid obstacles. Here, we assume that the complete map of the sensing field has been obtained before SenCar begins to collect data, which should include the location and shape information of obstacles in the sensing field. Then it is not difficult to adjust the basic moving path planning algorithm in Table 1 to avoid obstacles. For each candidate location of a turning point, SenCar will check if the line segment from the last turning point to it and the line segment from it to the next turning point are blocked by obstacles. If so, the candidate location is not eligible to be the turning point. Fig. 6 shows an example on how to check the eligibility of each possible location of the turning point. A new path from point A to B will be chosen from $A \rightarrow 1 \rightarrow B$ or $A \rightarrow 2 \rightarrow B$. Since the straight lines between A and 2, and between 2 and A are blocked by obstacles, 2 is not eligible to be a tuning point. The new path from A to B can only go through point 1.

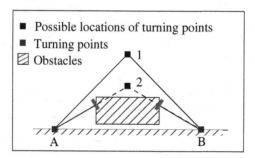

Fig. 6. Planning the moving path in the sensing field with obstacles. The line segments from A to 2 and from 2 to B are blocked by obstacles. Thus, 2 cannot be a turning point, while 1 is eligible to be a turning point.

4   Simulation Results

We have conducted extensive simulations to validate the algorithms we propose. In the simulation, we assume that a bunch of sensor nodes are densely deployed in the sensing field. Two-ray propagation model is used to describe the feature of the physical layer.

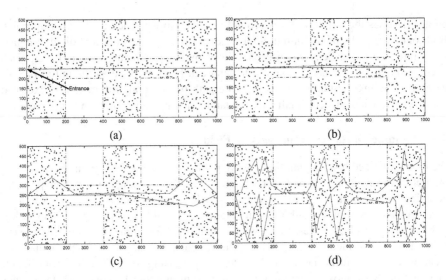

Fig. 7. SenCar starts the data gathering circle from the entrance of the building, collects data from sensors and returns to the entrance. (a) Initial layout of the network. (b) Layout and moving circle after Round 2. (c) Layout and moving circle after Round 4. (d) Layout and moving circle after Round 8.

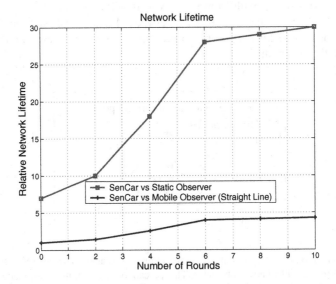

Fig. 8. The relative network lifetime of Scheme 3 compared to Scheme 1 and Scheme 2

With the maximum transmission power $0.858mw$, each node can communicate with other nodes as far as $40m$ away. The radio bandwidth is 200kbps. CBR traffic on the top of UDP is generated to measure the throughput. Each packet has a fixed size of 80 bytes, including header and payload. Let the grid length in the moving path planning algorithm be $10m$. Within each cluster, multi-hop polling protocol [14] is used as the inner-cluster protocol to avoid packets collision at the MAC layer. We evaluated the moving planning algorithm for both connected networks and disconnected networks.

4.1 Finding the Moving Circle in an Area with Obstacles

In this scenario, suppose that 800 sensors are densely deployed into a contaminated chemistry factory building to monitor the density of leaked chemicals. The map of the building and the initial layout of a bunch of sensors are shown in Fig. 7(a). The building consists of six $200m \times 200m$ large rooms, which are on both sides of a $1000m \times 100m$ aisle. The entire building is bounded by brick walls. SenCar has to move within the building. SenCar enters the building from the entrance, which is at the coordinates $(0m, 250m)$, collects data from all sensors, and returns to the entrance after a tour. We assume that the location information, connection patterns of sensors and the map of the building have been obtained during the deployment phase. Based on this information, SenCar calculates routes round by round by using the moving circle planning algorithm. As shown in Fig. 7, the initial moving circle consists two overlapped, straight-line moving paths, $(0m, 250m) \rightarrow (1000m, 250m)$ and $(1000m, 250m) \rightarrow (0m, 250m)$. Fig. 7(b), (c) and (d) show the moving circle after rounds 2, 4 and 8, respectively. From the figures, we can observe that, first, SenCar enters every room without hitting the walls of the building; second, as the number of rounds increases, SenCar moves zigzag around the building to get closer to the nodes. We next show that the movement of SenCar can balance the traffic load and prolong the network lifetime.

4.2 Network Lifetime

We now compare the network lifetime of the following three data gathering schemes: *Scheme 1*: A static observer placed in the center of the network (at point $(500m, 250m)$); *Scheme 2*: A mobile observer which can only move back and forth through the straight line between $(0m, 250m)$ and $(1000m, 250m)$; *Scheme 3*: SenCar which can move through a well-planned circle that starts and ends at point $(0m, 250m)$. For the network only containing static observer, we measured the optimal network lifetime by using the load balancing algorithm in [14]. We also evaluated the lifetime of the network, in which a mobile observer moves through straight lines. The optimal lifetime of the first two schemes is used as the performance reference for comparison purpose. The relative network lifetimes of *Scheme 3* compared to *Scheme 1* and *Scheme 2* are plotted in Fig. 8. From Fig. 8, we observe that the relative network lifetime ratios of SenCar compared to Scheme 1 and Scheme 2 keep increasing from rounds 1 to 10, and reach 29.8 and 4.5 at round 10, respectively. From this experiment, we can see that a mobile observer can prolong the network lifetime significantly compared to a static observer. Moreover, a well-planned moving path performs much better than a fixed straight line path for a mobile observer.

5 Conclusions

In this paper, we have proposed a new data collecting mechanism by introducing a mobile data observer, SenCar, to sensor networks. SenCar works like a mobile base station, starts the data gathering tour from the outside observer, traverses the entire sensor network, collects the data from nearby sensors, and then returns to the outside observer. We have showed that the moving path of SenCar can affect the network lifetime significantly. We presented a heuristic algorithm for planning the moving path/circle of SenCar and balancing traffic load in the network. By adopting a load balancing algorithm which finds the turning points and clusters the network recursively, network lifetime can be prolonged significantly. In addition, SenCar can avoid obstacles while moving. Our simulation results show that the proposed data gathering mechanism can prolong the network lifetime about 30 times compared to a network which has only a static observer, and about 4 times compared to a network whose mobile observer can only move along straight lines.

References

1. S. Chessa and P. Santi, "Crash faults identification in wireless sensor networks," *Computer Communications*, vol. 25, no. 14, pp. 1273-1282, 2002.
2. L. Schwiebert, S.K. S. Gupta and J. Weinmann, "Research challenges in wireless networks of biomedical sensors," *ACM MobiCom 2001*.
3. P. Juang, H. Oki, Y. Wang, M. Martonosi, L. Peh and D. Rubenstein, "Energy-efficient computing for wildlife tracking: Design tradeoffs and early experiences with zebranet," in *Architectural Support for Programming Languages and Operating Systems (ASPLOS)*, 2002.
4. T. Small and Z. Haas, "The shared wireless infostation model - a new ad hoc networking paradigm (or where there is a whale, there is a way)," *ACM MobiHoc 2003*.
5. I. Vasilescu, K. Kotay, D. Rus, M. Dunbabin and P. Corke, "Data collection, storage, and retrieval with an underwater sensor network," *Proc. of ACM Sensys*, 2005.
6. G. Asada, T. Dong, F. Lin, G. Pottie, W. Kaiser and H. Marcy, "Wireless integrated network sensors: low power systems on a chip," *European Solid State Circuits Conference*, 1998.
7. The Ultra Low Power Wireless Sensor Project, http://www-mtl.mit.edu/ jimg/project_top.html, 2004.
8. A Chakrabarty, A Sabharwal and B Aazhang, "Using predictable observer mobility for power efficient design of a sensor network," *Second International Workshop on Information Processing in Sensor Networks (IPSN)*, April 2003.
9. A. Pentland, R. Fletcher and A. Hasson, "Daknet: rethinking connectivity in developing nations," *IEEE Computer*, vol. 37, no. 1, pp. 78-83, January 2004.
10. D. Jea, A.A. Somasundara and M.B. Srivastava, "Multiple controlled mobile elements (data mules) for data collection in Sensor Networks," *2005 IEEE/ACM International Conference on Distributed Computing in Sensor Systems (DCOSS '05)*, June 2005.
11. A.A. Somasundara, A. Ramamoorthy, M.B. Srivastava, "Mobile element scheduling for efficient data collection in wireless sensor networks with dynamic deadlines," *IEEE Real Time Systems Symposium (RTSS)*, December 2004.
12. J.H. Chang and L. Tassiulas, "Energy conserving routing in wireless ad-hoc networks," *IEEE INFOCOM 2000*.

13. A. Bogdanov, E. Maneva and S. Riesenfeld, "Power-aware base station positioning for sensor networks," *IEEE INFOCOM 2004.*.

14. Z. Zhang, M. Ma and Y. Yang, "Energy efficient multi-hop polling in clusters of two-layered heterogeneous sensor networks," *19th IEEE International Parallel and Distributed Processing Symposium*, (IPDPS '05), Denver, 2005.

15. C. Guo, L.C. Zhong and J.M. Rabaey, "Low power distributed MAC for ad hoc sensor radio networks," *IEEE GLOBECOM 2001*

16. R.K. Ahuja, T.L. Magnanti and J.B. Orlin, *Network Flows: Theory, Algorithms, and Applications*, Prentice-Hall, 1993.

17. E.D. Kaplan, ed., *Understanding GPS – Principles and Applications*, Artech House, 1996.

18. S. Capkun, M. Hamdi and J.P. Hubaux, "GPS-free positioning in mobile ad-hoc networks," *Hawaii Int. Conf. on System Sciences*, Jan. 2001.

19. S.S. Skiena, *Algorithm Design Manual*, Springer-Verlag, pp. 319-322, 1997.

A Distributed Linear Least Squares Method for Precise Localization with Low Complexity in Wireless Sensor Networks

Frank Reichenbach[1], Alexander Born[2], Dirk Timmermann[1], and Ralf Bill[2]

[1] University of Rostock, Germany
Institute of Applied Microelectronics and Computer Engineering
{frank.reichenbach, dirk.timmermann}@uni-rostock.de
[2] University of Rostock, Germany
Institute for Geodesy and Geoinformatics
{alexander.born, ralf.bill}@uni-rostock.de

Abstract. Localizing sensor nodes is essential due to their random distribution after deployment. To reach a long network lifetime, which strongly depends on the limited energy resources of every node, applied algorithms must be developed with an awareness of computation and communication cost. In this paper we present a new localization method, which places a minimum computational requirement on the nodes but achieves very low localization errors of less than 1%. To achieve this, we split the complex least squares method into a less central precalculation and a simple, distributed subcalculation. This allows precalculating the complex part on high-performance nodes, e.g. base stations. Next, sensor nodes estimate their own positions by simple subcalculation, which does not exhaust the limited resources. We analyzed our method with three commonly used numerical techniques - normal equations, qr-factorization, and singular-value decomposition. Simulation results showed that we reduced the complexity on every node by more than 47% for normal equations. In addition, the proposed algorithm is robust with respect to high input errors and has low communication and memory requirements.

1 Introduction

The increasing miniaturization in the semiconductor field is leading to the evolution of very small and low-cost sensors [1]. Due to their small size they are strongly limited with respect to processor capacity, memory size and energy resources. Several thousands of such sensor nodes get into wireless contact with each other and form large ad hoc sensor networks. A wireless sensor network (WSN) will be placed around a field of interest or within an object. The sensor nodes are able to monitor different environmental parameters and to transmit them to a beacon or an infrastructure node. WSN enable new applications such as timely detection of wood fire or monitoring of artificial dikes.

P. Gibbons et al. (Eds.): DCOSS 2006, LNCS 4026, pp. 514–528, 2006.
© Springer-Verlag Berlin Heidelberg 2006

The resulting data are only meaningful when combined with the geographical position of the sensor. Possible positioning technologies are the Global Positioning System (GPS) or the Global System for Mobile Communication (GSM) [2],[3]. These systems are however, due to the size of the equipment, the high prices and the high energy requirements, unsuitable for miniaturized sensor nodes and could only be used for a small number of nodes [4].

In this paper we present a new approach to energy-saving determination of unknown coordinates with a high precision. Using this method, the calculations are split between the resource-limited sensor nodes and the high-performance base station.

This paper is structured as follows: In Section 2 we give a basic overview of the methods for positioning in wireless sensor networks. In Section 3 we describe the position estimation based on relationships to known points. Then, in Section 4, we examine the complexity of three classical solution techniques in order to compare them with our new method. Next, we present in Section 5 our new approach to split the least squares method with the aim to minimize the load on the sensor nodes. Furthermore, the new method is analyzed with respect to complexity, memory requirement and communication effort. After discussing the simulation results in Section 5, we finally conclude the paper with Section 6.

2   Related Work

For the above-mentioned reasons, existing positioning techniques (e.g. GPS) cannot be integrated on all sensor nodes. The number of nodes with known position has to be limited. These nodes are referred to here as beacons, with the remaining nodes classed as sensor nodes. For the positioning of the sensor nodes we distinguish between approximate and exact methods.

Approximative Localization. Many approximate approaches for the determination of sensor nodes exist in literature. These algorithms are resource-efficient but also result in higher positioning errors. Examples of such approaches are the hybrid methods [5], the Coarse Grained Localization [6], by using local coordinate systems [7], the Approximate Point in Triangulation-algorithm (APIT) [8], and the Weighted Centroid Localization (WCL) [9].

Exact Localization. In contrast to approximation methods, exact methods use the known beacon-positions and the distances to the sensor nodes in order to calculate their coordinates through the solution of non-linear equations. Using a minimum of three beacons (in two dimensions), the coordinates of the sensor nodes may be determined using intersection. The use of more than three beacons gives more information in the system and allows the refinement of the position and the detection and removal of outlying observations. The least squares method (LSM) is used for the solution of the simultaneous equations. The LSM produces accurate results, however it is complex and resource-intensive and therefore not feasible on resource-limited sensor nodes. Savvides et al. described methods to overcome these problems in [10]. Kwon et al. presented a distributed

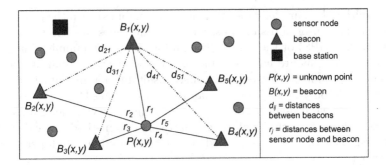

Fig. 1. Sensor network with one unknown point and beacons as reference points. Here beacon one was chosen as linearization tool.

solution using least squares whereby errors in acoustic measurements can be reduced [11]. Ahmed et al. published a new approach to combine the advantages of absolute and relative localization methods [12]. Moreover, Karalar et al. developed a low-energy system for positioning using least squares which may be integrated on individual sensor nodes [13]. A general overview of distributed positioning systems is given by Langendoen and Reijersin [14].

We demand exact localization methods that work on tiny sensor nodes with high limited energy resources. To achieve that, we transfer the complex calculations such as matrix multiplication, matrix inversion, and eigenvalue determination to the base station that can be e.g. a powerful desktop computer or a more efficient node in the network. Consequently, only simple calculations have to be executed on the sensor nodes. Additionally, we reduce the communication and memory requirements by optimizations of the proposed algorithm.

3   Background: Linearization and Least Squares Method

Estimating the position of an unknown point $P(x, y)$ requires in two-dimensions at least three known points (see Figure 1). With m known coordinates $B(x_i, y_i)$ and its distances r_i to them we obtain:

$$(x - x_i)^2 + (y - y_i)^2 = r_i^2 \quad (i = 1, 2, \ldots, m). \tag{1}$$

This system of equations must be linearized with either Taylor series [15] or a linearization tool [16]. Although the linearization tool is not as exact as the Taylor series, it requires no mathematical differentiation and it is suitable for a distributed implementation (discussed later in Section 5). Thus, we use the j'th equation of (1) as the linearization tool. By adding and subtracting x_j and y_j to all other equations this leads to:

$$(x - x_j + x_j - x_i)^2 + (y - y_j + y_j - y_i)^2 = r_i^2 \ (i = 1, 2, \ldots, j-1, j+1, \ldots, m). \tag{2}$$

With the distance r_j (r_i) that is the distance between the unknown point and the j'th (i'th) beacon and the distance d_{ij} that is the distance between beacon B_i and B_j this leads, after resolving and simplifying, to:

$$(x - x_j)(x_i - x_j) + (y - y_j)(y_i - y_j) = \frac{1}{2} \left[r_j^2 - r_i^2 + d_{ij}^2 \right] = b_{ij}. \qquad (3)$$

Because it is not important which equation we use as a linearization tool, $j = 1$ is sufficient. This is equal to choosing the first beacon and if $i = 2, 3, \ldots, m$ this leads to a linear system of equations with $m - 1$ equations and $n = 2$ unknowns.

$$
\begin{aligned}
(x - x_1)(x_2 - x_1) + (y - y_1)(y_2 - y_1) &= \tfrac{1}{2} \left[r_1^2 - r_2^2 + d_{21}^2 \right] = b_{21} \\
(x - x_1)(x_3 - x_1) + (y - y_1)(y_3 - y_1) &= \tfrac{1}{2} \left[r_1^2 - r_3^2 + d_{31}^2 \right] = b_{31} \\
&\vdots
\end{aligned}
$$

$$(x - x_1)(x_m - x_1) + (y - y_1)(y_m - y_1) = \tfrac{1}{2} \left[r_1^2 - r_m^2 + d_{m1}^2 \right] = b_{m1} \qquad (4)$$

This system of equations can be written in the matrix form:

$$\mathbf{A}\mathbf{x} = \mathbf{b} \qquad (5)$$

with:

$$
A = \begin{pmatrix}
x_2 - x_1 & y_2 - y_1 \\
x_3 - x_1 & y_3 - y_1 \\
\vdots & \vdots \\
x_m - x_1 & y_m - y_1
\end{pmatrix}, \mathbf{x} = \begin{pmatrix} x - x_1 \\ y - y_1 \end{pmatrix}, \mathbf{b} = \begin{pmatrix} b_{21} \\ b_{31} \\ \vdots \\ b_{m1} \end{pmatrix}. \qquad (6)
$$

This is the basic form that now has to be solved using the linear least squares method.

3.1   Solving the Linear Least Squares Problem

Due to the fact that overdetermined systems of equations with $m \gg n$ have not exact one solution for $\mathbf{A}\mathbf{x} = \mathbf{b}$, we have to apply the L2-norm [17]. This is also called the Euclidean Norm, which minimizes the sum of the squares:

$$\underset{x \in \Re^n}{Minimize} \ ||\mathbf{A}\mathbf{x} - \mathbf{b}||^2. \qquad (7)$$

To summarize (see Figure 2), linear systems of equations can be solved iteratively using Splitting techniques or directly with the normal equations or orthogonal factorization. Existing techniques are numerous but often the differences between them are small. For this reason, we focus our studies on three popular methods - normal equations, qr-factorization, and singular-value decomposition. For all others we recommend [18].

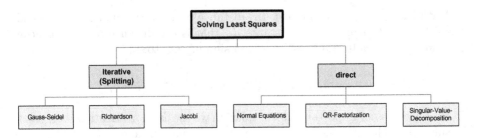

Fig. 2. Classification of common methods to solve a linear system of equations

Normal Equations. A trivial solution of the least squares problem is to re-convert after **x**. In this case, the unique solution of $A\mathbf{x} \approx \mathbf{b}$ is given by:

$$||A\mathbf{x} - \mathbf{b}||^2 \to A^T A\mathbf{x} = A^T\mathbf{b}. \tag{8}$$

Solving normal equations is a good choice if the linear system has many more equations than unknowns, i.e. $m >> n$, because after the multiplication $A^T A$ the result is only a quadratic $[n \times n]$-matrix. That decreases the following computation and makes it easy to implement in software. However, the numerical difficulties that can occur sometimes determine a completely wrong position, which leads to orthogonal techniques.

QR-Factorization. Orthogonal matrices transform vectors in different ways while they keep the length of the vector. Moreover, orthogonal matrices are invariant against the L2-norm, i.e. errors are not increased.

The qr-factorization transforms overdetermined linear systems of equations of the form $A\mathbf{x} \approx \mathbf{b}$ in a triangular system with the same solution, because it is:

$$||A\mathbf{x} - \mathbf{b}||^2 \to ||Q \begin{pmatrix} R_1 \\ 0 \end{pmatrix} \mathbf{x} - \mathbf{b}||^2 = ||\begin{pmatrix} R_1 \\ 0 \end{pmatrix} \mathbf{x} - Q^T\mathbf{b}||^2, \tag{9}$$

where Q is an orthogonal matrix (meaning that $Q^T Q = I$) and R_1 is an upper triangular matrix. This factorization is a standard method in numerics, is robust and stable to execute. In addition, the processing of rank defect matrices is possible.

Singular-Value Decomposition. A second method that we want to explain is the singular-value decomposition. If A has full rank and is a $[m \times n]$-matrix with $m > n$, then we can transform $A\mathbf{x} \approx \mathbf{b}$ in a diagonal system with the solution:

$$||A\mathbf{x} - \mathbf{b}||^2 \to ||U \begin{pmatrix} S_1 \\ 0 \end{pmatrix} V^T\mathbf{x} - \mathbf{b}||^2 = ||\begin{pmatrix} S_1 \\ 0 \end{pmatrix} V^T\mathbf{x} - U^T\mathbf{b}||^2 = ||\begin{pmatrix} S_1 \\ 0 \end{pmatrix} \mathbf{y} - U^T\mathbf{b}||^2, \tag{10}$$

where U is an orthogonal $[m \times m]$-matrix, V is an orthogonal $[n \times n]$-matrix and S is a diagonal matrix. The original algorithm has been implemented by Golub and Reinsch in [19]. This algorithm is also robust and stable to compute, but requires high computation effort due to root and eigenvalue operations.

4 Analysis: Complexity of the Methods

In the following, we will analyze the complexity of all three introduced methods. Although the literature offers numerous specifications, later we will reduce specific parts of the calculation. In order to mathematically define the complexity, we count the number of floating point operations (flops), which is commonly used in literature. The required number of computation cycles strongly depends on the hardware. Therefore, we count for every operation one flop whether it is an addition, subtraction, multiplication or division[1]. At this stage, we do not consider copying-operations in the memory, because this operation depends on the individual implementation of the algorithm.

As before, we will confine the explanation to two dimensions. Due to the linearization with a linearization tool the matrix A and the vector $\mathbf{b}$ have $(m-1)$-rows. For a clearer understanding we calculate with k-rows and substitute at the end: $m = k + 1$.

4.1 Complexity of the Normal Equations

The linear system of equations:

$$\mathbf{x} = \left(A^T A\right)^{-1} A^T \frac{1}{2}[r_1^2 - \mathbf{r}^2 + \mathbf{d}^2] \tag{11}$$

has to be solved. We divide the calculation into the following complexities.

1. Multiplying the $[n \times k]$-matrix A^T with the $[k \times n]$-matrix A leads to $\frac{n(n+1)}{2}$ flops[2].
2. The $[n \times n]$-matrix, resulting from 1., must be inverted[3] with a complexity of n^3.
3. The $[n \times n]$-matrix, resulting from 2., must be multiplied with the $[n \times k]$-matrix A^T, which costs $2n^2 k - nk$ flops. This leads to the precalculated Matrix A_p.
4. The matrix A_p must be multiplied with the k-vector $\mathbf{b}$. This step has a complexity of $2kn - n$ flops.
5. The calculation of $\mathbf{b}$ needs $5k + 1$ flops.

With $k = m - 1$ this leads to a total complexity of $15m - 5$ for the least squares method with m beacons and $n = 2$ unknowns.

4.2 Complexity of the QR-Factorization

Now, the complexity for the qr-factorization has to be studied. First the partial matrices of Q have to be determined, which in our case for $n = 2$ are limited to only two matrices; Q_1 and Q_2.

---

[1] It should be noted that in the arithmetic unit of a processor a division is a more complex operation than an addition. We will focus on theoretical analysis.

[2] Some operations can be saved by multiplying a transposed matrix with itself.

[3] The inversion of a matrix is very complex with n^3 flops (see [20]).

1. The calculation of Q_1 needs $\frac{5}{2}k^2 + \frac{9}{2}k - n + 5$ flops.
2. The calculation of Q_2 needs $\frac{5}{2}(k-1)^2 + \frac{9}{2}(k-1) - n + 5$ flops, because we do not need to consider the last line in the calculation.
3. The multiplication of Q_1Q_2A, where Q_1 and Q_2 have the size $k \times k$, needs $2k^3 + 3k^2 - 2k$ flops.
4. The calculation of $\mathbf{b}$ needs $5k + 1$ flops (see 4.1).
5. For the calculation of $Q_1^T\mathbf{b}$ it has to be considered, that only the upper two rows are needed for the multiplication. This leads to $8k - 4$ flops.
6. The calculation of $Rx \approx Q^T\mathbf{b}$, by back substitution, requires exactly 4 flops.

Summarized, the complexity of the qr-factorization is $4m^3 - 5m^2 + 13m - \frac{9}{2}$ flops.

4.3 Complexity of the SV-Decomposition

The sv-decomposition is more complicated than the previously discussed procedures, because a determination of eigenvalues is necessary, which has to be determined using several methods. In principal, $9n^3 + 8n^2k + 4k^2n$ flops are needed for the computation of U, V and S referring to [20]. With $n = 2$ this leads to $8k^2 + 24k + 48$ flops. Additionally, $\mathbf{x}$ must be determined with $S_1y = U_1^Tb$ and $\mathbf{x} = Vy$. Finally, this leads to a complexity of $8m^2 + 25m + 54$ flops for the svd.

Now, after discussing the standard methods, we will explain the new approach for distributing them in wireless sensor networks.

5 New Approach: Distributing Least Squares Problem

Linearizing non linear equations with a linearization tool has a significant advantage. All elements in matrix A are beacon positions $B_1(x, y)..B_m(x, y)$ only. Moreover, vector $\mathbf{b}$ consist of distances between the unknown sensor node and all beacons $r_1..r_m$ and distances $d_1..d_m$ between the first beacon and all others. Consequently, we split the complex computation into two parts - a less complex and a very simple part. First, we precalculate matrix A into a different form, which strongly depends on the solution method and will be discussed later. Then, with this precalculated form, a simple subcalculation starts. This splitting method works in a similar fashion with all three solution methods.

So far, in sensor networks it is desired to execute the entire localization algorithm on every node; completely distributed. With this assumption, the precalculation of a least squares method would be exactly the same on every sensor node, because matrix A is the same for every sensor node in a static network. This wastes limited resources and produces high redundancy.

Now, with our distributed approach, the high-performance base station executes the complex precalculation and the resource-aware sensor nodes estimate their own position with the simple subcalculation. In this subcalculation all sensor nodes use the same precalculated information combined with their individual measured distances to every beacon. This assumes that all beacons are able to communicate with every sensor node directly. This is difficult to achieve in real environments, due to obstacles and limited transmission ranges, but can be solved by multi-hopping techniques [21]. Thereby, beacons send packets over neighboring nodes hop by hop to the destination node. The number of hops is an indicator for a distance to the beacon. However, the focus in this paper is not the distance determination, but the localization process. Before we describe the methods in detail, we will postulate the entire algorithm.

Distributed Least Squares Algorithm

Step 0: Initialization Phase:
        All beacons send their position $B(x,y)$ to the base-
        station.
Step 1: Complex Precalculation Phase (central):
        Base station builds matrix A and vector d_p.
        Starting the complex precalculation (result strongly depends
        on the solution method).
Step 2: Communication Phase (distributed):
        Base station sends precalculated data and vector d_p
        to all sensor nodes.
Step 3: Simple Subcalculation Phase (distributed):
        Sensor nodes determine the distance to every beacon $r_1..r_m$.
        Sensor nodes receive the precalculated data and vector d_p,
        built vector b and estimate their own position P_{est} autonomously.

In the next section, we will adapt the algorithm to all three solution methods and analyze the results in detail.

5.1 Reduced Complexity

In Section 4.1 we already analyzed the complexity of all three solution methods. At this point it is important to know what we can save on the sensor nodes without complex precalculations regarding only the remaining subcalculation.

Normal Equations. We assume the constellation of the network in Figure 3 with sensor nodes, beacons, and a base station. On bases of (11) the base station precalculates $A_p = (A^T A)^{-1} A^T$ and $\mathbf{d}_p = \mathbf{d}^2$. The matrix A and vector $\mathbf{d}_p$ are sent to all sensor nodes. Together with the distances $\mathbf{r}$ to all beacons, which every sensor node must determine itself, the subcalculation starts:

$$\mathbf{x} = A_p \frac{1}{2}(r_1^2 - \mathbf{r}^2 + \mathbf{d}_p) \qquad (12)$$

This computation requires $8m - 11$ flops.

QR-Factorization. Here, the base station transmits the precalculated matrices Q, R, and also vector $\mathbf{d}_p$. With this, every sensor node reduces computation to the following:

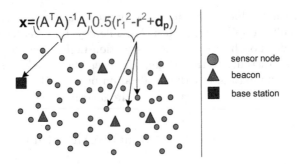

Fig. 3. Splitting the normal equations

1. Creating **b** needs $4k + 1$ flops (We use the substitution k=(m-1) again.).
2. Multiplying $\mathbf{y} = Q_1^T \mathbf{b}$, where Q_1^T is a $[2 \times k]$-matrix and **b** is a k-vector needs $8k - 4$ flops.
3. Solving $R\mathbf{x} \approx \mathbf{y}$ by back-substitution finally requires 4 flops.

Summing up, the subcalculation requires a reduced complexity of $12m - 11$.

SV-Decomposition. As the third method, the sensor nodes receive from the base station the matrices U, S, V and $\mathbf{d}_p$ and have to compute:

1. Creating **b** requires $4k + 1$ flops.
2. Solving $U_1^T \mathbf{b}$, where U_1^T is a $[2 \times k]$-matrix and **b** is a k-vector, results in the vector **z** and requires $4k - 2$ flops.
3. Then, **y** is calculated by back-substitution of $S_1 \mathbf{y} = \mathbf{z}$. Due to the two zero elements in S_1 this requires only 2 flops.
4. The last part of the calculation requires 6 flops, where **x** is determined by $\mathbf{x} = V\mathbf{y}$.

Adding all together leads to a reduced complexity of $8m + 1$ flops.

To compare all complexities Figure 4 was created. The complete solutions of the least squares method require much more operations than the reduced methods. Especially qr-factorization and sv-decomposition would exhaust the sensor nodes resources, because with only 50 beacons more than 10^4 floating point operations are required. In contrast, the normal equations are less complex. Obviously, all reduced calculations decrease the complexity and make the localization on resource aware sensor nodes feasible. However, a low communication is also demanded, which we will analyze next.

5.2 Communication Effort

As already described, communicating between sensor nodes is critical and must be minimized. Particularly, sending data over long distances stresses the energy capacity of sensor nodes. Communication between base station and beacons is less critical and must be preferred if possible. Therefore, we classify

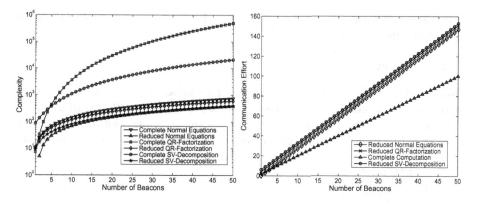

Fig. 4. Complexity of all complete and all reduced calculations in flops

Fig. 5. Communication effort of all reduced and the traditional method

communication in two phases. In an uncritical phase, all beacons send their positions to the base station. This causes no energy loss on the sensor nodes. Additionally, in a critical phase the base station sends precalculated information to the sensor nodes that have, in theory, to receive only. Practically, transmitting/sending is never lossless, due to errors in the transmission channel and protocols that require acknowledge packets etc. Furthermore, the base station cannot reach every sensor node in one hop, which demands multi-hopping over some nodes.

Normal Equations. Here, we focus on a theoretical comparison of the algorithms that is, for the moment, independent of protocol definitions and media access operations. Hence, every sensor node must receive the precalculated matrices and vector d_p. The communication effort directly depends on the used solution method.

At the normal equations the sensor node receives matrix A_p and vector $\mathbf{d}_p$, with $[n \cdot (m-1) + (m-1)]$ elements. This results in receiving $(3m-3)$ elements.

QR-Factorization. This requires transmitting Q_1^T with $[2 \cdot (m-1)]$ elements, matrix R with only two elements and also $\mathbf{d}_p$. Summarized, $(3m+1)$ elements must be send in the critical phase.

SV-Decomposition. By applying the sv-decomposition, $(3m+3)$ elements must be transmitted, because S_1 consists of two elements, U_1^T has $[(m-1) \cdot 2]$ elements, the quadratic matrix V consist of two elements and vector d_p of (m-1) elements.

We compared all communication efforts in Figure 5. The communication effort of all reduced methods is relatively low, comparing to the traditional method with much computation overhead[4]. The direct solution of the normal equations

---

[4] "Complete Computation" stands for the classical method, where all beacons send their positions directly to all sensor nodes. Thus, every sensor node must receive at least two positions to determine its own position that results in $2m$ elements.

minimizes communication. As an example, with 50 beacons not more than 100 elements must be received.

5.3 Memory Considerations

Normal Case. The reduced calculations must be feasible on sensor nodes with a very small memory, mostly not more than a few kilobyte RAM. In our case, the memory consuming operation is always the multiplication of $\mathbf{b}$ with the precalculated data. Without optimizations this would be for the three methods:

1. $A_p \cdot \frac{1}{2} \cdot (r_1^2 - \mathbf{r}^2 + \mathbf{d}_p)$
2. $Q_1^T \cdot \frac{1}{2} \cdot (r_1^2 - \mathbf{r}^2 + \mathbf{d}_p)$
3. $U_1^T \cdot \frac{1}{2} \cdot (r_1^2 - \mathbf{r}^2 + \mathbf{d}_p)$

In the worst case A_p, Q_1^T, U_1^T, and $\mathbf{r}$ plus $\mathbf{d}_p$ must be stored temporarily in memory before the execution on the sensor node can start. In more detail, $[2 \cdot (m - 1)] + (m - 1) + (m - 1) = (4m - 4)$ elements must be stored. On common microcontrollers, that are presently integrated on sensor node platforms, every element is stored in floating point representation as a 4 byte number. Accordingly, with $m = 100$ beacons, already 0.796 kb must be allocated, for localization only. Normally, the localization task is part of the middleware that has to execute many more tasks. Besides, temporary variables are needed that increases the memory consumption. Given these facts, we studied the critical operations in more detail and will describe optimizations in the next section.

Optimizations. In reality, input data for sensor nodes arrive in packets and will be disassembled into a serial data stream. Due to problems in the transmission channel (e.g. different paths or transmission errors) a sorted order of the incoming packets cannot be guaranteed. The data can arrive in an unsorted form and the calculation begins after receiving all data.

However, the reduced calculation has a further useful quality. Individual calculations of $A_p \cdot \mathbf{b}$ can be executed after the arrival of only some elements without collecting all data. Only one accumulator for the position $P_{est}(x)$ and one for the position $P_{est}(y)$ of the sensor node is needed. If we define $W = A_p, Q_1^T, U_1^T$ independently from the specific method, the following multiplication will always result:

$$\begin{pmatrix} w_{11} & \cdots & w_{1(m-1)} \\ w_{21} & \cdots & w_{2(m-1)} \end{pmatrix} \begin{pmatrix} b_1 \\ b_2 \\ \vdots \\ b_{(m-1)} \end{pmatrix} = \begin{pmatrix} w_{11} \cdot b_1 + \ldots + w_{1(m-1)} \cdot b_{(m-1)} \\ w_{21} \cdot b_1 + \ldots + w_{1(m-1)} \cdot b_{(m-1)} \end{pmatrix}.$$

With w_{ij} ($i = 1..2, j = 1..m - 1$) and b_s ($s = 1..m - 1$) the following assumptions can be made. If elements with $j = s$ are available, an immediate multiplication of $w_{ij} \cdot b_s$ and a subsequent accumulation in $P_{est}(x, y)$ is possible. The index i distinguishes into which accumulator it must be written; $P_{est}(x)$ at

$i = 1$ or $P_{est}(y)$ at $i = 2$. Finally, the optimized reduced calculation requires a worst-case calculation time of $(m - 1)/2$ if the elements arrive in reverse order.

To avoid the case of unsorted data it is also possible to send the elements w_{ij}, d_s in appropriate tuples, e.g. $w_{11}, d_1; w_{12}, d_2; \ldots; w_{ij}, d_s$. In best case, space in memory has then to be reserved for only a few temporary variables and two accumulators which reduces memory consumption to a minimum.

5.4   Example

The algorithm is intended for implementation and execution on a sensor platform. Due to this, we have represented the results in Table 1. We assume $m = 100$ beacons and $n = 2$ for the second dimension[5]. Furthermore, we assume that a floating point representation requires 4 byte of memory.

Table 1. Performance comparison with floating point numbers and 100 beacons. The communication effort and the memory capacity in the table refer to the reduced methods.

Algorithm	Full Complexity [flops]	Reduced Complexity [flops]	Savings [%]	Communication Effort [bytes]	Memory Capacity [bytes]
Normal Equation	1497	791	47.16	1188	≈ 1588
QR-Factorization	3951302	1201	99.97	1204	≈ 1588
SV-Decomposition	82556	803	99.03	1212	≈ 1588

The direct calculation by normal equations requires minimal computation whereas the calculation by the qr-factorization requires lowest data traffic. However, the normal equations are numerically instable and the sensitivity of the linear equalization problem can deteriorate. Remarkable are the saving of the calculation for the qr-factorization and the sv-decomposition, because the precalculation requires the largest expenditure. Summarized, this overall comparison shows the potential advantages of the new distributed approach.

5.5   Noisy Observations

Estimating the position basically requires beacon positions and distances. Due to various error influences, e.g. imprecise measuring of the signal of flight or defective GPS-coordinates, the applied algorithm must be robust to input errors. For this reason, we studied the behavior of the described methods in different simulations concerning noisy distances $\mathbf{r}$ and noisy beacon positions $B(x, y)$. To realize this simulation, we substituted the exact value by a chosen random value out of a Gaussian distribution $r_{appr}, B_{appr} \sim \mathcal{N}(\mu_{exact}, \sigma^2)$. The exact value was the arithmetic mean and the variance σ^2 was a parameter. We always simulated with 500 nodes, where 5% were beacons and the rest sensor nodes. We executed

---

[5] It must be considered here to add x_1 and y_1 to the final coordinates due to the linearization in (6).

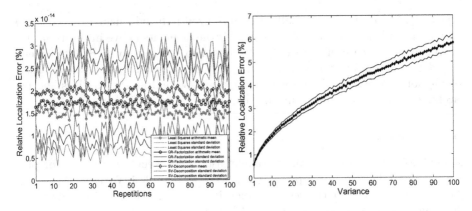

Fig. 6. Localization error with exact input   **Fig. 7.** Error with noisy distances

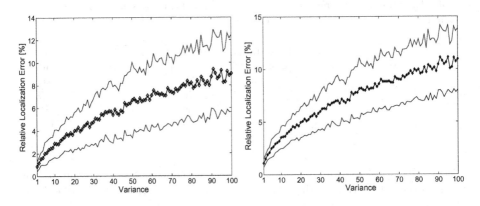

Fig. 8. Error with noisy beacon positions   **Fig. 9.** Error with complete noisy input

numerous series and averaged the results to avoid a strong influence of outliers. We created a test field with the size 100×100 where all nodes were placed by a uniform distribution. We determined in all simulations the averaged relative localization error.

In a first simulation we compared the achievable precision of the three methods that solves the least squares method. Figure 6 shows that with exact input the error for all three reduced methods ranges only in an interval of $0.5 \cdot 10^{-14}$. Following simulations with noisy input showed the same differences, leading us to continue simulating using only the normal equations.

Next, a simulation with noisy distances was executed (see Figure 7). An increasing variance of the Gaussian distribution resulted in an increasing relative localization error up to 6% at $\sigma^2 = 100$. The standard deviation is relatively small.

In more simulation series we studied the error for noisy beacon positions. This simulation describes the behavior of the algorithm in a slightly dynamic network, where the beacon positions can change after the precalculations are already executed and transmitted to the sensor nodes. The sensor nodes would determine new distances to the beacons but combine them with wrong beacon coordinates. Figure 8 shows the increasing error that rises over 8% at a variance of 100. The standard deviation is also higher compared to the previous simulation. Defective beacon positions influences the results strongly, but normally they are not the main problem. This means that the algorithm is able to manage slight changes in a dynamic network with an acceptable error.

In a last simulation we increased the variance for both, distances and beacon positions. Figure 9 shows the result. Here, the highest error occurs, as it was expected. However, these results show the robustness of the algorithm.

6   Conclusion

We have presented a new method for exact localization in resource-limited sensor networks by distributing the least squares method. Usually the calculation of this method is very complex with an increasing number of beacons. However, the use of the linearization tool enables us to split the complex calculation into a complex part, precalculated on the high-performance base station, and a very simple subcalculation on every sensor node. With low communication traffic the base station sends precalculated data to all sensor nodes. Sensor nodes must only receive data and compute their own position autonomously.

Simulations show that a complexity reduction of 99% (for qr-factorization and singular-value decomposition) and 47% (for normal equations), using 100 beacons, is achievable without increasing the communication requirements. Moreover, we described optimizations where the algorithm starts executing as soon as the first data arrive at the sensor node. This allowed high savings in the required memory capacity with only a few kilobytes of memory considered. Currently, we are studying the algorithm in extensive network simulations. In future, the implementation on a real sensor platform is planned.

Acknowledgment

This work was supported by the German Research Foundation under grant number TI254/15-1 and BI467/17-1 (keyword: Geosens). We appreciate comments given by Edward Nash, which helped us to improve this paper.

References

1. Akyildiz, I.F., Su, W., Sankarasubramaniam, Y., Cayirci, E.: Wireless sensor networks: A survey. Computer Networks **38** (2002) 393–422
2. Bill, R., Cap, C., Kohfahl, M., Mund, T.: Indoor and outdoor positioning in mobile environments a review and some investigations on wlan positioning. Geographic Information Sciences **10** (2004) 91–98
3. Gibson, J.: The mobile communications handbook. CRC Press (1996)

4. Min, R., Bhardwaj, M., Cho, S., Sinha, A., Shih, E., Wang, A., Chandrakasan, A.: Low-power wireless sensor networks. In: International Conference on VLSI Design. (2001) 205–210
5. Savarese, C., Rabaey, J., Langendoen, K.: Robust positioning algorithms for distributed ad-hoc wireless sensor networks. In: USENIX Technical Annual Conference. (2002) 317–327
6. Bulusu, N.: Gps-less low cost outdoor localization for very small devices. IEEE Personal Communications Magazine 7 (2000) 28–34
7. Capkun, S., Hamdi, M., Hubaux, J.P.: Gps-free positioning in mobile ad hoc networks. Cluster Computing 5 (2002) 157–167
8. Tian, H., Chengdu, H., Brian, B.M., John, S.A., Tarek, A.: Range-free localization schemes for large scale sensor networks. In: 9th annual international conference on Mobile computing and networking. (2003) 81–95
9. Blumenthal, J., Reichenbach, F., Timmermann, D.: Precise positioning with a low complexity algorithm in ad hoc wireless sensor networks. PIK - Praxis der Informationsverarbeitung und Kommunikation 28 (2005) 80–85
10. Savvides, A., Han, C.C., Srivastava, M.B.: Dynamic fine grained localization in ad-hoc networks of sensors. In: Seventh Annual ACM/IEEE International Conference on Mobile Computing and Networking. (2001) 166–179
11. Kwon, Y., Mechitov, K., Sundresh, S., Kim, W., Agha, G.: Resilient localization for sensor networks in outdoor environments. In: 25th IEEE International Conference on Distributed Computing Systems. (2005) 643–652
12. Ahmed, A.A., Shi, H., Shang, Y.: Sharp: A new approach to relative localization in wireless sensor networks. In: Second International Workshop on Wireless Ad Hoc Networking. (2005) 892–898
13. Karalar, T.C., Yamashita, S., Sheets, M., Rabaey, J.: An integrated, low power localization system for sensor networks. In: First Annual International Conference on Mobile and Ubiquitous Systems: Networking and Services. (2004) 24–30
14. Langendoen, K., Reijers, N.: Distributed localization in wireless sensor networks: A quantitative comparison. Computer Networks (Elsevier), special issue on Wireless Sensor Networks 43 (2003) 499–518
15. Niemeier, W.: Ausgleichsrechnung. de Gruyter (2002)
16. Murphy, W.S., Hereman, W.: Determination of a position in three dimensions using trilateration and approximate distances. (1999)
17. Gramlich, G.: Numerische Mathematik mit Matlab - Eine Einführung für Naturwissenschaftler und Ingenieure. dpunkt.verlag (2000)
18. Lawson, C.L., Hanson, R.: Solving Least Squares Problems. Englewood Cliffs, NJ: Prentice-Hall (1974)
19. Golub, G.H., Reinsch, C.: Singular Value Decomposition and Least Square Solutions, Linear Algebra, Volume II of Handbook for Automatic Computations. Springer Verlag (1971)
20. Golub, G.H., Loan, C.F.V.: Matrix Computations. The Johns Hopkins University Press (1996)
21. Niculescu, D., Nath, B.: Ad hoc positioning system (aps) using aoa. In: Proceedings of the IEEE Annu. Joint Conf. IEEE Computer and Communications Societies. (2003) 1734–1743

Consistency-Based On-line Localization in Sensor Networks

Jessica Feng, Lewis Girod, and Miodrag Potkonjak

4821 Boelter Hall, Los Angeles,
California 90034, USA
{jessicaf, miodrag}@cs.ucla.edu, girod@lecs.cs.ucla.edu

Abstract. We have developed a new on-line error modeling and optimization-based localization approach for sensor networks in the presence of distance measurement noise. The approach is solely based on the concept of consistency, and is developed specifically for the case of on-line localization, which refers to the situation when references are not available a priori. The localization problem is formulated as the task of maximizing the consistency between measurements and calculated distances. In addition, we also present a localized localization algorithm where a specified communication cost or the location accuracy is guaranteed while optimizing the other. We evaluated the approach in (i) both GPS-based and GPS-less scenarios; (ii) 1-D, 2-D and 3-D spaces, on sets of acoustic ranging-based distance measurements recorded by deployed sensor networks. The experimental evaluation indicates that localization of only a few centimeters is consistently achieved when the average and median distance measurement errors are more than a meter, even when the nodes have only a few distance measurements. The relative performance in terms of location accuracy compares favorably with respect to several state-of-the-art localization approaches. Finally, several insightful observations about the required conditions for accurate location discovery are deduced by analyzing the experimental results.

Keywords: Consistency, Location Discovery, Statistical Modeling.

1 Introduction

Sensor networks and pervasive computing systems form one of the fastest growing computer and networking research frontiers. Once the nodes that form a network or an infrastructure are deployed, invariably there is a need that each node discovers its position. Global position system (GPS) can greatly facilitates this task. However, due to obstacles such as trees and walls, the GPS system often does not lock to satellite signals. At the same time, GPS systems are relatively expensive and consume a significant amount of energy. Therefore, usually only a limited subset of nodes is equipped with GPS; other nodes deduce their locations by measuring distances between themselves. For this purpose, a variety of distance measurement technologies have been employed, including signal strength attenuation techniques, ultra wide band approaches, Doppler-assisted methods, carrier-phase-based measurements and acoustic signal-based techniques. The technologies differ significantly in terms of

P. Gibbons et al. (Eds.): DCOSS 2006, LNCS 4026, pp. 529–545, 2006.
© Springer-Verlag Berlin Heidelberg 2006

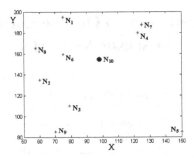

Fig. 1. Motivational example topology

Table 1. The distance measurements information

ID	LOCATION	REAL	GAUSSIAN	STAT 1	STAT 2
N_1	(75, 195)	45.893	45.791	56.697	44.193
N_2	(60, 135)	42.5	43.432	42.895	42.043
N_3	(79, 110)	48.654	78.066	49.008	39.964
N_4	(122.5, 180)	35.355	35.294	34.355	42.139
N_5	(150, 85)	87.5	86.362	56.988	87.479
N_6	(75, 159.4)	22.926	53.285	23.001	27.077
N_7	(125, 187.5)	42.573	42.938	43.837	41.992
N_8	(57.5, 165.4)	41.337	42.831	41.111	49.604
N_9	(70, 85)	75.208	71.427	87.449	74.574

Table 2. Solutions resulted using different error models (columns) based on different sets of measurements (rows)

	GAUSSIAN	STAT 1	STAT 2	CONSISTENCY
GAUSSIAN	0.0208	7.993	4.258	0.0424
STAT 1	8.179	0.0117	5.275	0.0315
STAT 2	7.658	6.042	0.0303	0.0396

maximum and minimum measuring range, resilience toward obstacles, power consumption, cost of deployment and power budget. Nevertheless they share a common denominator: distance measurements are prone to both small fluctuation and occasional large errors [1].

The *localization (location discovery) problem* can be defined in the following way. A total of N nodes, K of which ($K<<N$) have exact information about their positions. The measured distances, which are subject to errors, between M pairs of nodes are also available. The goal is to conclude the location (x_i, y_i) of each node i in such a way that $L(x_{ri}-x_i, y_{ri}-y_i)$ is minimized, where (x_{ri}, y_{ri}) is the actual location of i. Usually the targeted error norm L is L_1, $L2$, or $L\infty$.

It has been proven that location discovery problem is NP-complete [2]. It is also easy to see that the location discovery problem belongs to the class of nonlinear programs. A great variety of *centralized algorithms* (executed at a single place with the availability of the complete information about all measurements) and *localized algorithms* (executed by multiple nodes simultaneously and/or consecutively where each node has limited information provided by its neighbors) have been proposed. They range from iterative linearization and convex programming to conjugate direction-based and multiresolution search. [1][3] provide comprehensive surveys of state-of-the-art positioning designs and signal processing techniques. However, the effectiveness of these algorithms is constrained by the accuracy of the error model. There is a wide spectrum of available error models ranging from closed form parametric models to sophisticated kernel estimation-based non-parametric models. Nevertheless, none of them is a-priori applicable in new environments. The small example shown in Fig. 1 demonstrates the importance of the correct error model.

Consider 10 nodes $N_1,...,N_{10}$. We assume that the locations of the first nine nodes are available and error free. The topology of these 10 nodes is taken from a deployed network. The distances between the nodes are estimated based on the time-of-arrival of the acoustic signals. The traveling time of the acoustic signals is multiplied with the speed of the sound to estimate the distances between nodes – the *measured distances* [4][5]. Table 1 contains the information about the locations of the nine nodes (the second column); the *real/correct distances* obtained using the distance formula given the real locations of the nodes (the third column); the measured positions on two different days (the fifth and the sixth columns - STAT1 and STAT2). All measurements are in meters. In addition, the forth column shows the *simulated distances* generated under the widely used assumption of Gaussian noise model [6][7] on top of the real distances.

The goal is to locate N_{10} using the measured/simulated distances. We obtain the solution using the exhaustive search and following the maximum likelihood principle. Table 2 shows the results in term of location error, i.e. $(x_{r10}-x_{10}, y_{r10}-y_{10})$. The three rows indicate which set of measured/simulated distance measurements is used to derive N_{10}'s location (i.e. which type of error is in the distance measurements), and the four columns indicate the type of errors targeted by the maximum likelihood (i.e. the error model used as the optimization target). We see that when the correct type of errors is targeted, low location discrepancy is achieved, indicated by the bold italic numbers in Table 2. The average location error is between 1 and 3.3cm although some individual measurements have errors of more than 40m. However, when the errors in measurements and the optimization targeted error model do not match, the location error increases significantly. For example, when the Gaussian error model is assumed for the minimization of errors on the actually collected data - STAT1, the location error is more than 8m (8.179m). Even when the model obtained on one day is used as the optimization objective on another day, the resultant location error is still above 5m (6.042m and 5.275m). Therefore, we conclude that unless an accurate error model with respect to the measurements is targeted, accurate location discovery is not possible.

However, a simple condition of *pair-wise consistency* easily resolves this problem, at least for the example shown in Fig. 1. We say that a pair of measurements is pair-wise consistent if the longer measurement corresponds to the longer real distance. The formal definition of consistency is stated in Section 4. The last column in Table 2 shows the location errors yield based on the on-line localization. Regardless of what type of errors is in the distance measurements, the location error of N_{10} is always around 3cm. The final observation is that maximizing the percentage of consistent measurements can be easily mapped to nonlinear function minimization problem and solved using standard software [8].

We will try to demonstrate and statistically prove in the rest of the paper that the effectiveness of the pair-wise consistency modeling technique is not restricted to small instances of the problem. Our main technical goal is to demonstrate the effectiveness of the consistency-based formulation for location discovery in sensor networks where the location of each node is determined using information about the distances between limited number of communicating nodes (not limited to only beacons). We define the *calculated distance* between two nodes as the distance calculated by following the distance formula based on the locations proposed by the localization algorithm. We start by presenting the data acquisition process and restating the NP-completeness of the localization problem in Section 3. In Section 4, we demonstrate how to construct a monotonic continuous error model on-line when no golden standard (real distances) is available. The on-line model is based on the consistency between the measured and the calculated distances. The developed model is evaluated in two ways: i) using the standard statistical learn-and-test method; ii) evaluating the location accuracy when the models are served as the optimization objective..

2 Related Work

Wireless ad-hoc sensor networks (WASNs) are distributed embedded systems where each node combines sensing, computing, communication and storage capabilities. One of the fundamental tasks in WASNs is location discovery, which refers to the task where all the unknown-location nodes seek to determine the relative and/or absolute positions using the measured distances between different nodes. Such a distance can be measured by approaches include acoustic ranging methods [4][9][10], RSSI and RF proximity estimation [5][6], as well as algorithmic techniques [4][7]. In this paper, the set of distance measures we use as the demonstrative example was collected based on the line-of-sight acoustic signals [9][10].

Location discovery algorithms can be either centralized or localized [11]. Centralized algorithms assume that all the measured distances are forwarded to the center node, which then computes the location of each node using such information. Localized algorithms do not require the existence of the center node and allows each node to compute its position based on its local information by atomic multilateration, a method to estimate the location of a node if it is within the communication range of at least three beacons [12]. Iterative multilateration algorithm uses atomic multilateration as the primitive and treats an unknown-location node as a beacon once its location is resolved [13].

There are in general two different scenarios the location discovery problem is solved under. One of which assumes the measured distances between communicating nodes are available. Some of the recent work which are based on this assumption include [14][15][16]. Niculescu and Nath [16] propose a localization approach which is based on the basic idea of distance vector routing using only a fraction of beacons, with the assumption that each sensor node has some combination of ability to measure range, angle of arrival (AOF), orientation. They propose a lower bound for positioning error for a range/angle free algorithm, and examine the error characteristics of various classes of multihop ad-hoc positioning systems (APS)

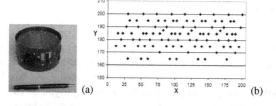

Fig. 2(a). A SH4 node. **2(b).** An example of the deployment topology.

algorithms. The localization method prop-osed by Galstyan et al. [17] is distributed and on-line, which means the localization process is conducted simultaneously with an application task. Sensor nodes use their geometric constraints induced by radio connectivity and sensing to decrease the uncertainty of positions. The performance of the algorithm is compared with the centralized (convex) programming. In addition to static networks, Hu and Evens [15] introduce the sequential Monte Carlo localization method for mobile networks, which exploit mobility to improve the accuracy and precision of positioning. A comprehensive study of the fundamental limitations and location accuracy bound for mobile positioning is presented in [18]. Sivavakeesar and Pavlou [19] propose an approach for hierarchically organized networks. By employing the dominating-set to perform periodic location updates on behalf of other nodes, the approach leads to less control traffic.

The second scenario which location discovery is solved under does not put any requirement on the availability of measured distances [20][21][22][23] [24]. He et al. [20] propose a range-free localization approach which performs the best when an irregular radio pattern and random node placement are considered. Yis et al. in [21] presents a localization method that uses the connectivity information (i.e. who is within the communication ranges of whom) to derive the positions of the unknown sensor nodes. Bruck et al. [22] study the localization problem in a 2-d and beacon-free environment, relying only on the local angle measurements. The approach determines a planar spanner of a unit disk graph, which can be used to generate a set of virtual coordinates. Finally, Viana et al. [23] argues that traditional localization schemes depend too much on the spatial distribution, which can lead to limitations that go against the principles of self-organization. The authors propose a location service for self-organizing networks that defines a logical multidimensional space, which is a strict mathematical representation of the network geographic space.

3 Preliminaries

In this section, we summarize all the necessary preliminaries for the derivation of pair-wise consistency-based on-line error models. We explain how the distance measurements were collected and discuss the relationship between the communication range verses the distance ranging range. Finally, we state the formulation for the localization problem as an optimization instance, especially as a nonlinear function minimization instance.

3.1 The Distance Measurements

The demonstrative example and all of our experimental results are conducted on sets of distance measurements that are collected using the acoustic signal detection-based ranging techniques. The number of deployed sensor nodes varies from 79 to 93, with the average being 90. The sensor nodes are custom designed based on an SH4 microprocessor running at 200MHz (Fig. 2(a)). The nodes were deployed at the Fort Leonard Wood Self Healing Minefield Test Facility, which measures 200m x 50m. The radio signal (communication) range is about 50m. Fig. 2(b) shows an example of the deployment topology. Each node is equipped with four independent speakers and microphones as the ranging tool. The distance between two nodes is obtained by timing the arrival of the acoustic signals [10]. Each node in the network takes turns to transmit the acoustic signals; all the nodes that receive the signals record the time of arrival and convert the time of flight to distance in meters. In total, there are 33 sets of distance measurements collected over the course of few days; each set consists of one round of acoustic signal transmission by all the nodes. For the sake of simplicity, we demonstrate the algorithms and techniques on a randomly selected subset of measurements, and we present the results for ten other randomly selected data sets in Section 6. The details on the experimental setup and the acoustic detection scheme used can be found in [9][10]. It is important to note that the techniques we propose in our study can be applied when no GPS devices are a-prior available, i.e., when no beacons are present in the network. In both on-line and the localized scenarios, the nonlinear function minimization formulation remains unchanged except that the coordinates of the beacons are now unknowns as well. Furthermore, the GPS-less localization can be viewed as the first step of conducting GPS-based localization. The nodes are first resolved as if no beacons are available (relative locations with respect to each other); then the absolute locations can be further derived given the ground truth of at least three nodes. We experimentally evaluated the GPS-less localization in Section 6.

From the communication point of view, we distinguish two types of the communications between a pair of nodes: (i) exchange of the acoustic signals for the purpose of distance ranging; (ii) and (ii) transmission and reception of radio signals (in terms of bytes) for the purpose of exchanging information. More specifically, we denote by L_i a set of nodes that receive node i's acoustic signals, therefore can estimate the distances between themselves to node i. Similarly, C_i denotes a set of nodes that receive the radio signals from i. We assume that the acoustic signal range (ASR) is independent from the radio signal range (RSR), which means that it is possible for a node i to have the distance estimate to another node j (i has received j's acoustic signals), while i can not exchange information with j (j is out of i's radio signal range), and vise versa. Furthermore, it is not necessary that all nodes in the network have the same ASR and RSR properties. This is a more realistic reflection of the actual deployed networks. For the sake of simplicity, we assume the ASR and the RSR are of the same range for demonstration, i.e. $L_i = C_i$ for each node i.

3.2 Location Discovery

The location discovery problem is traditionally formulated and solved as an optimization problem with the location error as the minimization objective. The basic intuition is that if the locations of the unknown nodes are resolved correctly, then the measured distances and the corresponding calculated distances should be of minimum discrepancy based on a specific/assumed error distribution. However, due to the environmental conditions and the natural imperfection of hardware devices, errors in measurements are inevitable for economically feasible systems. Most often, weighted L_1, L_2 or $L\infty$ norms of individual measurement error are adopted as the optimization target. Note that these norms implicitly assume a particular distribution of the measurement error. For example, the L_1 norm assumes the errors follow the uniform distribution. Maximizing the probabilities of certain error values occurring by following the Gaussian distribution of a particular variance is also a popular alternative. More recently, error models derived using statistical methods such as the kernel density estimation technique have also emerged [25]. In our study, we significantly enhance the application domain of using consistency for location discovery and its practical importance by developing on-line and distributed and localized approaches for consistency-based location discovery.

The location discovery problem can be formally stated as follows. In a k dimensional space, when we consider the homogeneous case where two sensor nodes i $(x_{1i}, x_{2i}, \ldots, x_{ki})$ and j $(x_{1j}, x_{2j}, \ldots, x_{kj})$ have measured distance d_{ij}, exactly one equation of the form of Equation (1) can be written where ε_{ij} denotes the discrepancy between the calculated distance and the measured distance.

$$\varepsilon_{ij} = \sqrt{\sum_{l=1}^{k} (x_{li} - x_{lj})^2} - d_{ij} \qquad (1)$$

After a set of equations that correspond to the pairs of nodes that have measured distances are written, where the unknown variables being the coordinates of the unknown nodes, the system of equations is then linearized and fed to a linear optimization mechanism. [13] provides a detailed procedure of how the system of equation is linearized. We formulated the location discovery problem in terms of a nonlinear function minimization instance where the objective function F has the form expressed in Equation (2). Function M can take the form of L_1, L_2 or $L\infty$ norms (F is subject to minimization), or the Gaussian distribution with various variance or the statistical error model constructed using the kernel density estimation technique (F is subject to maximization). In our study, M is the pair-wise consistency-based error model. Nonlinear programming is a direct extension of linear programming where the linear objective function is replaced by the nonlinear ones. Nonlinear programming has advantages in terms of computing power and formulation flexibility. The most important reason why we formulated the localization problem as a nonlinear programming is due to the NP-completeness of the localization problem [2].

$$F = M(\varepsilon_{ij}) \qquad (2)$$

where

$$\varepsilon_{ij} = \sqrt{\sum_{l=1}^{k} (x_{li} - x_{lj})^2} - d_{ij}$$

for pairs of nodes i and j that have measured distance d_{ij}. .

4 On-line Localization

In this section, we introduce an algorithm for location discovery that does not require the availability of the real distances nor the previously derived off-line error model. We start by presenting the two main concepts behind the approach: *on-line pair-wise consistency* and *hidden beacons*. Next, we explain how the problem can be solved as an instance of nonlinear function minimization. Finally, we describe a conceptually simple approach for simultaneous location discovery and construction of error model for the set of distance measurements.

We define consistency as the pair-wise relationship between two pairs of predicting and predicted variables. More specifically, two pairs $P_1(x_1, y_1)$ and $P_2(x_2, y_2)$ are consistent with respect to each other if and only if (Equation (3)).

$$((x_1 \geq x_2) \Rightarrow (y_1 \geq y_2) \vee (x_1 \leq x_2) \Rightarrow (y_1 \leq y_2)) \tag{3}$$

The location discovery problem can be formulated by using only the notion of pair-wise measurement consistency. The following objective function F (Equation (4)) measures to what extent a proposed solution by the optimization mechanism violates the consistency requirement. F is subject to minimization.

$$c_{ij} = \sqrt{(x_i - x_j)^2 + (y_i - y_j)^2} \, , \tag{4}$$
$$c_{kl} = \sqrt{(x_k - x_l)^2 + (y_k - y_l)^2} \, ,$$

for pairs of nodes i and j that have measured distance d_{ij}, & pairs of nodes k and l that have measured distance d_{kl}

$$\text{if } ((c_{ij} - c_{kl}) \cdot (d_{ij} - d_{kl}) < 0)$$
$$\quad F \mathrel{+}= [-((c_{ij} - c_{kl}) \cdot (d_{ij} - d_{kl}))]$$

For each pair of nodes i and j that have measured distance d_{ij}, c_{ij} is the calculated distance based the locations of i and j proposed by the optimization mechanism. Simply put, for all other pairs of nodes k and l that have measured distance d_{kl}, if d_{kl} is shorter or longer than d_{ij}, then c_{kl} should be also shorter or longer than c_{ij} respectively. Else, the pair is considered inconsistent, and a weight factor proportional to the inconsistency is imposed on the objective function. The objective function is to minimize the overall weight induced on the inconsistent pairs.

It is easy to see that the pair-wise consistency objective formulation shown in Equation (4) is not sufficient for the actual location discovery. This is so because any solution that has all distances between nodes multiplied by a factor Q_1 satisfies the consistency constraint equally well as the solution that has the distances between nodes multiplied by a factor Q_2 (Q_1 and Q_2 are arbitrary positive real numbers).

In order to overcome this problem, we introduce the concept of hidden beacons. The idea is simple but nevertheless is sufficient to fully resolve the scaling problem. We intentionally announce the locations of a small number of beacons is not available and include in the objective function that for each beacon, one more term that measures the difference between the real location of the beacon and the location proposed by the optimization mechanism. A significantly large weight factor is intentionally assigned in front of these terms in order to ensure that the hidden beacons are placed as closed as possible to their actual locations. Since the optimization mechanism has to satisfy the low discrepancy condition for hidden beacons, the proper scaling of all distances is consequently ensured.

At this point, it is sufficient to invoke the optimization algorithm that minimizes the following function $F2$ (Equation (5)), where F is specified in Equation (4):

$$F2 = F + \sum \varepsilon_s \qquad (5)$$

where $\varepsilon_s = \sqrt{(x_s{'}-x_s)^2 +(y_s{'}-y_s)^2}$, for all hidden beacons s

$(x_s{'}, y_s{'})$ is the location proposed by the optimization mechanism; (x_s, y_s) is the real location of beacon s.

The specified problem can be solved using variety of heuristic and probabilistic approaches such as simulated annealing, tabu search and genetic algorithms. However, in order to make our comparison consistent and to leverage on the power of nonlinear function minimization, we slightly modified the objective function $F2$ in the following way (Equation (6)):

$$
\begin{aligned}
&if\,((c_{ij}-c_{kl})\cdot(d_{ij}-d_{kl}) < 0) \qquad (6)\\
&\quad F\, += \,[-((c_{ij}-c_{kl})\cdot(d_{ij}-d_{kl})]
\end{aligned}
$$

else $F\, += H$

where H is a negative real constant

$F2 = F + \sum \varepsilon_s$

where $\varepsilon_s = \sqrt{(x_s{'}-x_s)^2 +(y_s{'}-y_s)^2}$

The motivation behind the alternation of the objective function is to provide nonlinear solver a continuous derivative in regions that are far from the final solution, so that the optimization can converge faster than in the case where the objective function is defined only as a binary function over the number of consistent pair of measurements.

Once we have the approach that produces the locations of all unknown nodes solely based on the consistency of the measurements and the locations of a small subset of beacons, it is straightforward to derive the error model for distance measurements (i.e. optimizing over equation (6) to yield an initial estimate of locations and distance errors and then switching to the distance error objective for further optimization). The error model is based on the measured distances and the corresponding calculated distances (distances derived based on the locations proposed by the on-line localization). Once the new error model is available, we can use this error model as the new objective function for the localization, and then consequently use the new resultant locations to construct the error model in an iterative fashion. This approach is reflected in all the experimental results presented in Section 6.

We have done two types of evaluation regarding the consistency-based error model (the PDF): i) whether the error model is a desirable optimization target; ii) whether the error model produces low location error when applied to the localization task. The results of the second criterion are presented in the experimental result section (Section 6). In terms of the first criterion, we claim that a desirable optimization objective should also follow the property of consistency, which means that an improvement in

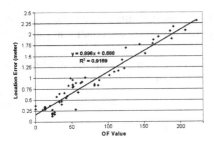

Fig. 3. The on-line objective function evaluation

terms of the objective function results to a smaller average location error. Fig. 3 shows the mean location errors for different objective function values. The average consistency of these pairs of the objective function value and the location error is 94.5%. In addition, the least linear squares fit results a slope of 0.896 and residual 0.916 (shown in the Figure).

5 Localized Algorithm

In this section, we present the localized localization algorithm that utilizes the on-line nonlinear function minimization-based formulation presented in Section 4. We start by stating the underlying abstractions and assumptions. Then we present the algorithm by discussing the dependencies between the centralized and the localized algorithms. The localization accuracy analysis and the communication cost studies are presented in Section 6.

From a network topology point of view, we assume stationary wireless sensor networks that are relatively densely deployed. A densely deployed network provides sufficient amount of data redundancy, which directly impacts the localization accuracy and the communication cost. We quantitatively study the tradeoffs between the average network connectivity, the location accuracy and the communication cost in Section 6. In addition, we assume that all the N deployed nodes are aware of the existence of their C_i and L_i (Section 3.1). However, node i is not aware of the properties (e.g. beacon or not, the connectivity) of its neighbors. The communication cost is computed by accumulating the number of transmitted bytes for the purpose of localization.

There are two goals in the localized location discovery: i) high location accuracy; ii) low communication cost. Depending on different availability of resources, circumstances and application requirements, the problem can be formulated by the standard primal-dual formulation. For example, one formulation is to have the location accuracy as the optimization objective while satisfying a specified communication cost as the constraint. The dual formulation in this case is optimizing the communication cost while satisfying a specified level of location accuracy. In our study, we optimize the location accuracy while keeping the communication cost under a specified level.

The basis of our localized algorithm is a series of invocations of the centralized algorithm with a specified parameter that limits the effective range of the algorithm. We first discuss the centralized algorithm. Fig. 4 presents the pseudocode of the centralized algorithm, which contains three main phases. During the first phase – *Level Discovery phase* (*LD-phase*), a central point of execution (CPE) (e.g. a gateway) initiates a breath first search (BFS) so that all the nodes in the

```
Centralized (Limit)
--------------------------------------------------------------------------------
1. The Gateway initiates the Level Discovery phase (LD-phase)
2. while (the current BFS level < Limit)
3.        level discovery messages propagate
4. The leaf nodes initiate the Measurement Gathering phase (MG-phase)
5. Optimization / Solving at the Gateway
6. The Gateway disseminates the results - Result Dissemination phase (RD-phase)
```

Fig. 4. Pseudocode of the centralized algorithm

```
1. A random node Initiates the Level Discovery phase (LD-phase)
   invoking Centralized (Level_Limit)
2. while (there exists boundary node i has not initiated the procedure)
3.        node i invokes Centralized (Level_Limit)
4. Procedure ends when no more messages progagate
```

Fig. 5. Pseudocode of the localized algorithm

network/cluster are aware of the shortest number of hops to the CPE (lines 1 to 3). The messages propagate during this phase have length one byte and each node (including the boundary/leaf nodes) has to broadcast exactly once in order to complete the BFS. The possible boundary nodes have to broadcast once in order to confirm they are indeed on the boundary of the network/cluster. For centralized algorithm, the *Limit* is set to a large constant so that the deepest BFS level is guaranteed to be smaller than *Limit*.

Once the BFS is completed, i.e. all nodes in the network are aware of their shortest hops to the CPE, the leaf nodes then initiate the *Measurement Gathering phase (MG-phase)* (line 4). During this phase, the neighborhood information and the distance measurements of all nodes are propagated back to the gateway through the shortest path identified in the first phase. For each node i, we assume B bytes have to be allocated for each of i's LD neighbors j, $j \in L_i$. In our study, $B = 3$, one byte for the j's ID, and two bytes for the distance measurement from i to j. The total number of bytes transmitted in this phase T is specified in Equation (7), where V_i is the level of node i:

$$T = \sum_{i}^{N} (V_i \cdot B) \tag{7}$$

Upon receiving all the distance measurements, the CPE invokes the optimization mechanism, which employs the on-line formulation as the optimization objective (line 5). Once the results are available, the CPE disseminates the information back to the nodes in the same fashion as in the first phase – *Result Dissemination phase (RD-phase)* (line 6). The only modification is that it is no longer necessary for the leaf nodes to further propagate the resulting information.

In the localized algorithm (Fig. 5), the key idea is to restrict the BFS not to expand to the entire network, but a limited area; and invoke the same centralized procedure with different CPEs. At the beginning, either a random node or the same CPE as in the centralized case starts the procedure (line 1). The parameter passed down to *Centralized( )* – *Level-Limit*, can be either specified by the user or statistically determined by analyzing the communication cost requirement. As shown in line 2 in Fig. 4, the BFS terminates when the BFS reaches the *Level-Limit*. All the boundary nodes created by the BFS (may or may not be the actual boundary nodes of the network/cluster) are future CPE candidates. After the first round, either one CPE candidate or multiple ones can initiate the centralized algorithm. The procedure terminates when all the possible boundary nodes have confirmed that it cannot reach

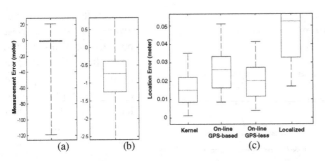

any more other new nodes, i.e., the actual boundary of the network. The termination of the procedure is marked by no more messages regarding localization propagate. We experimentally study how the comm.-unication cost of the localized algorithm scale to

Fig. 6(a). The measurement error (measured − real) boxplot. **6(b).** The measurement error boxplot zoom view. **6(c).** The boxplots of the location error comparison.

both size and the density of the network using simulation. The comprehensive results are presented in the following section.

6 Experimental Results

In this section, we experimentally evaluate the three consistency-based on-line localization algorithms: GPS-based and GPS-less centralized localization and the localized localization. In GPS-less localization, we first solve the instance without using any beacon information (obtain relative locations); then map the relative locations to the absolute positions using the available beacon information. The executions cross all four scenarios are done on a Pentium III 1200MHz processor. We conduct analysis of the localization algorithms in terms of the average connectivity, and the scalability in terms of network size, dimension and different types of measurement errors. In addition, we also present the results for 10 other randomly selected data sets. Finally, we compare the relative performance of the localization algorithms with a sample of previously published algorithms. All experiments are conducted based on the data produced by the deployed network (Section 3.1).

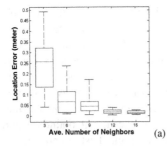
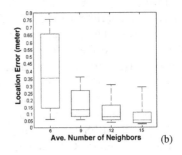

Fig. 7(a). The location error boxplots given different average connectivity for on-line GPS-less LD. **7(b).** The location error boxplots given different average connectivity for on-line GPS-based LD.

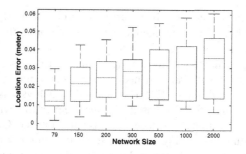

Fig. 8. The scalability study – location error boxplots given different network sizes

A good way to evaluate the overall effectiveness of both the objective function and the localization algorithm is to compare the input error (the distance measurements errors) and the resultant location errors. Fig. 6(a) and 6(b) present the boxplots of the distance measurement errors. The median of the measurement errors is -0.74m. In addition, the average measurement error is -6.73m. Fig. 6(c) compares the location errors of one off-line and three on-line localization algorithms: the centralized off-line algorithm with error model constructed using the kernel density estimation technique as the optimization objective [25]; the two on-line centralized localization algorithms: GPS-based and GPS-less; and finally the on-line localized algorithm. It is common to expect that the off-line localization algorithms outperform the on-line localization algorithms (as the plot indicates). However, we can conclude from the plot that our centralized on-line localization algorithms can achieve comparable results as the off-line approach, especially in the on-line GPS-less scenario. In addition, we conclude from the plots that without considering the beacons (GPS-less) yields better median location error than in the case of when beacons are available. Our explanation for this surprising finding is that the optimization has more degrees of freedom to alter each node's positions around in order to improve the objective function as opposed to when the beacons' positions are fixed. We compare the relative performance with a recent state-of-the-art literature [2] in terms of the ratio of the resultant location error and the input error (random noise). The authors introduced random noise which follows the Gaussian distribution with mean 0 and standard deviation 1cm, 5cm and 10cm. The resultant mean-square errors are 4.43cm, 14.39cm and 16.22cm respectively (e.g. the mean location errors are then 2.1cm, 3.8cm and 4.02cm respectively). Therefore, the corresponding ratio between the location error and the input error are 210%, 76% and 40.2% respectively. In our study, we consider the mean location error of the three algorithms and then normalize them against the mean input error (0.74m), the corresponding ratios are 2.35% (on-line GPS-based), 1.96% (on-line GPS-less), 3.7% (localized).

It is widely assumed that a high degree of connectivity of nodes results in smaller location errors. Fig. 7(a) and 7(b) show the boxplots of the location error distribution given different average number of LD neighbors for in both on-line GPS-less and localized scenarios. We see that while it is important to have more than minimally required three neighbors, once the number of neighbors per node is more than 10, one can expect very little further improvement. More importantly, the quality of the neighboring measurements matters much more than the sheer number of neighbors.

We have developed an integer linear programming (ILP)-based instance generator, which creates instances with random node placements while following a specified measurement error distribution [25]. The scalability analysis is conducted on the networks created using the ILP-based instance generator with the same error distribution as in the original instance. We use the on-line GPS-less localization approach for this study. From Fig. 8, we observe that initially the median location error increases by more than a factor of 2 when the network size doubles (79 nodes to 150 nodes). However, the increase diminishes with any further size increase. In addition, we observe that the location error distribution expands to a wider range as the network size grows. This is an expected consequence of the presence of large number of nodes. Our interpretation of this phenomenon is that some nodes have higher probability of getting 'lucky' and vise versa when the network size expands. It is interesting to note that no instances larger than 300 nodes are solved well using the centralized execution. Obviously the limit that can be addressed by the optimization software is reached (300 nodes). The instances larger than this critical point are solved by grouping 200 nodes consecutively and invoking the optimization in a distributed fashion.

In addition to network size, we also analyze the scalability in terms of dimensions and different types of errors in measurements. The study was done on the original network (79 nodes). Fig. 9(a) shows the location error boxplots when the localization

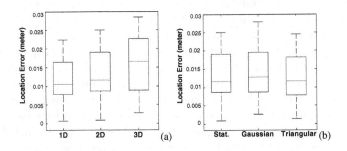

Fig. 9(a). The location error boxplots for different dimensions. **9(b).** The location error boxplots for different types of errors in measurements.

Table 3. Comparison of three previously published LD algorithms with our approaches

	AVE. LOCATION ERROR
	AVE. INPUT ERROR
N-HOP	5.15
ROBUST	4.43
GPS-BASED ON_LINE	0.00883
GPS-LESS ON-LINE	0.0066
LOCALIZED	0.0175

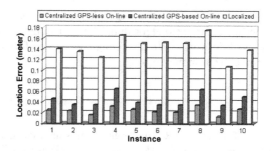

Fig. 10. The median location error comparison of the centralized off-line LD, the centralized on-line LD, and the localized LD across 10 independent data sets

is conducted in 1-d, 2-d and 3-d space using on-line GPS-based localization. It is interesting to note that in 3-d, the medium and the 75% percentile of the location error increased by almost 50% while the other percentiles have smaller fluctuations. In Fig. 9(b), we compare the performance on three sets of measurements that follow different types of error distribution. Stat. is the set of measurements we have obtained from the deployed networks. The other two sets of measurements are generated in simulation. On top of the real distances, random noise that i) follows the Gaussian distribution ($\mu=0$, $s=0.5$m) and ii) has triangular shape ($h=0.5$m, $b=0.5$m) are imposed. The mean location errors are within 15% of each other for all three sets of measurements. This finding supports that the consistency-based error model (as the optimization target) is effective regardless of the types of error distribution.

Furthermore, we examined the consistency of performance on all of the 33 data sets. For the sake of convenient visualization, Fig. 10 shows the results for 10 randomly selected instances where the number of neighbors is on average six per node. Centralized GPS-less off-line, centralized GPS- based on-line, and the localized algorithms are evaluated. Finally, we compare our localization algorithms with two previously published algorithms: N-hop multilateration [13] and robust positioning [2]. We compare the ratio of the average location error and the average input error (measurement error) as in the case of Fig. 6. In N-hop multilateration, the authors considered a 2cm white Gaussian error, and the average location error achieved in a network of 50 nodes with 10% initial beacons is 10.3cm. In robust positioning, an average of 4.43cm location error is achieved when 1cm of input noise is considered. Note that those two efforts conducted experiments using synthetic data in simulations, while our experiments are conducted based on the data produced by a deployed sensor network. The goal of this comparison is solely to demonstrate the performance quality of our algorithm, where our consistency-based on-line localization algorithms can achieve average location errors that are much smaller than the input noise.

7 Conclusion

We have developed a new on-line error modeling and location discovery approach that is solely based on the concept of consistency. The approach does not assume any a-priori knowledge about the error distribution and the optimization objective is the consistency between the measurements and the solutions provided by the optimization mechanism. We also present a localized algorithm that is based on several local invocations of the on-line centralized algorithm. The approach is evaluated using data produced from deployed networks. We also compared the performance with several other state-of-the-art localization methods.

References

1. G. Sun, J. Chen, W. Guo, and K.J.R. Liu, "Signal processing techniques in network-aided positioning," *IEEE Signal Processing Magazine*. 22(4), 2005, pp. 12–23.
2. D. Moore, J. Leonard, D. Rus, and S. Teller, "Robust distributed network localization with noisy range measurements," *International Conference on Embedded Networked Sensor Systems*, 2004, pp. 50–61.
3. A. Sayed, A. Tarighat, and N. Khajehnouri, "Network-based wireless location," *IEEE Signal Processing Magazine*. 22(4), 2005, pp. 24–40.
4. A. Ward, A. Jones, A. Hopper, "A new location technique for the active office," *IEEE Personal Communications*. 4(5), 1997, pp. 42–47.
5. P. Bahl, and V.N. Padmanabhan, "RADAR: an in-building RF-based user location and tracking system," *IEEE InfoCom*, 2000, pp.775–84.
6. J. Hightower, R. Want, G. Borriello, "SpotON: An Indoor 3d Location Sensing Technology Based on RF Signal Strength," CSE, University of Washington, WA, 2000.
7. C. Savarese, J. Rabaey, and K. Langendoen, "Robust positioning algorithms for distributed ad-hoc wireless sensor networks," *USENIX Technical Annual Conference*, 2002, pp. 317–327.
8. T. Hastie, R. Tibshirani, and J. Friedman, *The Elements of Statistical Learning: Data Mining, Inference, and Prediction,*. Springer-Verlag. NY, 2001.
9. W. Merrill, L. Girod, J. Elson, K. Sohrabi, F. Newberg, and W. Kaiser, "Autonomous position location in distributed, embedded, wireless systems," *IEEE CAS Workshop on Wireless Communications and Networking*. 2002.
10. W. Merrill, F. Newberg, L. Girod, and K. Sohrabi, "Battlefield ad-hoc LANs: a distributed processing perspective," *GOMACTech*, 2004.
11. N. Patwari, J.N. Ash, S. Kyperountas, A.O. Hero, R.L. Moses, and N.S. Correal, "Locating the nodes. *IEEE Signal Processing Magazine*, 22(4), 2005, pp. 54–69.
12. J. Feng, F. Koushanfar, and M. Potkonjak, "Localized algorithms for sensor networks," *Handbook of Sensor Networks: Compact Wireless and Wired Sensing Systems*. CRC Press, NY, 2004.
13. A. Savvides, C. Han, and M.B. Strivastava, "Dynamic fine-grained localization in ad-hoc networks of sensors," MobiCom'01, 2001, pp. 166–179.
14. X. Sheng, and Y.H. Hu, "Energy based acoustic source localization.," *ACM/IEEE IPSN'03*, 2003, pp. 285–300.
15. L. Hu, and E. Evans, "Localization for mobile sensor networks," *MobiCom'03*, 2003, pp. 45–57.
16. D. Niculescu, and B. Nath, "Error characteristics of ad hoc positioning systems (APS)," *MobiCom'04*, 2004, pp. 20–30.
17. A. Galstyan, B. Krishnamachari, K. Lerman, and S. Pattem, "Distributed online localization in sensor networks using a moving target," *ACM/IEEE IPSN'04*, 2004, pp. 61–70.
18. F. Gustafsson, and F. Gunnarsson, "Mobile positioning using wireless networks," *IEEE Signal Processing Magazine*, 22(4), 2005, pp. 41–53.

19. S. Sivavakeesar, and G. Pavlou, "Scalable location services for hierarchically organized mobile ad hoc networks," *MobiHoc'05*, 2005, pp. 217–228.
20. T. He, C. Huang, B.M. Blum, J.A. Stankovic, and T. Abdelzaher, "Range-free localization schemes for large scale sensor networks", *MobiHoc'03*, 2003, pp. 81–95.
21. Y. Shang, W. Ruml, Y. Zhang, and M.P.J. Fromherz, "Localization from mere connectivity,' *MobiHoc'03*, 2003, pp. 201–212.
22. J. Bruck, J. Gao, and A.Jiang, "Localization and routing in sensor networks by local angle information," *MobiHoc'05*, 2005, pp. 181–192.
23. A.C. Viana, M.D. de Amorim, S. Fdida, Y.Viniotis, and J.F. de Rezende, "Easily-managed and topology-independent location service for self-organizing networks," *MobiHoc'05*, 2005, pp. 193–204.
24. S, Guha, R. Murth, and E.G. Sirer, "Sextant: a unified node and event localization framework using non-convex constraints," *MobiHoc'05*, 2005, pp. 206–216.
25. J. Feng, and M. Potkonjak, "Location discovery using data-driven statistical error modeling," To appear in *IEEE InfoCom'06*, 2006.

The Robustness of Localization Algorithms to Signal Strength Attacks: A Comparative Study

Yingying Chen, Konstantinos Kleisouris, Xiaoyan Li, Wade Trappe,
and Richard P. Martin

Department of Computer Science and Wireless Information Network Laboratory
Rutgers University, 110 Frelinghuysen Rd, Piscataway, NJ 08854
{yingche, kkonst, xili, rmartin}@cs.rutgers.edu,
trappe@winlab.rutgers.edu

Abstract. In this paper, we examine several localization algorithms and eval-uate their robustness to attacks where an adversary attenuates or amplifies the signal strength at one or more landmarks. We propose several performance met-rics that quantify the estimator's precision and error, including Hölder metrics, which quantify the variability in position space for a given variability in signal strength space. We then conduct a trace-driven evaluation of several point-based and area-based algorithms, where we measured their performance as we applied attacks on real data from two different buildings. We found the median error de-graded gracefully, with a linear response as a function of the attack strength. We also found that area-based algorithms experienced a decrease and a spatial-shift in the returned area under attack, implying that precision increases though bias is introduced for these schemes. We observed both strong experimental and the-oretic evidence that all the algorithms have similar average responses to signal strength attacks.

1 Introduction

Secure localization is important for distributed sensor systems because the position of sensor nodes is a critical input for many sensor network tasks, such as tracking, monitor-ing and geometric-based routing. However, assuring the validity of localization results is not straight-forward because these algorithms rely on physical measurements that can be affected by non-cryptographic attacks. Although there has been recent research on securing localization, to date there has been no study on the robustness of localiza-tion algorithms to physical attacks. In this paper, we investigate the susceptibility of a wide range of signal strength localization algorithms to attacks on the Received Sig-nal Strength (RSS). RSS is an attractive basis for localization because all commodity radio technologies, such as 802.11, 802.15.4, and Bluetooth provide it, and thus the same algorithms can be applied across different platforms. Also, using RSS allows the localization system to reuse the existing communication infrastructure, rather than re-quiring the additional cost needed to deploy specialized localization infrastructure, such as ceiling-based ultrasound, GPS, or infrared methods.

In this work, we investigate the response of several localization algorithms to unan-ticipated power losses and gains, i.e. attenuation and amplification attacks. In these

P. Gibbons et al. (Eds.): DCOSS 2006, LNCS 4026, pp. 546–563, 2006.
© Springer-Verlag Berlin Heidelberg 2006

attacks, the attacker modifies the RSS of a sensor node or landmark, for example, by placing an absorbing or reflecting material around the node. Specifically, we investigate point-based and area-based RF fingerprinting algorithms, whereby a database of collected RF fingerprints are measured at several landmarks for an initial set of locations. In order to evaluate the robustness of these algorithms, we provide a generalized characterization of the localization problem, and then present several performance metrics suitable for quantifying performance. We present a new family of metrics, which we call Hölder metrics, for quantifying the susceptibility of localization algorithms to perturbations in signal strength readings. We use worst-case and average-case versions of the Hölder metric, which describe the maximum and average variability as a function of changes in the RSS. We then experimentally evaluate the performance of a wide variety of localization algorithms after applying attenuation and amplification attacks to real data measured from two different office buildings.

Using experimentally observed localization performance, we found that the error for a wide variety of algorithms scaled with surprising similarity under attack. The single exception was the Bayesian Networks algorithm, which degraded slower than the others in response to attacks against a single landmark. In addition to our experimental observations, we found a similar average-case response of the algorithms using our Hölder metrics. However, we observed that methods which returned an average of likely positions had less variability and are thus less susceptible than other methods.

We also observed that all algorithms degraded gracefully, experiencing linear scaling in localization error as a function of the amount of loss or gain (in dB) an attack introduced. This observation applied to various statistical descriptions of the error, leading us to conclude that no algorithm "collapses" in response to an attack. This is important because it means that, for all the algorithms we examined, there is no tipping point at which an attacker can cause gross errors. In particular, we found the mean error of most of the algorithms for both buildings scaled between 1.3-1.8 ft/dB when all the landmarks were attenuated simultaneously, and 0.5-0.8 ft/dB when attenuating a single landmark. We also showed experimentally that RSS can be easily attenuated by 15 dB, and that, as a general rule of thumb, very simple signal strength attacks can lead to localization errors of 20-30 ft.

Finally, we conducted a detailed evaluation of area-based algorithms as this family of algorithms return a set of potential locations for the transmitter. Thus, it is possible that these algorithms might return a set with a larger area in response to an attack and could have less precision (or more uncertainty) under attack. However, we found all three of our area-based algorithms shifted the returned areas rather than increased returned area. Further, one of the algorithms, the Area Based Probability (ABP) scheme, significantly shrank the size of the returned area in response to very large changes in signal strength.

The rest of this paper is organized as follows. We first discuss related work in Section 2. Next, in Section 3 we give an overview of the algorithms used in our study and discuss how signal strength attacks can be performed. In Section 4, we provide a formal model of the localization problem as well as introduce the metrics that we use in this paper. We then examine the performance of the algorithms through an experimental study in Section 5, and discuss the Hölder metrics for these algorithms in Section 6. Finally, we conclude in Section 7.

2   Related Work

In general, localization algorithms can be categorized as: range-based vs. range-free, scene matching, and aggregate or singular. The range-based algorithms involve distance estimation to landmarks using the measurement of various physical properties like RSS [1], Time Of Arrival (TOA) [2] and Time Difference Of Arrival (TDOA) [3]. Rather than use precise physical property measurements, range-free algorithms use coarser metrics like connectivity [4] or hop-counts [5] to landmarks to place bounds on candidate positions. In scene matching approaches, a radio map of the environment is constructed, either by measuring actual samples, using signal propagation models, or some combination of the two. A node then measures a set of radio properties (often just the RSS of a set of landmarks), the *fingerprint*, and attempts to match these to known location(s) on the radio map. These approaches are almost always used in indoor environments because signal propagation is extensively affected by reflection, diffraction and scattering, and thus ranging or simple distance bounds cannot be effectively employed. Matching fingerprints to locations can be cast in statistical terms [6,7], as a machine-learning classifier problem [8], or as a clustering problem [9]. Finally, a third dimension of classification extends to aggregate or singular algorithms. Aggregate approaches use collections of many nodes in the network in order to localize (often by flooding), while localization of a node in singular methods only requires it to communicate to a few landmarks. For example, algorithms using optimization [10] or multidimensional scaling [4] require many estimates between nodes.

Recently, it has been recognized that there are many non-cryptographic attacks that can affect localization performance. For example, wormhole attacks tunnel through a faster channel to shorten the observed distance between two nodes [11]. Compromised nodes may delay response messages to disrupt distance estimation [12] and compromised landmarks may even broadcast completely invalid information [13]. Physical barriers can directly distort the physical property used by localization. [12] provided a thorough survey of potential attacks to various localization algorithms based on their underlying physical properties.

Secure localization algorithms have been proposed to address these attacks. [14] uses a distance bounding protocol [15, 16] to upperbound the distance between two nodes. Location estimation (via multilateration) with distances from the bounding protocol can be verified against these bounds and any inconsistency will then indicate attack. [17] uses hidden and mobile base stations to localize and verify location estimate. Since such base station locations are hard for attackers to infer, it is hard to launch an attack, thereby providing extra security. [18] uses both directional antenna and distance bounding to achieve security. Compared to all these methods, which employ location verification and discard location estimate that indicates under attack, [13] and [12] try to eliminate the effect of attack and still provide good localization. [12] makes use of the data redundancy and robust statistical methods to achieve reliable localization in the presence of attacks. [13] proposes to detect attacks based on data inconsistency from received beacons and to use a greedy search or voting algorithm to eliminate the malicious beacon information.

In our work, we focus only on fingerprinting algorithms that use RSS, and provide an investigation into the feasibility of signal strength attacks as well as the susceptibility

Table 1. Algorithms under study

Algorithm	Abbreviation	Description
Area-Based		
Simple Point Matching	SPM	Maximum likelihood matching of the RSS to an area using thresholds.
Area Based Probability	ABP-α	Bayes rule matching of the RSS to an area probabilistically bounded by the confidence level α%.
Bayesian Network	BN	Returns the most likely area using a Bayesian network approach.
Point-Based		
RADAR	R1	Returns the closest record in the Euclidean distance of signal space.
Averaged RADAR	R2	Returns the average of the top 2 closest records in the signal map.
Gridded RADAR	GR	Applies RADAR using an interpolated grid signal map.
Highest Probability	P1	Applies maximum likelihood estimation to the received signal.
Averaged Highest Probability	P2	Returns the average of the top 2 likelihoods.
Gridded Highest Probability	GP	Applies likelihoods to an interpolated grid signal map.

of fingerprinting algorithms to such attacks. Of previous work, only [12] proposed a possible solution to the fingerprint-based localization, but the susceptibility of different fingerprinting methods was not completely investigated.

3   Algorithms and Signal Strength Attacks

In this paper we are only concerned with localization algorithms that employ signal strength measurements. There are several ways to classify localization schemes that use signal strength: range-based schemes, which explicitly involve the calculation of distances to landmarks; and RF fingerprinting schemes whereby a radio map is constructed using prior measurements, and a device is localized by referencing this radio map. For this study, we focus on indoor localization schemes, and therefore we restrict our attention to RF fingerprinting methods, which have had more success for indoor environments. RF fingerprinting methods can be further broken down into two main categories: point-based methods, and area-based methods.

Point-based methods return an estimated point as a localization result. A primary example of a point-based method is the RADAR scheme [9]. Variations of RADAR, such as Averaged RADAR and Gridded RADAR have been proposed in [19]. On the other hand, area-based algorithms return a *most likely* area in which the true location resides. Two examples of area-based localization algorithms are the Area Based Probability (ABP) method [19] and the Bayesian Networks method [20]. One of the major advantages of area-based methods compared to point-based methods is that they return a region, which has an increased chance of capturing the transmitter's true location.

For this paper, we have selected a representative set of algorithms from each class of RF fingerprinting schemes for conducting our analysis. The algorithms we have selected are presented in Table 1. Although there are a variety of other fingerprinting localization algorithms that may be studied, our results are general and can be applied to other point-based and area-based methods. More details for these algorithms can be found in [9, 19, 20].

To attack signal-strength based localization systems, an adversary must attenuate or amplify the RSS readings. This can be done by applying the attack at the transmitting device, e.g. simply placing foil around the 802.11 card; or by directing the attack at the landmarks. For example, we may steer the lobes and nulls of an antenna to target select landmarks. A broad variety of attenuation attacks can be performed by introducing materials between the landmarks and sensors [12]. We measured the effect of different materials on the RF propagation when inserted between the landmarks and the sensors. Figure 1 shows the experimental results. These materials are easy to access and attacks utilizing these kind of materials can be simply performed with low cost. Based upon the results in Figure 1, we see that there is a linear relationship between the unattacked signal strength and the attacked signal strength in dB for various materials. The linear relationship suggests that there is an easy way for an adversary to control the effect of his/her attack on the observed signal strength.

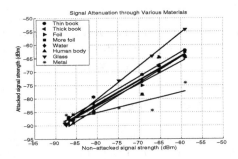

Fig. 1. Signal attenuation when going through a barrier

In the rest of this paper, we will use the linear attenuation model to describe the effect of an attack on the RSS readings at one or more landmarks. The resulting attacked readings are then used to study the consequent effects on localization for the algorithms surveyed above. In particular, in this study, we apply our attacks to individual landmarks, which might correspond to placing a barrier directly in front of a landmark, as well as to the entire set of landmarks, which corresponds to placing a barrier around the transmitting device. Similar arguments can be made for amplification attacks, whereby barriers are removed between the source and receivers. Although there are many different and more complex signal strength attack methods that can be used, we believe their effects will not vary much from the linear signal strength attack model we use in this paper, and note that such sophisticated attacks could involve much higher cost to perform.

4 Measuring Attack Susceptibility

The aim of a localization attack is to perturb a set of signal strength readings in order to have an effect on the localization output. When selecting a localization algorithm, it is desirable to have a set of metrics by which we can quantify how susceptible a localization algorithm is to varying levels of attack by an adversary. In this section, we shall provide a formal specification for an attack, and present several measurement tools for quantifying the effectiveness of an attack.

4.1   A Generalized Localization Model

In order to begin, we need to specify a model that captures a variety of RF-fingerprinting localization algorithms. Let us suppose that we have a domain D in two-dimensions, such as an office building, over which we wish to localize transmitters. Within D, a set of n landmarks have been deployed to assist in localization. A wireless device that transmits with a fixed power in an isotropic manner will cause a vector of n signal strength readings to be measured by the n landmarks. In practice, these n signal strength readings are averaged over a sufficiently large time window to remove statistical variability. Therefore, corresponding to each location in D, there is an n-dimensional vector of signal readings $\mathbf{s} = (s_1, s_2, \cdots, s_n)$ that resides in a range R.

This relationship between positions in D and signal strength vectors defines a fingerprint function $F : D \rightarrow R$ that takes our real world position (x, y) and maps it to a signal strength reading $\mathbf{s}$. F has some important properties. First, in practice, F is not completely specified, but rather a finite set of positions (x_j, y_j) is used for measuring a corresponding set of signal strength vectors $\mathbf{s}_j$. Additionally, the function F is generally one-to-one, but is not onto. This means that the inverse of F is a function G that is not well-defined: There are holes in the n-dimensional space in which R resides for which there is no well-defined inverse.

It is precisely the inverse function G, though, that allows us to perform localization. In general, we will have a signal strength reading $\mathbf{s}$ for which there is no explicit inverse (e.g. perhaps due to noise variability). Instead of using G, which has a domain restricted to R, we consider various pseudo-inverses G_{alg} of F for which the domain of G_{alg} is the complete n-dimensional space. Here, the notation G_{alg} indicates that there may be different *algorithmic* choices for the pseudo-inverse. For example, we shall denote G_R to be the RADAR localization algorithm. In general, the function G_{alg} maps an n-dimensional signal strength vector to a region in D. For point-based localization algorithms, the image of G_{alg} is a single point corresponding to the localization result. On the other hand, for area-based methods, the localization algorithm G_{alg} produces a set of likely positions.

An attack on the localization algorithm is a perturbation to the correct n-dimensional signal strength vector $\mathbf{s}$ to produce a corrupted n-dimensional vector $\tilde{\mathbf{s}}$. Corresponding to the uncorrupted signal strength vector $\mathbf{s}$ is a correct localization result $\mathbf{p} = G_{alg}(\mathbf{s})$, while the corrupted signal strength vector produces an attacked localization result $\tilde{\mathbf{p}} = G_{alg}(\tilde{\mathbf{s}})$. Here, $\mathbf{p}$ and $\tilde{\mathbf{p}}$ are set-valued and may either be a single point or a region in D.

4.2   Attack Susceptibility Metrics

We wish to quantify the effect that an attack has on localization by relating the effect of a change in a signal strength reading $\mathbf{s}$ to the resulting change in the localization result $\mathbf{p}$. We shall use $\mathbf{p}_0$ to denote the correct location of a transmitter, $\mathbf{p}$ to denote the estimated location (set) when there is no attack being performed, and $\tilde{\mathbf{p}}$ to denote the position (set) returned by the estimator after an attack has affected the signal strength. There are several performance metrics that we will use:

Estimator Distance Error: An attack will cause the magnitude of $\mathbf{p}_0 - \tilde{\mathbf{p}}$ to increase. For a particular localization algorithm G_{alg} we are interested in the statistical characterization of $\|\mathbf{p}_0 - \tilde{\mathbf{p}}\|$ over all possible locations in the building. The characterization of $\|\mathbf{p}_0 - \tilde{\mathbf{p}}\|$ depends on whether a point-based method or an area-based method is used, and can be described via its mean and distributional behavior. For a point-based method, we may measure the cumulative distribution (cdf) of the error $\|\mathbf{p}_0 - \tilde{\mathbf{p}}\|$ over the entire building. For area-based methods, we replace $\tilde{\mathbf{p}}$, which is a set, with its median (along the x and y dimensions separately). Thus, for area-based metrics, we calculate the CDF of the distance between the median of the estimated locations $\tilde{\mathbf{p}}_{med}$ and the true location, i.e. $\|\mathbf{p}_0 - \tilde{\mathbf{p}}_{med}\|$.

The CDF provides a complete statistical specification of the distance errors. It is often more desirable to look at the average behavior of the error. For point-based methods, the average distance error is simply $E[\|\mathbf{p}_0 - \tilde{\mathbf{p}}\|]$, which is just the average of $\|\mathbf{p}_0 - \tilde{\mathbf{p}}\|$ over all locations. Area-based methods allow for more options in defining the average distance error. First, for a particular value of $\mathbf{p}_0$, $\tilde{\mathbf{p}}$ is a set of points. For each $\mathbf{p}_0$, we get a collection of error values $\|\mathbf{p}_0 - \mathbf{q}\|$, as $\mathbf{q}$ varies over points in $\tilde{\mathbf{p}}$. For each $\mathbf{p}_0$, we may extract the minimum, 25th percentile, median, 75th percentile, and maximum. These quartile values of $\|\mathbf{p}_0 - \mathbf{q}\|$ are then averaged over the different positions $\mathbf{p}_0$.

Estimator Precision: An area-based localization algorithm returns a set $\mathbf{p}$. For localization, precision refers to the size of the returned estimated area. This metric quantifies the average value of the area of the localized set $\mathbf{p}$ over different signal strength readings $\mathbf{s}$. Generally speaking, the smaller the size of the returned area, the more precise the estimation is. When an attack is conducted, it is possible that the precision of the answer $\tilde{\mathbf{p}}$ is affected.

Precision vs. Perturbation Distance: The perturbation distance is the quantity $\|\mathbf{p}_{med} - \tilde{\mathbf{p}}_{med}\|$. The precision vs. perturbation distance metric depicts the functional dependency between precision and increased perturbation distance.

Hölder Metrics: In addition to error performance, we are interested in how dramatically the returned results can be perturbed by an attack. Thus, we wish to relate the magnitude of the perturbation $\|\mathbf{s} - \tilde{\mathbf{s}}\|$ to its effect on the localization result, which is measured by $\|G_{alg}(\mathbf{s}) - G_{alg}(\tilde{\mathbf{s}})\|$. In order to quantify the effect that a change in the signal strength space has on the position space, we borrow a measure from functional analysis [21], called the Hölder parameter (also known as the Lipschitz parameter) for G_{alg}. The Hölder parameter H_{alg} is defined via

$$H_{alg} = \max_{\mathbf{s},\mathbf{v}} \frac{\|G_{alg}(\mathbf{s}) - G_{alg}(\mathbf{v})\|}{\|\mathbf{s} - \mathbf{v}\|}.$$

For continuous G_{alg}, the Hölder parameter measures the maximum (or worst-case) ratio of variability in position space for a given variability in signal strength space. Since the traditional Hölder parameter describes the worst-case effect an attack might have, it is natural to also provide an average-case measurement of an attack, and therefore we introduce the average-case Hölder parameter

$$\overline{H}_{alg} = \text{avg}_{\mathbf{s},\mathbf{v}} \frac{\|G_{alg}(\mathbf{s}) - G_{alg}(\mathbf{v})\|}{\|\mathbf{s} - \mathbf{v}\|}.$$

These parameters are only defined for continuous functions G_{alg}, and many localization algorithms are not continuous. For example, if we look at G_R for RADAR, the result of varying a signal strength reading is that it will yield a *stair-step* behavior in position space, i.e. small changes will map to the same output and then suddenly, as we continue changing the signal strength vector, there will be a change to a new position estimate (we have switched over to a new Voronoi cell in signal space). In reality, this behavior does not concern us too much, as we are merely concerned with whether adjacent Voronoi cells map to close positions. We will revisit this issue in Section 6. Finally, we emphasize that Hölder metrics measure the perturbability of the returned results, and do not directly measure error.

5   Experimental Results

In this section we present our experimental results. We first describe our experimental method. Next, we examine the impact of attacks on the RSS to localization error when attacking all landmarks simultaneously as well as single-landmark attacks. We then quantify the algorithms' linear responses to RSS changes. Finally, we present a precision study that investigates the impact of attacks on the returned areas for area-based algorithms.

5.1   Experimental Setup

Figure 2 shows our experimental set up. The floor map on the left, (a) is the 3rd floor of the CoRE building at Rutgers, which houses the computer science department and has an area of 200x80ft (16000 ft^2). The other floor shown in (b) is an industrial research laboratory (we call the Industrial Lab), which has an area of 225x144ft (32400 ft^2). The stars are the training points, the small dots are testing points, and the larger squares are the landmarks, which are 802.11 access points. Notice that the 4 CoRE landmarks are more co-linear than the 5 landmarks in the Industrial Lab.

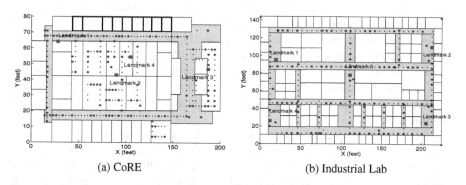

(a) CoRE   (b) Industrial Lab

Fig. 2. Deployment of landmarks and training locations on the experimental floors

For both attenuation and amplification attacks, we ran the algorithms but modified the RSS of the testing points. We altered the RSS by +/-5 dB to +/-25 dB, in increments of 5 dB. We experimented with different ways to handle signals that would fall below the detectable threshold of -92 dBm for our cards. We found that substituting the minimal signal (-92 dBm) produced about the same localization results and did not require changing the algorithms to special case missing data.

We experimented different training set sizes, including 35, 115, 225, 253 and 286 points. Although there are some small differences, we found that the behavior of the algorithms matches previous results and varied little after using 115 training points, and we thus used a training set size of 115 for this study.

5.2   Localization Error Analysis

In this section, we analyze the estimator distance error through the statistical characterization of $\|\mathbf{p}_0 - \tilde{\mathbf{p}}\|$ by presenting the error CDFs of all the algorithms as a function of attenuation and amplification attacks. The CDF provides a complete statistical specification of the distance errors.

Figure 3(a) shows the normal performance of the algorithms for the CoRE building and (e) shows the results for the Industrial Lab. For the area-based algorithms, the median tile error is presented, as well as the minimum and maximum tile errors for ABP-75. As in previous work, the algorithms all obtain similar performance, with the exception of BN which slightly under-performs the other algorithms.

Figures 3(b) and 3(c) show the error CDFs under simultaneous landmark attenuation attacks of 10 and 25 dB for CoRE, respectively, while Figure 3(f) and 3(g) show the similar results in the industrial lab. First, bulk of the curves shift to the right by roughly equal amounts: no algorithm is qualitatively more robust than the others. Comparing the two buildings, the results show that the industrial lab errors are slightly higher for attacks at equal dB, but again, qualitatively the impact of the building environment is not very significant.

Figures 3(d) and 3(h) show the error CDFs for the CoRE and Industrial Lab under a 10 dB amplification attack. The results are qualitatively symmetric with respect to the outcome of the 10 dB attenuation attack. We found that, in general, comparing amplifications to attenuations of equal dB, the errors were qualitatively the same.

An interesting feature is that the minimum error for APB-75 also shifts to the right by roughly the same amount as the other curves. Figures 3(a) and 3(e) show that, in the non-attacked case, the minimum tile error for ABP-75 is quite small, meaning that the localized node is almost always within or very close to the returned area. However, under attacks, the closest part of the returned area moves away from the true location at the same rate as the median tile. We observed similar effects for the SPM and BN algorithms.

Next, we examine attacks against a single landmark. We found attacks against certain landmarks had a much higher impact than against others in the CoRE building. Figure 4(a) and 4(b) show the difference in the error CDF by comparing attacks of landmarks 1 and 2. Figure 2(a) shows that landmark 1 is at the southern end of the building, while landmark 2 is in the center and is close to landmark 4. The tail of the curves in Figure 4(a) are much worse than for 4(b), showing that when landmark 1 is attacked significantly more high errors are returned. we observed a similar effect for amplification attacks.

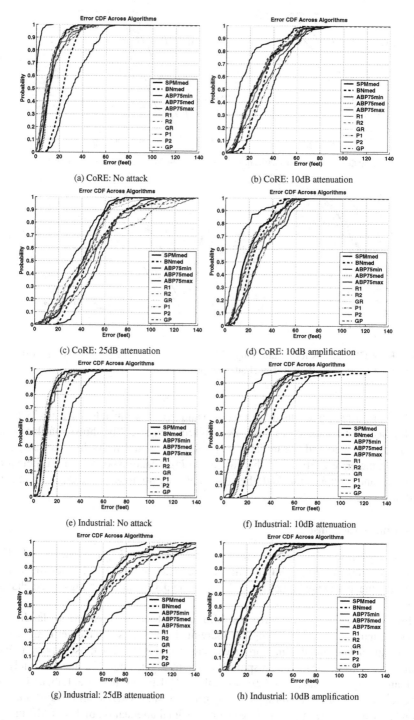

(a) CoRE: No attack

(b) CoRE: 10dB attenuation

(c) CoRE: 25dB attenuation

(d) CoRE: 10dB amplification

(e) Industrial: No attack

(f) Industrial: 10dB attenuation

(g) Industrial: 25dB attenuation

(h) Industrial: 10dB amplification

Fig. 3. Error CDF across localization algorithms when attacks are performed on all the landmarks

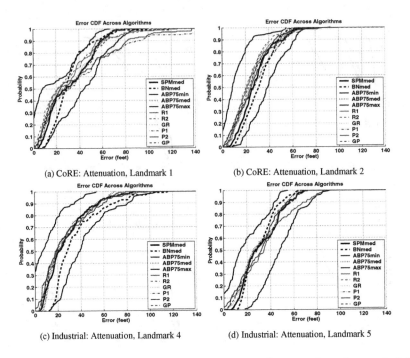

Fig. 4. Error CDF across localization algorithms when attacks are performed on an individual landmark. The attack is 25dB of signal attenuation.

The Industrial Lab results in Figures 4(c) and (d) show much less sensitivity to landmark placement compared to the CoRE building. Figure 2(b) shows that landmark 5 is centrally located and we initially suspected this would result in attack sensitivity. However, the error CDFs show that the remaining 4 landmarks provide sufficient coverage: as landmark 5 is attacked, the error CDFs are not much different from attacking landmark 4.

5.3   Linear Response

In this section, we show that the average distance error, $E[\|\mathbf{p}_0 - \tilde{\mathbf{p}}\|]$, of all the algorithms scales in a linear way to attacks: the localization error changes linearly with respect to the amount of signal strength change in dB (recall it is a log-scaled change in power).

Figure 5 plots the median error vs. RSS attenuation for simultaneous landmark attacks in Figure 5(a) and 5(d), and for individual landmarks in the other figures. Points are measured data, and the lines are linear least-squares fits. The most important feature is that, in all cases, the median responses of all the algorithms fits a line extremely well, with an average R^2-statistic of 0.98 for both the CoRE and Industrial Lab, and a worse-case R^2 of 0.94 for both buildings. Comparing the slopes across all the algorithms, we found a mean change in positioning error vs. signal attenuation of 1.55 ft/dB under simultaneous attacks with a minimum of 1.3 ft/dB and maximum of 1.8 ft/dB. For the single landmark attack, the slope was substantially less, 0.64 ft/dB, although BN degrades consistently less than the other algorithms at 0.44 ft/dB. The linear fit results are quite important as it means that no algorithm has a cliff where the average positioning error suffers a catastrophic failure under attack. Instead, it remains proportional to the severity of the attack.

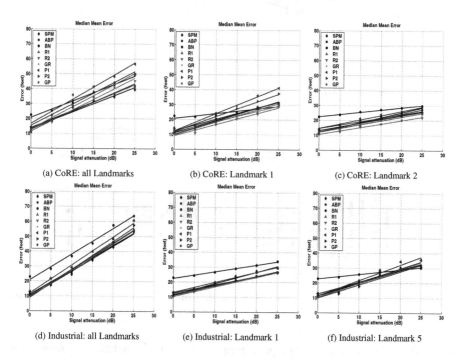

Fig. 5. Median mean error across localization algorithms under attenuation attack

While the median error characterizes the overall response to attacks, it does not address whether an attacker can cause a few, large errors. We examined the response of the maximum error as a function of the strength of the attack, i.e. how the 100th percentile error scales as a function of the change in dB. We note that this characterization is not the same as, nor is directly related to, the Hölder metrics. Those metrics define the rates of change between physical and signal space within the localization function itself, while here we characterize the change in the estimator error to the change in signal, i.e. $\|p_0 - \tilde{p}\|/\|s - v\|$.

Figure 6 plots the worst-case error for each algorithm as a function of signal dB for the CoRE building. The figure shows that almost all the responses are again linear, with least-squares fits of R^2 values of 0.84 or higher, though SPM does not have a linear response. The second important point is the algorithms' responses vary, falling into three groups. BN, R1 and R2 are quite poor, with the worse case error scaling at about 4 ft/dB. P1 and P2, are in a second class, scaling at close to 3 ft/dB. The gridded algorithms, GP and GR, as well as ABP-75 fair better, scaling at 2 ft/dB or less. Finally,

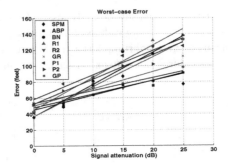

Fig. 6. Maximum error as a function of attack strength for CoRE

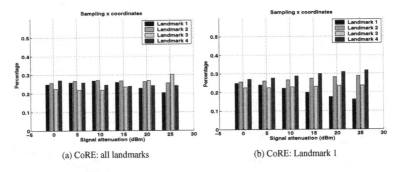

(a) CoRE: all landmarks   (b) CoRE: Landmark 1

Fig. 7. Contribution of each Landmark during sampling in the BN algorithm under attenuation attacks

SPM is in a class by itself, with a poor linear fit (R^2 of 0.61) and the maximum error topping out at about 85 ft after 15 dB of attack.

Examining the error CDFs and the maximum errors, we can see that most of the localizations move fairly slowly in response to an attack, at about 1.5 ft/dB. However, for some of the algorithms, particularly BN, R1 and R2, the top part of the error CDF moves faster, at about 4 ft/dB. What this means is that, for a select few points, an attacker can cause more substantial errors of over 100 ft. However, at most places in the building, an attack can only cause errors with much less magnitude.

Figure 5 show that BN is more robust compared to other algorithms for individual landmark attacks. Recall BN uses a Monte-Carlo sampling technique (Gibbs sampling) to compute the full joint-probability distribution for not just the position coordinates, but also for every node in the Bayesian network. Under a single landmark attack we found the network reduces the contribution of network nodes directly affected by the attacked landmark to the full joint-probability distribution while increasing other landmarks' contributions. In effect, the network "discounts" the attacked landmark's contribution to the overall joint-density because the attacked data from that landmark is highly unlikely given the training data.

To show this effect we developed our own Gibbs sampler so that we could observe the relative contributions of each node in the Bayesian network to the final answer.

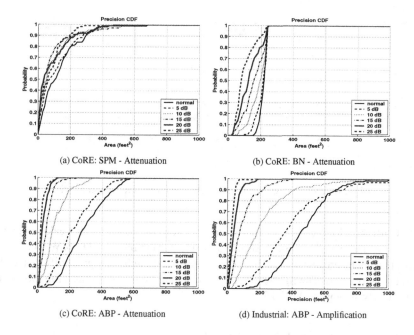

(a) CoRE: SPM - Attenuation

(b) CoRE: BN - Attenuation

(c) CoRE: ABP - Attenuation

(d) Industrial: ABP - Amplification

Fig. 8. Analysis of precision CDF across area-based algorithms. The attack is performed on all the landmarks.

Figure 7 shows the percentage contribution for each landmark to overall joint-density. For instance, in CoRE, the contribution of each landmark starts almost uniformly. When Landmark 1 under attack, the contribution of Landmark 1 goes from 0.25 down to 0.15.

5.4 Precision Study

In this section, we examine the area-based algorithms' precision in response to attacks. Figure 8 shows the CDF of the precision (i.e. size of the returned area) for different area-based algorithms under attack for all the landmarks in CoRE and Industrial Lab. We found the algorithms did not become less precise in response to attacks, but rather, the algorithms tended to shift and shrink the returned areas. Figure 8(a) shows a small average shrinkage for SPM in the CoRE building, and likewise, 8(b) shows a similar effect for BN.

ABP-75 had the most dramatic effect. Figures 8(c) and 8(d) show the precision versus the attack strength for both buildings. The shrinkages are quite substantial. We found that, under attack, the probability densities of the tiles shrank to small values that were located on a few tiles– reflecting the fact that an attack causes there not to be a likely position to localize a node. We also found that this effect held for amplification attacks, as is shown in Figure 8(d). The shrinking precision behavior may be useful for attack detection, although a full characterization of how this effect occurs remains for future work.

Examining this effect further, Figure 9 presents the precision vs. the attack strength, with a least squares line fit. Figure 9(a) shows the effect when attacking all landmarks

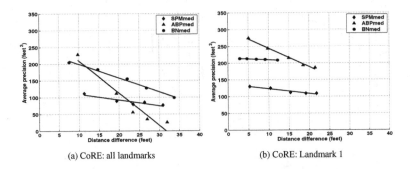

Fig. 9. Precision vs. perturbation distance under attenuation attack

on the CoRE building. Figure 9(b) shows a downward trend, but much weaker, when attacking one landmark. We observed similar results for the Industrial Lab. We see mostly linear changes in precision in response to attacks, although with great differences between the algorithms. The figures show that the decrease in precision as function of dB is particularly strong for ABP-75.

6   Discussion About Hölder Metrics

In the previous section we examined the experimental results, and looked at the performance of several localization algorithms in terms of error and precision. We now focus on the performance of these localization algorithms in terms of the Hölder metrics. The Hölder metrics measure the variability of the *returned* answer in response to changes in the signal strength vectors.

We first discuss the practical aspects of measuring H and $\overline{H}$ for different algorithms. In Section 4, the Hölder parameters are defined by calculating the maximum and average over the entire n-dimensional signal strength space. In practice, it is necessary to perform a sampling technique to measure H and $\overline{H}$. Additionally, as noted earlier, the definition of H and $\overline{H}$ are only suitable for (Hölder) continuous functions G_{alg}. In reality, several localization algorithms, such as RADAR, are not continuous and involve the tessellation of the signal strength space into Voronoi cells V_j, and thus only a discrete set of localization results are produced (image of V_j under G_{alg}). Hence, for any $s \in V_j$ we have $G_R(s) = (x_j, y_j)$. Unfortunately, for neighboring Voronoi cells, we may take $s \in V_j$ and $v \in V_i$ such that they are arbitrarily close (i.e. $\|s - v\| \to 0$), while $\|G_R(s) - G_R(v)\| \neq 0$. In such a case, the formal calculation of H and $\overline{H}$ is not possible. However, for our purposes, we are only interested in measuring the notion of adjacency of Voronoi cells in signal space yielding *close* localization results. Thus, our calculation of H and $\overline{H}$ is only performed over the centroids of the various Voronoi cells for localization algorithms that tessellate of signal strength space.

The Hölder parameters for the different localization algorithms are presented in Table 2. Examining these results, there are several important observations that can be made. First, if we examine the results for $\overline{H}$ we see that, for each building, all of the algorithms have very similar $\overline{H}$ values. Hence, we may conclude that the average vari-

Table 2. Analysis of (worst-case) H and (average-case) $\overline{H}$

Algorithms	CoRE: H	LAB: H	CoRE: $\overline{H}$	LAB: $\overline{H}$
Area-Based				
SPM	23.7646	11.0659	1.8856	2.3548
ABP-75	20.0347	23.0652	1.8548	2.3424
BN	31.7324	14.9168	2.0595	2.5873
Point-Based				
R1	36.2400	20.7846	1.9750	2.3677
R2	19.8586	8.7313	1.9138	2.3058
GR	35.9880	20.6886	1.9691	2.3628
P1	20.8832	20.7846	1.9793	2.3683
P2	19.8586	8.7313	1.9178	2.3058
GP	21.8303	20.6886	1.9649	2.2882

ability of the returned localization result to a change in the signal strength vector is roughly the same for all algorithms. This is an important result as it means, regardless of which RF fingerprinting localization system we deploy, the average susceptibility of the returned results to an attack is essentially identical.

However, if we examine the results for H, which reflects the worst-case susceptibility, then we see that there are some differences across the algorithms. First, comparing H and $\overline{H}$ for both point-based and area-based algorithms, we see that the worst-case variability can be much larger than the average variability. Additionally, the point-based methods appear to cluster. Notably, RADAR (R1) and Gridded Radar (GR) have similar performance across both CoRE and the Industrial Lab, while averaged RADAR (R2) and averaged Highest Probability (P2) have similar performance across both buildings. A very interesting phenomena is observed by looking at the algorithms that returned an average of likely locations (R2 and P2). Across both buildings these algorithms exhibited less variability compared to other algorithms. This is to be expected as averaging is a smoothing operation, which reduces variations in a function. This observation suggests that R2 and P2 are more robust from a worst-case point-of-view than other point-based algorithms.

7   Conclusion

In this paper, we analyzed the robustness of RF-fingerprinting localization algorithms to attacks that target signal strength measurements. We first examined the feasibility of conducting amplification and attenuation attacks, and observed a linear dependency between non-attacked signal strength and attacked signal strength readings for different barriers placed between the transmitter and a landmark receiver. We provided a set of performance metrics for quantifying the effectiveness of an attenuation/amplification attack. Our metrics included localization error, the precision of area-based algorithms, and a new family of metrics, called Hölder metrics, that quantify the variability of the returned answer versus change in the signal strength vectors. We conducted a trace-driven evaluation of several point-based and area-based localization algorithms where the linear attack model was applied to data measured in two different office buildings. We found that the localization error scaled similarly for all algorithms under attack. Further, we found that, when attacked, area-based algorithms did not experience a degradation in precision although they experienced degradation in accuracy. We then examined the variability of the localization results under attack by measuring the Hölder metrics. We found that all algorithms had similar average variability, but those methods returned the average of a set of most likely positions exhibited less variability. This result suggests that the average susceptibility of the returned results to an attack is essentially identical across point-based and area-based algorithms, though it might be desirable to employ either area-based methods or point-based methods that perform averaging in order to lessen the worst-case effect of a potential attack.

References

1. Hightower, J., Vakili, C., Borriello, G., Want, R.: (Design and calibration of the spoton ad-hoc location sensing system) (unpublished).
2. Enge, P., Misra, P.: Global Positioning System: Signals, Measurements and Performance. Ganga-Jamuna Pr (2001)
3. Priyantha, N., Chakraborty, A., Balakrishnan, H.: The cricket location-support system. In: Proceedings of the ACM International Conference on Mobile Computing and Networking (MobiCom). (2000)
4. Shang, Y., Ruml, W., Zhang, Y., Fromherz, M.P.J.: Localization from mere connectivity. In: Proceedings of the Fourth ACM International Symposium on Mobile Ad-Hoc Networking and Computing (MobiHoc). (2003)
5. Niculescu, D., Nath, B.: Ad hoc positioning system (APS). In: Proceedings of the IEEE Global Telecommunications Conference (GLOBECOM). (2001) 2926–2931
6. Youssef, M., Agrawal, A., Shankar, A.U.: WLAN location determination via clustering and probability distributions. In: Proceedings of IEEE PerCom'03, Fort Worth, TX (2003)
7. Roos, T., Myllymaki, P., H.Tirri: A Statistical Modeling Approach to Location Estimation. IEEE Transactions on Mobile Computing 1(1) (2002)
8. Battiti, R., Brunato, M., Villani, A.: Statistical Learning Theory for Location Fingerprinting in Wireless LANs. Technical Report DIT-02-086, University of Trento, Informatica e Telecomunicazioni (2002)
9. Bahl, P., Padmanabhan, V.N.: Radar: An in-building rf-based user location and tracking system. In: Proceedings of the IEEE International Conference on Computer Communications (INFOCOM). (2000)
10. Doherty1, L., Pister, K.S.J., ElGhaoui, L.: Convex position estimation in wireless sensor networks. In: Proceedings of the IEEE International Conference on Computer Communications (INFOCOM). (2001)
11. Hu, Y., Perrig, A., Johnson, D.: Packet leashes: a defense against wormhole attacks in wireless networks. In: Proceedings of the IEEE International Conference on Computer Communications (INFOCOM). (2003)
12. Li, Z., Trappe, W., Zhang, Y., Nath, B.: Robust statistical methods for securing wireless localization in sensor networks. In: Proceedings of the Fourth International Symposium on Information Processing in Sensor Networks (IPSN 2005). (2005)
13. Liu, D., Ning, P., Du, W.: Attack-resistant location estimation in sensor networks. In: Proceedings of the Fourth International Symposium on Information Processing in Sensor Networks (IPSN 2005). (2005)
14. Capkun, S., Hubaux, J.P.: Secure positioning of wireless devices with application to sensor networks. In: Proceedings of the IEEE International Conference on Computer Communications (INFOCOM). (2005)

15. Brands, S., Chaum, D.: Distance-bounding protocols (1994)
16. Sastry, N., Shankar, U., Wagner, D.: Secure verification of location claims. In: Proceedings of the 2003 ACM workshop on wireless security. (2003) 1–10
17. Capkun, S., Hubaux, J.: (Securing localization with hidden and mobile base stations) to appear in Proceedings of IEEE Infocom 2006.
18. Lazos, L., Poovendran, R., Capkun, S.: Rope: robust position estimation in wireless sensor networks. In: Proceedings of the Fourth International Symposium on Information Processing in Sensor Networks (IPSN 2005). (2005) 324–331
19. Elnahrawy, E., Li, X., Martin, R.P.: The limits of localization using signal strength: A comparative study. In: Proceedings of the First IEEE International Conference on Sensor and Ad hoc Communcations and Networks (SECON 2004). (2004)
20. Madigan, D., Elnahrawy, E., Martin, R., Ju, W., Krishnan, P., Krishnakumar, A.S.: Bayesian indoor positioning systems. In: Proceedings of the IEEE International Conference on Computer Communications (INFOCOM). (2005) 324–331
21. Lang, S.: Real and Functional Analysis. Springer (1993)

Author Index

Lecture Notes in Computer Science

For information about Vols. 1–3956

please contact your bookseller or Springer

Vol. 3998: T. Calamoneri, I. Finocchi, G.F. Italiano (Eds.), Algorithms and Complexity. XII, 394 pages. 2006.

Vol. 3997: W. Grieskamp, C. Weise (Eds.), Formal Approaches to Software Testing. XII, 219 pages. 2006.

Vol. 3996: A. Keller, J.-P. Martin-Flatin (Eds.), Self-Managed Networks, Systems, and Services. X, 185 pages. 2006.

Vol. 3995: G. Müller (Ed.), Emerging Trends in Information and Communication Security. XX, 524 pages. 2006.

Vol. 3994: V.N. Alexandrov, G.D. van Albada, P.M.A. Sloot, J. Dongarra (Eds.), Computational Science – ICCS 2006, Part IV. XXXV, 1096 pages. 2006.

Vol. 3993: V.N. Alexandrov, G.D. van Albada, P.M.A. Sloot, J. Dongarra (Eds.), Computational Science – ICCS 2006, Part III. XXXVI, 1136 pages. 2006.

Vol. 3992: V.N. Alexandrov, G.D. van Albada, P.M.A. Sloot, J. Dongarra (Eds.), Computational Science – ICCS 2006, Part II. XXXV, 1122 pages. 2006.

Vol. 3991: V.N. Alexandrov, G.D. van Albada, P.M.A. Sloot, J. Dongarra (Eds.), Computational Science – ICCS 2006, Part I. LXXXI, 1096 pages. 2006.

Vol. 3990: J. C. Beck, B.M. Smith (Eds.), Integration of AI and OR Techniques in Constraint Programming for Combinatorial Optimization Problems. X, 301 pages. 2006.

Vol. 3989: J. Zhou, M. Yung, F. Bao, Applied Cryptography and Network Security. XIV, 488 pages. 2006.

Vol. 3987: M. Hazas, J. Krumm, T. Strang (Eds.), Location- and Context-Awareness. X, 289 pages. 2006.

Vol. 3986: K. Stølen, W.H. Winsborough, F. Martinelli, F. Massacci (Eds.), Trust Management. XIV, 474 pages. 2006.

Vol. 3984: M. Gavrilova, O. Gervasi, V. Kumar, C.J. K. Tan, D. Taniar, A. Laganà, Y. Mun, H. Choo (Eds.), Computational Science and Its Applications - ICCSA 2006, Part V. XXV, 1045 pages. 2006.

Vol. 3983: M. Gavrilova, O. Gervasi, V. Kumar, C.J. K. Tan, D. Taniar, A. Laganà, Y. Mun, H. Choo (Eds.), Computational Science and Its Applications - ICCSA 2006, Part IV. XXVI, 1191 pages. 2006.

Vol. 3982: M. Gavrilova, O. Gervasi, V. Kumar, C.J. K. Tan, D. Taniar, A. Laganà, Y. Mun, H. Choo (Eds.), Computational Science and Its Applications - ICCSA 2006, Part III. XXV, 1243 pages. 2006.

Vol. 3981: M. Gavrilova, O. Gervasi, V. Kumar, C.J. K. Tan, D. Taniar, A. Laganà, Y. Mun, H. Choo (Eds.), Computational Science and Its Applications - ICCSA 2006, Part II. XXVI, 1255 pages. 2006.

Vol. 3980: M. Gavrilova, O. Gervasi, V. Kumar, C.J. K. Tan, D. Taniar, A. Laganà, Y. Mun, H. Choo (Eds.), Computational Science and Its Applications - ICCSA 2006, Part I. LXXV, 1199 pages. 2006.

Vol. 3979: T.S. Huang, N. Sebe, M.S. Lew, V. Pavlović, M. Kölsch, A. Galata, B. Kisačanin (Eds.), Computer Vision in Human-Computer Interaction. XII, 121 pages. 2006.

Vol. 3978: B. Hnich, M. Carlsson, F. Fages, F. Rossi (Eds.), Recent Advances in Constraints. VIII, 179 pages. 2006. (Sublibrary LNAI).

Vol. 3977: N. Fuhr, M. Lalmas, S. Malik, G. Kazai (Eds.), Advances in XML Information Retrieval and Evaluation. XII, 556 pages. 2006.

Vol. 3976: F. Boavida, T. Plagemann, B. Stiller, C. Westphal, E. Monteiro (Eds.), Networking 2006. Networking Technologies, Services, and Protocols; Performance of Computer and Communication Networks; Mobile and Wireless Communications Systems. XXVI, 1276 pages. 2006.

Vol. 3975: S. Mehrotra, D.D. Zeng, H. Chen, B.M. Thuraisingham, F.-Y. Wang (Eds.), Intelligence and Security Informatics. XXII, 772 pages. 2006.

Vol. 3973: J. Wang, Z. Yi, J.M. Zurada, B.-L. Lu, H. Yin (Eds.), Advances in Neural Networks - ISNN 2006, Part III. XXIX, 1402 pages. 2006.

Vol. 3972: J. Wang, Z. Yi, J.M. Zurada, B.-L. Lu, H. Yin (Eds.), Advances in Neural Networks - ISNN 2006, Part II. XXVII, 1444 pages. 2006.

Vol. 3971: J. Wang, Z. Yi, J.M. Zurada, B.-L. Lu, H. Yin (Eds.), Advances in Neural Networks - ISNN 2006, Part I. LXVII, 1442 pages. 2006.

Vol. 3970: T. Braun, G. Carle, S. Fahmy, Y. Koucheryavy (Eds.), Wired/Wireless Internet Communications. XIV, 350 pages. 2006.

Vol. 3969: Ø. Ytrehus (Ed.), Coding and Cryptography. XI, 443 pages. 2006.

Vol. 3968: K.P. Fishkin, B. Schiele, P. Nixon, A. Quigley (Eds.), Pervasive Computing. XV, 402 pages. 2006.

Vol. 3967: D. Grigoriev, J. Harrison, E.A. Hirsch (Eds.), Computer Science – Theory and Applications. XVI, 684 pages. 2006.

Vol. 3966: Q. Wang, D. Pfahl, D.M. Raffo, P. Wernick (Eds.), Software Process Change. XIV, 356 pages. 2006.

Vol. 3965: M. Bernardo, A. Cimatti (Eds.), Formal Methods for Hardware Verification. VII, 243 pages. 2006.

Vol. 3964: M. Ü. Uyar, A.Y. Duale, M.A. Fecko (Eds.), Testing of Communicating Systems. XI, 373 pages. 2006.

Vol. 3963: O. Dikenelli, M.-P. Gleizes, A. Ricci (Eds.), Engineering Societies in the Agents World VI. XII, 303 pages. 2006. (Sublibrary LNAI).

Vol. 3962: W. IJsselsteijn, Y. de Kort, C. Midden, B. Eggen, E. van den Hoven (Eds.), Persuasive Technology. XII, 216 pages. 2006.

Vol. 3960: R. Vieira, P. Quaresma, M.d.G.V. Nunes, N.J. Mamede, C. Oliveira, M.C. Dias (Eds.), Computational Processing of the Portuguese Language. XII, 274 pages. 2006. (Sublibrary LNAI).

Vol. 3959: J.-Y. Cai, S. B. Cooper, A. Li (Eds.), Theory and Applications of Models of Computation. XV, 794 pages. 2006.

Vol. 3958: M. Yung, Y. Dodis, A. Kiayias, T. Malkin (Eds.), Public Key Cryptography - PKC 2006. XIV, 543 pages. 2006.